Advances in Intelligent Systems and Computing

Volume 358

Series editor

Janusz Kacprzyk, Polish Academy of Sciences, Warsaw, Poland
e-mail: kacprzyk@ibspan.waw.pl

About this Series

The series "Advances in Intelligent Systems and Computing" contains publications on theory, applications, and design methods of Intelligent Systems and Intelligent Computing. Virtually all disciplines such as engineering, natural sciences, computer and information science, ICT, economics, business, e-commerce, environment, healthcare, life science are covered. The list of topics spans all the areas of modern intelligent systems and computing.

The publications within "Advances in Intelligent Systems and Computing" are primarily textbooks and proceedings of important conferences, symposia and congresses. They cover significant recent developments in the field, both of a foundational and applicable character. An important characteristic feature of the series is the short publication time and world-wide distribution. This permits a rapid and broad dissemination of research results.

More information about this series at http://www.springer.com/series/11156

Hoai An Le Thi · Ngoc Thanh Nguyen
Tien Van Do
Editors

Advanced Computational Methods for Knowledge Engineering

Proceedings of 3rd International Conference on Computer Science, Applied Mathematics and Applications – ICCSAMA 2015

Editors
Hoai An Le Thi
LITA - UFR MIM
University of Lorraine - Metz
France

Ngoc Thanh Nguyen
Institute of Informatics
Wrocław University of Technology
Wrocław
Poland

Tien Van Do
Department of Networked Systems
and Services
Budapest University of Technology
and Economics
Budapest
Hungary

ISSN 2194-5357 ISSN 2194-5365 (electronic)
Advances in Intelligent Systems and Computing
ISBN 978-3-319-17995-7 ISBN 978-3-319-17996-4 (eBook)
DOI 10.1007/978-3-319-17996-4

Library of Congress Control Number: 2015937023

Springer Cham Heidelberg New York Dordrecht London

Printed on acid-free paper

Springer International Publishing AG Switzerland is part of Springer Science+Business Media
(www.springer.com)

Preface

This volume contains the extended versions of papers presented at the 3th International Conference on Computer Science, Applied Mathematics and Applications (ICCSAMA 2015) held on 11-13 May, 2015 in Metz, France. The conference is co-organized by Laboratory of Theoretical and Applied Computer Science (University of Lorraine, France), Analysis, Design and Development of ICT systems (AddICT) Laboratory (Budapest University of Technology and Economics, Hungary), Division of Knowledge Management Systems (Wroclaw University of Technology, Poland), School of Applied Mathematics and Informatics (Hanoi University of Science and Technology, Vietnam), and in cooperation with IEEE SMC Technical Committee on Computational Collective Intelligence.

The aim of ICCSAMA 2015 is to bring together leading academic scientists, researchers and scholars to discuss and share their newest results in the fields of Computer Science, Applied Mathematics and their applications. These two fields are very close and related to each other. It is also clear that the potentials of computational methods for knowledge engineering and optimization algorithms are to be exploited, and this is an opportunity and a challenge for researchers.

After the peer review process, 36 papers have been selected for including in this volume. Their topics revolve around Computational Methods, Optimization Techniques, Knowledge Engineering and have been partitioned into 5 groups: Mathematical Programming and Optimization: theory, methods and software; Operational Research and Decision making; Machine Learning, Data Security, and Bioinformatics; Knowledge Information System; and Software Engineering.

It is observed that the ICCSAMA 2013, 2014 and 2015 clearly generated a significant amount of interaction between members of both communities on Computer Science and Applied Mathematics, and we hope that these discussions have seeded future exciting development at the interface between computational methods, optimization and engineering.

The materials included in this book can be useful for researchers, Ph.D. and graduate students in Optimization Theory and Knowledge Engineering fields. It is the hope of the editors that readers can find many inspiring ideas and use them to their research. Many such challenges are suggested by particular approaches and models presented in individual chapters of this book. We would like to thank all authors, who contributed to the success of the conference and to this book. Special thanks go to the members of the Steering and Program Committees for their contributions to keeping the high quality of the selected papers. Cordial thanks are due to the Organizing Committee members for their efforts and the organizational work. Finally, we cordially thank Prof. Janusz Kacprzyk and Dr. Thomas Ditzinger from Springer for their supports.

March 2015

Hoai An Le Thi
Ngoc Thanh Nguyen
Tien Van Do

Organization

ICCSAMA 2015 is co-organized by Laboratory of Theoretical and Applied Computer Science (University of Lorraine, France), Analysis, Design and Development of ICT systems (AddICT) Laboratory (Budapest University of Technology and Economics, Hungary), Division of Knowledge Management Systems (Wroclaw University of Technology, Poland), School of Applied Mathematics and Informatics (Hanoi University of Science and Technology, Vietnam), and in cooperation with IEEE SMC Technical Committee on Computational Collective Intelligence.

Organizing Committee

Conference Chair

Hoai An Le Thi — LITA–University of Lorraine, France

Conference Co-Chair

Ngoc Thanh Nguyen — Wroclaw University of Technology, Poland
Tien Van Do — Budapest University of Technology and Economics, Hungary

Publicity Chair

Hoai Minh Le — LITA–University of Lorraine, France

Members

Hoai Minh Le — University of Lorraine, France
Quang Thuy Le — Hanoi University of Science and Technology, Vietnam

Quang Thuan Nguyen	Hanoi University of Science and Technology, Vietnam
Duy Nhat Phan	University of Lorraine, France
Anh Son Ta	Hanoi University of Science and Technology, Vietnam
Thi Thuy Tran	University of Lorraine, France
Xuan Thanh Vo	University of Lorraine, France

Program Committee

Program Chair

Hoai An Le Thi	LITA–University of Lorraine, France

Program Co-Chair

Tao Pham Dinh	National Institute for Applied Sciences, Rouen, France
Hung Son Nguyen	Warsaw University, Poland
Duc Truong Pham	University of Birmingham, UK

Members

Alain Bui	University of Versailles Saint-Quentin-en-Yvelines, France
Minh Phong Bui	Eotvos Lorand University, Hungary
Thi Thu Ha Dao	University of Versailles Saint-Quentin-en-Yvelines, France
Nam Hoai Do	Budapest University of Technology and Economics, Hungary
Tien Van Do	Budapest University of Technology and Economics, Hungary
Thanh Nghi Do	Can Tho University, Vietnam
Quang Thuy Ha	Vietnam National University, Vietnam
Ferenc Hain	Budapest College of Communications and Business, Hungary
Chi Hieu Le	University of Greenwich, UK
Hoai Minh Le	University of Lorraine, France
Nguyen Thinh Le	Humboldt Universität zu Berlin, Germany
Hoai An Le Thi	University of Lorraine, France
Marie Luong	University Paris 13, France
Van Sang Ngo	University of Rouen, France
Anh Linh Nguyen	Warsaw University, Poland

External Referees

Sponsoring Institutions

University of Lorraine (UL), France
Laboratory of Theoretical and Applied Computer Science, UL, France

Budapest University of Technology and economics, Hungary,
Division of Knowledge Management Systems, Wroclaw University of Technology, Poland
School of Applied Mathematics and Informatics, Hanoi University of Science and Technology, Vietnam
Springer
Technical Committee on Computational Collective Intelligence
Vietnam Journal of Computer Science

Contents

Part I: Mathematical Programming and Optimization: Theory, Methods and Software

A Cutting Plane Approach for Solving Linear Bilevel Programming Problems 3
Almas Jahanshahloo, Majid Zohrehbandian

A Direct Method for Determining the Lower Convex Hull of a Finite Point Set in 3D 15
Thanh An Phan, Thanh Giang Dinh

A Hybrid Intelligent Control System Based on PMV Optimization for Thermal Comfort in Smart Buildings 27
Jiawei Zhu, Fabrice Lauri, Abderrafiaa Koukam, Vincent Hilaire

DC Approximation Approach for ℓ_0-minimization in Compressed Sensing 37
Thi Bich Thuy Nguyen, Hoai An Le Thi, Hoai Minh Le, Xuan Thanh Vo

DC Programming and DCA Approach for Resource Allocation Optimization in OFDMA/TDD Wireless Networks 49
Canh Nam Nguyen, Thi Hoai Pham, Van Huy Tran

DC Programming and DCA for a Novel Resource Allocation Problem in Emerging Area of Cooperative Physical Layer Security 57
Thi Thuy Tran, Hoai An Le Thi, Tao Pham Dinh

Scheduling Problem for Bus Rapid Transit Routes 69
Quang Thuan Nguyen, Nguyen Ba Thang Phan

Part II: Operational Research and Decision Making

Application of Recently Proposed Metaheuristics to the Sequence Dependent TSP 83
Samet Tonyali, Ali Fuat Alkaya

Comparative study of Extended Kalman Filter and Particle Filter for Attitude Estimation in Gyroless Low Earth Orbit Spacecraft 95
Nor Hazadura Hamzah, Sazali Yaacob, Hariharan Muthusamy, Norhizam Hamzah

Graph Coloring Tabu Search for Project Scheduling 107
Nicolas Zufferey

Quality of the Approximation of Ruin Probabilities Regarding to Large Claims 119
Aicha Bareche, Mouloud Cherfaoui, Djamil Aïssani

Part III: Machine Learning, Data Security, and Bioinformatics

An Improvement of Stability Based Method to Clustering 129
Minh Thuy Ta, Hoai An Le Thi

A Method for Building a Labeled Named Entity Recognition Corpus Using Ontologies 141
Ngoc-Trinh Vu, Van-Hien Tran, Thi-Huyen-Trang Doan, Hoang-Quynh Le, Mai-Vu Tran

A New Method of Virus Detection Based on Maximum Entropy Model 151
Nhu Tuan Nguyen, Van Huong Pham, Ba Cuong Le, Duc Thuan Le, Thi Hong Van Le

A Parallel Algorithm for Frequent Subgraph Mining 163
Bay Vo, Dang Nguyen, Thanh-Long Nguyen

Combining Random Sub Space Algorithm and Support Vector Machines Classifier for Arabic Opinions Analysis 175
Amel Ziani, Nabiha Azizi, Yamina Tlili Guiyassa

Efficient Privacy Preserving Data Audit in Cloud 185
Hai-Van Dang, Thai-Son Tran, Duc-Than Nguyen, Thach V. Bui, Dinh-Thuc Nguyen

Incremental Mining Class Association Rules Using Diffsets 197
Loan T.T. Nguyen, Ngoc Thanh Nguyen

Mathematical Morphology on Soft Sets for Application to Metabolic Networks . . . 209
Mehmet Ali Balcı, Ömer Akgüller

Molecular Screening of Azurin-Like Anticancer Bacteriocins from Human Gut Microflora Using Bioinformatics . . . 219
Van Duy Nguyen, Ha Hung Chuong Nguyen

Non-linear Classification of Massive Datasets with a Parallel Algorithm of Local Support Vector Machines . . . 231
Thanh-Nghi Do

On the Efficiency of Query-Subquery Nets with Right/Tail-Recursion Elimination in Evaluating Queries to Horn Knowledge Bases . . . 243
Son Thanh Cao

Parallel Multiclass Logistic Regression for Classifying Large Scale Image Datasets . . . 255
Thanh-Nghi Do, François Poulet

Statistical Features for Emboli Identification Using Clustering Technique . . . 267
Najah Ghazali, Dzati Athiar Ramli

Twitter Sentiment Analysis Using Machine Learning Techniques . . . 279
Bac Le, Huy Nguyen

Video Recommendation Using Neuro-Fuzzy on Social TV Environment . . . 291
Duc Anh Nguyen, Trong Hai Duong

Part IV: Knowledge Information System

A Two-Stage Consensus-Based Approach for Determining Collective Knowledge . . . 301
Van Du Nguyen, Ngoc Thanh Nguyen

Context in Ontology for Knowledge Representation . . . 311
Asmaa Chebba, Thouraya Bouabana-Tebibel, Stuart H. Rubin

Designing a Tableau Reasoner for Description Logics . . . 321
Linh Anh Nguyen

Granular Floor Plan Representation for Evacuation Modeling . . . 335
Wojciech Świeboda, Maja Nguyen, Hung Son Nguyen

Integrated Assessment Model on Global-Scale Emissions of Air Pollutants . . . 345
Thanh Binh Nguyen

Query-Subquery Nets with Stratified Negation . . . 355
Son Thanh Cao

Part V: Software Engineering

Distributed Hierarchy of Clusters in the Presence of Topological Changes .. 369
François Avril, Alain Bui, Devan Sohier

Native Runtime Environment for Internet of Things 381
Valentina Manea, Mihai Carabas, Lucian Mogosanu, Laura Gheorghe

Searching for Strongly Subsuming Higher Order Mutants by Applying Multi-objective Optimization Algorithm 391
Quang Vu Nguyen, Lech Madeyski

Some Practical Aspects on Modelling Server Clusters 403
Nam H. Do, Thanh-Binh V. Lam

Erratum to: A Direct Method for Determining the Lower Convex Hull of a Finite Point Set in 3D E1
Thanh An Phan, Thanh Giang Dinh

Author Index 415

Part I

Mathematical Programming and Optimization: Theory, Methods and Software

A Cutting Plane Approach for Solving Linear Bilevel Programming Problems

Almas Jahanshahloo and Majid Zohrehbandian*

Department of Mathematics, Karaj Branch, Islamic Azad University, Alborz, Iran
{Almasj63,zohrebandian}@yahoo.com

Abstract. Bilevel programming (BLP) problems are hierarchical optimization problems having a parametric optimization problem as part of their constraints. From the mathematical point of view, the BLP problem is NP-hard even if the objectives and constraints are linear. This paper proposes a cutting plane approach to solve linear BLP problem which is the simplest case of BLP problems. Our approach is based on the idea that is commonly used in computational mathematics: solving a relaxation problem that is easier to solve and giving a tight approximation by introduction of cutting planes. Therefore, by exploring the theoretical properties of linear BLP, we extend the cutting plane approach for solving linear BLP problems. Numerical examples are provided to illustrate the approach.

Keywords: Linear Bilevel Programming Problem, Quadratic Programming, Cutting plane, Simplex algorithm pivot.

1 Introduction

Bilevel programming (BLP) problems are mathematical optimization problems having a hierarchical structure, where the set of all variables is partitioned between two vectors x (upper-level or leader decision variables) and y (lower level or follower decision variables), and y is to be chosen as an optimal solution of a second mathematical programming problem parameterized in x.

From a historical point of view, BLP problem, which is closely related to the economic problem of Stackelberg (1952) in the field of game theory, was first introduced by Bracken and McGill (1973). Since that time there have been broad interests to this problem both from the practical and the theoretical points of view and many interesting papers have introduced solution algorithms, theoretical properties, and selected applications of BLP problems.

Fortuny-Amat and McCarl (1981), Anandalingam and White (1990), and Anandalingam and Friesz (1992) discussed the importance effect of BLP problem in interpretation of some economic problems. Ding (2012) and Ding et al. (2012) introduced some equilibrium existence theorems for the multi-leader-follower generalized multiobjective games. Dorsch (2013) studied generalized Nash equilibrium problems and

* Corresponding author.

H.A. Le Thi et al. (eds.), *Advanced Computational Methods for Knowledge Engineering*,
Advances in Intelligent Systems and Computing 358, DOI: 10.1007/978-3-319-17996-4_1

bilevel optimization side by side. This perspective comes from the crucial fact that both problems heavily depend on parametric issues.

Anh et al. (2012) presented a gradient-type algorithm for solving bilevel variational inequalities, which is another field, related to the BLP problems. Bao et al. (2007), Dinh et al. (2010), Ye and Zhu (2010), and Ye (2011) mentioned related developments in semiinfinite and multiobjective bilevel programs. Houska (2013) presented a novel sequential convex bilevel programming algorithm for semiinfinite programming and Jayswal et al. (2013) derived duality results for such problems.

As it proposed by Ye and Zhu (1995, 2010), generalized differentiation plays a fundamental role in the study of BLP problems. Mordukhovich et al. (2012) derived new results on generalized differentiation in variational analysis and applied them to deriving necessary optimality conditions for the optimistic version of BLP problems. By using an upper estimate of Clarke subdifferential of value function in variational analysis, Kohli (2012) developed new necessary KKT type optimality conditions for optimistic BLP problem with convex lower level problem. Lin and Chuang (2010) applied variational inclusion problem on metric spaces to study BLP problems and mathematical programs with equilibrium constraint (MPEC).

Harker and Pang (1988) and Judice and Faustino (1992) used the relationship between BLP and MPEC problems to propose solution algorithm for them. Liou and Yao (2005) introduced the group decision problems in a BLP structure, and its corresponding optimization with equilibrium constraints. Dempe and Dutta (2012) and Allende and Still (2013) studied mathematical programs with complementarity constraints (MPCC) and bilevel optimization side by side, where the lower level problem in a BLP can be replaced by the KKT or the FJ-condition, and this leads to a special structured MPEC problem. Finally, Lv et al. (2008, 2010) transformed the inverse optimal value problem into a corresponding nonlinear BLP problem equivalently.

The technical report by Fulop (1993) about relationship of linear BLP problem and multiple objective linear programming (MOLP) is widely used for introducing solution algorithms for BLP problem. Based on this relationship, Glackin et al. (2009) proposed an algorithm that uses simplex pivots. Strekalovsky et al. (2010), Mersha and Dempe (2011), and Calvete et al. (2012) all discussed various direct search algorithms on efficient extreme points of a convex polyhedron produced by relaxation on BLP problem. Candler and Townsley (1982), Bialas and Karwan (1982, 1984), and Shi et al. (2005) used extreme point enumeration for solving BLP problem.

Ren et al. (2013) proposed a hybrid algorithm to solve nonlinear BLP problems. Etoa (2010) used an enumeration sequential quadratic programming algorithm for solving convex quadratic BLP problems. Bard and Falk (1982), Bard and Moore (1992) and Hansen et al. (1992), discussed various branch-and-bound methods for solving BLP problems. Finally, Meng et al. (2012) presented an objective penalty function with two penalty parameters for BLP inequality constrained under the convexity assumption to the lower level problem.

Moreover, for some theoretical papers on the structure of the feasible set and optimality conditions for BLP problems, see Henrion and Surowiec (2011), Dempe and Zemkoho (2011, 2013), Suneja and Kohli (2011), Gadhi and Dempe (2012), Jongen and Shikhman (2012), Ruuska et al. (2012), Dempe et al. (2013), and Liu et al. (2014). And see Huijun et al. (2008), Sonia et al. (2008), Xu et al. (2012), Fang et al. (2013), Sun et al. (2013), and Yao et al. (2013), for some selected applications of BLP problems. Finally, we refer the readers to the books by Bard (1998), Migdalas et al. (1998) and Dempe (2002), and the bibliographies therein for theory, applications and solution methods of general BLP problems, and to two survey articles by Colson et al. (2005, 2007), which can give a comprehensive overview over BLP problems.

In this paper and based on cutting plane approach, we propose an algorithm for solving linear BLP problems. The rest of the paper is structured as follows. In section 2 and after reviewing some well-known concepts about BLP problems, we introduce our solution algorithm for solving linear BLPs. In section 3, we use numerical examples to explore the proposed algorithm. Finally Section 4 concludes the paper.

2 Solving Linear BLP Problems

The majority of research on BLP problems has centered on linear BLP. This paper addresses linear BLP problem of the form (1), and we intend to give a cutting plane idea for attacking it. To this end, we address optimistic linear BLP problem where the functions F(x,y) and f(x,y) are the upper-level (leader) and lower-level (follower) objective functions, respectively, and it is assumed that, whenever the optimal solution set of follower is not a singleton, the leader is allowed to select one of the y^* element in the optimal set, which suits him best.

$$\begin{aligned}
\min_{x,y} \quad & F(x,y) = d_1^T x + d_2^T y \\
s.t. \quad & A_1 x + A_2 y \le b_1 \\
& x \ge 0 \\
& \min_{y} \quad f(x,y) = c_1^T x + c_2^T y \\
& s.t. \quad A_3 x + A_4 y \le b_2 \\
& \qquad\quad y \ge 0
\end{aligned} \tag{1}$$

Definition 1. Extended region of the linear BLP problem (1) is defined as follows:

$$ER = \{(x,y) \mid A_1 x + A_2 y \le b_1, A_3 x + A_4 y \le b_2, x, y \ge 0\}$$

Definition 2. The follower rational reaction set is defined as follows:

$$FRR(x) = \{y \mid y \in \arg\min\{f(x,z): \quad (x,z) \in ER\}\}$$

Definition 3. Feasible region of linear BLP problem (1) is defined as follows:

$$FR1 = \{(x,y) \mid (x,y) \in ER, y \in FRR(x)\}$$

Based on these definitions, the linear BLP problem (1) can be written as minimization of F(x,y) on feasible set FR1, where the set FR1 is a subset of frontier points of ER and then, due to linearity of leader objective function, F(x,y), the optimal solution of model (1) is settle on one of the extreme points of the set ER. Based on this property and to find global solutions of linear BLP problem (1), we can implicitly enumerate the extreme points of the constraint region ER.

Theorem. Optimal objective value of the quadratic programming problem (2) is nonnegative, and every feasible solution of it with zero objective value is corresponding to a feasible solution of linear BLP (1) with leader objective value equals to h*.

$$\begin{aligned} \min_{x,y,\lambda,h} \quad & c_1^T x + c_2^T y - \lambda^T (b_2 - A_3 x) \\ s.t. \quad & (x,y) \in ER \\ & A_4^T \lambda \geq c_2, \quad \lambda \geq 0 \\ & d_1^T x + d_2^T y - h = 0 \end{aligned} \tag{2}$$

Proof. It is straightforward based on a model introduced in Dempe (2002), page 47.

Hence, feasible solutions of model (1) are corresponding to the optimal solutions with zero objective value of model (2). We refer to the feasible set of (2) as FR2.

These properties provide theoretical foundation for us to introduce our algorithm. In other words, by evaluating leader objective function F(x,y) on extreme points of extended region ER and using simplex pivots, we can introduce suitable cuts and construct a restricted set S, contains optimal solution of original problem, as a subset of extended region ER. Then, the linear BLP (1) can be written as minimization of F(x,y) on restricted set S.

We begin by minimizing F(x,y) on extended region ER to find an initial extreme point. If the optimal point (x^0,y^0) is feasible for linear BLP problem, the linear BLP is easy and (x^0,y^0) solves it. But, if $(x^0, y^0) \notin FR1$, we produce a suitable cut and add it to the constraints set ER to produce a restricted set S of ER, which it still includes optimal solution of linear BLP (1). Then, in k-th iteration of the algorithm, due to feasibility or infeasibility of the optimal point (x^k,y^k) and based on F(x,y) and $F(x^k,y^k)$, we produce a suitable cut and add it to the restricted set S. Then we pivot to an adjacent extreme point of the restricted set S. The above idea can be accomplished by the following algorithm:

Algorithm (Cutting Plane Approach for Solving Linear BLP Problem (1))

Step 0. Put $S = ER, \quad E = \{\}, \quad \underline{z} = -\infty, \quad \bar{z} = +\infty$.

Step 1. Compute $(x^*, y^*) = \arg\min\{d_1^T x + d_2^T y \mid (x,y) \in ER\}$. If $y^* \in FRR(x^*)$, then (x*,y*) is the global optimum for linear BLP (1). Otherwise, put $E = E \cup \{(x^*, y^*)\}$, and $\underline{z} = d_1^T x^* + d_2^T y^*$. Compute $(x^*, y^*) = \arg\min\{z = c_1^T x + c_2^T y - \lambda^T (b_2 - A_3 x) \mid (x, y, \lambda, h) \in FR2\}$. If z*>0 then linear BLP (1) is infeasible and we can terminate the algorithm, but if z*=0 then we have a feasible solution $(\bar{x}, \bar{y}) = (x^*, y^*)$ and we can set the upper bound as $\bar{z} = h = d_1^T x^* + d_2^T y^*$.

Step 2. If $\underline{z} = \bar{z}$ then go to step 7. Otherwise, firstly, choose an extreme point $(\hat{x}, \hat{y})$ in S which is adjacent to (x*,y*) and it is not on the hyperplane $d_1^T x + d_2^T y = h$, where $h = d_1^T x^* + d_2^T y^*$. If there is not such a point in S, enumerate the extreme points on the hyperplane $d_1^T x + d_2^T y = h$ for finding such a point $(\hat{x}, \hat{y})$. Put $E = E \cup \{(\hat{x}, \hat{y}\}$ and go to step 3. If there is not such a point in S, go to step 7.

Step 3. Compute

$(x^*, y^*) = \arg\min\{z = c_1^T x + c_2^T y - \lambda^T (b_2 - A_3 x) \mid h \le d_1^T \hat{x} + d_2^T \hat{y}, (x, y, \lambda, h) \in FR2\}$. Note that, for solving this problem, it is better to begin from the current point $(\hat{x}, \hat{y})$, because if it is a feasible solution of linear BLP (1), we can easily determine suitable values for h and λ, to introduce the optimal solution of this quadratic problem.
Step 4. If z*=0 go to step 5; Otherwise, go to step 6.
Step 5. We have a feasible solution (x*,y*) for linear BLP (1) and optimum value for it is less than $h = d_1^T x^* + d_2^T y^*$. Then, put

$S = S \cap \{d_1^T x + d_2^T y \le h\}$, $(\bar{x}, \bar{y}) = (x^*, y^*)$, $\bar{z} = h$, and go to step 2.

Step 6. We have an infeasible solution $(\hat{x}, \hat{y})$ for linear BLP (1) and Optimum value for linear BLP (1) is more than $h = d_1^T \hat{x} + d_2^T \hat{y}$. Put $(x^*, y^*) = (\hat{x}, \hat{y})$,

$\underline{z} = d_1^T \hat{x} + d_2^T \hat{y}$, $S = S \cap \{d_1^T x + d_2^T y \ge h\}$, and go to step 2 where $(x^*, y^*) = (\hat{x}, \hat{y})$.

Step 7. Put $z^* = \bar{z}$, $(x^*, y^*) = (\bar{x}, \bar{y})$ and stop; (x*,y*) is the global optimum for linear BLP (1).

3 Numerical Example

Example 1. Consider the linear BLP (3) introduced in Glackin et al. (2009), where the extended region ER is equal to the convex hull of the extreme points (0,1,3), (0.5,1.5,2), (0,2,2), (0,1,0), (1,2,0) and (0,3,0). Note that the Glackin et al's method can't use for solving the linear BLPs where the upper level constraints depends on y (follower's decision variables).

$$\begin{aligned} \min_{x_1, x_2, y} \quad & -2x_1 + 4x_2 + 3y \\ s.t. \quad & x_1 - x_2 \le -1 \\ & x_1, x_2 \ge 0 \\ & \min_{y} \quad -y \\ & \quad x_1 + x_2 + y \le 4 \\ & \quad 2x_1 + 2x_2 + y \le 6 \\ & \quad y \ge 0 \end{aligned} \tag{3}$$

Step 1 of cutting plane algorithm yields (x^*, y^*)=(0,1,0) which is not feasible for model (3). Put $\underline{z} = -2x_1^* + 4x_2^* + 3y^* = -4$. We arbitrarily choose extreme point (0,1,3) which is adjacent to (0,1,0). Step 3 yields z*=0 and optimal solution (0,1,3) which is feasible for model (3). Put $\bar{z} = -2x_1^* + 4x_2^* + 3y^* = 13$, $(\bar{x}_1, \bar{x}_2, \bar{y}) = (0,1,3)$. By adding the constraint $-2x_1 + 4x_2 + 3y \le \bar{z} = 13$ to the constraint set S, the extreme point (0,2,2) will be remove from S. Then, we choose extreme point (0.5,1.5,2) which is adjacent to (0,1,3). Step 3 yields z*=0 and optimal solution (x^*, y^*)= (0.5,1.5,2). Put $\bar{z} = -2x_1^* + 4x_2^* + 3y^* = 11$, $(\bar{x}_1, \bar{x}_2, \bar{y}) = (0.5,1.5,2)$. The extreme point (0,3,0) and all of the extreme points on the additional constraint $-2x_1+4x_2+3y=13$ will be remove from S , by addition of the constraint $-2x_1+4x_2+3y\le\bar{z}=11$ to the constraint set S. Adjacent to (0.5,1.5,2), we arbitrarily choose extreme point (1,2,0). Step 3 yields z*=0 and (x^*,y^*)=(1,2,0). Put $\bar{z}=-2x_1^*+4x_2^*+3y^*=6$, $(\bar{x}_1,\bar{x}_2,\bar{y})=(1,2,0)$. By adding the constraint $-2x_1+4x_2+3y\le\bar{z}=6$ to the constraint set S, the extreme point (0.5,1.5,2) will be remove from S and S will be the convex hull of extreme points (0,1,0), (1,2,0) and two additional extreme points (0,1,2) and (0.1.5,0). By enumerating the two adjacent extreme points (0,1,2) and (0.1.5,0), we find that the only adjacent extreme point with the property $-2x_1+4x_2+3y<6$, is (0,1,0) which is in E and we study it beforehand. Hence, the algorithm has been terminated and the global optimal solution of model (3) is (x^*,y^*)= (1,2,0) with objective value equals to z^*=6.

Example 2. Consider the linear BLP (4) introduced in Shi et al. (2005). They mentioned that by using the k-th best approach, and after enumeration of 34 extreme points, global optimal solution occurs at x*= (1.45, 1.25, 18.55, 4.63, 4.05, 4.85, 6.66, 12.45, 0.38, 15.25, 0.22) with leader objective value F* = 607.44. But, this solution is infeasible for linear BLP (4)! Because it violates the last constraint of the leader (i.e. $x_1+x_{11}\ge 7.9$). However, by substitution of KKT conditions instead of follower problems, model (4) can be transform to model (2). Finally, by execution of the proposed cutting plane algorithm in this paper, in step 1, we have z*>0 and we conclude that the problem is infeasible.

Despite of infeasibility of this problem, let us to execute the other steps of the proposed cutting plane algorithm, which produces a sequence of infeasible adjacent extreme points: x^{*1}= (1.45, 3.04, .71, .58, 4.05, 29.95, .72, 6.15, 6.28, 4.40, 69.05) with leader objective value F*=230.7464, x^{*2}= (1.66, 3.49, .52, .67, 3.84, 29.74, .83, 5.94, 7.22, 4.19, 68.84) with leader objective value F*=244.0582, x^{*3}= (3.48, 7.31, 1.08, 1.39, 2.02, 27.92, 1.74, 4.12, 15.12, 2.37, 67.02) with leader objective value F*=362.6635, x^{*4}= (4.43, 9.31, 1.37, 1.77, 1.07, 29.97, 2.22, 3.17, 14.17, 1.42, 66.07) with leader objective value F*=426.0158, x^{*5}= (4.43, 9.31, 1.37, 1.77, 1.07, 29.97, 2.22, 18.3, 14.17, 1.42, 66.07) with leader objective value F*=895.1080.

Then, we have x^{*6}= (5.5, 11.55, 1.71, 2.2, 0, 25.9, 2.75, 17.19, 13.1, 1.76, 65) with leader objective value F*=965.6255, x^{*7}= (5.97, 12.53, 1.85, 2.39, 0, 25.43, 2.98, 16.7, 12.63, 1.9, 64.53) with leader objective value F*=991.8267, x^{*8}= (7, 11.5, 2.17, 2.8, 0, 24.4, 3.5, 15.63, 11.6, 2.24, 63.5) with leader objective value F*=1051.250, x^{*9}= (7, 11.5, 2.17, 2.8, 0, 0, 3.5, 15.63, 11.6, 2.24, 63.5) with leader objective value F*=1075.406, x^{*10}= (7, 0, 2.17, 2.8, 0, 0, 3.5, 15.63, 11.6, 2.24, 63.5) with leader objective value F*=1081.156, x^{*11}= (7, 0, 2.17, 22.4, 0, 0, 3.5, 15.63, 11.6, 2.24, 63.5) with leader objective value F*=1094.876, x^{*12}= (7, 0, 13, 22.4, 0, 0, 3.5, 15.63, 11.6, 2.24, 63.5) with leader objective value F*=1127.366, x^{*13}= (7, 0, 13, 22.4, 0, 0, 3.5, 15.63, 11.6, 9.7, 63.5) with leader objective value F*=1130.350, x^{*14}= (7, 0, 13, 22.4, 0, 0, 3.5, 15.63, 11.6, 9.07, 1.05) with leader objective value F*=1144.714, x^{*15}= (7, 0, 13, 22.4, 0, 0, 3.5, 15.63, 1.82, 9.7, 1.05) with leader objective value F*=1147.647. Finally, because $\underline{z}=1147.647$ and $\bar{z}=+\infty$, the linear BLP (4) is infeasible.

Note that, Shi et al. (2005) introduced infeasible value F* = 607.44 after 34 enumeration and we showed in the 5-th iteration of the algorithm that there is not any feasible solution for linear BLP (4) with F*=895. In other words, the execution of K-th best approach needs lots of enumeration to produce the point x^{*15} which have the maximum infeasible value for leader objective function on the extended region ER.

$$
\begin{aligned}
\min_{x_1,\cdots,x_{11}} \quad & 80x_1 - 0.5x_2 + 3x_3 + 0.7x_4 - 10x_5 \\
& -0.99x_6 + 13x_7 + 31x_8 - 0.3x_9 + 0.4x_{10} - 0.23x_{11} \\
s.t. \quad & x_1 + x_2 \le 18.5, \quad -0.31x_1 + x_3 \ge 0, \quad 3.2x_1 - x_4 \ge 0, \\
& x_1 + x_5 \ge 5.5, \quad -0.25x_1 - x_6 \le 0, \quad 4.6x_1 - x_7 \ge 0, \\
& 0.5x_1 + 0.48x_8 \le 11, \quad 4.34x_1 - x_9 \ge 0, \quad 0.32x_1 - x_{10} \le 0 \\
& x_1 + x_{11} \ge 7.9, \quad x_1,\cdots,x_{11} \ge 0 \\
& \min_{x_2} \ 0.9x_1 + 3x_2 \\
& s.t. \ 2.1x_1 - x_2 \ge 0, \quad x_1 + x_2 \ge 2.7, \quad x_1, x_2 \ge 0 \\
& \min_{x_3} \ 3x_1 - 2.2x_3 \\
& s.t. \ x_1 + x_3 \le 20, \quad 0.8x_1 + 0.91x_3 \ge 1.8, \quad x_1, x_3 \ge 0 \\
& \min_{x_4} \ 0.4x_1 - x_4 \\
& s.t. \ x_1 + x_4 \le 31.5, \quad -0.4x_1 + x_4 \ge 0, \quad x_1, x_4 \ge 0 \\
& \min_{x_5} \ 0.7x_1 + 21x_5 \\
& s.t. \ 0.3x_1 + 0.4x_5 \le 8.5, \quad -2.8x_1 + x_5 \le 0, \quad x_1, x_5 \ge 0 \\
& \min_{x_6} \ 10x_1 + 0.67x_6 \\
& s.t. \ x_1 + x_6 \le 31.4, \quad x_1 + x_6 \ge 6.3, \quad x_1, x_6 \ge 0 \\
& \min_{x_7} \ 2x_1 - 3x_7 \\
& s.t. \ 2x_1 + x_7 \le 17.5, \quad -0.5x_1 + x_7 \ge 0, \quad x_1, x_7 \ge 0 \\
& \min_{x_8} \ 0.75x_1 - 20.5x_8 \\
& s.t. \ -8.6x_1 + x_8 \le 0, \quad x_1 + x_8 \ge 7.6, \quad x_1, x_8 \ge 0 \\
& \min_{x_9} \ 0.3x_1 + 6.7x_9 \\
& s.t. \ x_1 + x_9 \le 18.6, \quad 0.26x_1 - x_9 \le 0, \quad x_1, x_9 \ge 0 \\
& \min_{x_{10}} \ 0.65x_1 - 3.2x_{10} \\
& s.t. \ x_1 + x_{10} \le 16.7, \quad x_1 + x_{10} \ge 5.85 \quad x_1, x_{10} \ge 0 \\
& \min_{x_{11}} \ x_1 + 0.56x_{11} \\
& s.t. \ x_1 + x_{11} \le 70.5, \quad -0.15x_1 + x_{11} \ge 0, \quad x_1, x_{11} \ge 0
\end{aligned}
\tag{4}
$$

4 Conclusion

Our cutting plane algorithm can be used effectively to solve linear BLP problems having a small number of leader/follower variables, such as the problems that arise in many practical applications. This cutting plane based algorithm is most effective when the number of the extreme points of the extended region ER is small because it performs a partial enumeration of those points. However, in the worst case, the proposed algorithm moves on a sequence of increasing infeasible vertices and decreasing feasible vertices of the set ER, on one of the paths from the best infeasible vertex to the best feasible vertex with respect to the leader objective function. Then, the number of enumerated vertices, in the worst case, is much less than the explicit enumeration methods like the k-th best approach.

Moreover, the proposed algorithm can solve general optimistic linear BLPs, while a lot of proposed methods in literature, only can use for solving linear BLPs for which the leader problem only depends on the leader's decision variables.

Finally, the performance of our cutting plane algorithm might be improved by using more sophisticated methods to find a feasible solution of linear BLP or to optimize a quadratic function over the linear constraints.

References

1. Allende, G.B., Still, G.: Solving bilevel programs with the KKT-approach. Mathematical Programming, Ser. A 138(1-2), 309–332 (2013)
2. Anandalingam, G., Friesz, T.L.: Hierarchical optimization. Annals of Operations Research 34 (1992)
3. Anandalingam, G., White, D.: A solution method for the linear static Stackelberg problem using penalty functions. IEEE Trans. Automat. Contr. 35, 1170–1173 (1990)
4. Anh, P.N., Kim, J.K., Muu, L.D.: An extra gradient algorithm for solving bilevel pseudomonotone variational inequalities. Journal of Global Optimization 52, 627–639 (2012)
5. Bao, T.Q., Gupta, P., Mordukhovich, B.S.: Necessary conditions in multiobjective optimization with equilibrium constraints. Journal of Optimization Theory and Applications 135, 179–203 (2007)
6. Bard, J.: Practical Bilevel Optimization. Kluwer Academic, Dordrecht (1998)
7. Bard, J., Falk, J.: An explicit solution to the multi-level programming problem. Comput. Oper. Res. 9, 77–100 (1982)
8. Bard, J., Moore, J.: An algorithm for the discrete bilevel programming problem. Nav. Res. Logist. 39, 419–435 (1992)
9. Bialas, W., Karwan, M.: On two-level optimization. IEEE Transactions on Automatic Control 27, 211–214 (1982)
10. Bialas, W., Karwan, M.: Two level linear programming. Management Science 30, 1004–1020 (1984)
11. Bracken, J., McGill, J.: Mathematical programs with optimization problems in the constraints. Operations Research 21, 37–44 (1973)
12. Calvete, H.I., Gale, C., Dempe, S., Lohse, S.: Bilevel problems over polyhedral with extreme point optimal solutions. Journal of Global Optimization 53, 573–586 (2012)

13. Candler, W., Townsley, R.: A linear twolevel programming problem. Computers and Operations Research 9, 59–76 (1982)
14. Colson, B., Marcotte, P., Savard, G.: Bilevel programming: A survey. 4OR 3, 87–107 (2005)
15. Colson, B., Marcotte, P., Savard, G.: An overview of bilevel optimization. Annals of Operations Research 153, 235–256 (2007)
16. Dempe, S.: Foundations of Bilevel Programming. Kluwer Academic, Dordrecht (2002)
17. Dempe, S., Dutta, J.: Is bilevel programming a special case of a mathematical program with complementarity constraints? Mathematical Programming, Ser. A 131, 37–48 (2012)
18. Dempe, S., Gadhi, N., Zemkoho, A.B.: New Optimality Conditions for the Semivectorial Bilevel Optimization Problem. Journal of Optimization Theory and Applications 157(1), 54–74 (2013)
19. Dempe, S., Zemkoho, A.B.: The Generalized Mangasarian-Fromowitz Constraint Qualification and Optimality Conditions for Bilevel Programs. Journal of Optimization Theory and Applications 148, 46–68 (2011)
20. Dempe, S., Zemkoho, A.B.: The bilevel programming problem: reformulations, constraint qualifications and optimality conditions. Mathematical Programming, Ser. A 138(1-2), 447–473 (2013)
21. Ding, X.P.: Existence and iterative algorithm of solutions for a class of bilevel generalized mixed equilibrium problems in Banach spaces. Journal of Global Optimization 53(3), 525–537 (2012)
22. Ding, X.P., Liou, Y.C., Yao, J.C.: Existence and algorithms for bilevel generalized mixed equilibrium problems in Banach spaces. Journal of Global Optimization 53(2), 331–346 (2012)
23. Dinh, N., Mordukhovich, B.S., Nghia, T.T.A.: Subdifferentials of value functions and optimality conditions for DC and bilevel infinite and semi-infinite programs. Mathematical Programming 123, 101–138 (2010)
24. Dorsch, D., Jongen, H.T., Shikhman, V.: On Intrinsic Complexity of Nash Equilibrium Problems and Bilevel Optimization. Journal of Optimization Theory and Applications 159(3), 606–634 (2013)
25. Eichfelder, G.: Multiobjective bilevel optimization. Mathematical Programming, Ser. A 123, 419–449 (2010)
26. Etoa, J.B.E.: Solving convex quadratic bilevel programming problems using an enumeration sequential quadratic programming algorithm. Journal of Global Optimization 47, 615–637 (2010)
27. Fang, S., Guo, P., Li, M., Zhang, L.: Bilevel Multiobjective Programming Applied to Water Resources Allocation. Mathematical Problems in Engineering, Article ID 837919 2013 (2013)
28. Fortuny-Amat, J., McCarl, B.: A representation and economic interpretation of a two-level programming problem. J. Oper. Res. Soc. 32, 783–792 (1981)
29. Fulop, J.: On the equivalence between a linear bilevel programming problem and linear optimization over the efficient set. Technical Report WP93-1, Laboratory of Operations Research and Decision Systems, Computer and Automation Institute, Hungarian Academy of Sciences (1993)
30. Gadhi, N., Dempe, S.: Necessary Optimality Conditions and a New Approach to Multiobjective Bilevel Optimization Problems. Journal of Optimization Theory and Applications 155, 100–114 (2012)

31. Glackin, J., Ecker, J.G., Kupferschmid, M.: Solving Bilevel Linear Programs Using Multiple Objective Linear Programming. Journal of Optimization Theory and Applications 140, 197–212 (2009)
32. Hansen, P., Jaumard, B., Savard, G.: New branch-and-bound rules for linear bilevel programming. SIAM J. Control Optim. 13, 1194–1217 (1992)
33. Harker, P., Pang, J.: Existence of optimal solutions to mathematical programs with equilibrium constraints. Oper. Res. Lett. 7, 61–64 (1988)
34. Henrion, R., Surowiec, T.: On calmness conditions in convex bilevel programming. Applicable Analysis 90(6), 951–970 (2011)
35. Houska, B., Diehl, M.: Nonlinear robust optimization via sequential convex bilevel programming. Mathematical Programming, Ser. A 142(1-2), 539–577 (2013)
36. Huijun, S., Ziyou, G., Jianjun, W.: A bi-level programming model and solution algorithm for the location of logistics distribution centers. Applied Mathematical Modelling 32, 610–616 (2008)
37. Jayswal, A., Prasad, A.K., Ahmad, I., Agarwal, R.P.: Duality for semi-infinite programming problems involving (H_p,r)-invex functions. Journal of Inequalities and Applications 2013, 200 (2013)
38. Jongen, H.T., Shikhman, V.: Bilevel optimization: on the structure of the feasible set. Mathematical Programming, Ser. B 136, 65–89 (2012)
39. Judice, J., Faustino, A.: A sequential LCP method for bilevel linear programming. Annals of Operations Research 34, 89–106 (1992)
40. Kohli, B.: Optimality Conditions for Optimistic Bilevel Programming Problem Using Convexifactors. J. Optim. Theory. Appl. 152, 632–651 (2012)
41. Lin, L.J., Chuang, C.S.: Saddle Point Problems, Bilevel Problems, and Mathematical Program with Equilibrium Constraint on Complete Metric Spaces. Journal of Inequalities and Applications, Article ID 306403 2010 (2010)
42. Liou, Y.C., Yao, J.C.: Bilevel Decision via Variational Inequalities. Computers and Mathematics with Applications 49, 1243–1253 (2005)
43. Liu, B., Wan, Z., Chen, J., Wang, G.: Optimality conditions for pessimistic semivectorial bilevel programming problems. Journal of Inequalities and Applications 2014, 41 (2014)
44. Lv, Y., Chena, Z., Wan, Z.: A penalty function method based on bilevel programming for solving inverse optimal value problems. Applied Mathematics Letters 23, 170–175 (2010)
45. Lv, Y., Hua, T., Wan, Z.: A penalty function method for solving inverse optimal value problem. Journal of Computational and Applied Mathematics 220, 175–180 (2008)
46. Meng, Z., Dang, C., Shen, R., Jiang, M.: An Objective Penalty Function of Bilevel Programming. Journal of Optimization Theory and Applications 153, 377–387 (2012)
47. Mersha, A.G., Dempe, S.: Direct search algorithm for bilevel programming problems. Comput. Optim. Appl. 49, 1–15 (2011)
48. Migdalas, A., Pardalos, P.M., Varbrand, P.: Multilevel Optimization: Algorithms and Applications. Kluwer Academic Publishers, Dordrecht (1998)
49. Mordukhovich, B.S., Nam, N.M., Phan, H.M.: Variational Analysis of Marginal Functions with Applications to Bilevel Programming. Journal of Optimization Theory and Applications 152, 557–586 (2012)
50. Ren, A., Wang, Y., Jia, F.: A Hybrid Estimation of Distribution Algorithm and Nelder-Mead Simplex Method for Solving a Class of Nonlinear Bilevel Programming Problems. Journal of Applied Mathematics, Article ID 2013 378568 (2013)
51. Ruuska, S., Miettinen, K., Wiecek, M.M.: Connections Between Single-Level and Bilevel Multiobjective Optimization. J. Optim. Theory. Appl. 153, 60–74 (2012)

52. Shi, C., Zhang, G., Lu, J.: The Kth-Best Approach for Linear Bilevel Multi-follower Programming. Journal of Global Optimization 33, 563–578 (2005)
53. Sonia, K.A., Puri, M.C.: Bilevel time minimizing transportation problem. Discrete Optimization 5, 714–723 (2008)
54. Sun, X., Lu, H.P., Chu, W.J.: A Low-Carbon-Based Bilevel Optimization Model for Public Transit Network. Mathematical Problems in Engineering, Article ID 374826 2013 (2013)
55. Suneja, S.K., Kohli, B.: Optimality and Duality Results for Bilevel Programming Problem Using Convexifactors. Journal of Optimization Theory and Applications 150, 1–19 (2011)
56. Stackelberg, H.: The theory of market economy. Oxford University Press, Oxford (1952)
57. Strekalovsky, A.S., Orlov, A.V., Malyshev, A.V.: On computational search for optimistic solutions in bilevel problems. Journal of Global Optimization 48, 159–172 (2010)
58. Xu, J., Tu, Y., Zeng, Z.: A Nonlinear Multiobjective Bilevel Model for Minimum Cost Network Flow Problem in a Large-Scale Construction Project. Mathematical Problems in Engineering, Article ID 463976 2012 (2012)
59. Yao, L., Xu, J.: A Class of Expected Value Bilevel Programming Problems with Random Coefficients Based on Rough Approximation and Its Application to a Production-Inventory System. Abstract and Applied Analysis, Article ID 312527 2013 (2013)
60. Ye, J.J.: Necessary optimality conditions for multiobjective bilevel programs. Math. Oper. Res. 36(1), 165–184 (2011)
61. Ye, J.J., Zhu, D.L.: Optimality conditions for bilevel programming problems. Optimization 33, 9–27 (1995)
62. Ye, J.J., Zhu, D.L.: New necessary optimality conditions for bilevel programs by combining MPEC and the value function approach. SIAM J. Optim. 20, 1885–1905 (2010)

A Direct Method for Determining the Lower Convex Hull of a Finite Point Set in 3D

Thanh An Phan[1,2] and Thanh Giang Dinh[2,3]

[1] Institute of Mathematics, Vietnam Academy of Science and Technology, Hanoi, Vietnam
[2] CEMAT, Instituto Superior Técnico, University of Lisbon, Lisbon, Portugal
[3] Department of Mathematics, Vinh University, Vinh City, Vietnam
dtgiang@math.ist.utl.pt

Abstract. Determining the convex hull, its lower convex hull, and Voronoi diagram of a point set is a basic operation for many applications of pattern recognition, image processing, and data mining. To date, the lower convex hull of a finite point set is determined from the entire convex hull of the set. There arises a question "How can we determine the lower convex hull of a finite point set without relying on the entire convex hull?" In this paper, we show that the lower convex hull is wrapped by lower facets starting from an extreme edge of the lower convex hull. Then a direct method for determining the lower convex hull of a finite point set in 3D without the entire convex hull is presented. The actual running times on the set of random points (in the uniform distribution) show that our corresponding algorithm runs significantly faster than the incremental convex hull algorithm and some versions of the gift-wrapping algorithm.

Keywords: Convex Hull, Extreme Edge, Gift-wrapping Algorithm, Lower Convex Hull, Pattern Recognition, Voronoi Diagram.

1 Introduction

The convex hull and its lower convex hull of a finite point set in 3D have important applications in computer graphics, computer aided design, pattern recognition, and tomography (see [1] and [4]). The convex hull and its lower convex hull can be used to obtain characteristic shape measurements to enable objects to be classified by automatic pattern recognition. They can be constructed by many known algorithms, for example, the divide-and-conquer algorithm and the incremental algorithm (see [9]) and versions of the gift-wrapping algorithm (see [2], [10], [11]). Experiments have revealed that algorithms to compute the convex hull of n points in 3D would appear to have a worst-case complexity of $O(n^2)$, and not of $O(n\log n)$ as was previously thought (discovered by Day in [3]). The 3D convex hull of a set of points is also an intermediate stage in several other important geometric algorithms, for example, Delaunay triangulations and Voronoi diagram generation. By the way Voronoi diagram is a tool in pattern recognition. For example, face recognition based on fusion of Voronoi diagrams (see [4] and [6]). Determining the Voronoi diagram of a finite point set is equivalent to determining the lower convex hull of the corresponding point set (see [8]). To date, the

The original version of this chapter was revised: Acknowledgement section has been updated. The erratum to this chapter is available at 10.1007/978-3-319-17996-4_37

H.A. Le Thi et al. (eds.), *Advanced Computational Methods for Knowledge Engineering*,
Advances in Intelligent Systems and Computing 358, DOI: 10.1007/978-3-319-17996-4_2

lower convex hull of a finite point set is determined from the entire convex hull of the set. There arises the question "How can we determine the lower convex hull of a finite point set without relying on the entire convex hull?"

In this paper, we use a direct method to solve this question. Firstly, some properties of lower facets of the lower convex hull of a finite point set are presented (Propositions 2 - 3). Secondly, we show that a lower convex hull can be wrapped by lower facets starting from an extreme edge of the lower convex hull (Proposition 1 and Theorem 1). As a result, an entire convex hull, and therefore the convex hull algorithms mentioned above, are not really necessary for determining the lower convex hull. That is also the reason why our algorithm (which is introduced in Section 4.2) for determining the lower convex hull of a finite point set runs significantly faster than any known incremental algorithm (implemented in [9]) and some versions of the gift-wrapping algorithm, e.g., [2] (Table 1 and Figure 3). Moreover, we deal with the degenerate case in which four or more points are coplanar in 3D. In Section 5.2, we describe an application to compute Voronoi diagrams on a sphere (the problem is introduced by H. S. Na et al. in [7]), by determining Delaunay triangulation via the corresponding lower convex hull. The running times described by Table 2 and Figure 3 show the efficiency of our algorithm.

2 Preliminaries

Before starting the analysis, we recall some definitions and properties. For any points p, q in space, set $[p, q] := \{(1-\lambda)p + \lambda q : 0 \le \lambda \le 1\}$ and denote pq ($[p, q]$, respectively) the directed line (directed segment, respectively) through the points p and q. Given three points $p = (p_x, p_y, p_z), q = (q_x, q_y, q_z), t = (t_x, t_y, t_z)$, and a vector normal $\overrightarrow{n} = (n_x, n_y, n_z)$ to the plane (denoted by (p, q, t)) through three points p, q, t, all the points on the plane (p, q, t) satisfy the equation $xn_x + yn_y + zn_z = d$, where $d := q_x n_x + q_y n_y + q_z n_z$.

Definition 1. *(see [9]) Suppose that $n_z \neq 0$ and $w = (w_x, w_y, w_z)$. If $w_z > w_z^* := (d - w_x n_x - w_y n_y)/n_z$ ($w_z < w_z^*$, respectively) then w is called above (below, respectively) the plane.*

For $p, q, t, w \in \mathbb{R}^3$, set

$$V(p, q, t, w) = \begin{vmatrix} p_x & p_y & p_z & 1 \\ q_x & q_y & q_z & 1 \\ t_x & t_y & t_z & 1 \\ w_x & w_y & w_z & 1 \end{vmatrix}. \quad (1)$$

Then, for $\overrightarrow{n} = (n_x, n_y, n_z) = \overrightarrow{pq} \times \overrightarrow{pt}$,

$$V(p, q, t, w) = d - n_x w_x - n_y w_y - n_z w_z. \quad (2)$$

Let P be a finite set of points in $\mathbb{R}^3$. Then the convex hull CH(P) of P is a convex polytope ([[5], pp. 43-47]). A convex polytope is described by means of its boundary which consists of faces. Each face of a convex polytope is a convex set (that is, a lower-dimensional convex polytope), a k-face denotes a k-dimensional face (that is, a face whose affine hull has dimension k). If a convex polytope is 3-dimensional, its 2-faces are called facets, its 1-faces are called edges, and its 0-faces are vertices (see [[10], page 97]). If no four points p, q, t, w of P are coplanar, then all the facets of CH(P) are triangular. Let f be a triangular facet of CH(P) and let p, q, and t be three vertices of f. Then the facet f has an orientation in the sense that CH(P) lies in one side of the plane containing f and the other side is completely empty. In this case we assume that $V(p, q, t, w)$ is positive for all $w \in P \setminus \{p, q, t\}$. It is shown that this is equivalent to p, q, t form a counter-clockwise circuit when viewed from the side away from w (see [[9], page 23]).

Definition 2. *(see [[8], page 80]) A lower facet is a triangle whose three vertices belong to P such that all points of P are on or above the plane passing these vertices. Lower convex hull, denoted by $CH_L(P)$, consists of all lower facets of the convex hull. We call a facet of $CH(P)$ which is not a lower facet is an upper facet.*

We denote the triangular face containing three ordered vertices $p, q, t \in P$ by (p, q, t) or (e, t), with $e := [p, q]$. We choose $\overrightarrow{n} = \overrightarrow{pq} \times \overrightarrow{pt}$ as the respective normal vector of the plane passing (p, q, t). Given edge $[p, q]$ of CH$_L(P)$, to determine if three points $p, q, t \in P$ form a lower facet, we determine if w is above the plane passing (p, q, t) for all $w \in P \setminus \{p, q, t\}$. Four points p, q, t, w are considered be coplanar if

$$|V(p, q, t, w)| = 0. \tag{3}$$

The problem of dealing with the degenerate case in which four or more points are coplanar in 3D will be solved in this paper.

3 Lower Convex Hull of a Finite Point Set in 3D

Take $P = \{p_i = (x_i, y_i, z_i) \in \mathbb{R}^3,\ i = 0, 1, \ldots, n-1\}, n \geq 4$. In this section, we determine an edge of CH$_L(P)$. Set $P' = \{p'_0, p'_1, \ldots, p'_{n-1}\}$ with the coordinates $(0, y_i, z_i), i = 0, 1, \ldots, n-1$ (this is the projection of P in the direction parallel to the x-coordinate axis onto the plane $0yz$). We view the plane $0yz$ from $x \approx +\infty$ in 3D and Oy axis is from right to left. Choose $p'_0 \in P'$ as follows

$$\text{On the plane } 0yz, \text{ find the rightmost lowest point of } P' \text{ and label it } p'_0. \tag{4}$$

It means that we find the lowest point firstly. In case there are several with the same minimum z-coordinate, we will choose the rightmost of the lowest points, i.e., the one with smallest y-coordinate. We now consider the following three cases:

- (A_1) If $p'_i = p'_0$ for all $i = 1, 2, ..., n-1$ then all points of P are collinear, so there doesn't exist CH$_L(P)$.
- (A_2) If p'_i belongs to a straight line $\mathcal{L}$ for all $i = 0, 1, ..., n-1$ then all points of P are coplanar and the plane which contains P is perpendicular to Oyz. We consider two cases of the straight line $\mathcal{L}$: if $\mathcal{L}$ is parallel to Oz, there doesn't exist CH$_L(P)$; otherwise, we have a degenerate case CH$_L(P)$ = CH(P), where CH(P) is the 2D convex hull of P.
- (A_3) P' contains at least three non-collinear points. Then,

$$\text{take } p'_1 \ (p'_{n-1}, \text{ respectively) such that all other points of } P' \text{ are}$$
$$\text{on or to the left (right, respectively) of } p'_0p'_1 \ (p'_0p'_{n-1}, \text{respectively).} \quad (5)$$

It was known that $[p_0, p_1]$ and $[p_0, p_{n-1}]$ are edges of CH(P) (see [[9], page 69], [[2], page 979]). Now we will show that at least one of them is also an edge of CH$_L(P)$.

From now we assume that (A_3) holds. Let $q' \in \{p'_1; p'_{n-1}\}$, then $[p'_0, q']$ is an extreme edge of the 2D convex hull of P' on the plane Oyz. If $q' := p'_1$ ($q' := p'_{n-1}$, respectively) then we denote $q = p_1$ ($q = p_{n-1}$, respectively). Because we view the plane Oyz from $x \approx +\infty$ in 3D and Oy axis is from right to left, we can choose $q' \in \{p'_1; p'_{n-1}\}$ satisfies at least one of following conditions:

- $p_{0y} < q_y$ and all other points of P' are on or to the right of p'_0q',
- $p_{0y} > q_y$ and all other points of P' are on or to the left of p'_0q', (6)

where $p'_0 = (0, p_{0y}, p_{0z}), q' = (0, q_y, q_z)$ (see Fig. 1).

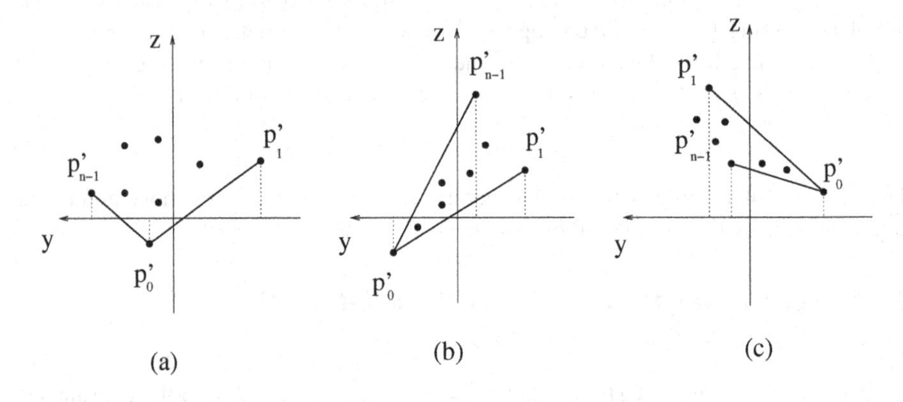

Fig. 1. The rightmost lowest point of P' on the plane Oyz is p'_0. We can choose $q' \in \{p'_1; p'_{n-1}\}$ satisfies at least one of conditions (6). In detail, case (a) $q' := p'_1$ or $q' := p'_{n-1}$, case (b) $q' := p'_1$ and case (c) $q' := p'_{n-1}$.

Proposition 1. *Suppose that $p'_0, q' \in P'$ are determined by conditions (4) and (6). Then, $[p_0, q]$ is an edge of CH$_L(P)$.*

Proof. Let Π be a plane through two points p'_0, q' and Π is parallel to the x-coordinate axis, then $p_0, q \in \Pi$. By the hypothesis that P' contains at least three non-collinear points, we have $P \not\subset \Pi$. It follows that there exists a point $u \in P \setminus \Pi$. Firstly we prove that u is above the plane Π for all the points $u \in P \setminus \Pi$.

A vector normal to plane Π is $\overrightarrow{n}^{\Pi} = (0, q_z - p_{0z}, p_{0y} - q_y)$. It follows that all the points on plane Π satisfy the equation: $xn^{\Pi}_x + yn^{\Pi}_y + zn^{\Pi}_z = d^{\Pi}$, where $n^{\Pi}_x = 0, n^{\Pi}_y = q_z - p_{0z}, n^{\Pi}_z = p_{0y} - q_y, d^{\Pi} = (q_z - p_{0z})p_{0y} + (p_{0y} - q_y)p_{0z}$.

For all $u = (u_x, u_y, u_z) \in P \setminus \Pi$, let $u\bar{u}$ be a directed line that is parallel to Oz axis. Since $n^{\Pi}_z \neq 0$, the plane Π is not parallel to Oz axis. Hence, $\Pi \cap u\bar{u} \neq \emptyset$. Take $u^* := \Pi \cap u\bar{u}$ then $u^*_z = \frac{(u_y - p_{0y})(q_z - p_{0z})}{q_y - p_{0y}} + p_{0z}$.

By (6), we can suppose that $p_{0y} < q_y$, and $u' = (0, u_y, u_z)$ is to the right of the directed line p'_0q', for all $u \in P \setminus \Pi$. Since u' is to the right of p'_0q', we have $\frac{(u_y - p_{0y})(q_z - p_{0z})}{q_y - p_{0y}} + p_{0z} < u_z$.

It means that $u^*_z < u_z$ and thus u is above the plane Π for all the points $u \in P \setminus \Pi$.

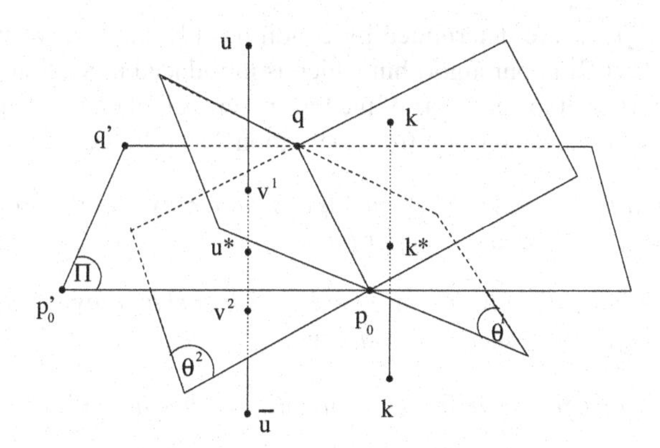

Fig. 2. The plane Π is perpendicular to Oyz and through p'_0, q'. For $u \in P\setminus\Pi$, let $u\bar{u}$ be a directed line that is parallel to Oz axis and we define $u^* := \Pi \cap u\bar{u}$; $v^i := \Theta^i \cap u\bar{u}$, ($i = 1$ or 2).

On the other hand, since $[p'_0, q']$ is an extreme edge of the 2D convex hull of P' on the plane Oyz, $[p_0, q]$ is an edge of CH(P) as shown in [[9], page 69].

Let $\Theta^1 = (p_0, q, t)$ and $\Theta^2 = (p_0, k, q)$ be two distinct facets of CH(P), then $t, k \in P \setminus \Pi$. Hence, $s(t) > 0$ and $s(k) > 0$.

Denote $\overrightarrow{n}^i = (n^i_x, n^i_y, n^i_z)$ the vector normal to plane Θ^i, ($i = 1, 2$), where $\overrightarrow{n}^1 = \overrightarrow{p_0q} \times \overrightarrow{p_0t}$, $\overrightarrow{n}^2 = \overrightarrow{p_0k} \times \overrightarrow{p_0q}$. We note that

$$\begin{aligned} n^1_x &= (q_y - p_{0y})(t_z - p_{0z}) - (t_y - p_{0y})(q_z - p_{0z}) = s(t) > 0, \\ n^2_z &= -(q_x - p_{0x})(k_y - p_{0y}) + (k_x - p_{0x})(q_y - p_{0y}). \end{aligned} \tag{7}$$

We now assume that Θ^1, Θ^2 are two upper facets of CH(P). Then, at least one of n^1_z and n^2_z is non-zero. Suppose without loss of generality that $n^1_z \neq 0$. Hence, the plane Θ^1 is not parallel to Oz axis. It implies that $\Theta^1 \cap u\bar{u} \neq \emptyset$. Take $v^1 := \Theta^1 \cap u\bar{u}$ then $v^1_z = \frac{n^1_x(p_{0x} - u_x) + n^1_y(p_{0y} - u_y)}{n^1_z} + p_{0z}$. Since $\Theta^1 = (p_0, q, t)$ is a facet of CH(P), u is on the positive side of Θ^1 for all $u \in P$, $V(p_0, q, t, u) = d^1 - n^1_x u_x - n^1_y u_y - n^1_z u_z > 0$. Moreover, since Θ^1 is a upper facet of CH(P), $v^1_z > u_z$ for all $u \in P\setminus\Pi$ (by Definition 2), meaning $(d^1 - n^1_x u_x - n^1_y u_y)/n^1_z > u_z$ (by Definition 1). Therefore, we get $n^1_z > 0$. On the other hand, $u^*_z < u_z$ implies that $v^1_z > u^*_z$ for all $u \in P \setminus \Pi$. For all the points $u \in P \setminus \Pi$, we have $v^1_z - u^*_z = \frac{n^1_x[(p_{0x} - u_x)(q_y - p_{0y}) + (q_x - p_{0x})(u_y - p_{0y})]}{n^1_z(q_y - p_{0y})}$. Choose $u = k$ and thanks to $v^1_z > u^*_z$ for all $u \in P \setminus \Pi$, then $v^1_z - k^*_z = \frac{-n^1_x n^2_z}{n^1_z(q_y - p_{0y})} > 0$, where $k^* := \Pi \cap k\bar{k}$ and the directed line $k\bar{k}$ is parallel to Oz axis (since the plane Π is not parallel to Oz axis, $\Pi \cap k\bar{k} \neq \emptyset$). By (7) and $p_{0y} < q_y$, $n^1_z > 0$, we see that $n^2_z < 0$.

Since $\Theta^2 = (p_0, k, q)$ is a facet of CH(P), $V(p_0, k, q, w) = d^2 - n_x^2 w_x - n_y^2 w_y - n_z^2 w_z > 0$, for all $w \in P$. By Definition 1 and $n_z^2 < 0$, w is above the plane Θ^2. This is a contradiction with the assumption that Θ^2 is an upper facets of CH(P). Using the arguments similar to the case $n_z^2 \neq 0$, we conclude that Θ^1 or Θ^2 is a lower facet, thus $[p_0, q]$ is an edge of CH$_L(P)$. □

In this paper, if p_0', q' are determined by conditions (4) and (6), we use $[p_0, q]$ as the first edge of CH$_L(P)$ in our algorithm which is introduced in Section 4.2. We now present some properties of lower facets of the lower convex hull of a finite point set. We can come to an immediate conclusion from Definition 2 as follows.

Proposition 2. *In a lower convex hull, an edge is shared by exactly either two lower facets or by one lower facet and one upper facet.*

Definition 3. *An edge of the lower convex hull which is shared by exactly one lower facet and one upper facet is called a bound-edge.*

Proposition 3. *For* $a, b, p \in P$, *if* $\vec{n} = (n_x, n_y, n_z) = \vec{ab} \times \vec{ap}$ *and* $n_z \neq 0$, *then*

(i) If (a, b, p) *is a facet of CH*(P) *and* $n_z < 0$ *then it is a lower facet.*
(ii) If $n_z > 0$ *then* (a, b, p) *is not a lower facet.*

Proof. It follows directly from (2). □

4 Algorithm for Determining the Lower Convex Hull of a Finite Point Set

Recall that $P = \{p_i = (x_i, y_i, z_i),\ i = 0, 1, \ldots, n-1\}, n \geq 4$. If $v \in P$ is indexed i (i.e., $v = p_i$) then set $v_{next} := p_{i+1}$. Let $\mathcal{E}_L(P)$ be the set of edges of CH$_L(P)$. By Proposition 1, if p_0', q' are determined by conditions (4) and (6) then $\mathcal{E}_L(P) \neq \emptyset$. If there exist four or more coplanar points in P then facets of CH(P) are possible convex polygons which have more three edges. Based on Proposition 3, the following procedure either determines a lower facet through e or returns p_0 or p_{n-1}.

4.1 Determining Lower Facets

Given $P \subset \mathbb{R}^3$, *edge* $e := [a, b]$ *of* CH$_L(P)$. Procedure **LF**(e, P) (stands for **L**ower **F**acets) *either determines a set of points* $LP = \{p_{l1}, ..., p_{lm}\} \subset P$ *such that the polygon* $\langle p_{l1}, ..., p_{lm}\rangle$ *is a lower facet through e or returns* $LP = \{p_0\}$ *or* $LP = \{p_{n-1}\}$.

1: **procedure LF**(e, P)
2: Take $p_l \in P$. $l := 0$.
3: **while** $p_l \in P$ **do**
4: Consider (e, p_l) with $p_l = (p_{l_x}, p_{l_y}, p_{l_z})$. Assume that $\vec{ab} \times \vec{ap_l} = (n_{l_x}, n_{l_y}, n_{l_z})$ and $d := p_{l_x} n_x + p_{l_y} n_y + p_{l_z} n_z$. Consider an array $CP := \emptyset$.
5: **while** $p_l = a$ or $p_l = b$ **do**
6: $l \leftarrow l + 1$

7: **if** $p_{l+1} = p_0$ and ($p_l = a$ or $p_l = b$) **return** p_{n-1} ▷ $LP = \{p_{n-1}\}$
8: **if** $n_{l_z} < 0$
9: **while** $v \in P \setminus \{a, b, p_l\}$ **do**
10: **if** $n_{l_x} v_x + n_{l_y} v_y + n_{l_z} v_z = d$
11: $CP \leftarrow CP \cup \{v\}$
▷ *CP is the set of coplanar points of $\{a, b, p_l\}$*
12: **else**
13: **if** $n_{l_x} v_x + n_{l_y} v_y + n_{l_z} v_z < d$ ▷ *due to Proposition 3 (i), v is above the plane passing (e, p_l)*
14: $v \leftarrow v_{next}$
15: **else**, $l \leftarrow l + 1$
16: **if** $p_l = p_0$ **return** p_0 ▷ $LP = \{p_0\}$
17: **else, goto 4**
18: **else**, $l \leftarrow l + 1$ ▷ *due to Proposition 3 (ii), (e, p_l) is not a lower facet*
19: **if** $p_{l+1} = p_0$ and ($p_l = a$ or $p_l = b$) **return** p_{n-1} ▷ $LP = \{p_{n-1}\}$
20: **if** $p_l \neq p_0$ **goto 4**
21: **else, return** p_0 ▷ $LP = \{p_0\}$
22: $CP \leftarrow CP \cup \{p_l, a, b\}$.
23: $LP \leftarrow \text{CH}(CP)$ ▷ $\text{CH}(CP)$ *is the 2D convex hull of point set CP*
24: **return** LP.
25: **end procedure**

The main algorithm for finding all the facets of $\text{CH}_L(P)$ is modified from the gift-wrapping algorithm (see [10], page 134) as follows.

4.2 Main Algorithm and Correctness

In the following, all edges of a newly obtained facet are put into a file named $\mathcal{E}_L(P)$ of all edges which are candidates for being used in steps. Let us denote by F_e and F'_e two faces sharing the edge e. Note that each e of $\mathcal{E}_L(P)$ is a candidate if either F_e or F'_e, but not both, has been generated in the gift-wrapping exploration. The lower convex hull is wrapped by lower facets starting from an extreme edge determined by (6) of the lower convex hull.

Input: $P := \{p_i = (x_i, y_i, z_i) \in \mathbb{R}^3, i = 0, 1, \ldots, n-1\}, n \geq 4$.
Output: Set $\mathcal{Q}$ of all the faces of $\text{CH}_L(P)$.
1: Set $P' := \{p'_0, p'_1, \ldots, p'_{n-1}\}$ with $p'_i = (0, y_i, z_i), i = 0, 1, \ldots, n-1$.
2: Take p'_0 determined by (4).
3: **If** $p'_i = p'_0$ for all $i = 1, 2, ..., n-1$ **then** there doesn't exist $\text{CH}_L(P)$ ▷ *by* (A_1)
4: **else**
5: **if** p'_i belongs to a straight line $\mathcal{L}$ for all $i = 0, 1, ..., n-1$ **then**
6: **if** $\mathcal{L} \parallel Oz$ **then** there doesn't exist $\text{CH}_L(P)$ ▷ *by* (A_2)
7: **else** $\text{CH}_L(P) = \text{CH}(P)$ ▷ $\text{CH}(P)$ *is the 2D convex hull of P*
8: **else** Take p'_1 (or p'_{n-1}, respectively) determined by (6);
9: Set $e := [p_0, p_1]$ (or $e := [p_0, p_{n-1}]$, respectively);
10: Consider a queue $\mathcal{Q} := \emptyset$ and a file $\mathcal{E}_L(P) := \emptyset$;
11: Call **LF**(e, P) to obtain a set of points $LP = \{p_{l1}, ..., p_{lm}\}$ such that
12: the polygon $F_e = \langle p_{l1}, ..., p_{lm} \rangle$ is a lower facet through e;

13: Push edges of F_e into $\mathcal{E}_L(P)$;
14: Push F_e into $\mathcal{Q}$;
15: **while** ($\mathcal{Q} \neq \emptyset$) **do**
16: **begin** Extract F_e from the front of $\mathcal{Q}$;
17: $T :=$ edges of F_e;
18: **for each** $e \in T \cap \mathcal{E}_L(P)$ **do** ▷ *e is a candidate*
19: **begin** call **LF**(e, P) to obtain a set of points
20: $LP = \{p_{l1}, ..., p_{lm}\}$ such that the polygon
21: $\langle p_{l1}, ..., p_{lm}\rangle$ is a lower facet through e;
22: **if** e is not a bound-edge **then** ▷ *i.e.,* $\langle p_{l1}, ..., p_{lm}\rangle$
▷ *is the lower facet* F'_e *sharing e with* F_e
23: insert into $\mathcal{E}_L(P)$ all edges of F'_e not yet present
24: and delete all those already present;
25: push F'_e into $\mathcal{Q}$;
26: **end**;
27: output F_e;
28: **end**;

Theorem 1. *The algorithm described in Section 4.2 determines* $CH_L(P)$ *in* $O(nk)$ *time, where* n *is the number of points of* P *and* k *is the number of facets of* $CH_L(P)$.

Proof. Under the assumptions (4) and (6), there exists the edge e_1 which belongs to $\mathcal{E}_L(P)$. The procedure **LF**(e_1, P) determines a lower facet $F = \langle p_{l1}, ..., p_{lm}\rangle$ through the edge e_1. Consider next candidate $e_2 = [p_{li}, p_{l\,i+1}]$. If this edge is not a bound-edge then the procedure determines a lower facet F' sharing e_2 with F. Otherwise, e_2 is a bound-edge. Then $e_3 = [p_{l\,i+1}, p_{l\,i+2}]$ or $e_3 = [p_{l\,i+2}, p_{l\,i+1}]$ is considered as a next candidate in the algorithm in Section 4.2. The process is similar with e_3. Hence, the algorithm determines all faces of the lower convex hull. We can compute the 2D convex hull in line 23 of the procedure **LF**(e, P) by using Graham's algorithm. It follows that this procedure runs in $O(n\log n)$ time. Note that the gift-wrapping algorithm for finding CH(P) runs in $O(nk)$ time, where n is the number of points of P and k is the number of facets of CH(P) (see [[10], pp. 134-136]). Because the works from line 10 to the end of the new algorithm are the same with ones of the gift-wrapping algorithm except the call **LF**(e, P), the algorithm for finding CH$_L(P)$ runs in $O(nk)$ time (where n is the number of points of P and k is the number of facets of CH$_L(P)$) after the works from line 1 to line 10 (which take $O(n)$ time). We conclude the algorithm determines the lower convex hull in $O(nk)$ time. □

5 Numerical Experiments

5.1 Determining the Lower Convex Hull of a Finite Point Set in 3D

Given a set of n points $P \subset \mathbb{R}^3$. We compute the 2D convex hull in line 23 of the procedure **LF**(e, P) by using Graham's algorithm implemented in [[9], page 77]. The actual running times of known convex hull algorithms for computing the entire convex hull then computing the lower convex hull are slower than our algorithm (computing the lower convex hull without relying on the entire convex hull). Table 1 and Fig. 3

show that our new algorithm runs significantly faster than the incremental algorithm (implemented in [9]) and a version of the gift-wrapping algorithm [2]. Note that the speedup is only shown for large n, as visible from Table 1. For the comparison to be meaningful, three implementations use the same code for file reading. The algorithms are implemented in C and compiled by the GNU Compiler Collection under Ubuntu Linux 10.4 and are executed on platform Intel(R) Core(TM) 4, CPU 3.07 GHz and 1.8 GB RAM.

Table 1. The actual running times (time in sec.) with the input is $P = \{p_i = (x_i, y_i, z_i), i = 0, 1, \ldots, n-1\} \subset \mathbb{R}^3$, where the set $\{(x_i, y_i), i = 0, 1, \ldots, n-1\}$ is randomly positioned in the interior of a square of size 100×100 (in the uniform distribution)

Number of points	Find $\mathrm{CH}_L(P)$ without entire convex hull Our Algorithm 4.2	Find $\mathrm{CH}_L(P)$, relying on entire convex hull	
		Incremental Algorithm	Modified Gift-wrapping Algorithm
1000	0.009839	0.020649	0.027743
10000	0.075452	0.624167	0.526191
50000	0.607473	13.301419	3.265638
70000	0.611422	25.906851	4.222786
100000	1.449733	55.162455	9.663950
200000	3.598728	281.797376	22.046477

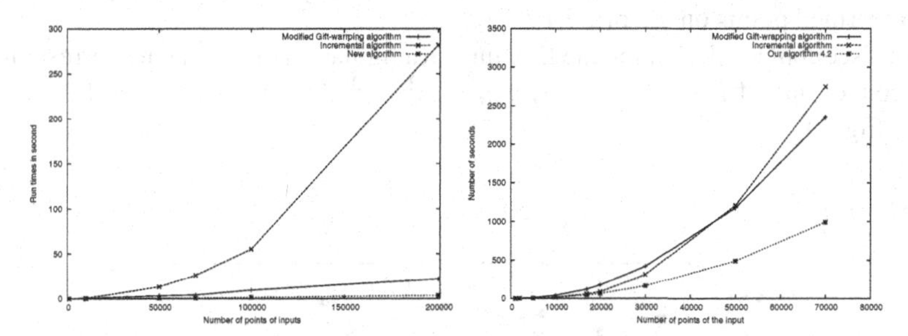

(a) The actual running times of Table 1 (b) The actual running times of Table 2

Fig. 3. The actual running times described in Table 1 and Table 2 show that our new algorithm (i.e., Algorithm 4.2) runs significantly faster than the incremental algorithm (implemented in [9]) and a version of the gift-wrapping algorithm [2]

5.2 Application for Computing Voronoi Diagrams on a Sphere

Let $P^* = \{p_i^* = (x_i, y_i) \in \mathbb{R}^2, i = 0, 1, \ldots, n-1\}, n \geq 4$. Take $P = \{p_i = (x_i, y_i, z_i) \in \mathbb{R}^3, i = 0, 1, \ldots, n-1\}$ such that $z_i = x_i^2 + y_i^2$, $i = 0, 1, \ldots, n-1$.

Theorem 2. *(see [[9], page 186]) The Delaunay triangulation of a set of points in two dimensions is precisely the projection to the $xy-$plane of the lower convex hull of the transformed points in three dimensions, transformed by mapping upwards to the paraboloid $z = x^2 + y^2$.*

Based on Theorem 2, we obtain the following sequential algorithm (see [8], [9], [10], etc) for the connection between the lower convex hull and Delaunay triangulation (in step 2 of this algorithm, we use Algorithm 4.2 for constructing the lower convex hull).

Given: Set $P^* = \{p_i^* = (x_i, y_i) \in \mathbb{R}^2,\ i = 0, 1, \ldots, n-1\}, n \geq 4.$
Find: Delaunay triangulation of P^*.

1. Create set $P = \{p_i = (x_i, y_i, z_i) \in \mathbb{R}^3,\ i = 0, 1, \ldots, n-1\}$ such that $z_i = x_i^2 + y_i^2,\ i = 0, 1, \ldots, n-1$.
2. Construct the lower convex hull $\mathrm{CH}_L(P)$ of P in $\mathbb{R}^3$.
3. Project all lower faces of $\mathrm{CH}_L(P)$ in the direction parallel to the $z-$coordinate axis onto the plane Oxy (containing P^*) and return the resulting Delaunay triangulation.

Given a set of points U on a unit sphere $S := \{(x, y, z) \in \mathbb{R}^3 : x^2 + y^2 + z^2 = 1\}$, the spherical Voronoi diagram $SV(U)$ of U has the same combinatorial structure as the convex hull of U (introduced by Na et al. in [7]). Let us denote by T the tangent plane to S at the point opposite to $\sigma = (0, 0, -1) \in S$, i.e., $T := \{(x, y, z) \in \mathbb{R}^3 : z = 1\}$. In [7], an important step in the algorithm for determining $SV(U)$ is computing the planar Voronoi diagram (or Delaunay triangulation) of suitably transformed points on T, with input is points on S. In detail, the point $\tau \in U$ is projected (transformed) onto the plane T from the fixed point σ (see Fig. 4). Hence, the transformed point $\tau^* \in T$ is computed by formula: $\tau = (x, y, z) \mapsto \tau^* = (\frac{2x}{z+1}, \frac{2y}{z+1}, 1)$. We denote P^* the set of all transformed points on the plane T.

In this section, we determine the Delaunay triangulation of P^* via the corresponding lower convex hull of $P = \{(x_i, y_i, z_i) : z_i = x_i^2 + y_i^2, (x_i, y_i) \in P^*, i = 0, 1, \ldots, n-1\}$ (see Fig. 5).

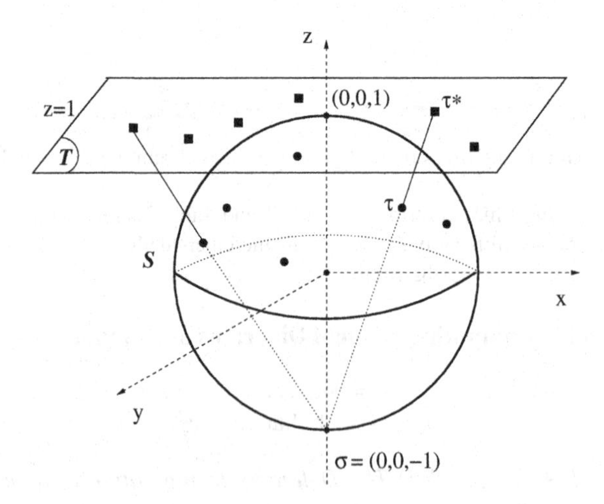

Fig. 4. The point τ (on the sphere S) is transformed onto the plane T from the fixed point σ, where T is the tangent plane to S at the point opposite to $\sigma = (0, 0, -1)$

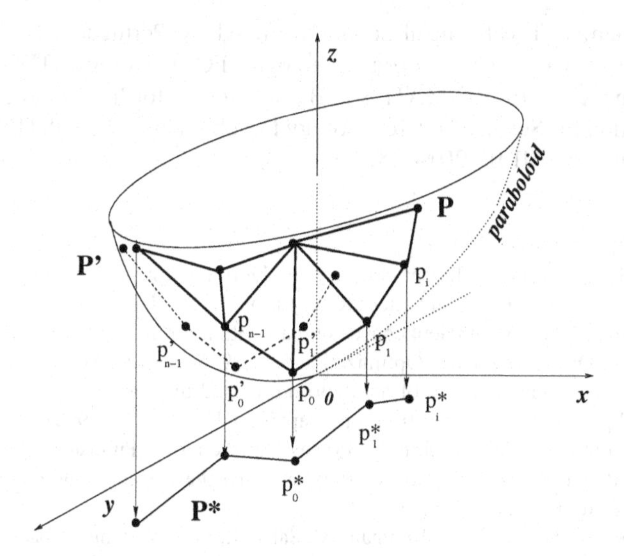

Fig. 5. P' (P^*, respectively) is the projection of P in the direction parallel to the x-coordinate axis (z-coordinate axis, respectively) onto the plane $0yz$ ($0xy$, respectively). If $[p'_0, p'_1]$ ($[p'_0, p'_{n-1}]$, respectively) is determined by (6), then $[p_0, p_1]$ ($[p_0, p_{n-1}]$, respectively) is an edge of $\mathrm{CH}_L(P)$.

Table 2 and Fig. 3 show that our new algorithm (i.e., Algorithm 4.2) runs significantly faster than the incremental algorithm (implemented in [9]) and a version of the gift-wrapping algorithm [2]. For the comparison to be meaningful, three implementations use the same code for file reading.

Table 2. The actual running times (time in sec.) of algorithms to determine the Delaunay triangulation of $P^* = \{(2x_i/(z_i+1), 2y_i/(z_i+1)) : (x_i, y_i, z_i) \in S,\ i = 0, 1, \ldots, n-1\}$ with the input is $U = \{p_i = (x_i, y_i, z_i) : x_i^2 + y_i^2 + z_i^2 = 1,\ i = 0, 1, \ldots, n-1\} \subset \mathbb{R}^3$, where the set $\{(x_i, y_i),\ i = 0, 1, \ldots, n-1\}$ is randomly positioned in the interior of a square of size 1×1 (in the uniform distribution, the error 10^{-13})

Number of points	Find $\mathrm{CH}_L(P)$ without entire convex hull Our Algorithm 4.2	Find $\mathrm{CH}_L(P)$, relying on entire convex hull Incremental Algorithm	Modified Gift-wrapping Algorithm
1000	0.131589	0.101180	0.323113
2000	0.522748	0.442554	1.372205
5000	3.543717	3.458963	9.963105
10000	14.996426	15.315926	39.254659
17000	45.839407	57.076066	120.063742
20000	65.911719	86.668035	175.638411
30000	164.105265	310.215363	417.372167

Acknowledgments. The financial supports offered by Portuguese National Funds through Fundação para a Ciência e a Tecnologia (FCT), Portugal, TWAS Research Grants Programme 13-054 RG/MATHS in Basic Sciences for Individuals, and the National Foundation for Science and Technology Development (NAFOSTED) Vietnam under grant number 101.01-2014.28.

References

1. Akl, S.G., Toussaint, G.: Efficient convex hull algorithms for pattern recognition applications. In: 4th Int'l Joint Conf. on Pattern Recognition, Kyoto, Japan, pp. 483–487 (1978)
2. An, P.T., Trang, L.H.: An efficient convex hull algorithm for finite point sets in 3D based on the Method of Orienting Curves. Optimization 62(7), 975–988 (2013)
3. Day, A.M.: An implementation of an algorithm to find the convex hull of a set of three-dimensional points. ACM Transactions on Graphics 9(1), 105–132 (1990)
4. Luo, D.: Pattern Recognition and Image Processing. Woodhead Publishing (1998)
5. McMullen, P., Shephard, G.C.: Convex Polytopes and the Upper Bound Conjecture. Cambridge University Press, Cambridge (1971)
6. Meethongjan, K., Dzulkifli, M., Rehman, A., Saba, T.: Face recognition based on fusion of Voronoi diagram automatic facial and wavelet moment invariants. International Journal of Video & Image Processing and Network Security 10(4), 1–8 (2010)
7. Na, H.S., Lee, C.N., Cheong, O.: Voronoi diagrams on the sphere. Computational Geometry 23(2), 183–194 (2002)
8. Okabe, A., Boots, B., Sugihara, K.: Spatial Tessellations: Concepts and Applications of Voronoi Diagrams, 2nd edn. John Wiley & Sons Ltd (2000)
9. O'Rourke, J.: Computational Geometry in C, 2nd edn. Cambridge University Press (1998)
10. Preparata, F.P., Shamos, M.I.: Computational Geometry - An Introduction, 2nd edn. Second Edition. Springer, New York (1988)
11. Sugihara, K.: Robust gift wrapping for the three-dimensional convex hull. Journal of Computer and System Sciences 49, 391–407 (1994)

A Hybrid Intelligent Control System Based on PMV Optimization for Thermal Comfort in Smart Buildings

Jiawei Zhu, Fabrice Lauri, Abderrafiaa Koukam, and Vincent Hilaire

IRTES-SET, UTBM, 90010 Belfort Cedex, France
{jiawei.zhu,fabrice.lauri,abder.koukam,vincent.hilaire}@utbm.fr

Abstract. With the fast development of human society, on one hand, environmental issues have drawn incomparable attention, so energy efficiency plays a significant role in smart buildings; on the other hand, spending more and more time in buildings leads occupants constantly to improve the quality of life there. Hence, how to manage devices in buildings with the aid of advanced technologies to save energy while increase comfort level is a subject of uttermost importance. This paper presents a hybrid intelligent control system, which is based on the optimization of the predicted mean vote, for thermal comfort in smart buildings. In this system, the predicted mean vote is adopted as the objective function and after employing particle swarm optimization the near-optimal temperature preference is set to a proportional-integral-derivative controller to regulate the indoor air temperature. In order to validate the system design, a series of computer simulations are conducted. The results indicate the proposed system can both provide better thermal comfort and consume less energy comparing with the other two intelligent methods: fuzzy logic control and reinforcement learning control.

Keywords: thermal comfort, smart building, intelligent control, particle swarm optimization, energy.

1 Introduction

United Nations Environment Programme [1] indicates that buildings use about 40% of global energy, 25% of global water, 40% of global resources, and they emit approximately 1/3 of Green House Gas (GHG) emissions. With the development of human society, environmental issues have drawn more and more attention. In this background, buildings can offer a great potential for achieving significant GHG emission reductions in different countries. Furthermore, energy consumption in buildings can be reduced by using advanced technologies and management. On the other hand, people spend greater part of their time in buildings. As the quality of life in home is increasingly considered as of paramount importance, many people constantly seek to improve comfort in their living spaces. Meanwhile, the popularization of the concept of home office makes the productivity in smart buildings economically significant. How to manage buildings in a proper

H.A. Le Thi et al. (eds.), *Advanced Computational Methods for Knowledge Engineering*,
Advances in Intelligent Systems and Computing 358, DOI: 10.1007/978-3-319-17996-4_3

way to improve energy efficiency and comfort level while reducing pollution at the same time is therefore a subject of uttermost importance.

Corresponding to the increasing demands for environment, comfort, energy, and productivity, intelligent computing and control methods are applied for improving comfort conditions in smart buildings thanks to the dramatically rapid development of information technologies. Widespread utilization of low-power, high-capacity, small form-factor computing devices, the proliferation of powerful but low-cost sensors and actuators, and ubiquitous networking technologies make the intelligent control more easily come true.

Many techniques have been used for controlling thermal comfort in smart buildings. Authors in [2] present a model predictive controller, which uses both weather forecast and thermal model , applied to the inside temperature control of real buildings. However, on one hand, it is difficult to obtain accurate thermal models for a variety of buildings. On the other hand, thermal comfort does not only relate to the environmental conditions like room temperature, but also factors about people themselves. In order to describe thermal comfort formally, P.O. Fanger [3] invents a thermal comfort model called Predictive Mean Vote index (PMV). Based on this model, Dounis and Manolakis [4] design a fuzzy logic system, which sets PMV and the ambient temperature as input variables while the heating power as output variables, to regulate the space heating system. But as the authors indicated, the proposed expert system works well only if the knowledge embedded in its rule base is sound. It means that a tuning and optimizing process is needed at the later stage in order to have good results, which is time consuming. In the work of K. Dalamagkidis *et al.*[5], they develop a reinforcement learning controller to achieve thermal comfort in buildings with minimal energy consumption. The reinforcement learning signal is a function of the thermal comfort and the energy consumption. However, the convergence speed of reinforcement learning in this application is very slow.

In this work, we propose a novel hybrid intelligent control system to maintain thermal comfort for occupants. This system utilizes Proportional-Integral-Derivative (PID) to control the heating system regularly at the lower level while employs Particle Swarm Optimization (PSO) to compute near-optimal setpoint inversely from PMV model based on present environmental and personal conditions at the higher level. The contribution of this work is threefold: firstly, the thermal comfort problem is formally described; secondly, a novel hybrid control method is proposed, with which occupants' thermal comfort can be improved and energy consumption can be reduced; thirdly, by conducting experiments and comparing with fuzzy logic control and reinforcement learning control, the better performance of our proposed method is proved. The rest of this paper is organized as follows. Section 2 mathematically describes the thermal comfort problem. Section 3 presents the proposed method in detail. Experimental results and analysis are given in section 4. Finally, we conclude in section 5.

2 Problem Description

2.1 Building Thermal Model

The room temperature is affected not only by auxiliary heating / cooling systems and electric appliances, but also by the solar radiation and the ambient temperature. According to [6], the heat balance of a building can be expressed as

$$\phi_h(t) + \phi_s(t) = \phi_t(t) + \phi_c(t) \tag{1}$$

where ϕ_h is the heat supplied by all internal heat sources; ϕ_s is the heat gained by solar radiation; ϕ_t is the heat loss through external contact; ϕ_c is the heat retained by the building.

In order to analyze the thermal dynamics of a building, we can consider it as a thermal network, which is analogous to an electric circuit network by regarding heat flows as electric current, temperature as voltage and treating thermal transmittance (U) and thermal capacitance (C) as electric resistance and electric capacitance respectively, as shown in Fig.1. In the figure, there are two nodes which represent the temperature inside and outside the room, and six sub-circuits which indicate physical components of the room, including room air, internal partitions, ceiling, floor and external walls.

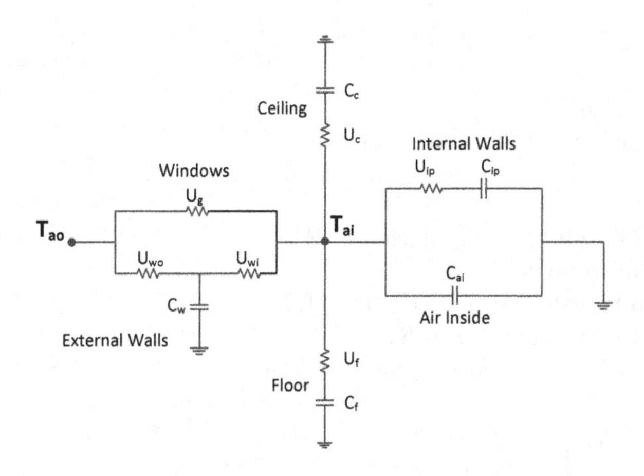

Fig. 1. Thermal Network

Before deriving state-space equations of the building, two definitions should be mentioned:

$$\phi = A \times U \times (T_1 - T_2) \tag{2}$$

where ϕ is the heat transfer in watts, A is the area in square metres, U is the thermal transmittance, T_1 is the temperature on one side of an object and T_2 is the temperature on the other side of the object.

$$C = \frac{\Delta Q}{\Delta T} \tag{3}$$

where C is the thermal capacitance, ΔQ is the change of heat and ΔT is the change of temperature.

Now the thermal system of the building depicted in Fig.1 can be expressed by Equations (4) - (8), except the sub-circuit of windows, because of its so tiny thermal mass that we assume windows have not the property of thermal capacitance to store heat:

$$\frac{dT_w}{dt} = \frac{A_w}{C_w}\Big[U_{wi}(T_{ai} - T_w) + U_{wo}(T_{ao} - T_w)\Big] \tag{4}$$

$$\frac{dT_f}{dt} = \frac{A_f}{C_f}\left[\frac{pQ_s}{A_f} + U_f(T_{ai} - T_f)\right] \tag{5}$$

$$\frac{dT_c}{dt} = \frac{A_c}{C_c}\Big[U_c(T_{ai} - T_c)\Big] \tag{6}$$

$$\frac{dT_{ip}}{dt} = \frac{A_{ip}}{C_{ip}}\left[\frac{(1-p)Q_s}{A_{ip}} + U_{ip}(T_{ai} - T_{ip})\right] \tag{7}$$

$$\begin{aligned}\frac{dT_{ai}}{dt} = \frac{1}{C_{ai}}\Big[&Q_p + Q_e + (A_g U_g + U_v)(T_{ao} - T_{ai}) \\ &+ A_w U_{wi}(T_w - T_{ai}) + A_f U_f(T_f - T_{ai}) \\ &+ A_c U_c(T_c - T_{ai}) + A_{ip} U_{ip}(T_{ip} - T_{ai})\Big]\end{aligned} \tag{8}$$

In above equations,

- Q_p heat supplied by the heating system in W,
- Q_e heat gained by using electrical equipments in W,
- Q_s solar radiation through glazing in W,
- T temperature in K,
- U thermal transmittance in $W/(m^2 \cdot K)$,
- C thermal capacitance in J/K,
- p fraction of solar radiation entering floor.

Subscripts:

- w external wall,
- wi inside part of external wall,
- wo outside part of external wall,
- f floor,
- c ceiling,
- ip internal partition,
- ao outdoor air,
- ai indoor air,
- v ventilation
- g glazing

Above equations can be stacked using the state-space notation:

$$\dot{\mathbf{x}} = \mathbf{A}\mathbf{x} + \mathbf{B}\mathbf{u} \tag{9}$$

where $\dot{\mathbf{x}}$ is a vector of derivatives of temperatures of external walls(T_w), floor(T_f), ceiling(T_c), internal partitions(T_{ip}) and air inside(T_{ai}), $\mathbf{A}$, $\mathbf{B}$ are matrices of coefficients, $\mathbf{x}$ is a vector of states and $\mathbf{u}$ is the input vector, including Q_p, Q_e, Q_s and T_{ao}. The area of each component of the building is known after choosing a physical building model, and the properties of different building materials can be obtained from [7].

2.2 Thermal Comfort

Thermal comfort is the condition of mind that expresses satisfaction with the thermal environment and is assessed by subjective evaluation [8]. The Predictive Mean Vote index (PMV) derived by P.O. Fanger [3] stands among the most recognized thermal comfort models, which predicts the mean value of the votes of a large group of persons on the 7-point thermal sensation scale, based on the heat balance of the human body. The 7-point thermal sensation scale separately indicates -3 for cold, -2 for cool, -1 for slightly cool, 0 for neutral, +1 for slightly warm, +2 for warm, and +3 for hot. Moreover, the ISO recommends maintaining PMV at level 0 with a tolerance of 0.5 as the best thermal comfort. The calculation of PMV is non-linear and non-convex [10]. It is affected by six factors, four environmental-dependent: air temperature T_{ai}, mean radiant temperature (MRT) T_{mrt}, air velocity v_{air}, and relative humidity h_r and two personal-dependent: occupant's clothing insulation I_{cl} and activity level M, as given below:

$$PMV = F(T_{ai}, T_{mrt}, v_{air}, h_r, I_{cl}, M) \tag{10}$$

In our problem, based on other parameters and an optimal PMV value we need to inversely calculate the optimal air temperature as a proper preference for the heating controller. This can be mathematically expressed as:

$$T_{ai} = G(PMV^*, T_{mrt}, v_{air}, h_r, I_{cl}, M) \tag{11}$$

3 Proposed Method

In order to solve the thermal comfort problem presented in the last section, we propose a novel hybrid intelligent control method, as can be seen in Fig.2. In this figure, occupants set PMV preference (recall that 0 represents comfort), and according to the present sensing data from the environment, the Inverse PMV Optimization module can provide a proper indoor temperature setpoint. Based on this setpoint and the sensing indoor temperature, the PID controller can calculate an applicable heating power for the space heating system.

In the Inverse PMV Optimization module, it has been proved that the PMV calculation is non-linear and non-convex [10], so classic optimization tools are

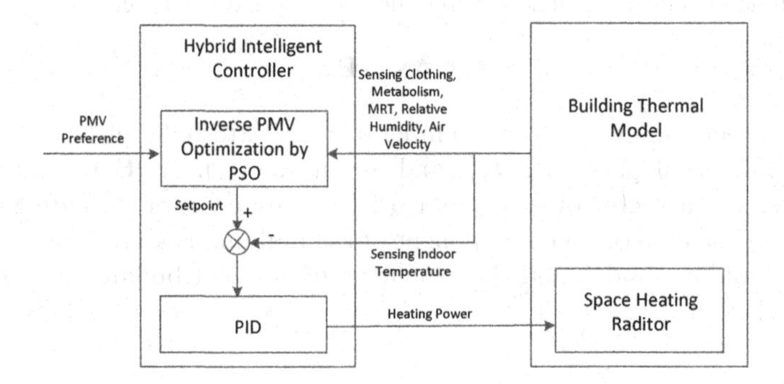

Fig. 2. Hybrid Intelligent Controller Diagram

not suitable. In this research, we use PSO to search for the near-optimal solutions. Particle Swarm Optimization (PSO), which is derived from simulating social behavior as a stylized representation of the movement of organisms in a bird flock or fish school, is originally proposed and developed by [11,12]. It is a metaheuristic algorithm that has been turned out to be powerful to solve the complex non-linear and non-convex optimization problems [13]. Moreover, it has several other advantages, such as fewer parameters to adjust, easier to escape from the local optimal solutions, and so on.

In PSO, a population of candidate solutions, here dubbed particles that include position vector $\mathbf{x}$ and velocity vector $\mathbf{v}$, is randomly generated around the search-space initially. After that they are iteratively updated to simulate the movement around the search-space according to mathematical formulae over the particles' position and velocity, as expressed below:

$$\begin{aligned} \mathbf{v}_i^{k+1} =& w \cdot \mathbf{v}_i^k + c_1 \cdot rand() \cdot ({\mathbf{p_{best}}}_i^k - \mathbf{x}_i^k) \\ & + c_2 \cdot rand() \cdot ({\mathbf{g_{best}}}^k - \mathbf{x}_i^k) \end{aligned} \tag{12}$$

$$\mathbf{x}_i^{k+1} = \mathbf{x}_i^k + \mathbf{v}_i^{k+1} \tag{13}$$

where w is inertia weight, c_1 and c_2 are acceleration constants, $rand()$ generates random value between 0 and 1, ${\mathbf{p_{best}}}_i^k$ is the local best position, ${\mathbf{g_{best}}}^k$ is global best position, i is the particle index, and k is the iteration index. For Equation 12, the first part expresses particle's inertia of previous action, the second part reflects particle's cognition that stimulates the particle to minish errors, and the last part is called social part, which indicates the information sharing and cooperation among particles.

At the lower level of the proposed method, we use PID to regulate the heating power regularly. There are three reasons that we choose it: firstly due to the huge thermal mass of the building, the indoor temperature change is a rather slow process, that means it does not need an exquisite control method; secondly the work of Paris *et al.*[14] has proved that PID is already good enough to regulate the indoor temperature comparing with other hybrid methods such as PID-FLC and PID-MPC; thirdly, PID can be implemented easily which only needs three scalar parameters. A typical equation, that describes a PID regulator, is the following:

$$Q_p(t) = K_p e(t) + K_d \frac{de(t)}{dt} + K_i \int e(t)dt \tag{14}$$

where K_p is the proportional parameter, K_d is the derivative parameter, and K_i is the integral parameter.

4 Experimentation

In experiments, we compare our proposed hybrid intelligent control approach, denoted as HIC, with fuzzy logic control (FLC) and reinforcement learning control (RLC). In PSO part of HIC, inertia weight $w = 0.4$, acceleration constants $c_1 = c_2 = 2$, particle number is 30, and maximum iteration number is 150. In PID part, proportional gain is 2000, integral gain is 0.5, and derivative gain is 0. The maximum power of the electric heating radiator is $2000W$.

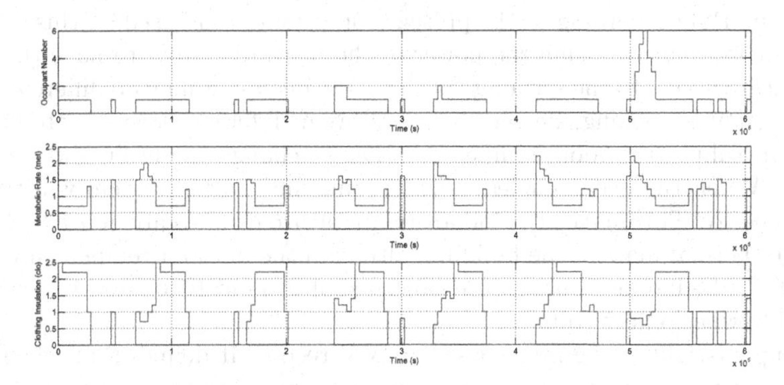

Fig. 3. Simulation of Occupant's Number, Metabolic Rate, and Clothing Insulation of One Week

The simulations of occupant number, metabolic rate, and clothing insulation are shown in Fig. 3, which depict the general life of a person who works or studies during the daytime, has lunch at home sometimes, and returns home cooking, taking exercise, watching TV, etc. in the evenings regularly on weekdays, and invites friends to have a party at weekend.

Fig. 4 show the simulation results of our proposed method. The simulation environment is assumed in Oslo with a duration of one week in winter from 08/01/2014 00:00 to 14/01/2014 11:59, 604800 seconds in total. The ambient

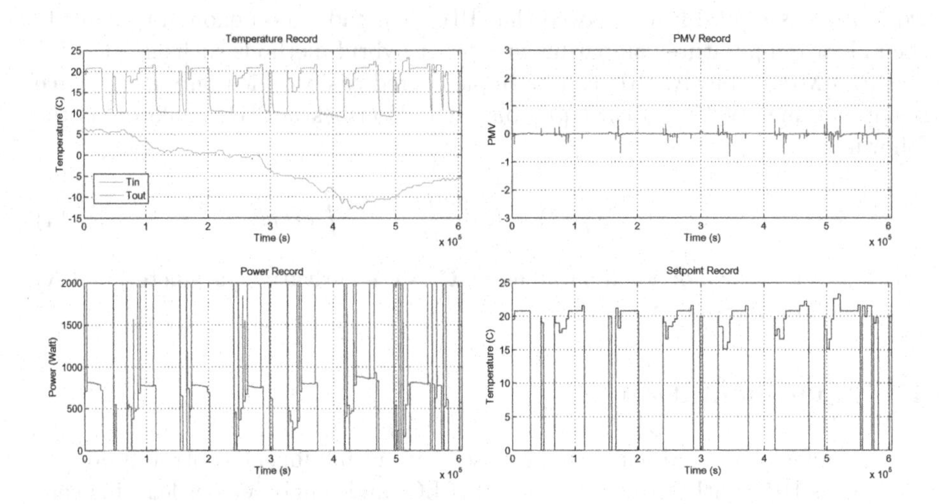

Fig. 4. Simulation Results of HIC Method

weather data is obtained from [15]. The upper-left subfigure records the variations of indoor and outdoor air temperature, in yellow line and blue line respectively. When the occupant is at home, the system will compute an optimal setpoint by PSO according to the present conditions, and based on this setpoint the PID controller can tune the power of the heating system to have the room temperature access to the setpoint. Because the occupant may do different activities like sitting, cooking, sleeping, etc. and wear different clothes with different insulation values, the room temperature has to change to obtain good thermal comfort. When the occupant leaves the room, the heating system will turn off to save energy. This makes the indoor temperature drop to about $10°C$. Due to the huge thermal mass of the building, although the outdoor temperature drops below $0°C$, the indoor temperature can be kept around $10°C$, and this is called thermal flywheel effect [16].

The upper-right subfigure shows the PMV record. It defines that when there is no one in the room, the PMV is set to 0. From the subfigure, it can be seen that the PMV can be kept between +0.5 and -0.5 to assure excellent thermal comfort, except a few minutes' slight cool or warm feeling, which are caused by (1) leaving and then returning home in extremely cold weather, in which case the heating system is unable to warm the room instantly; and (2) the arriving of a mount of visitors from whom the metabolism and thermal radiation make the room over warm. The two bottom subfigures indicate the output heating power and the intelligent setpoint picking during this simulation.

Fig. 5 is the simulation results using FLC and RLC separately. By comparing these 4 figures with Fig. 4, it can be found that our proposed method can provide much better thermal comfort than these two methods, for they sometimes overshoot and have considerable vibrations. For FLC, the performance is

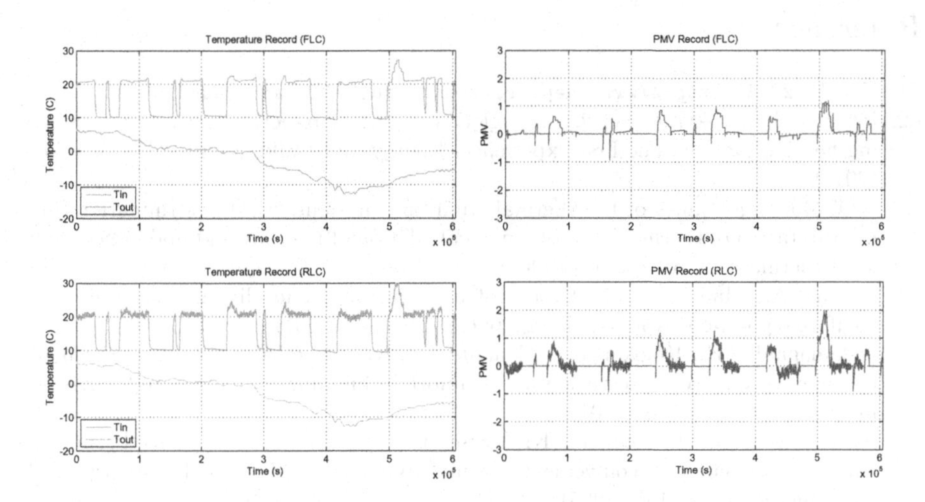

Fig. 5. Simulation Results of FLC Method and RLC Method

tightly related to the designing of membership functions and rules, which are often done empirically. Through optimizing these factors, the results may be better, but due to the complexity of PMV calculation it is hard for this expert system to outperform our proposed method. For RLC, because we discretize the action space, it causes the indoor temperature and PMV value vibrate continually within some extent. By using some techniques may handle this problem, but it is time consuming. The total energy consumed by employing our HIC method is $2.4817 \times 10^8 J$, while they are $2.7393 \times 10^8 J$ and $2.7236 \times 10^8 J$ for FLC and RLC respectively.

5 Conclusion

Nowadays, smart buildings not only mean providing a more comfortable living environment, but also dedicate to reduce the energy cost and environmental pollution. In order to achieve this goal, this paper proposes a hybrid intelligent control system. In detail, a traditional Proportional-Integral-Derivative (PID) controller is used to directly regulate the heating system, while a Particle Swarm Optimization (PSO) method is employed to compute near-optimal setpoints for the PID controller based on present environmental and personal conditions. The advantages of this approach are threefold: (1) better thermal comfort, (2) less energy cost, (3) easier to implement. The experimental results have proved these points and indicated that it outperforms the other two control methods: fuzzy logic control and reinforcement learning control.

References

1. UNEP (2014), http://www.unep.org/sbci/AboutSBCI/Background.asp
2. Privara, S., Siroky, J., Ferkl, L., Cigler, J.: Model predictive control of a building heating system: The first experience. Energy and Buildings 43(2C3), 564–572 (2011)
3. ISO7730: Ergonomics of the thermal environment - analytical determination and interpretation of thermal comfort using calculation of the pmv and ppd indices and local thermal comfort criteria (2005)
4. Dounis, A., Manolakis, D.: Design of a fuzzy system for living space thermal-comfort regulation. Applied Energy 69(2), 119–144 (2001)
5. Dalamagkidis, K., Kolokotsa, D., Kalaitzakis, K., Stavrakakis, G.: Reinforcement learning for energy conservation and comfort in buildings. Building and Environment 42(7), 2686–2698 (2007)
6. Achterbosch, G., de Jong, P., Krist-Spit, C., van der Meulen, S., Verberne, J.: The development of a comvenient thermal dynamic building model. Energy and Buildings 8(3), 183–196 (1985)
7. ASHRAE: ASHRAE Handbook: Fundamentals. American Society of Heating, Refrigerating, and Air-Conditioning Engineers (2005)
8. Ashrae: ANSI/ ASHRAE Standard 55-2004, Thermal Comfort Conditions for Human Occupancy. American Society of Heating, Air-Conditioning, and Refrigeration Engineers, Inc. (2004)
9. Bermejo, P., Redondo, L., de la Ossa, L., Rodríguez, D., Flores, J., Urea, C., Gámez, J.A., Puerta, J.M.: Design and simulation of a thermal comfort adaptive system based on fuzzy logic and on-line learning. Energy and Buildings 49, 367–379 (2012)
10. Ma, Y.: Model Predictive Control for Energy Efficient Buildings. University of California, Berkeley (2013)
11. Kennedy, J., Eberhart, R.: Particle swarm optimization. In: Proceedings of the IEEE International Conference on. Neural Networks, vol. 4, pp. 1942–1948 (November 1995)
12. Shi, Y., Eberhart, R.: A modified particle swarm optimizer. In: The 1998 IEEE International Conference on Evolutionary Computation Proceedings, pp. 69–73. IEEE World Congress on Computational Intelligence (May 1998)
13. Lin, F.J., Teng, L.T., Lin, J.W., Chen, S.Y.: Recurrent functional-link-based fuzzy-neural-network-controlled induction-generator system using improved particle swarm optimization. IEEE Transactions on Industrial Electronics 56(5), 1557–1577 (2009)
14. Paris, B., Eynard, J., Grieu, S., Talbert, T., Polit, M.: Heating control schemes for energy management in buildings. Energy and Buildings 42(10), 1908–1917 (2010)
15. NOAA: National Climatic Data Center. National Oceanic And Atmospheric Administration (2014)
16. Tsilingiris, P.: Thermal flywheel effects on the time varying conduction heat transfer through structural walls. Energy and Buildings 35(10), 1037–1047 (2003)

DC Approximation Approach for ℓ_0-minimization in Compressed Sensing

Thi Bich Thuy Nguyen, Hoai An Le Thi, Hoai Minh Le, and Xuan Thanh Vo

Laboratory of Theoretical and Applied Computer Science (LITA)
UFR MIM, University of Lorraine, Ile du Saulcy, 57045 Metz, France
{thi-bich-thuy.nguyen,hoai-an.le-thi,
minh.le,xuan-thanh.vo}@univ-lorraine.fr

Abstract. In this paper, we study the effectiveness of some non-convex approximations of ℓ_0-norm in compressed sensing. Using four continuous non-convex approximations of ℓ_0-norm, we reformulate the compressed sensing problem as DC (Difference of Convex functions) programs and then DCA (DC Algorithm) is applied to find the solutions. Computational experiments show the efficiency and the scalability of our method in comparison with other nonconvex approaches such as iterative reweighted schemes (including reweighted ℓ_1 and iterative reweighted least-squares algorithms).

Keywords: Compressed Sensing, Sparse Recovery, $\ell_0 - norm$, DC Programming, DCA.

1 Introduction

Compressed Sensing or Compressive Sensing (CS), was introduced by Donoho [11] and Candes et al. [6], is a new signal processing technique. It provides a framework for efficiently acquiring a signal and after that recovering it from very few measurements when the interested signal is very sparse in some basis. The number of measurements needed to be stored is far lower than the Shannon-Nyquist rate while maintaining the essential information.

In this paper, we consider the fundamental issue in CS, which is signal recovery and study the effect of the sensing matrix on the recovery result. The problem can be stated as follows. Given a sensing (measurement) matrix $A \in \mathbb{R}^{m \times n}$ ($m << n$) and a measurement vector $b \in \mathbb{R}^m$, we seek to recover the signal $x \in \mathbb{R}^n$ such that $Ax = b$. Without further information, it is impossible to recover x from y since this linear system is highly underdetermined, and has therefore infinitely solutions. However, if we assume that the vector x is highly sparse, then the situation dramatically changes as it will be outlined. The approach for a recovery procedure that comes first to mind is to search for the sparsest vector x which is consistent with the measurement vector $y = Ax$. This leads to solving the ℓ_0-miminization problem

$$\min \{||x||_0 : \ Ax = b\} . \tag{1}$$

H.A. Le Thi et al. (eds.), *Advanced Computational Methods for Knowledge Engineering*,
Advances in Intelligent Systems and Computing 358, DOI: 10.1007/978-3-319-17996-4_4

Since ℓ_0–norm is intrinsically combinatorial, solving problem (1) is NP-hard in general. In order to overcome this issue, in several works, one approximates the ℓ_0–norm by a continuous, convex or nonconvex function.

The representative convex method is to replace the ℓ_0–norm by the ℓ_1–norm. Then the resulting problem which is called basis pursuit (BP)([3]) has the form

$$\min\{||x||_1 : Ax = b\}, \tag{2}$$

where $||x||_1 = \sum_{i=1}^{n} |x_i|$ is the ℓ_1–norm of the vector $x \in \mathbb{R}^n$.

The ℓ_1 minimization problem (2) is a convex optimization problem and thus tractable. However, to recovery exactly the signal, it requires some situations. It has been proved in [8] that nonconvex minimizations are able to recover sparsity in a more efficient way and require fewer measurements than BP.

Nonconvex continuous approaches were extensively developed in which the ℓ_0 term is approximated by a continuous, nonconvex function. Some approximations proposed to approximate the ℓ_0-norm. The first one is piecewise exponential (PiE) function, which was developed by Bradley and Mangasarian ([2]), studied later in the works of Mohimani et al. ([17], [18]) and applied to CS in the works of Rinaldi et al. ([33]). Logarithmic approximation and ℓ_p-norm with $p < 1$ were studied by Rao and Kreutz-Delgado [35] and Fu [14]. In [8] and [9] Chartrand and Yin considered the case of $0 \leq p \leq 1$ and applied to compressed sensing. Other very used approximations are Smoothly Clipped Absolute Deviation (SCAD) function ([13]), the capped–ℓ_1 (CaP) function ([32]) (CaP) and the piecewise linear (PiL) approximation ([38]),...

Dealing with these approximations, in the context of compressed sensing, several algorithms have been developed such as iteratively reweighted ℓ_1 (IRL1) ([5],[15],[39]), iteratively reweighted least-squares (IRLS) ([8], [9], [10],[19]), Successive Linear Approximation (SLA) ([34], [33]), Local Linear Approximation (LLA)([41]), Two-stage ℓ_1 ([37]), Adaptive Lasso ([40]), Local Quadratic Approximation (LQA) algorithm ([41],[13]), Difference of Convex functions Algorithm (DCA) ([16],[26]), proximal alternating linearized minimization algorithm (PALM)([1]),... Recently, Le Thi et al. [27] gave a rigorous study on DC (Difference of Convex functions) approximation approaches for sparse optimization on both theoretical and algorithmic aspects. In their work, a unifying DC approximation, including all standard approximations, of the ℓ_0–norm is proposed. Furthermore, DCA schemes developed in that paper cover all standard nonconvex algorithms for dealing with ℓ_0-norm.

In this paper, we study four DC approximations of ℓ_0-norm involving PiE, SCAD, CaP, PiL that appeared as the best approximations in [27]. With these approximation, we then reformulate the problem (1) as DC programs and apply DCA to solve the approximation problems. To evaluate the efficiency of our approach, we perform comparisons with two standard nonconvex algorithms including reweighted-ℓ_1 ([5]) and IRLS-ℓ_p ([19]) that are in fact special versions of DCA ([27]) using logarithmic (log) and ℓ_p approximations respectively.

The remainder of the paper is organized as follows. The section 2 introduces briefly DC Programming and DCA. In section 3, approximation approach based on DCA for solving the compressed sensing problem is described. The computational results with some data sets are reported in section 4. Finally, section 5 concludes the paper.

2 DC Programming and DCA

DC programming and DCA constitute the backbone of smooth/nonsmooth nonconvex programming and global optimization. They address the problem of minimizing a function f which is the difference of two convex functions on the whole space $\mathbb{R}^d$ or on a convex set $C \subset \mathbb{R}^d$. Generally speaking, a DC program is an optimization problem of the form :

$$\alpha = \inf\{f(x) := g(x) - h(x) : x \in \mathbb{R}^d\} \qquad (P_{dc})$$

where g, h are lower semi-continuous proper convex functions on $\mathbb{R}^d$. Such a function f is called a DC function, and $g - h$, a DC decomposition of f, while the convex functions g and h are DC components of f. If at least one of the DC components is polyhedral convex, (P_{dc}) is called a polyhedral DC program.

The idea of DCA is simple: each iteration l of DCA approximates the concave part $-h$ by its affine majorization (that corresponds to taking $y^l \in \partial h(x^l)$) and solves the resulting convex problem (that is equivalent to determining a point $x^{l+1} \in \partial g^*(y^l)$ where g^* is the conjugate function of the convex function g).

The generic DCA scheme is shown below.

DCA scheme

Initialization:

Let $x^0 \in \mathbb{R}^d$ be a best guess, $l = 0$.

Repeat

- Calculate $y^l \in \partial h(x^l)$
- Calculate $x^{l+1} \in \arg\min\{g(x) - \langle y^l, x\rangle : x \in \mathbb{R}^d\}$ (P_l)
- $l = l + 1$

Until convergence of $\{x^l\}$.

Convergence properties of DCA and its theoretical basis can be found in ([20],[21],[22]). For instance it is important to mention that

- DCA is a descent method without linesearch: the sequence $\{g(x^l) - h(x^l)\}$ is decreasing.
- If the optimal value α of problem (P_{dc}) is finite and the infinite sequences $\{x^l\}$ and $\{y^l\}$ are bounded, then every limit point x^* of the sequence $\{x^l\}$ is a critical point of $g - h$, i.e. $\partial h(x^*) \cap \partial g(x^*) \neq \emptyset$.
- DCA has a *linear convergence* for general DC programs and has *finite convergence* for polyhedral DC programs.

For a complete study of DC programming and DCA the reader is referred to [20],[21],[22], and the references therein.

It is worth to note that the construction of DCA involves the convex DC components g and h but not the DC function f itself. Moreover, a DC function f has infinitely many DC decompositions $g - h$ which have a crucial impact on the qualities (speed of convergence, robustness, efficiency, globality of computed solutions,...) of DCA. The solution of a nonconvex program by DCA must be composed of two stages: the search of an *appropriate* DC decomposition and that of a *good* initial point. DCA has been successfully applied to numerous and various nonconvex optimization problems to which it quite often gave global solutions and proved to be more robust and more efficient than related standard methods (see [36],[16],[24],[23],[31],[26],[29],[30],[28] and the list of references at `http://www.lita.univ-lorraine.fr/~lethi/index.php/en/research/dc-programming-and-dca.html`).

3 DC Approximation Approach for Solving Problem (1)

The ℓ_0-norm results in a combinatorial optimization problem, and hence is not practical for large scale problems. We will replace ℓ_0-norm by an approximation function such that the approximate problem of (1) can be expressed as a DC program to which DCA is applicable.

Define the step function $s : \mathbb{R} \mapsto \mathbb{R}$ by $s(t) = 1$ for $t \neq 0$ and $s(t) = 0$ for $t = 0$. Then for $x \in \mathbb{R}^n$ we have $\|x\|_0 = \sum_{i=1}^n s(x_i)$.

Let $\varphi : \mathbb{R} \mapsto \mathbb{R}$ be a continuous function that approximates s, i.e. $\varphi(t) \approx s(t)$ for all $t \in \mathbb{R}$. Then for $x \in \mathbb{R}^n$ we have

$$\|x\|_0 \approx \sum_{i=1}^n \varphi(x_i).$$

Using this approximation, we can formulate the approximate problem of the compressed sensing problem (1) in the form

$$\min \left\{ F(x) := \sum_{i=1}^n \varphi(x_i) : x \in \Omega \right\}, \tag{3}$$

where $\Omega = \{x \in \mathbb{R}^n : Ax = b\}$.

Suppose that φ can be expressed as a DC function of the form

$$\varphi(t) = g(t) - h(t), \quad \forall t \in \mathbb{R}, \tag{4}$$

where g and h are convex functions on $\mathbb{R}$. Then the problem (3) can be expressed as a DC program as follows

$$\min\{G(x) - H(x) : x \in \mathbb{R}^n\}, \tag{5}$$

where

$$G(x) = \chi_\Omega(x) + \sum_{i=1}^n g(x_i), \quad H(x) = \sum_{i=1}^n h(x_i)$$

are clearly convex functions on $\mathbb{R}^n$.

DCA applied to (5) consists of computing, at each iteration l,
- $y^l \in \partial H(x^l)$ that is equivalent to $y_i^l \in \partial h(x_i^l)$ for all $i = 1, \ldots, n$.
- Compute x^{l+1} by

$$\begin{aligned} x^{l+1} &\in \arg\min\{G(x) - \langle y^l, x\rangle : x \in \mathbb{R}^n\} \\ &= \arg\min\left\{\sum_{i=1}^{n} g(x_i) - \langle y^l, x\rangle : Ax = b\right\}. \end{aligned}$$

Since Ω is a polyhedral convex set, if either g or h is polyhedral then (5) is a polyhedral DC program. In such case, DCA applied to (5) benefits from convergence properties of polyhedral DC programs. In summary, DCA for solving (5) can be described as follows.

DCA-App
Input: matrix $A \in \mathbb{R}^{m\times n}$ $(m << n)$, $b \in \mathbb{R}^m$
Output: the sparsest vector $x \in \mathbb{R}^n$ such that $Ax = b$
Initialization: Let $x^0 \in \mathbb{R}^n$ be a best guess, $\epsilon > 0$ be a small tolerance, $l = 0$
Repeat:
1. Compute $y_i^l \in \partial h(x_i^l)$, $\forall i = 1, \ldots, n$.
2. Compute x^{l+1} as a solution to the convex program

$$\min\left\{\sum_{i=1}^{n} g(x_i) - \langle y^l, x\rangle : Ax = b\right\}. \qquad (P_k)$$

3. $l = l + 1$
Until $\|x^l - x^{l-1}\|/(1 + \|x^l\|) < \epsilon$ or $|F(x^l) - F(x^{l-1})|/(1 + F(x^l)) < \epsilon$.

In this paper, we consider four approximation functions including piecewise exponential (PiE) function ([2]), the Smoothly Clip Absolute Deviation (SCAD) function ([13]), capped-ℓ_1 (CaP) function ([31]), and the piecewise linear approximation (PiL) recently proposed in [25]. The formulation of these functions is summarized in Table 1. Note that among these functions, SCAD is not actually an approximation of the step function s, but it becomes an approximation of s if we scale it with the factor $\frac{2}{(a+1)\lambda^2}$ that does not affect the solution set of the problem (3). DC decompositions of these functions are described in Table 2. These approximations except PiL, together with their DC decomposition, have been applied to find sparse classifiers in SVMs ([24], [31], [23]).

The implementation of Algorithm DCA-App according to each specific approximation function differs from each other in the computation of $y_i^l \in \partial h(x_i^l)$ in the step 1 and the subproblem in the step 2. The computation of $y_i^l \in \partial h(x_i^l)$ is given in Table 2. The subproblem in case of approximations PiE, SCAD, CaP has the form

$$\min\left\{\lambda\|x\|_1 - \langle y^l, x\rangle : Ax = b\right\}$$

Table 1. DC approximation functions

Name	Formula of $\varphi(t),\ t \in \mathbb{R}$
PiE ([2])	$1 - e^{-\lambda\lvert t\rvert},\ \lambda > 0$
SCAD ([13])	$\begin{cases} \lambda\lvert t\rvert & \text{if } \lvert t\rvert \leq \lambda \\ -\frac{t^2 - 2\alpha\lambda\lvert t\rvert + \lambda^2}{2(\alpha-1)} & \text{if } \lambda < \lvert t\rvert \leq \alpha\lambda \\ \frac{(\alpha+1)\lambda^2}{2} & \text{if } \lvert t\rvert > \alpha\lambda \end{cases}$ $,\ \alpha > 2, \lambda > 0$
CaP ([31])	$\min\{1, \lambda\lvert t\rvert\},\ \lambda > 0$
PiL ([25])	$\min\left\{1, \max\left\{0, \frac{\lvert t\rvert - a}{b - a}\right\}\right\},\ 0 < a < b$

$$\Leftrightarrow \min\left\{\lambda\sum_{i=1}^{n} t_i - \langle y^l, x\rangle : Ax = b, \quad -t_i \leq x_i \leq t_i \ \forall i = 1, \ldots, n\right\},$$

which is a linear program.

And the subproblem in case of approximation PiL has the form

$$\min\left\{\frac{1}{b-a}\sum_{i=1}^{n} \max(a, |x_i|) - \langle y^l, x\rangle : Ax = b\right\}$$

$$\Leftrightarrow \min\left\{\frac{1}{b-a}\sum_{i=1}^{n} t_i - \langle y^l, x\rangle : Ax = b, \quad -t_i \leq x_i \leq t_i, a \leq t_i \ \forall i = 1, \ldots, n\right\},$$

which is also a linear program.

Table 2. DC decomposition $\varphi = g - h$ and calculation of ∂h. The notation $\text{sgn}(t)$ denotes sign of t

Name	$g(t)$	$h(t)$	$\overline{t} \in \partial h(t)$
PiE	$\lambda\lvert t\rvert$	$\lambda\lvert t\rvert - 1 + e^{-\lambda\lvert t\rvert}$	$\text{sgn}(t)(\lambda\lvert t\rvert - e^{-\lambda\lvert t\rvert})$
SCAD	$\lambda\lvert t\rvert$	$\begin{cases} 0 & \text{if } \lvert t\rvert \leq \lambda \\ \frac{(\lvert t\rvert-\lambda)^2}{2(\alpha-1)} & \text{if } \lambda \leq \lvert t\rvert \leq \alpha\lambda \\ \lambda\lvert t\rvert - \frac{(\alpha+1)\lambda^2}{2} & \text{otherwise} \end{cases}$	$\begin{cases} 0 & \text{if } \lvert t\rvert \leq \lambda \\ \text{sgn}(t)\frac{\lvert t\rvert-\lambda}{\alpha-1} & \text{if } \lambda < \lvert t\rvert < \alpha\lambda \\ \text{sgn}(t)\lambda & \text{otherwise} \end{cases}$
CaP	$\lambda\lvert t\rvert$	$\max\{1, \lambda\lvert t\rvert\} - 1$	$\begin{cases} 0 & \text{if } \lvert t\rvert \leq \frac{1}{\lambda} \\ \text{sgn}(t)\lambda & \text{otherwise} \end{cases}$
PiL	$\frac{\max\{a, \lvert t\rvert\}}{b-a}$	$\frac{\max\{b, \lvert t\rvert\}}{b-a} - 1$	$\begin{cases} 0 & \text{if } \lvert t\rvert \leq b \\ \frac{\text{sgn}(t)}{b-a} & \text{otherwise} \end{cases}$

Observe that for all considered approximation functions, we always have g is polyhedral convex, so (5) is a polyhedral DC program. Thus, DCA-App applied to these approximations has finite convergence ([20],[21],[22]). Some other convergence properties are stated in the following Theorem.

Theorem 1 (Convergence properties of DCA-App)

i) *DCA generates a sequence* $\{x^l\}$ *such that the sequence* $\{F(x^l)\}$ *is monotonously decreasing.*
ii) *If* x^* *is a limit point of* $\{x^l\}$ *then* x^* *is a critical point of* (5).
iii) *The sequence* $\{x^l\}$ *converges to* x^* *after a finite number of iterations. The point* x^* *is almost always a local minimizer of the corresponding optimization problems. Especially,*
- *In CaP approximation, if* $x_i^* \notin \{-\frac{1}{\lambda}; \frac{1}{\lambda}\}$, $\forall i = 1, \ldots, n$ *then* x^* *is a local minimizer of optimization problem.*
- *In PiL approximation, if* $x_i^* \notin \{-a; a\}$, $\forall i = 1, \ldots, n$ *then* x^* *is a local minimizer of optimization problem.*

Proof. (i) and (ii) are direct consequences of the convergence properties of DCA for general DC programs. Below, we are going to prove iii).

In the case of CaP approximation, the second DC component, says H, is a polyhedral convex function. Moreover, if the condition $x_i^* \notin \{-\frac{1}{\lambda}; \frac{1}{\lambda}\}$ $\forall i = 1, \ldots, n$ holds, H is differentiable at x^*. Then using the convergence property of DCA for polyhedral DC programs ([20],[22]), we deduce that x^* is local minimizer of optimization problem. Moreover, since a polyhedral convex function is almost always differentiable, say, it is differentiable everywhere except on a set of measure zero, we can say that x^* is almost always a local minimizer of this problem. The proof is similar for PiL since its second DC component H is also polyhedral convex and is differentiable if the condition $x_i^* \notin \{-a; a\}$ $\forall i = 1, \ldots, n$ holds.

In the case of PiE (resp. SCAD) approximation, the first DC component of the problem, says G, is a polyhedral convex function, so is G^*. Hence, the dual DC program of the original problem is also a polyhedral DC program. Thus, if y^* is the limit point of the sequence $\{y^l\}$ generated by DCA-App, we can say that y^* is almost always a local minimizer of the dual DC program of the optimization problem. On another hand, according to the property of transportation of local minimizers in DC programming, we have the following ([20],[22]): let y^* be a local minimizer of the dual program and $x^* \in \partial G^*(y^*)$. If H is differentiable at x^* then x^* is a local minimizer of the problem. Combining this property with the facts that y^* is almost always a local minimizer of the dual DC program of and H is differentiable everywhere, we conclude that x^* is almost always a local minimizer of optimization problem. The proof is then complete. □

4 Numerical Experiments

In this section, we study the efficiency of our proposed approach in comparison with Reweighted-ℓ_1 algorithm [7] and IRLS-ℓ_p algorithm [19] with $p = 0.5$. All algorithms have been implemented in the Visual C++2008, run on a PC Intel i5CPU650, 3.2 GHz of 4GB RAM. CPLEX 12.3 is used for solving the convex linear problems. The initial point is taken at $x^0 = \mathbf{0}$.

In our experiments, we will study how the sensing matrix affects on the results of signal recovery problem. Two important properties of A which were most widely known are *Restricted Isometry Property* (RIP) and *coherence* or *mutual coherence*. For more details we refer the reader to ([4]) and ([12]).

In this paper, we will test on three scenarios. In the first test, we consider a random Gaussian matrix which is incoherent and has small RIP constants with high probability. In the second test, we also consider the random Gaussian matrix but with fewer measurements and in the last test with the matrix coherence.

The test protocol is executed as follows: we sample a random $m \times n$ matrix A and generate a target signal $x \in \mathbb{R}^n$ with $||x||_0 = k$. The k nonzero spike positions are chosen randomly, and the nonzero values are chosen randomly from a zero-mean unit-variance Gaussian distribution. We then compute the measurement $b = Ax$ and apply each solver to produce a reconstruction x^* of x. The reconstruction is considered as success if the relative error $\frac{||x-x^*||}{||x||} \leq 10^{-2}$. We run 100 independent trials and record the corresponding success rates at various sparsity levels $k \in \{5, \ldots, 35\}$.

The parameter for Reweighted-ℓ_1 is chosen as $\epsilon = 0.1$. For IRLS-ℓ_p, we chose $p = 0.5$. For our algorithms, the parameters are set to $\lambda = 0.05$ (for PiL, SCAD, and CaP), $\alpha = 40$ (for SCAD), $a = 0.001$, $b = 0.101$ (for PiL).

In our first experiment, we test on RIP matrix. Set $m = 64, n = 256$, a random $m \times n$ matrix A with i.i.d. Gaussian entries is sampled. The sparsity level k is taken in $\{10, \ldots, 35\}$.

The experiment results are given in Figure 1.

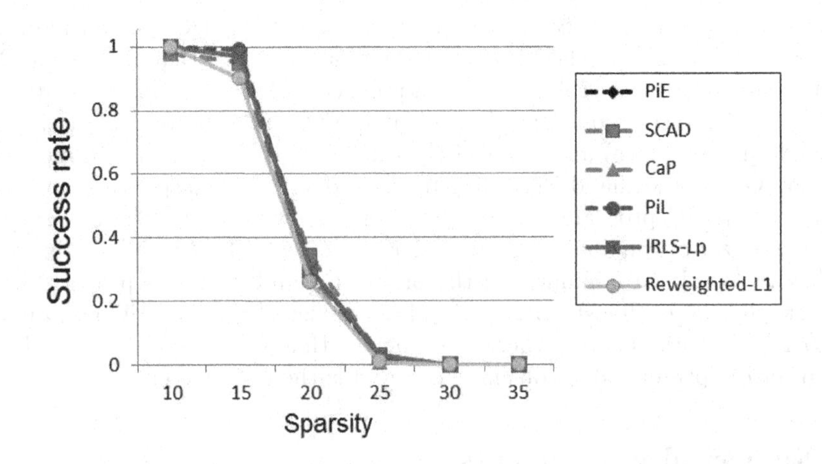

Fig. 1. Success rates using incoherent sensing matrix, $m = 64, n = 256$

In figure 1, the six algorithms have the same performances but DCA-App-SCAD and DCA-App-PiL are slightly better than the others and all of five algorithms outperform Reweighted ℓ_1.

In the second experiment, we focus on the random Gaussian matrix A but it has m very smaller than n. The size of A is $m \times n = 100 \times 2000$. We also test 100 times with the sparsity of x belonging to interval $[5, 35]$ with step–size is 5. The results are shown in Figure 2.

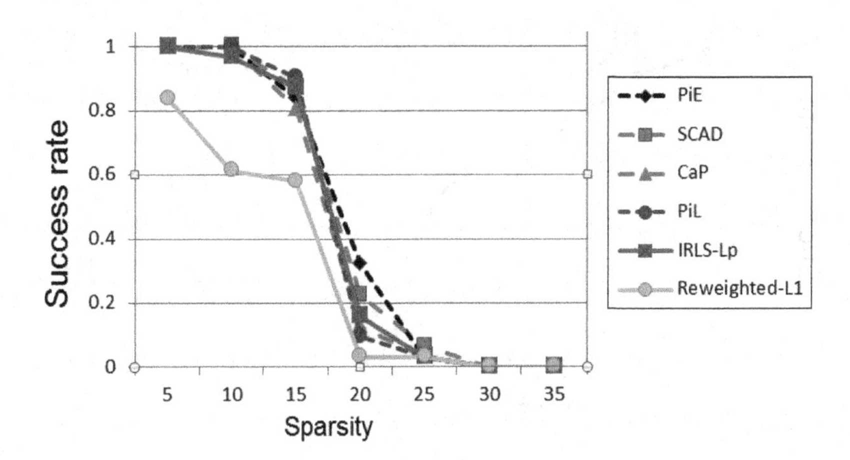

Fig. 2. Success rates using incoherent sensing matrix, $m = 100, n = 2000$

In this example, there are differences between the results of these algorithms. It can be seen in Figure 2 that when k is small, i.e. x is very sparse, all the algorithms have exact recovery with high rate except Reweighted ℓ_1. But when k increases, the success rate decreases. The best performance in this case belongs to DCA-App-PiE and DCA-App-SCAD, the followings are IRLS-ℓ_p and DCA-App-CaP and DCA-App-PiL. Reweighted ℓ_1 has the lowest success rate in this test.

In the third experiment, we consider the case matrix A is highly coherent. Specifically, A is a random partial discrete cosine transform (DCT) matrix with size $m \times n = 100 \times 2000$ and its columns are computed by

$$A_i = \frac{1}{\sqrt{m}} \cos(2i\pi\xi_i/F), \quad \forall i = 1, \ldots, n \tag{6}$$

where ξ_i's are random vector uniformly distributed in $([0, 1]^m)$, and $F \in \mathbb{N}$ is a *refinement factor.*

The number F is closely related to the conditioning of A in the sense that larger F corresponds to large coherence of A. In our test, coherence of A is greater than 0.9999. Note that the *mutual coherence* $\mu(A)$ of a matrix A measures the similarity between the matrix columns. If $\mu(A) = 1$, it implies the existence of two parallel atoms, and this causes confusion in the reconstruction of sparse coefficients. The sparsity of x is also in the set $\{5, 10, \ldots, 35\}$. Figure 3 presents the results of our experiment.

In figure 3, when k is small, DCA-App-PiL and DCA-App-SCAD have the highest success rate but when the sparsity of x is greater, DCA-App-SCAD and DCA-App-PiE have the best performance. IRLS-ℓ_p and Reweighted ℓ_1 give the success rate lower than the others.

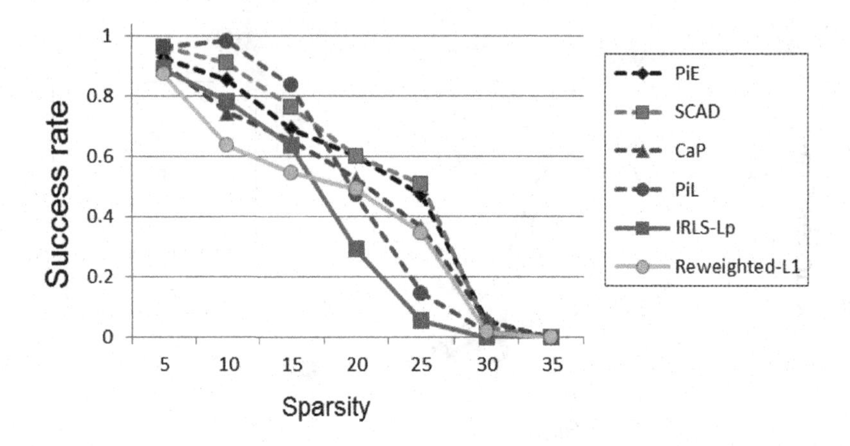

Fig. 3. Success rates using highly coherent sensing matrix, $m = 100, n = 2000$

5 Conclusions

In this paper, we have presented a DC approximation approach based on DC programming and DCA for compressed sensing. Using four approximation functions of ℓ_0–norm, we reformulated the compressed sensing problem as DC programs and then used DCA to solve the resulting problems. In our experiments, we have compared 6 algorithms: DCA-App-PiE, DCA-App-SCAD, DCA-App-CaP, DCA-App-PiL, Reweighted-ℓ_1 and IRLS-ℓ_p, which are working with ℓ_0–norm. We performed three experiments in which matrix A is incoherent and highly coherent. The experiments showed the efficiency of each approximation and also the power of DCA. It is observed that DCA-App-PiL gives the best performance in all tests even in case A is highly coherent. The following is DCA-App-SCAD. DCA-App-PiE and DCA-App-CaP are both good when A is incoherent or when x is very sparse. All DCA based algorithms have higher success rate than IRLS-ℓ_p and Reweighted-ℓ_1 when A is highly coherent.

References

1. Attouch, H., Bolte, J., Redont, P., Soubeyran, A.: Proximal alternating minimization and projection methods for nonconvex problems: An approach based on the Kurdyka-Lojasiewicz inequality. Mathematics of Operations Research 35(2), 438–457 (2010)
2. Bradley, P.S., Mangasarian, O.L.: Feature Selection via concave minimization and support vector machines. In: Proceeding of International Conference on Machina Learning ICML 1998 (1998)

3. Chen, S., Donoho, D.L., Saunders, M.: Atomic decomposition by basis pursuit. SIAM Journal on Scientific Computing 20(1), 33–61 (1998)
4. Candès, E.J., Tao, T.: Near-optimal signal recovery from random projections: universal encoding strategies? IEEE Transaction Information Theory 52(12), 5406–5425 (2006)
5. Candès, E.J., Wakin, M.B., Boyd, S.: Enhancing Sparsity by Reweighted l1 Minimization. Journal of Fourier Analysis and Applications 14(5), 877–905 (2008); special issue on sparsity
6. Candès, E.J., Romberg, J., Tao, T.: Robust Uncertainty Principles: Exact Signal Reconstruction From Highly Incomplete Frequency Information (2006)
7. Candés, E.J., Paige, A.: Randall: Highly Robust Error Correction by Convex Programming. IEEE Transactions Information Theory Information Theory 54(7), 2829–2840 (2008)
8. Chartrand, R.: Exact Reconstruction of Sparse Signals via Nonconvex Minimization. IEEE Signal Process. Lett. 14(10), 707–710 (2007)
9. Chartrand, R., Yin, W.: Iteratively Reweighted Algorithms for Compressive Sensing. In: IEEE International Conference on Acoustics, Speech, and Signal Processing (2008)
10. Daubechies, I., DeVore, R., Fornasier, M., Güntük, C.: Iteratively reweighted least squares minimization for sparse recovery. Commun. Pure Appl. Math. 63, 1–38 (2010)
11. Donoho, D.L.: Compressed sensing. IEEE Trans. Inform. Theory 52(4), 1289–1306 (2006)
12. Donoho, D.L., Xiaoming, H.: Uncertainty principles and ideal atomic decomposition. IEEE Transactions on Information Theory 47(7), 2845–2862 (2001)
13. Fan, J., Li, R.: Variable selection via nonconcave penalized likelihood and its oracle properties. J. Amer. Stat. Ass. 96(456), 1348–1360 (2001)
14. Fu, W.J.: Penalized regressions: The bridge versus the Lasso. Journal of Computational and Graphical Statistics 7, 397–416 (1998)
15. Foucart, S., Lai, M.: Sparsest solutions of underdetermined linear systems via ℓ_q-minimization for $0 < q \leq 1$, Appl. Comput. Harmon. Anal. 26, 395–407 (2009)
16. Gasso, G., Rakotomamonjy, A., Canu, S.: Recovering sparse signals with a certain family of nonconvex penalties and DC programming. IEEE Transactions on Signal Processing 57(12), 4686–4698 (2009)
17. Mohimani, G.H., Babaie-Zadeh, M., Jutten, C.: Fast Sparse Representation Based on Smoothed ℓ^0 Norm. In: Davies, M.E., James, C.J., Abdallah, S.A., Plumbley, M.D. (eds.) ICA 2007. LNCS, vol. 4666, pp. 389–396. Springer, Heidelberg (2007)
18. Mohimani, H., Babaie-Zadeh, M., Jutten, C.: A fast approach for overcomplete sparse decomposition based on smoothed L0 norm. IEEE Transactions on Signal Processing 57(1), 289–301 (2009)
19. Lai, M.-J., Xu, Y., Yin, W.: Improved Iteratively reweighted least squares for unconstrained smoothed ℓ_p minimization. SIAM J. Numer. Anal. 51(2), 927–957 (2013)
20. Pham Dinh, T., Le Thi, H.A.: Convex analysis approach to DC programming: Theory, algorithms and applications. Acta Math. Vietnamica 22(1), 289–357 (1997)
21. Le Thi, H.A., Pham Dinh, T.: DC Optimization Algorithm for Solving The Trust Region Problem. SIAM Journal on Optimization 8(2), 476–505 (1998)
22. Le Thi, H.A., Pham Dinh, T.: The DC (difference of convex functions) Programming and DCA revisited with DC models of real world nonconvex optimization problems. Annals of Operations Research 133, 23–46 (2005)

23. Le Thi, H.A., Van Nguyen, V., Ouchani, S.: Gene Selection for Cancer Classification Using DCA. In: Tang, C., Ling, C.X., Zhou, X., Cercone, N.J., Li, X. (eds.) ADMA 2008. LNCS (LNAI), vol. 5139, pp. 62–72. Springer, Heidelberg (2008)
24. Le Thi, H.A., Le Hoai, M., Nguyen, V.V., Pham Dinh, T.: A DC Programming approach for feature selection in support vector machines learning. Adv. Data Analysis and Classification 2(3), 259–278 (2008)
25. Le Thi, H.A.: A new approximation for the ℓ_0–norm. Research report LITA EA 3097, University of Lorraine, France (2012)
26. Le Thi, H.A., Nguyen Thi, B.T., Le, H.M.: Sparse signal recovery by difference of convex functions algorithms. In: Selamat, A., Nguyen, N.T., Haron, H. (eds.) ACIIDS 2013, Part II. LNCS, vol. 7803, pp. 387–397. Springer, Heidelberg (2013)
27. Le Thi, H.A., Pham Dinh, T., Le, H.M., Vo, X.T.: DC approximation approaches for sparse optimization. European Journal of Operational Research 244(1), 26–46 (2015)
28. Le, H.M., Le Thi, H.A., Nguyen, M.C.: Sparse Semi-Supervised Support Vector Machines by DC Programming and DCA. Neurocomputing (November 27, 2014), (published online), doi:10.1016/j.neucom.2014.11.051,
29. Le Thi, H.A., Nguyen, M.C., Pham Dinh, T.: A DC programming approach for finding Communities in networks. Neural Computation 26(12), 2827–2854 (2014)
30. Le Thi, H.A., Vo, X.T., Pham Dinh, T.: Feature Selection for linear SVMs under Uncertain Data: Robust optimization based on Difference of Convex functions Algorithms. Neural Networks 59, 36–50 (2014)
31. Ong, C.S., Le Thi, H.A.: Learning sparse classifiers with difference of convex functions algorithms. Optimization Methods and Software 28(4), 830–854 (2013)
32. Peleg, D., Meir, R.: A bilinear formulation for vector sparsity optimization. Signal Processing 88(2), 375–389 (2008) ISSN 0165–1684
33. Rinaldi, F.: Concave programming for finding sparse solutions to problems with convex constraints. Optimization Methods and Software 26(6), 971–992 (2011)
34. Rinaldi, F., Schoen, F., Sciandrone, M.: Concave programming for minimizing the zero-norm over polyhedral sets. Comput. Opt. Appl. 46(3), 467–486 (2010)
35. Rao, B.D., Kreutz-Delgado, K.: An affine scaling methodology for best basis selection. IEEE Trans. Signal Processing 47, 87–200 (1999)
36. Thiao, M., Pham Dinh, T., Le Thi, H.A.: DC Programming Approach for a Class of Nonconvex Programs Involving *lo* Norm. In: Le Thi, H.A., Bouvry, P., Pham Dinh, T. (eds.) MCO 2008. CCIS, vol. 14, pp. 348–357. Springer, Heidelberg (2008)
37. Zhang, T.: Some sharp performance bounds for least squares regression with regularization. Ann. Statist. 37, 2109–2144 (2009)
38. Zhang, C., Shao, Y., Tan, J., Deng, N.: Mixed-norm linear support vector machine. Neural Computing and Applications 23(7-8), 2159–2166 (2013)
39. Zhao, Y., Li, D.: Reweighted l1-Minimization for Sparse Solutions to Underdetermined Linear Systems. SIAM J. Opt. 22(3), 1065–1088 (2012)
40. Zou, H.: The adaptive lasso and its oracle properties. J. Amer. Stat. Ass. 101, 1418–1429 (2006)
41. Zou, H., Li, R.: One-step sparse estimates in nonconcave penalized likelihood models. The Annals of Statistics 36(4), 1509–1533 (2008)

DC Programming and DCA Approach for Resource Allocation Optimization in OFDMA/TDD Wireless Networks*

Canh Nam Nguyen**, Thi Hoai Pham, and Van Huy Tran

School of Applied Mathematics and Informatics,
Hanoi University of Science and Technology, Vietnam
{nam.nguyencanh,hoai.phamthi,huy.tranvan}@hust.edu.vn

Abstract. The next generation broadband wireless networks deploy OFDM/OFDMA as the enabling technologies for broadband data transmission with QoS capabilities. Many optimization problems arise in the conception of such a network. This article studies an optimization problem in resource allocation. We build mathematical model for the considered problem and then propose a new approach for it resolution. Our approach bases on DC programming and DCA. Preliminary numerical results will be reported to show the efficiency of the proposed method.

Keywords: DC programming, DCA, Pure 0-1 programming, Branch-and-Bound, WiMAX.

1 Introduction

OFDMA transmission technique is becoming the preferred technology for the fourth generation broadband wireless networks such as WiMAX. OFDMA is a combination of TDMA and FDMA schemes, in which user's data can be transmitted simultaneously in time domain and frequency domain. From a network operator point of view, it is very important to utilize the channel resources effectively because the available radio resources are limited, while the revenue should be maximized. From a user point of view, it is more important to have a fair resource allocation so that their data sessions are not in outage situation and the requested quality of service is guaranteed. Generally, there has been a strong motivation to improve the channel capacity by increasing the spectral efficiency while providing fairness and meeting Quality of Service (QoS) requirements of all users simultaneously [7,1].

Thus two functionalities should be deployed in the MAC layer of the OFDMA wireless systems in order to ensure spectral efficiency, fairness and QoS, namely the scheduler and the resource allocation functions. In this article we consider a resource allocation problem in an OFDMA/TDD frame. Until now there are

* This research is funded by Vietnam National Foundation for Science and Technology Development (NAFOSTED) under grant number 101.01-2013.19.

** Corresponding author.

H.A. Le Thi et al. (eds.), *Advanced Computational Methods for Knowledge Engineering*,
Advances in Intelligent Systems and Computing 358, DOI: 10.1007/978-3-319-17996-4_5

several efforts in research community attempting to maximize the efficiency in an OFDMA/TDD frame [2,8]. However in OFDMA/TDD schemes, we find that the research on these previous mentioned issues is not sufficient. We then propose a new approach to solve this problem.

The considered problem is first formulated as a 0-1 linear programming problem. We then suggest a Branch and Bound scheme to solve to obtained problem. An efficient algorithm appears when we combine a local algorithm in Difference of Convex (DC) programming called DC Algorithm (DCA). To this purpose, we first apply the theory of exact penalization in DC [3,6] as that of minimizing a DC function over a polyhedral convex set. The resulting problem is then handled by DCA introduced by Pham Dinh Tao in 1985 and extensively developed since 1994 by Le Thi Hoai An and Pham Dinh Tao (see [9,4,5] and reference therein).

The article is organized as follows. After the problem statement at the beginning of section 2, we will introduce its mathematical formulation. Section 3 deals with DC reformulation and application of DCA to the obtained problem. We report computational experiments in section 4. The article will ends with some conclusions.

2 Problem Statement

Consider an OFDMA/TDD frame which contains K users, M sub-chanels and N time-slots. Each user will attain his maximum efficiency if he is allocated suitable resources, i.e., with right sub-channel and in right time. But there are, of course, many conflicts within users, see Fig. 1.

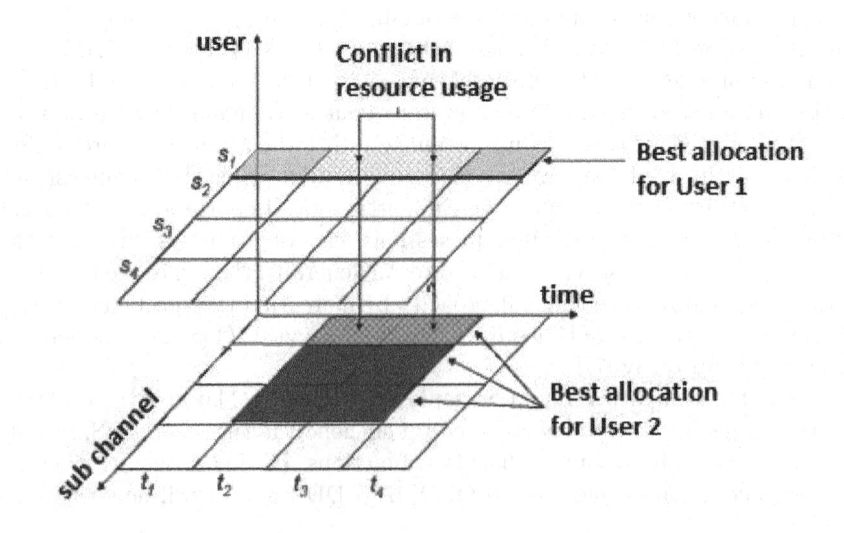

Fig. 1. OFDMA/TDD frame in which resource conflict occurs so often

Let $b_{ijk}, 1 \leq i \leq M, 1 \leq j \leq N, 1 \leq k \leq K$, be the number of bit data that user k can send if he is provided i sub-channel and time-slot j, our efficiency measure. Our problem is to allocate the resources in a frame such that maximize the efficiency. Nevertheless such an allocation should satisfy the following conditions

- At a slot of time and for a specific sub-channel there is at most one user (to avoid conflict)
- Resource allocation for users should be done in sharp of rectangle (IEEE802.16e standard in WiMAX network), see Fig. 2.

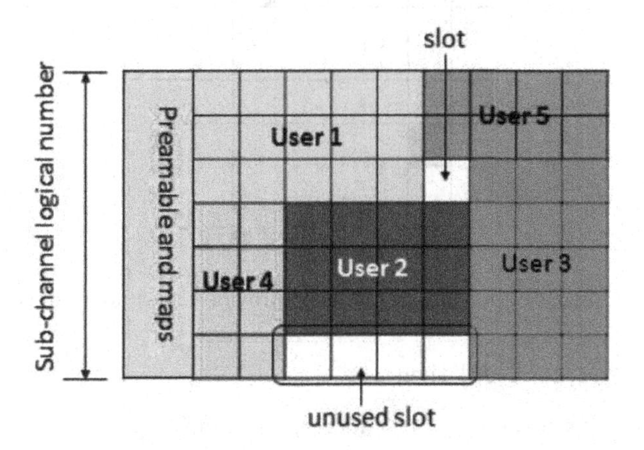

Fig. 2. Rectangle design in a frame

Mathematical Formulation

In the follows we will formulate the above problem as a pure 0-1 linear programming.

If we denote $x_{ijk} \in \{0, 1\}$ for $1 \leq i \leq M, 1 \leq j \leq N, 1 \leq k \leq K$ with convention

$$x_{ijk} = \begin{cases} 1 \text{ if user } k \text{ is provided sub-channel } i \text{ in time-slot } j \\ 0 \text{ otherwise} \end{cases}, \tag{1}$$

then the total data that can be transfered is

$$f(x) := \sum_{k=1}^{K} \sum_{i=1}^{N} \sum_{j=1}^{M} b_{ijk} x_{ijk}. \tag{2}$$

In order to avoid the conflict, at a time-slot j and in one sub-channel i there is at most one user, we introduce the following constraint

$$\sum_{k=1}^{K} x_{ijk} \leq 1 \quad \forall i = \overline{1, M}, \forall j = \overline{1, N}. \tag{3}$$

In order to get the rectangle allocation we use the well-known technique in optimization called "Big-M". Because a user should be allocated resource in a rectangle form so if user k is assigned for two nodes in the frame then all the node in rectangle made by those two nodes will be assigned for this user, see the Fig. 3. On the other hand, we have some conditions like if $x_{i_1 j_1 k^*} = x_{i_2 j_2 k^*} = 1$

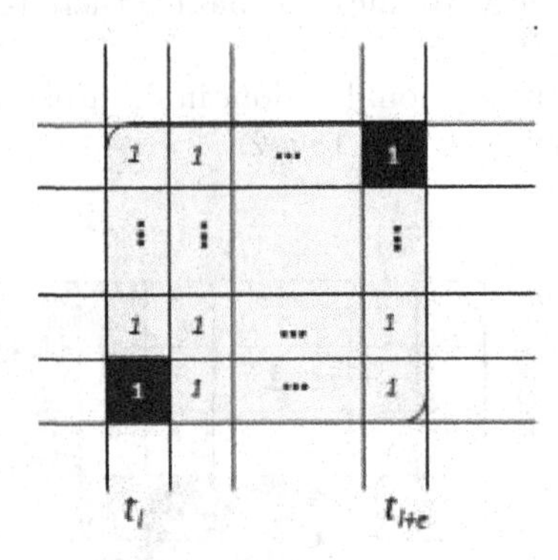

Fig. 3. Allocation for one user if he is provided twos nodes in the frame

for some $k^* \in \{1, 2, \ldots, K\}$ then

$$x_{ijk^*} = 1, \quad \forall \min\{i_1, i_2\} \leq i \leq \max\{i_1, i_2\}, \min\{j_1, j_2\} \leq j \leq \max\{j_1, j_2\}.$$

This condition will be represented by the following constraints

$$(|i_1 - i_2| + 1)(|j_1 - j_2| + 1)(x_{i_1 j_1 k} + x_{i_2 j_2 k} - 1) - \sum_{i \in I_{i_1 i_2}} \sum_{j \in J_{j_1 j_2}} x_{ijk} \leq 0, \quad (4)$$

where

$$I_{i_1 i_2} = \{i : \min\{i_1, i_2\} \leq i \leq \max\{i_1, i_2\}\},$$
$$J_{j_1 j_2} = \{j : \min\{j_1, j_2\} \leq j \leq \max\{j_1, j_2\}\}.$$

Finally we arrive in the optimization problem with the objective function (2) and the constraints (3), (4) and (1).

Since the number of variables and number of constraints increase so fast with respect to the number of users, number of sub-channels and number of time-slots, we then deal with a problem in the form of a large-scale pure 0-1 linear program.

3 A Global Optimization Based on DC Programming Approach

We will solve the obtained optimization problem by the well known global method, Branch and Bound. Linear relaxation is applied in the bounding procedures of our algorithm as in classical approaches. The algorithm is truly speed up, can solve the problem in fact, when we combine DCA in our Branch and Bound scheme.

3.1 DC Reformulation

By using the well known results concerning the exact penalty [3,6] we will formulate the considered pure 0-1 problem in the form of concave minimization programming. Clearly, the set D of feasible points x determined by the system of constraints $\{(3),(4),(1)\}$ is a nonempty, bounded polyhedral convex set in $\mathbb{R}^n$ with $n = K + M + N$. Our problem can be expressed in the form:

$$\alpha = \min \left\{c^T x : x \in D, x \in \{0,1\}^n\right\}. \tag{5}$$

Let us consider function p defined by $p(x) = \sum_{i=1}^{n} \min\{x_i, 1 - x_i\}$.

Set $K := \{x \in D : x \in [0,1]^n\}$. It is clear that p is concave and finite on K, $p(x) \geq 0$ for all $x \in K$, and

$$\{x \in D : x \in \{0,1\}^n\} = \{x \in K : p(x) \leq 0\}.$$

Hence Problem (5) can be rewritten as

$$\alpha = \min \{c^T x : x \in K, p(x) \leq 0\} \tag{6}$$

From Theorem 1 below we get, for a sufficiently large number τ $(\tau > \tau_0)$, the equivalent concave minimization problem to (5) is :

$$\min \{c^T x + \tau p(x) : x \in K\}. \tag{7}$$

Theorem 1. *(Theorem 1, [3]) Let K be a nonempty bounded polyhedral convex set, f be a finite concave function on K and p be a finite nonnegative concave function on K. Then there exists $\tau_0 \geq 0$ such that the following problems have the same solution set:*

$$(P_t) \qquad \alpha(t) = \inf\{f(x) + \tau p(x) : x \in K\}$$
$$(P) \qquad \alpha \quad = \inf\{f(x) : x \in K, p(x) \leq 0\}.$$

More precisely if the vertex set of K, denoted by $V(K)$, is contained in $\{x \in K, p(x) \leq 0\}$, then $\tau_0 = 0$, otherwise $\tau_0 = \min\left\{\frac{f(x)-\alpha(0)}{S} : x \in K, p(x) \leq 0\right\}$, where $S := \min\{p(x) : x \in V(K), p(x) > 0\} > 0$. □

Clearly that (7) is a DC program with

$$g(x) = \chi_K(x) \quad \text{and} \quad h(x) = -c^T x - \tau \sum_{i=1}^{n} \min\{x_i, 1 - x_i\} \tag{8}$$

where $\chi_K(x) = 0$ if $x \in K$, $+\infty$, otherwise (the indicator function on K).

3.2 DCA for Solving Problem (7)

In this section we investigate a DC programming approach for solving (7).
A DC program is that of the form

$$\alpha := \inf\{f(x) := g(x) - h(x) \mid x \in \mathbb{R}^n\}, \tag{9}$$

with g, h being lower semicontinuous proper convex functions on $\mathbb{R}^n$, and its dual is defined as

$$\alpha := \inf\{h^*(y) - g^*(y) \mid y \in \mathbb{R}^n\}, \tag{10}$$

where $g^*(y) := \sup\{\langle x, y\rangle - g(x) \mid x \in \mathbb{R}^n\}$, the conjugate function of g.
Based on local optimality conditions and duality in DC programming, the DCA consists in the construction of two sequences $\{x^k\}$ and $\{y^k\}$, candidates to be optimal solutions of primal and dual programs respectively. These two sequences $\{x^k\}$ and $\{y^k\}$ are determined in the following way

$$y^k \in \partial h(x^k); \quad x^{k+1} \in \partial g^*(y^k). \tag{11}$$

It should be reminded that if either g or h is polyhedral convex, then (9) is called a polyhedral DC program for which DCA has a finite convergence ([9,10,5]). That is the case for DCA applied to (7).

DCA for Solving (7)

By the definition of h, a subgradient $y \in \partial h(x)$ can be chosen :

$$y \in \partial h(x) \Rightarrow y = -c + \begin{cases} \tau & \text{if } x_i \geq 0.5 \\ -\tau & \text{otherwise.} \end{cases} \tag{12}$$

Algorithm 1 (DCA applied to (7))
Let $\epsilon > 0$ be small enough and x^0. Set $k \leftarrow 0$; $er \leftarrow 1$.
while $er > \epsilon$ **do**
Compute $y^k \in \partial h(x^k)$ via (12).
Solve the linear program: $\min\{-\langle y^k, x\rangle : x \in K\}$ to obtain x^{k+1}
$er \leftarrow \min\left\{||x^{k+1} - x^k||, |f(x^{k+1}) - f(x^k)|\right\}$
$k \leftarrow k + 1$
endwhile

Remark 1. The application of DCA to Problem (7) leads to solve a series of linear programming problems.

The convergence of Algorithm **1** can be summarized in the next theorem whose proof is essentially based on the convergence theorem of a DC polyhedral program ([9,5]).

Theorem 2. *(Convergence properties of Algorithm* 1*)*

(i) *Algorithm* 1 *generates a sequence* $\{x^k\}$ *contained in* $V(K)$ *such that the sequence* $\{g(x^k) - h(x^k)\}$ *is decreasing.*
(ii) *If at iteration* r *we have* $x^r \in \{0,1\}^n$, *then* $x^k \in \{0,1\}^n$ *for all* $k \geq r$.
(iii) *The sequence* $\{x^k\}$ *converges to* $x^* \in V(K)$ *after a finite number of iterations. The point* x^* *is a critical point of Problem (7) (i.e.* $\partial g(x^*) \cap \partial h(x^*) \neq \emptyset$*). Moreover if* $x_i^* \neq \frac{1}{2}$ *for* $i = 1, \ldots, n$, *then* x^* *is a local solution to Problem (7).*

4 Computational Experiments

The algorithms were implemented on a PC Intel Core i3, CPU 2.2GhZ, 4G RAM, in C++. To solve the Linear Programming problems in relaxation procedure and in DCA, we used CLP solver, a very famous open source solver from COIN-OR (`www.coin-or.org`).

To test the performance of the algorithms as well as the efficiency of DCA, we randomly generated 10 examples, with number of users, number of sub-channels and number of slots of time varying from 3 to 6 which are quite large. For all test problems, we always obtained the ϵ-optimal solution, with $\varepsilon \leq 0.05$.

In Table 1 we compare the performance of two algorithms: the Branch and Bound algorithm (BB) and the combined DCA-Branch and Bound (DCA-BB). We limit the algorithms in number of iteration by 10^5, i.e, after 10^5 number of iterations the problem will be considered as failed in finding an optimal solution.

Some notations used in Table 1 :

#Var :	Number of variables	UB :	Upper bound
#Constr :	Number of constraints	LB :	Lower bound
#Iter :	Number of iterations		

Table 1. Numerical results

Prob	m	n	k	#Var	#Constr	B&B				DCA - B&B			
						#Iter	UB	LB	time(s)	#Iter	UB	LB	time(s)
P1	3	3	4	36	280	5	4981	4981	1.4	2	4981	4981	0.9
P2	4	4	3	48	1215	11	2267	2267	0.5	2	2267	2267	0.3
P3	4	4	4	64	1615	5	16528	16500	1.4	2	16543	16500	1
P4	3	3	6	54	416	39	40	40	8.8	17	40	40	1.4
P5	4	4	5	80	2015	797	876753	876624	25.4	95	876654	876534	23.4
P6	4	4	6	96	2415	49	5962	5950	65.6	38	5962	5871	35.4
P7	5	5	4	100	6024	123	9564	9514	792.4	37	9587	9432	76.8
P8	6	6	3	108	13055	137	3757	3743	1023.2	33	3789	3641	151.4
P9	5	4	6	120	4699	-	-	-	-	399	12367	12356	1500
P10	5	5	5	125	7524	-	-	-	-	278	98794	98765	1809

- : Failed in finding optimal solution

From the results we can see that the combination DCA-BB is more efficient than the BB specially when we deal with large scale problems. The reason is that DCA can give good solution for problems in any dimension. This gives us a very potential approach when we face up real life problems.

Conclusion. We have presented a DC programming approach and DCA for solving a resource allocation optimization problem in wireless network. The method is based on DCA and branch-and-bound algorithm. Computational results show that DCA is really efficient and at the same time it can give an integer solution while working in a continuous domain.

References

1. Ali-Yahiya, T., Beylot, A.-L., Pujolle, G.: Downlink resource allocation strategies for OFDMA based mobile WiMAX. Telecommunication Systems 44(1-2), 29–37 (2010)
2. Bacioccola, A., Cicconetti, C., Lenzini, L., Mingozzi, E.A.M.E., Erta, A.A.E.A.: A downlink data region allocation algorithm for IEEE 802.16e OFDMA. In: Proc. 6th Int. Conf. Information, Communications and Signal Processing, Singapore, pp. 1–5 (2007)
3. Le Thi, H.A., Pham Dinh, T., Le, D.M.: Exact Penalty in DC Programming. Vietnam Journal of Mathematics 27(2), 1216–1231 (1999)
4. Le Thi, H.A., Pham Dinh, T.: DC Programming: Theory, Algorithms and Applications. In: The First International Workshop on Global Constrained Optimization and Constraint Satisfaction (Cocos 2002), Valbonne-Sophia Antipolis, France, October 2-4, 28 pages (2002)
5. Le Thi, H.A., Pham Dinh, T.: The DC (Difference of Convex functions) Programming and DCA revisited with DC models of real world nonconvex optimization problem. Annals of Operations Research 133, 23–46 (2005)
6. Le Thi, H.A., Pham Dinh, T., Huynh Van, N.: Exact penalty and error bounds in DC programming. Journal of Global Optimization 52(3), 509–535 (2012)
7. Rodrigues, E.B., Casadevall, F.: Control of the trade-off between resource efficiency and user fairness in wireless networks using utility-based adaptive resource allocation. IEEE Communications Magazine 49(9), 90–98 (2011)
8. Rodrigues, E.B., Casadevall, F., Sroka, P., Moretti, M., Dainelli, G.: Resource Allocation and Packet Scheduling in OFDMA-based Cellular Networks. In: 4th International Conference on Cognitive Radio Oriented Wireless Networks and Communications, pp. 1–6 (2009)
9. Pham Dinh, T., Le Thi, H.A.: Convex analysis approach to DC programming: Theory, Algorithms and Applications. Acta Mathematica Vietnamica, dedicated to Professor Hoang Tuy on the occasion of his 70th birthday 22(1), 289–355 (1997)
10. Pham Dinh, T., Le Thi, H.A.: DC optimization algorithms for solving the trust region subproblem. SIAM J. Optimization 8, 476–505 (1998)

DC Programming and DCA for a Novel Resource Allocation Problem in Emerging Area of Cooperative Physical Layer Security

Thi Thuy Tran[1], Hoai An Le Thi[1], and Tao Pham Dinh[2]

[1] Laboratory of Theoretical and Applied Computer Science (LITA)
UFR MIM, University of Lorraine, Ile du Saulcy, 57045 Metz, France
[2] Laboratory of Mathematics, INSA-Rouen
University of Normandie, 76801 Saint-Etienne-du-Rouvray Cedex, France
{hoai-an.le-thi,thi-thuy.tran}@univ-lorraine.fr,
pham@insa-rouen.fr

Abstract. In this paper, we consider the problem of resource allocation in emerging area of cooperative physical layer security, whose objective function is of the difference of two convex functions, called DC function, and whose some constraints are coupling. We propose a novel DC decomposition for the objective function and use DCA (DC algorithms) to solve this problem. The main advantage of the proposed DC decomposition is that it leads to strongly convex quadratic sub-problems that can be easily solved by both distributed and centralized methods. The numerical results show the efficiency of this DC decomposition compared with the one proposed in ([2]). The DC decomposition technique in this article may be applied for a wide range of multiuser DC problems in communication systems, signal processing and networking.

Keywords: Resource allocation, physical layer, DC programming and DCA.

1 Introduction and Related Works

In communication systems, it is quite popular to encounter nonconvex optimization problems involving large scale data. This requires to have efficient, fast, scalable algorithms for dealing with them to meet the reality requirements such as running time, confidential matter, service quality.... DC programming and DCA (DC algorithms) will be an appropriate choice for such cases. It is due to the fact that almost all of the challenging nonconvex optimization problems in general and the challenging nonconvex problems in communication system in particular can be reformulated as DC programs. Furthermore, many evidences have also shown that DCA outperforms other algorithms in a lot of cases ([14],[15],[11], [12],...). In fact, there are more and more researchers using DC programming and DCA as an innovative approach to nonconvex programming because of its effectiveness ([23],[9],[1],[21],...). Nevertheless, the results obtained from DCA

H.A. Le Thi et al. (eds.), *Advanced Computational Methods for Knowledge Engineering*,
Advances in Intelligent Systems and Computing 358, DOI: 10.1007/978-3-319-17996-4_6

heavily depend on how the problem is decomposed into a DC program. In other words, finding a suitable DC decomposition plays an essential role in achieving a good result for DC program as well as optimizing the running time of computer.

Power control and resource allocation techniques for cellular communication systems have been recently studied in ([3],[8],...). To keep up with the pace of the dramatic developments of communication networks and computing systems, resource allocation nowadays continues to be the fundamental challenge to meet the increasing demand of users in terms of both quantity and quality. The aim of resource allocation is to efficiently assign the limited available resources among users while it still satisfies users' service demand. Many tools proposed recently, including optimization theory, control theory, game theory, and auction theory have been used to model and solve a wide range of practical resource allocation problems. As a result, resource allocation in communication system is an urgent research topic that has vast applications. It is vital to develop advanced and effective resource allocation techniques for guaranteeing the optimal performance of these systems and networks.

Physical layer security has been considered as a promising technique to enable the exchange of confidential messages over a wireless medium with the presence of unauthorized eavesdroppers, without relying on higher-layer encryption. The fundamental principle behind physical layer security is to exploit the inherent randomness of noise or the friendly jammer and communication channels to limit the amount of information that can be extracted by a fraudulent receiver. With appropriately designed coding and transmit precoding schemes in addition to the exploitation of any available channel state information, physical layer security schemes enable secret communication over a wireless medium without the aid of an encryption key ([17]). Recently, cooperative transmissions using friendly jammers to enhance physical layer security has drawn increasing attention ([4],[6], [16],[22],[5]).

In this paper, we consider the resource allocation problem in emerging area of cooperative physical layer security, whose objective function is of DC-type and whose some constraints are coupling. The purpose of this problem is to find an effective way to allocate the power of sources from users as well as from friendly jammers in order to maximize the sum of their secrecy rates. This model is introduced and solved by three authors Alberth Alvarado, Gesualdo Scutari, and Jong-Shi Pang from the United State, using their novel decomposition technique proposed in ([2]). They have formulated the system design as a game where the legitimate users are players who cooperate with the jammers to maximize their own secrecy rate. We will propose here a different DC decomposition technique for this problem, which can result in strongly convex quadratic subproblems with separate variables, which is simple to solve by both centralized and distributed methods.

2 A Novel Resource Allocation Problem in the Emerging Area of Cooperative Physical Layer Security

We take into account a wireless communication system comprised of Q transmitter and receiver pairs-the legitimate users, J friendly jammers, and a single eavesdropper. OFDMA transmissions are assumed for the legitimate users over flat-fading and quasi-static (constant within the transmission) channels. Let us denote H_{qq}^{SD}, H_{jq}^{JD}, H_{je}^{JE}, H_{qe}^{SE}, respectively as the channel gain of the legitimate source-destination pair q, the channel gain between the transmitter of jammer j and the receiver of user q, the channel gain between the transmitter of jammer j and the receiver of the eavesdropper, the channel gain between the source of user q and the eavesdropper. We also assume CSI of the eavesdropper's (cross-) channels; this is a common assumption in PHY security literature; see, e.g., [7]. CSI on the eavesdropper's channel can be obtained when the eavesdropper is active in the network and its transmissions can be monitored.

We follow the CJ paradigm, in which the friendly jammers cooperate with the users by giving a proper interference profile to mask the eavesdropper. The power allocation of source q is denoted by p_q; p_{jq}^J is the fraction of power allocated by friendly jammer j over the channel used by user q. $\mathbf{p}_q^J \triangleq (p_{jq}^J)$ is the power profile allocated by all the jammers over the channel of user q. The power budget of user q and jammer j do not exceed P_q and P_j^J, respectively.

From information theoretical assumption, the maximum achievable rate (i.e. the rate at which information can be transmitted secretly from the source to its intended destination) on link q is

$$r_{qq}(p_q, \mathbf{p}_q^J) \triangleq \log\left(1 + \frac{H_{qq}^{SD} p_q}{\sigma^2 + \sum_{j=1}^{J} H_{jq}^{JD} p_{jq}^J}\right). \tag{1}$$

Similarly, the rate on the channel between source q and the eavesdropper is

$$r_{qe}(p_q, \mathbf{p}_q^J) \triangleq \log\left(1 + \frac{H_{qe}^{SE} p_q}{\sigma^2 + \sum_{j=1}^{J} H_{je}^{JE} p_{jq}^J}\right). \tag{2}$$

The secrecy rate of user q is then (see, e.g., [7])

$$r_q^s \triangleq [r_{qq}(p_q, \mathbf{p}_q^J) - r_{qe}(p_q, \mathbf{p}_q^J)]_+. \tag{3}$$

Problem Formulation: The system design is formulated as a game where the legitimate users are the players who cooperate with the jammers to maximize their own secrecy rate. More generally, each user q seeks together with the jammers the tuple $(p_q, \mathbf{p}_q^J)$ satisfying the following optimization problem:

$$
\begin{aligned}
\text{Max}_{\ (p_q,\mathbf{p}_q^J)_q\geq 0} \quad & r(\mathbf{p},\mathbf{p}^J) \triangleq \sum_{q=1}^{Q} r_q^s \\
\text{s.t.} \quad & p_q \leq P_q, \qquad \forall q=1,\ldots,Q, \\
& \sum_{r=1}^{Q} p_{jr}^J \leq P_j^J, \qquad \forall j=1,\ldots,J.
\end{aligned} \tag{4}
$$

Note that

$$
r_{qq}(p_q,\mathbf{p}^J) \geq r_{qe}(p_q,\mathbf{p}^J) \Longleftrightarrow \left[\begin{array}{l} p_q = 0 \\ \frac{H_{qq}^{SD}}{\sigma^2+\sum_{j=1}^{J} H_{jq}^{JD} p_{jq}^J} \geq \frac{H_{qe}^{SE}}{\sigma^2+\sum_{j=1}^{J} H_{je}^{JE} p_{jq}^J} \end{array}\right.
$$

$$
\Longleftrightarrow \left[\begin{array}{l} p_q = 0 \\ \sum_{j=1}^{J}(H_{qq}^{SD}H_{je}^{JE} - H_{qe}^{SE}H_{jq}^{JD})p_{jq}^J + (H_{qq}^{SD} - H_{qe}^{SE})\sigma^2 \geq 0. \quad (*) \end{array}\right.
$$

If (*) is violated, it means that $r_q^s = 0$, i.e., the secrecy rate of the players equal to zero, which is insignificant since players try to maximize their secrecy rate. Therefore, we can ignore the feasible players' strategy profiles not satisfying (*). It leads us to solving the following smooth problem:

$$
\begin{aligned}
\text{Max}_{\ (p_q,\mathbf{p}_q^J)_q\geq 0} \quad & r(\mathbf{p},\mathbf{p}^J) \triangleq \sum_{q=1}^{Q}[r_{qq}(p_q,\mathbf{p}_q^J) - r_{qe}(p_q,\mathbf{p}_q^J)] \\
\text{s.t.} \quad & p_q \leq P_q, \qquad \forall q=1,\ldots,Q, \\
& \sum_{r=1}^{Q} p_{jr}^J \leq P_j^J, \qquad \forall j=1,\ldots,J \\
& \sum_{j=1}^{J}(H_{qq}^{SD}H_{je}^{JE} - H_{qe}^{SE}H_{jq}^{JD})p_{jq}^J + (H_{qq}^{SD} - H_{qe}^{SE})\sigma^2 \geq 0 \qquad \forall q=1,\ldots,Q.
\end{aligned} \tag{5}
$$

Instead of solving the problem (5), we will solve the equivalent problem as follows.

$$
\begin{aligned}
\text{Min}_{\ (p_q,\mathbf{p}_q^J)_q\geq 0} \quad & r_1(\mathbf{p},\mathbf{p}^J) \triangleq \sum_{q=1}^{Q}[-r_{qq}(p_q,\mathbf{p}_q^J) + r_{qe}(p_q,\mathbf{p}_q^J)] \\
\text{s.t.} \quad & p_q \leq P_q, \qquad \forall q=1,\ldots,Q, \\
& \sum_{r=1}^{Q} p_{jr}^J \leq P_j^J, \qquad \forall j=1,\ldots,J \\
& \sum_{j=1}^{J}(H_{qq}^{SD}H_{je}^{JE} - H_{qe}^{SE}H_{jq}^{JD})p_{jq}^J + (H_{qq}^{SD} - H_{qe}^{SE})\sigma^2 \geq 0 \qquad \forall q=1,\ldots,Q.
\end{aligned} \tag{6}
$$

3 DC Programming and DCA for Solving the Problem (6)

3.1 A Brief Introduction of DC Programming and DCA

DC Programming and DCA constitute the backbone of smooth/nonsmooth nonconvex programming and global optimization. They are introduced by Pham Dinh Tao in 1985 in their preliminary form and extensively developed by Le Thi Hoai An and Pham Dinh Tao since 1994 to become now classic and more and more popular. DCA is a continuous primal dual subgradient approach. It is based on local optimality and duality in DC programming in order to solve standard DC programs which are of the form

$$\alpha = \inf\{f(x) := g(x) - h(x) : \ x \in \mathbb{R}^n\}, \quad (P_{dc})$$

with $g, h \in \Gamma_0(\mathbb{R}^n)$, which is a set of lower semi-continuous proper convex functions on $\mathbb{R}^n$. Such a function f is called a DC function, and $g - h$, a DC decomposition of f, while the convex functions g and h are DC components of f. A constrained DC program whose feasible set C is convex always can be transformed into an unconstrained DC program by adding the indicator function of C to the first DC component.

Recall that, for a convex function ϕ, the subgradient of ϕ at x_0, denoted as $\partial\phi(x_0)$, is defined by

$$\partial\phi(x_0) := \{y \in \mathbb{R}^n : \phi(x) \geq \phi(x_0) + \langle x - x_0, y\rangle, \forall x \in \mathbb{R}^n\}$$

The main principle of DCA is quite simple, that is, at each iteration of DCA, the convex function h is approximated by its affine minorant at $y^k \in \partial h(x^k)$, and it leads to solving the resulting convex program.

$$\begin{aligned} y^k \quad &\in \partial h(x^k) \\ x^{k+1} &\in \arg\min_{x \in \mathbb{R}^n} \{g(x) - h(x^k) - \langle x - x^k, y^k\rangle\}. \quad (P_k) \end{aligned}$$

The computation of DCA is only dependent on DC components g and h but not the function f itself. Actually, there exist infinitely many DC decompositions corresponding to each DC function and they generate various versions of DCA. Choosing the appropriate DC decomposition plays a key role since it influences on the properties of DCA such as convergence speed, robustness, efficiency, globality of computed solutions,...DCA is thus a philosophy rather than an algorithm. For each problem we can design a family of DCA based algorithms. To the best of our knowledge, DCA is actually one of the rare algorithms for nonsmooth nonconvex programming which allow to solve large-scale DC programs. DCA was successfully applied for solving various nonconvex optimization problems, which quite often gave global solutions and is proved to be more robust and more efficient than related standard methods ([18,19,20] and the list of reference in ([10]).

This is a DCA generic scheme:

- **Initialization.** Choose an initial point x^0. $0 \longleftarrow k$
- **Repeat.**
  Step 1. For each k, x^k is known, computing $y^k \in \partial h(x^k)$.
  Step 2. Calculating $x^{k+1} \in \partial g^*(y^k)$
  where $\partial g^*(y^k) = \arg\min_{x \in \mathbb{R}^n} \{g(x) - h(x^k) - \langle x - x^k, y^k \rangle : x \in C\}$
  Step 3. $k \leftarrow k+1$
- **Until** stopping condition is satisfied.

The convergence properties of DCA and its theoretical basis is analyzed and proved completely in ([18], [13],[19]). Some typical important properties of DCA are worth being recalled here.

i) DCA is a descent method without line search but with global convergence : the sequences $\{g(x^k) - h(x^k)\}$ and $\{h^*(y^k) - g^*(y^k)\}$ are decreasing.

ii) If the optimal value α of DC program is finite and the infinite sequences $\{x^k\}$ and $\{y^k\}$ are bounded, then every limit point x^* (resp. y^*) of sequence $\{x^k\}$ (resp. $\{y^k\}$) is a critical point of $(g-h)$ (resp. $(h^* - g^*)$), i.e. $\partial g(x^*) \cap \partial h(x^*) \neq \emptyset$ (resp. $\partial h^*(y^*) \cap \partial g^*(x^*) \neq \emptyset$)

iii) DCA has a linear convergence for DC programs.

iv) DCA has a finite convergence for polyhedral DC programs

3.2 DC Programming and DCA for the Problem (6)

The New DC Decomposition for the Objective Function of (6)
For any value of ρ, the objective function of the problem (6) can be written in the form:

$$r_1(\mathbf{p}, \mathbf{p}^J) = G(\mathbf{p}, \mathbf{p}^J) - H(\mathbf{p}, \mathbf{p}^J),$$

where

$$G(\mathbf{p}, \mathbf{p}^J) = \rho \left(\sum_{q=1}^{Q} p_q^2 + \sum_{j=1}^{J} \sum_{q=1}^{Q} p_{jq}^{J2} \right),$$

$$H(\mathbf{p}, \mathbf{p}^J) = \left[\rho \left(\sum_{q=1}^{Q} p_q^2 + \sum_{j=1}^{J} \sum_{q=1}^{Q} p_{jq}^{J2} \right) - \sum_{q=1}^{Q} \left(-r_{qq}(p_q, \mathbf{p}_q^J) + r_{qe}(p_q, \mathbf{p}_q^J) \right) \right].$$

Proposition 1. *If $\rho \geq \frac{M^2}{\sigma^4}\sqrt{1 + 2J + 4J^2}$, then both $G(\mathbf{p}, \mathbf{p}^J)$ and $H(\mathbf{p}, \mathbf{p}^J)$ are convex, where $M = \max_{q=1,\ldots,Q; j=1,\ldots,J} \{H_{qe}^{SE}, H_{qq}^{SD}, H_{jq}^{JD}, H_{je}^{JE}\}$.*

Proof. We have

$$\nabla r_1(\mathbf{p}, \mathbf{p}^J) = \left[G_1(\mathbf{p}, \mathbf{p}^J) = \left(\frac{\partial r_1}{\partial p_q} \right)_{q=1,\ldots,Q}, G_2(\mathbf{p}, \mathbf{p}^J) = \left(\frac{\partial r_1}{\partial p_{jq}^J} \right)_{\substack{j=1,\ldots,J \\ q=1,\ldots,Q}} \right]^T,$$

where

$$\frac{\partial r_1}{\partial p_q} = \frac{H_{qe}^{SE}}{\sigma^2 + H_{qe}^{SE} p_q + A} - \frac{H_{qq}^{SD}}{\sigma^2 + H_{qq}^{SD} p_q + B},$$

$$\frac{\partial r_1}{\partial p_{jq}^J} = \frac{H_{je}^{JE}}{\sigma^2 + H_{qe}^{SE} p_q + A} - \frac{H_{jq}^{JD}}{\sigma^2 + H_{qq}^{SD} p_q + B} - \frac{H_{je}^{JE}}{\sigma^2 + A} + \frac{H_{jq}^{JD}}{\sigma^2 + B},$$

with $A = \sum_{k=1}^{J} H_{ke}^{JE} p_{kq}^J$, $B = \sum_{k=1}^{J} H_{kq}^{JD} p_{kq}^J$.

Firstly, we show that there exists a constant L such that

$$\|\nabla r_1(\mathbf{p}, \mathbf{p}^J) - \nabla r_1(\widehat{\mathbf{p}}, \widehat{\mathbf{p}}^J)\|_2 \leq L \|(\mathbf{p}, \mathbf{p}^J) - (\widehat{\mathbf{p}}, \widehat{\mathbf{p}}^J)\|_2.$$

We have

$$\begin{aligned} &\|\nabla r_1(\mathbf{p}, \mathbf{p}^J) - \nabla r_1(\widehat{\mathbf{p}}, \widehat{\mathbf{p}}^J)\|_2^2 \\ &= \|\nabla G_1(\mathbf{p}, \mathbf{p}^J) - \nabla G_1(\widehat{\mathbf{p}}, \widehat{\mathbf{p}}^J)\|_2^2 + \|\nabla G_2(\mathbf{p}, \mathbf{p}^J) - \nabla G_2(\widehat{\mathbf{p}}, \widehat{\mathbf{p}}^J)\|_2^2. \end{aligned}$$

$$\begin{aligned} &\|\nabla G_1(\mathbf{p}, \mathbf{p}^J) - \nabla G_1(\widehat{\mathbf{p}}, \widehat{\mathbf{p}}^J)\|_2^2 \\ &= \sum_{q=1}^{Q} \left(\frac{H_{qe}^{SE}}{\sigma^2 + H_{qe}^{SE} p_q + A} - \frac{H_{qe}^{SE}}{\sigma^2 + H_{qe}^{SE} \widehat{p}_q + A_1} + \frac{H_{qq}^{SD}}{\sigma^2 + H_{qq}^{SD} \widehat{p}_q + B_1} \right. \\ &\qquad \left. - \frac{H_{qq}^{SD}}{\sigma^2 + H_{qq}^{SD} p_q + B} \right)^2 \\ &= \sum_{q=1}^{Q} \left(\frac{H_{qe}^{SE^2} (\widehat{p}_q - p_q) + \sum_{j=1}^{J} H_{qe}^{SE} H_{je}^{JE} (\widehat{p}_{jq}^J - p_{jq}^J)}{(\sigma^2 + H_q e^{SE} p_q + A)(\sigma^2 + H_{qe}^{SE} \widehat{p}_q + A_1)} \right. \\ &\qquad \left. + \frac{H_{qe}^{SD^2} (p_q - \widehat{p}_q) + \sum_{j=1}^{J} H_{qq}^{SD} H_{jq}^{JD} (p_{jq}^J - \widehat{p}_{jq}^J)}{(\sigma^2 + H_{qe}^{SD} p_q + B)(\sigma^2 + H_{qq}^{SD} \widehat{p}_q + B_1)} \right)^2, \end{aligned}$$

with $A_1 = \sum_{k=1}^{J} H_{ke}^{JE} \widehat{p}_{kq}^J$; $B_1 = \sum_{k=1}^{J} H_{kq}^{JD} \widehat{p}_{kq}^J$.

Applying the Cauchy-Schwartz inequality and noting that all denominators of the fractions above are greater than σ^2 and $M = \max_{\substack{q=1,\dots,Q \\ j=1,\dots,J}} \{H_{qe}^{SE}, H_{qq}^{SD}, H_{jq}^{JD}, H_{je}^{JE}\}$, we obtain

$$\|\nabla G_1(\mathbf{p}, \mathbf{p}^J) - \nabla G_1(\widehat{\mathbf{p}}, \widehat{\mathbf{p}}^J)\|_2^2 \leq \tfrac{4M^4}{\sigma^8}(1 + J) \|(\mathbf{p}, \mathbf{p}^J) - (\widehat{\mathbf{p}}, \widehat{\mathbf{p}}^J)\|_2^2. \quad (**)$$

We also have $\|\nabla G_2(\mathbf{p}, \mathbf{p}^J) - \nabla G_2(\widehat{\mathbf{p}}, \widehat{\mathbf{p}}^J)\|_2^2$

$$\begin{aligned} &= \sum_{j=1}^{J} \sum_{q=1}^{Q} \left[\frac{H_{je}^{JE}}{\sigma^2 + H_{qe}^{SE} p_q + A} - \frac{H_{je}^{JE}}{\sigma^2 + H_{qe}^{SE} p_q + A_1} + \frac{H_{je}^{JE}}{\sigma^2 + A_1} - \frac{H_{je}^{JE}}{\sigma^2 + A} + \right. \\ &\qquad \left. + \frac{H_{jq}^{JD}}{\sigma^2 + H_{qq}^{SD} \widehat{p}_q + B_1} - \frac{H_{jq}^{JD}}{\sigma^2 + H_{qq}^{SD} p_q + B} + \frac{H_{jq}^{JD}}{\sigma^2 + B} - \frac{H_{jq}^{JD}}{\sigma^2 + B_1} \right]^2 \end{aligned}$$

$$= \sum_{j=1}^{J}\sum_{q=1}^{Q}\left[\left(\frac{H_{je}^{JE}H_{qe}^{SE}}{M_1M_2} - \frac{H_{jq}^{JD}H_{qq}^{SD}}{M_3M_4}\right)(\widehat{p}_q - p_q)\right.$$
$$\left. + \sum_{k=1}^{J}\left(\frac{H_{je}^{JE}H_{ke}^{JE}}{M_1M_2} - \frac{H_{je}^{JE}H_{ke}^{JE}}{M_3M_4} - \frac{H_{jq}^{JD}H_{kq}^{JD}}{M_5M_6} + \frac{H_{jq}^{JD}H_{kq}^{JD}}{M_7M_8}\right)(\widehat{p}_{kq}^J - p_{kq}^J)\right]^2.$$

where $M_i, (i = 1, ..., 8)$ are the denominators of the fractions in the above expression, respectively. Applying the Cauchy-Schwarz inequality we also gain the following inequality $\|\nabla G_2(\mathbf{p}, \mathbf{p}^J) - \nabla G_2(\widehat{\mathbf{p}}, \widehat{\mathbf{p}}^J)\|_2^2 \le \frac{4M^4}{\sigma^8}J(1+4J)\|(\mathbf{p}, \mathbf{p}^J) - (\widehat{\mathbf{p}}, \widehat{\mathbf{p}}^J)\|_2^2$. (***)

Adding (**) to (***) and then taking the square root of both sides, we obtain $\|\nabla r_1(\mathbf{p}, \mathbf{p}^J) - \nabla r_1(\widehat{\mathbf{p}}, \widehat{\mathbf{p}}^J)\|_2 \le \frac{2M^2}{\sigma^4}\sqrt{1+2J+4J^2}\|(\mathbf{p}, \mathbf{p}^J) - (\widehat{\mathbf{p}}, \widehat{\mathbf{p}}^J)\|_2$. This inequality results in $\nabla^2(r_1) \preceq \frac{2M^2}{\sigma^4}\sqrt{1+2J+4J^2}I$. Furthermore, $\nabla^2 H = 2\rho I - \nabla^2 r_1$, thus $\nabla^2 H \succeq 0$ if $\rho \ge \frac{M^2}{\sigma^4}\sqrt{1+2J+4J^2}$, which ensures that $H(\mathbf{p}, \mathbf{p}^J)$ is a convex function. In addition, $G(\mathbf{p}, \mathbf{p}^J)$ is also convex for all positive values of ρ. Therefore, with $\rho \ge \frac{M^2}{\sigma^4}\sqrt{1+2J+4J^2}$, $G(\mathbf{p}, \mathbf{p}^J) - H(\mathbf{p}, \mathbf{p}^J)$ is a DC decomposition of the objective of (6), which is different from that in ([2]). As a result, we obtain a new DC program as follows.

$$\begin{aligned}
\text{Min}\ & r_1(\mathbf{p}, \mathbf{p}^J) \triangleq G(\mathbf{p}, \mathbf{p}^J) - H(\mathbf{p}, \mathbf{p}^J)\\
\text{s.t.}\ & (p_q, \mathbf{p}_q^J)_q^Q \ge 0,\\
& p_q \le P_q, \qquad \forall q = 1, \ldots, Q,\\
& \sum_{r=1}^{Q} p_{jr}^J \le P_j^J, \qquad \forall j = 1, \ldots, J\\
& \sum_{j=1}^{J}(H_{qq}^{SD}H_{je}^{JE} - H_{qe}^{SE}H_{jq}^{JD})p_{jq}^J + (H_{qq}^{SD} - H_{qe}^{SE})\sigma^2 \ge 0, \quad \forall q = 1, \ldots, Q
\end{aligned}$$

Following the generic DCA scheme described in Section 3.1, DCA applied on (6) is given by the algorithm below.

DCA Scheme for DC Program (6)

- **Initialization.** Choose an initial point $x^{0.} = (\mathbf{p}^0, \mathbf{p}^{J,0}), 0 \longleftarrow k$.
- **Repeat.**
  Step 1. For each $k, x^k = (\mathbf{p}^k, \mathbf{p}^{J,k})$ is known, compute $y^k = (\overline{\mathbf{p}}^k, \overline{\mathbf{p}}^{J,k}) = \nabla H(x^k)$ with

$$\nabla H(x) = \begin{bmatrix} \left(2\rho p_q - \frac{H_{qe}^{SE}}{\sigma^2 + H_{qe}^{SE}p_q + A} + \frac{H_{qq}^{SD}}{\sigma^2 + H_{qq}^{SD}p_q + B}\right)_{q=1,\ldots,Q} \\ \left(2\rho p_{jq}^J - \frac{H_{je}^{JE}}{\sigma^2 + H_{qe}^{SE}p_q + A} + \frac{H_{je}^{JE}}{\sigma^2 + A} + \frac{H_{jq}^{JD}}{\sigma^2 + H_{qq}^{SD}p_q + B} - \frac{H_{jq}^{JD}}{\sigma^2 + B}\right)_{\substack{j=1,\ldots,J\\ q=1,\ldots,Q}} \end{bmatrix},$$

where $A = \sum_{k=1}^{J} H_{ke}^{JE}p_{kq}^J; B = \sum_{k=1}^{J} H_{kq}^{JD}p_{kq}^J$.

Step 2. Find $x^{k+1} = (\mathbf{p}^{k+1}, \mathbf{p}^{J,k+1})$ by solving the following convex subproblem.

$$\text{Min} \quad \rho\left(\sum_{q=1}^{Q} p_q^2 + \sum_{j=1}^{J}\sum_{q=1}^{Q} p_{jq}^{J2}\right) - < y^k, x >$$

s.t.

$$x = (x_q)_{q=1}^{Q} = (p_q, \mathbf{p}_q^J)_{q=1}^{Q} \geq 0$$
$$p_q \leq P_q, \qquad \forall q = 1, \ldots, Q$$
$$\sum_{r=1}^{Q} p_{jr}^J \leq P_j^J, \qquad \forall j = 1, \ldots, J$$

$$\sum_{j=1}^{J} (H_{qq}^{SD} H_{je}^{JE} - H_{qe}^{SE} H_{jq}^{JD}) p_{jq}^J + (H_{qq}^{SD} - H_{qe}^{SE})\sigma^2 \geq 0 \ \forall q = 1, ..., Q. \quad (7)$$

Step 3. $k \leftarrow k + 1$.

- **Until** the stopping condition is satisfied.

4 Numerical Results

4.1 Datasets

The position of the users, jammers, and eavesdropper are randomly generated within a square area; the channel gain $H_{qq}^{SD}, H_{qe}^{SE}, H_{jq}^{JD}, H_{je}^{JE}$ are Rayleigh distributed with mean equal to one and normalized by the distance between the transmitter and the receiver. The results are collected only for the channel realizations satisfying condition (7).

4.2 Setting Parameters and Stopping Criteria

In this paper, we present some experiment results obtained from DCA and make a comparision with those obtained from SCA mentioned in ([2]). Therefore, for the fair comparison, all parameters are set in the same way as in ([2]). All users and jammers have the same power budget, i.e. $P_q = P_j^J = P$ and we set $\mathbf{snr} = \frac{P}{\sigma^2} = 10dB$. The number of jammers, J, is set to equal $[Q/2]$.

In Table 1, the initial points are randomly generated five times and then they are shared for all algorithms in the paper. In Table 2, the initial points are also randomly generated five times but separately for each algorithm. In each table, the average of the results gained corresponding to each initial point is chosen as the final value of SSR and the standard deviation is also accompanied (that is indicated in parenthesis). The best values of SSR are collected and reported in the SSR Best column. The DCA-based algorithms are terminated when at least one of the following criteria are satisfied:

Table 1. System secrecy rate (SSR) versus number Q of legitimate users

Q	DCA			SCA		
	SSR	SSR Best	CPU(s)	SSR	SSR Best	CPU(s)
10	**5.236** (0.0107)	**5.248**	6.2	4.915 (0.011)	4.923	194.8
20	13.013 (0.055)	13.104	8.2	**13.199** (0.052)	**13.205**	677.2
30	**15.907**(0.040)	**15.943**	14.1	15.901 (0.043)	15.909	1861.1
40	**19.167** (0.008)	**19.179**	22.4	18.974 (0.007)	18.982	2615.7
50	**28.714** (0.038)	**28.744**	25.7	26.719 (0.025)	26.720	4000.3

Table 2. System secrecy rate (SSR) versus number Q of legitimate users

Q	DCA			SCA		
	SSR	SSR Best	CPU(s)	SSR	SRR Best	CPU(s)
10	**5.242** (0.004)	**5.249**	5.6	4.921(0.015)	4.923	166.5
20	13.091 (0.027)	13.134	7.6	**13.196** (0.106)	**13.385**	952.9
30	**15.900** (0.023)	**15.926**	12.8	15.856 (0.019)	15.878	1635.4
40	**19.128** (0.052)	**19.199**	20,6	18.975 (0,007)	18.985	2650.8
50	**28.742** (0.018)	**28.759**	20.6	26.719 (0.012)	26.727	3424.3

- The absolute value of the difference of the System Secrecy Rate (SSR) in two consecutive iterations becomes smaller than $1e-5$.
- The norm of the difference of two tuples $(\mathbf{p}, \mathbf{p}^J)$ in two consecutive iterations becomes smaller than $1e-5$.

4.3 Numerical Results and Comments

In Table 1 and Table 2, we compare the value of SSR versus the number Q of legitimate users obtained by DCA and SCA, respectively. These tables also show the running times of both algorithms versus the number of legitimate users. In Table 3, we demonstrate the dependence of SSR on $\mathtt{snr}$ in the case of 10 users. In

Table 3. System secrecy rate (SSR) versus the various values of $\mathtt{snr}$ in the case of 10 users

	snr	10	20	30	40
DCA	SSR	5.245	6.419	6.638	6.481
SCA	SSR	4.954	6.122	6.410	6.453

general, the numerical results show that SSR achieved by all algorithms tends to increase with the number of users as well as snr. It can be observed from the both Table 1 and Table 2 that the DCA yields system secrecy rates better than those of SCA in almost all of cases while it is much less expensive. More detailed, though the proposed DCA is a centralized approach, its consuming times are less than those of SCA that is implemented as a distributed approach, for all the datasets. Table 3 expresses the comparison of the gains from both algorithms corresponding to the different values of snr in the case of 10 users. This table shows that the increase of snr tends to lead to the rise of SSR. The gains of DCA is always better than those of SCA.

5 Conclusions

In this paper, we have presented a new DC decomposition and the corresponding DCA-based algorithm for the resource allocation problem in physical layer. The experimental results on some datasets have shown the robustness as well as the efficiency of DCA based on this new DC decomposition in term of both the running time and the objective value. It provides more evidences to show that DC programming and DCA is an efficient and robust approach for solving the nonconvex optimization problems in the wide range of areas. In addition, the technique of decomposing DC proposed in this paper can be also applied for the various optimization problems and it is especially useful if one wants to solve these problems in distributed ways to take advantage of all CPUs in the computers at the same time as well as help to separate a large scale problem into smaller scale ones.

References

1. Al-Shatri, A., Weber, T.: Achieving the maximum sum rate using dc programming in cellular networks. IEEE Trans. Signal Process 60(3), 1331–1341 (2012)
2. Alvarado, A., Scutari, G., Pang, J.-S.: A new decomposition method for multiuser dc-programming and its application. IEEE Transactions on Signal Processing 62(11), 2984–2998 (2014)
3. Chiang, M.: Geometric programming for communication systems. Foundations and Trends in Communication and Information Theory 2(1-2), 1–154 (2005)
4. Dong, L., Han, Z., Petropulu, A., Poor, H.: Improving wireless physical layer security via cooperating relays. IEEE Trans. Signal Process 58(3), 1875–1888 (2010)
5. Han, Z., Marina, N., Debbah, M., Hjorungnes, A.: Physical layer security game: intersaction between source, eavesdropper, and friendly jammer. EURASIP J. Wireless Commun. Netw 2009(1), 1–11 (2009)
6. He, X., Yener, A.: Cooperative jamming: The tale of friendly interference for secrecy, pp. 65–88. Spinger (2010)
7. Jorswieck, E., Wolf, A., Gerbracht, S.: Secrecy on the Physical Layer in Wireless Networks, vol. 20. INTECH (2010)
8. Julian, D., Chiang, M., O'Neill, D., Boyd, S.P.: QoS and fairness constrained convex optimization of resource allocation for wireless cellular and ad hoc networks. In: IEEE INFOCOM 2002, pp. 477–486 (June 2002)

9. Kha, H.H., Tuan, H.D., Nguyen, H.H.: Fast global optimal power allocation in wireless network by local dc programming. IEEE Trans. on Wireless Communications 11(2), 510–512 (2012)
10. Le Thi, H.A.: Dc programming and dca, http://lita.sciences.univ-metz.fr/lethi
11. Le Thi, H.A., Nguyen, M.C., Dinh, T.P.: A dc programming approach for finding communities in networks. Neural Computation 26(12), 2827–2854 (2014)
12. Le Thi, H.A., Nguyen, M.C., Pham Dinh, T.: Self-organizing maps by difference of convex functions optimization. Data Mining and Knowledge Discovery 28, 1336–1365 (2014)
13. Le Thi, H.A., Pham Dinh, T.: The DC (difference of convex functions) programming and DCA revisited with DC models of real world nonconvex optimization problems. Ann. Oper. Resp. 133, 23–46 (2005)
14. Le Thi, H.A., Pham Dinh, T., Le, H.M., Vo, X.T.: Dc approximation approaches for sparse optimization. European Journal of Operational Research 244(1), 26–46 (2015)
15. Le Thi, H.A., Vo, X.T., Pham Dinh, T.: Feature selection for linear svms under uncertain data: Robust optimization based on difference of convex functions algorithms. Neural Networks 59, 36–50 (2014)
16. Li, J., Petropulu, A., Weber, S.: On cooperative relaying schemes for wireless physical layer security. IEEE Trans. Signal Process 59(10), 4985–4997 (2011)
17. Mutherjee, A., Fakoorian, S.A.A., Huang, J., Swindlehurst, A.L.: Principle of physical layer security in multiuser wireless networks: A survey. IEEE Communication Survey and Tutorials 16(3), 1550–1573 (2014)
18. Pham Dinh, T., Le Thi, H.A.: Convex analysis approach to dc programming: Theory, algorithms and applications. Acta Mathematica Vietnamica 22(1), 289–357 (1997)
19. Pham Dinh, T., Le Thi, H.A.: optimization algorithms for solving the trust region subproblem. SIAM J. Optimization 8 (1998)
20. Pham Dinh, T., Le Thi, H.A.: Recent advances on DC programming and DCA. To appear in Transactions on Computational Collective Intelligence 37 (2013)
21. Piot, V., Geist, M., Pietquin, O.: Difference of convex functions programming for reinforcement learning. In: Advances in Neural Information Processing Systems (2014)
22. Stanojev, I., Yener, A.: Improving secrecy rate via spectrum leasing for friendly jamming. IEEE Trans. Inf. Forensics Security 12(1), 134–145 (2013)
23. Vucic, N., Schubert, M.: Dc programming approach for resource allocation in wireless networks. In: IEEE, Proceedings of the 8th International Symposium on Modeling and Optimization in Mobile, Ad Hoc and Wireless Networks, pp. 380–386 (2010)

Scheduling Problem for Bus Rapid Transit Routes

Quang Thuan Nguyen and Nguyen Ba Thang Phan

School of Applied Mathematics and Informatics,
Hanoi University of Science and Technology, Vietnam
thuan.nguyenquang@hust.edu.vn,
phanbathang125692@gmail.com

Abstract. Bus Rapid Transit (BRT) system plays an important role in public transport. In order to improve BRT operation quality, the headway optimization and scheduling combination are considered. Based on an existing model, this work proposes a modifying model that overcomes the limitation of the old one. The problem is then solved by a genetic algorithm - based method. Experimental results show that total cost is significantly decreasing after using scheduling combination strategies. The optimal frequency of BRT vehicles is also determined.

Keywords: Bus Rapid Transit, Genetic Algorithm, Scheduling.

1 Introduction

Bus Rapid Transit (BRT) can be defined as a corridor in which buses operate on a dedicated right-of-way such as a bus-way or a bus lane reserved for buses on an arterial road or freeway. It can be understood as a rapid mode of transportation that can combine the quality of rail transit and the flexibility of buses [5]. BRT may be an efficient transportation service, especially for cities in developing countries where there are high transit-dependent populations and financial resources are limited [9]. Nowadays, many cities all over the world have developed the BRT system such as Bogota, Delhi, Guangzhou, Jakarta, ...

To improve the quality of BRT service or bus systems in general, it is absolutely scheduled. One usually schedules the frequency of each bus route, namely, the headway - the time between two consecutive buses leaving the initial station - is calculated. Another scheduling problem is to plan when BRT vehicles should stop at certain stations. This problem is called the problem of scheduling combination for buses. Most researches focused on traditional bus scheduling [4], [13], [3], [1], [11], [8]. For BRT systems, there are fewer researches. The frequency of BRT route is studied by Bai et al. in 2007 [2] or by Liang et al. in 2011 [6]. Scheduling combination for BRT is considered in [7]. Sun et al. in 2008 investigated the frequency and scheduling combination for a BRT route in the mean time [12]. The solution methods were usually heuristic such as taboo search methods [2], genetic algorithms [6], [12].

According to the BRT vehicle operation form and stops number, the scheduling is regularly divided to three forms: normal scheduling, zone scheduling and express scheduling. With the normal scheduling form, vehicles run along routes and stop at every station from the initial stop to the end. The zone scheduling is defined as vehicles

H.A. Le Thi et al. (eds.), *Advanced Computational Methods for Knowledge Engineering*,
Advances in Intelligent Systems and Computing 358, DOI: 10.1007/978-3-319-17996-4_7

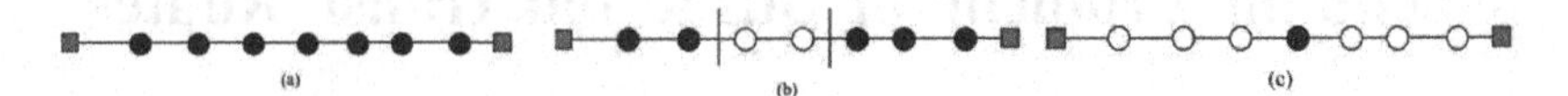

Fig. 1. (a) Normal schedule; (b) Zone schedule; (c) Express schedule

only run on high-traffic volume zone while for the express scheduling, vehicles only stop at certain stations with the large demand of passengers (Figure 1).

It is clear that the design and assignment of BRT vehicles to suitable scheduling forms are transport planners' very important tasks. Sun et al. proposed a genetic algorithm (GA)- based method to give the optimal headway and a suitable scheduling combination so that the total cost (including the waiting time, traveling time, operation BRT cost ...) is minimized [12]. The authors were based on the assumption that a BRT vehicle is not allowed to cross another BRT vehicle. This leads to an unrealistic thing when an express BRT vehicle may cross a normal one.

Our work improves the limitation above by rebuilding a mathematical model. The modifying model is then solved by a genetic algorithm-based method. We also applied this model to a BRT route in Hanoi. The experimented route is being constructed and expectedly used at the end of 2015. The promising experimental results make a valuable contribution of this work.

The next section is devoted to describe BRT systems and the mathematical problem. Section 3 presents a brief of the solution method. The case study is dedicated in Section 4. The paper is ended by some conclusions and acknowledgements.

2 Optimization Model of BRT Systems

Consider a BRT system with the following assumptions:

i) BRT vehicles are completely prioritized, i.e. they never stop due to traffic lights;
ii) BRT vehicles run at constant speed, namely, the running time between two any stations does not change;
iii) The headway is fixed in the study period;
iv) The passenger arrival rate is uniform and unchanged in the given period;
v) The duration for stop and acceleration and deceleration are fixed;

We are using some notations as follows:

i– BRT vehicle ith i $(i = 1, 2, \cdots, M)$, in which M is the total number of operating vehicles in the study period;

j– stop jth on the BRT route $(j = 1, 2, \cdots, N)$, in which N is the total number of stops on the route;

l– scheduling form: $l = 1$ means the normal scheduling, $l = 2$ shows the zone scheduling and $l = 3$ assigns to the express scheduling;

$\delta^l_{i,j}$– binary variable, for the scheduling form l, gets the value of 1 if vehicle i stops at j, otherwise is 0;

$\delta^l_{i,jk}$– binary variable, for the scheduling form l, gets the value of 1 if vehicle i stops at both j and k, otherwise is 0;

T– studied period;

T_0 – dwelling duration at every stop;
c – acceleration and deceleration duration;
h – headway;
$a_{i,j}$ - the arrival time vehicle i at stop j;
$d_{i,j}$ - the departure time vehicle i at stop j;
t_j - the running time of vehicles between stop $j-1$ and j;
$r_{j,k}$ - passenger arrival rate at stop j who want to go to stop $k(k > j)$;
R_j - arrival rate at stop j: ($R_j = \sum_{k=j+1}^{N} r_{j,k}$);
$A_{i,j}$ - the number of alighting passengers at stop j from vehicle i;
$B_{i,j}$ - the number of boarding passengers at stop j from vehicle i;
$L_{i,j}$ - the number of passengers on BRT when vehicle i runs from j to $j+1$;
C_1 - weight of cost of waiting passenger;
C_2 - weight of cost of on-board passenger;
C_3 - weight of operation cost of vehicles.

Ij – BRT vehicle ith actually leaving stop j; At the initial station, the order of BRT vehicles leaving the station is $1, 2, ..., i, ..., M$, respectively. Since the situation of crossing among vehicles, the order of BRT vehicles leaving the station $j \neq 1$ is probably not the same to the original order. To avoid confusion, in the next, the actual order of vehicle ith is denoted by Ith (the big ith). To determine exactly the order of vehicles leaving stop j, we can base on the value of $d_{i,j}$.

$h_{I,j}$ - headway between vehicle I and $I-1$ at stop j;
$s_{I,jk}$- passenger number wanting to stop k, but missing vehicle I when it leaves stop j;
$S_{I,j}$ - total number of passengers missing vehicle I at stop j;
$W_{I,jk}$ - passenger number wanting to go from stop j to stop k by vehicle I.

Input

- A BRT route having N stations, a fleet of M BRT vehicles;
- Three scheduling forms: Normal, Zone and Express;
- Running time between two consecutive stations t_{ij};
- Dwelling time T_0 and acceleration/deceleration time c;
- Matrix of passenger arrival rate r_{jk};
- Studied period T.

Output

- Headway h and the assignment of each vehicle to certain scheduling form;
- Total cost.

Objective Function

The objective is to minimize the total cost f containing three terms $f = f_1 + f_2 + f_3$. In particular,

- $f_1 = C_1.\sum_{i=1}^{M}\sum_{j=1}^{N}[\frac{R_j.h_{I,j}^2}{2} + S_{I-1,j}.h_{I,j}]$,

where the first term is the average passenger waiting time for vehicle I at stop j in the duration of $h_{I,j}$ and the second term is the waiting time for vehicle I of the passengers missing vehicle $I-1$;

- $f_2 = C_2\sum_{i=1}^{M}\sum_{j=1}^{N}[L_{i,j-1} - A_{i,j}].\delta_{i,j}^l.T_0 + C_2\sum_{i=1}^{M}\sum_{j=1}^{N-1} L_{i,j}[t_{j+1} + (\delta_{i,j}^l + \delta_{i,j+1}^l).c]$,

where the first term is the total waiting time of on-board passengers when vehicles stop and the second one is the travel time of on-board passengers.

- $f_3 = C_3\sum_{i=1}^{M}\sum_{j=1}^{N}\delta_{i,j}^l.T_0 + C_3\sum_{i=1}^{M}\sum_{j=1}^{N-1}[t_{j+1} + (\delta_{i,j}^l + \delta_{i,j+1}^l).c]$,

that is the total cost of operation BRT vehicles.

Constraints
(1) Time Constraints
The arrival time and departure time at stop 1 are equal for every vehicle:

$$a_{i,1} = d_{i,1} = (i-1).h, \qquad i = \overline{1,\ M}. \tag{1}$$

The arrival time of vehicle i at stop j is equal to the sum of the departure time of that vehicle at stop $j-1$ and the running time and acceleration/deceleration time:

$$a_{i,j} = d_{i,j-1} + t_j + (\delta_{i,j-1}^l + \delta_{i,j}^l).c. \tag{2}$$

At stop j, the departure time of vehicle i is the sum of the its arrival time and the dwelling time:

$$d_{i,j} = a_{i,j} + \delta_{i,j}^l.T_0. \tag{3}$$

The headway at stop j between vehicle I and $I-1$:

$$h_{I,j} = d_{I,j} - d_{I-1,j}. \tag{4}$$

(2) Passenger Number Constraints
The number of passengers waiting vehicle I at stop j is composed of the number of passengers missing vehicle $I-1$ and a new coming passenger number:

$$W_{I,jk} = s_{I-1,jk} + r_{j,k}.h_{I,j}. \tag{5}$$

The passenger number wanting to go to stop k but missing vehicle I at stop j depends on the passenger number waiting vehicle I and that vehicle I has plan to stop at j and k or not:

$$s_{I,jk} = W_{I,jk}(1 - \delta_{I,jk}^l). \tag{6}$$

The passenger number missing vehicle I at stop j is the sum of passengers missing vehicle I at stop j and wanting to all k after j:

$$S_{I,j} = \sum_{k=j+1}^{N} s_{I,jk}. \tag{7}$$

The alighting passenger number of vehicle i at stop j equals the sum of boarding passenger number at all stop before j:

$$A_{i,j} = \delta^l_{i,j}.\sum_{k=1}^{j-1} W_{i,kj}.\delta^l_{i,kj}. \tag{8}$$

The boarding passenger number of vehicle i at stop j equals the sum of alighting passenger number at all stops after j:

$$B_{i,j} = \delta^l_{i,j}.\sum_{k=j+1}^{N} W_{i,jk}.\delta^l_{i,jk}. \tag{9}$$

The number of on-board passengers of vehicle i from stop j to stop $j+1$ equals the one from the stop $j-1$ to j plus boarding passengers at stop j minus alighting passengers at stop j:

$$L_{i,j} = L_{i,j-1} + B_{i,j} - A_{i,j}. \tag{10}$$

In summary, the scheduling combination problem can be formulated as follows: $\min\{f_1 + f_2 + f_3\}$, subject to the constraints (1)-(4) and (5)-(10). This problem is in the form of 0-1 integer non-linear programming problem, that is so difficult.

3 A GA-Based Solution Method

Genetic algorithm (GA) is a branch of evolutionary computation in which one imitates the biological processes of reproduction and natural selection to solve for the fittest solutions. GA allows one to find solutions to problems that other optimization methods cannot handle due to a lack of continuity, derivatives, linearity, or other features. The general schema of GA is in Figure 2 (refer to [10] for more detail).

Coding

The most important procedure in GA is the coding step. In our problem, each one of three scheduling forms are easily encoded by using a 2-bit binary number: 01 represents the normal scheduling, 10 is the zone scheduling and 11 is assigned to the express scheduling. The length of chromosomes depends on the frequency or the headway. For instance, consider the study period $T = 15$ minutes with the headway $h = 3$ minutes. That means the number of vehicles is operated $M = 5$. If an individual has the chromosome of 11 10 00 01 10 then it is understood as that 5 vehicles are assigned to 11, 10, 00, 01 and 10, respectively.

Genetic Operators

Three main operators in a GA schema are Selection, Crossing and Mutation.

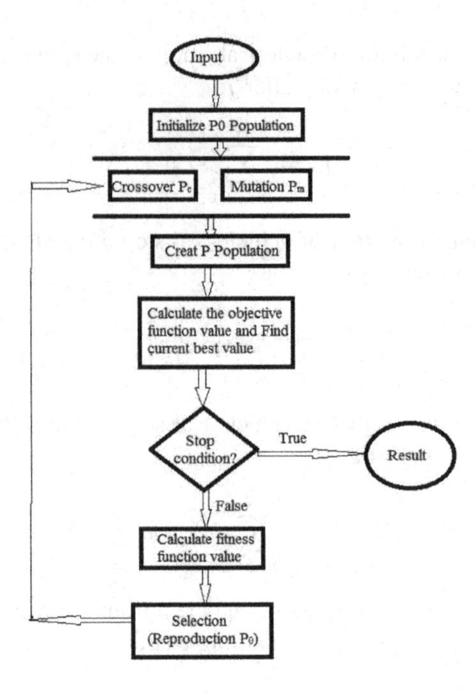

Fig. 2. GA schema

i) Selection:
Given a population at certain generation, Selection determine which individuals are chosen for the next generation. We use Roulette gambling law for this step [13].
ii) Crossing:
Two individuals cross at a certain position. For example, the crossing of two head genes of 00 00 00 00 00 and three tail genes of 11 11 11 11 11 is 00 00 11 11 11. The crossing rate is set to P_c.
iii) Mutation:
Given a mutation rate of P_m. The individual is mutated at a random position, namely, "0" becomes "1" or vise versa.

4 A Case Study in Hanoi

Hanoi, the capital of Vietnam, is planning to have 8 BRT routes by 2030. At the present, Kim Ma-Yen Nghia BRT route is being constructed and expectedly used in the end of 2015. We apply the model to this route. The route map is showed in Figure 3. The BRT route is investigated with the following parameters:

i) The study period is $T = 1h$ and the number of stops is $N = 23$;
ii) $T_0 = 30s$ and $c = 15s$;
iii) $C_1 = 0.2$ and $C_2 = 0.3$ and $C_3 = 0.5$;
iv) The average speed is $22.5km/h = 375m/min$. Based on the distance and the speed, the running time is determined in Table 1.

Fig. 3. Kim Ma- Yen Nghia BRT route

We test the demand of transport in two scenarios: the normal hour and the rush hour. The algorithm is coded in C and run on a medium-configured laptop.

Normal Hour Scenario

The matrix of passenger arrival rates for the normal hour is given in Figure 4. This matrix has the size of 23×23. The number located at line j and colume k is the passenger rate who want to go from stop j to stop k.

0	4.2	1.8	3.5	0.1	2.0	2.5	2.9	0.9	1.3	1.5	0.6	4.6	3.2	2.8	1.2	4.2	4.6	4.3	2.8	3.7	4.2	0.5
0	0	0.3	0.4	4.3	3.3	2.2	1.7	1.9	4.6	4.8	2.7	2.2	3.9	2.0	1.3	1.8	5.0	3.6	4.5	0.4	1.2	2.3
0	0	0	3.4	2.4	1.5	4.2	1.2	0.4	1.9	4.8	4.5	1.3	0.8	3.8	1.0	2.4	4.2	3.0	2.9	1.7	3.6	4.1
0	0	0	0	4.3	3.9	0.7	4.1	4.3	1.5	4.9	4.7	0.6	4.1	3.0	2.1	0.1	0.7	0.2	4.4	4.9	3.0	2.4
0	0	0	0	0	3.5	0.5	0.7	4.1	1.7	2.7	3.2	0.9	4.5	4.0	2.7	2.4	3.8	3.9	1.9	3.3	3.0	4.2
0	0	0	0	0	0	3.4	1.6	4.0	0.9	0.5	3.1	2.8	0.7	2.4	3.7	2.2	4.6	2.5	2.3	2.1	3.0	2.8
0	0	0	0	0	0	0	2.4	4.8	1.3	3.7	4.1	1.2	3.7	0.6	1.8	0.6	2.5	3.2	0.3	0.1	0.1	4.2
0	0	0	0	0	0	0	0	2.5	1.7	3.1	0.8	4.2	0.8	3.8	0.8	3.8	0.4	3.4	4.6	1.0	1.0	0.9
0	0	0	0	0	0	0	0	0	2.2	3.9	2.3	4.7	0.7	3.1	1.4	1.9	0.1	4.2	1.3	0.6	1.1	1.0
0	0	0	0	0	0	0	0	0	0	2.5	3.8	4.9	3.4	4.6	4.2	0.3	0.1	4.2	3.7	2.5	2.1	4.7
0	0	0	0	0	0	0	0	0	0	0	2.2	4.9	5.0	1.9	3.5	3.2	3.5	0.4	5.0	1.9	3.9	0.1
0	0	0	0	0	0	0	0	0	0	0	0	3.9	2.8	1.8	2.9	4.4	4.9	3.4	0.8	2.2	1.1	1.8
0	0	0	0	0	0	0	0	0	0	0	0	0	1.4	1.5	1.0	1.7	3.6	0.2	0.1	5.0	2.0	0.7
0	0	0	0	0	0	0	0	0	0	0	0	0	0	4.9	0.4	2.5	0.9	4.5	1.0	4.0	0.3	4.6
0	0	0	0	0	0	0	0	0	0	0	0	0	0	0	3.6	4.4	4.4	2.4	3.8	1.5	0.4	4.9
0	0	0	0	0	0	0	0	0	0	0	0	0	0	0	0	0.1	0.9	1.9	3.1	4.7	4.9	3.2
0	0	0	0	0	0	0	0	0	0	0	0	0	0	0	0	0	4.0	4.9	1.0	0.8	2.3	2.3
0	0	0	0	0	0	0	0	0	0	0	0	0	0	0	0	0	0	3.9	4.3	3.9	3.0	4.1
0	0	0	0	0	0	0	0	0	0	0	0	0	0	0	0	0	0	0	0.8	0.9	4.2	1.6
0	0	0	0	0	0	0	0	0	0	0	0	0	0	0	0	0	0	0	0	3.9	0.7	1.2
0	0	0	0	0	0	0	0	0	0	0	0	0	0	0	0	0	0	0	0	0	0.3	3.5
0	0	0	0	0	0	0	0	0	0	0	0	0	0	0	0	0	0	0	0	0	0	2.3
0	0	0	0	0	0	0	0	0	0	0	0	0	0	0	0	0	0	0	0	0	0	0

Fig. 4. Matrix of passenger arrival rates for the normal hour (people/min)

Table 1. Running time on link

Link	Distance (metres)	Time (minutes)
Yen Nghia - 01	1589	4.237
01 - 02	511	1.363
02 - 03	512	1.365
03 - 04	548	1.461
04 - 05	503	1.341
05 - 06	417	1.112
06 - 07	470	1.253
07 - 08	630	1.680
08 - 09	550	1.467
09 - 10	410	1.093
10 - 11	590	1.573
11 - 12	585	1.560
12 13	565	1.507
13 14	720	1.920
14 - 15	640	1.707
15 - 16	625	1.667
16 - 17	1130	3.013
17 - 18	665	1.773
18 - 19	840	2.240
19 - 20	720	1.920
20 - 21	770	2.053
21 - Kim Ma	710	1.893

Based on the passenger arrival rates, three scheduling forms are chosen as in Table 2. As showed in Table 2, for the normal scheduling a vehicle catches/takes passengers at every stations; for the zone scheduling, it does passengers at $12/23$ and it does passengers at $7/23$ for the express scheduling.

Table 2. Three scheduling forms ("1" means stop and "0" is non-stop)

Stop	01	02	03	04	05	06	07	08	09	10	11	12	13	14	15	16	17	18	19	20	21	22	23
Nomal	1	1	1	1	1	1	1	1	1	1	1	1	1	1	1	1	1	1	1	1	1	1	1
Zone	1	0	0	1	1	1	0	0	0	1	1	1	0	1	1	0	0	1	1	0	0	0	1
Express	1	0	0	1	1	0	0	0	0	0	1	0	0	0	1	0	0	1	0	0	0	0	1

The GA is run with the parameters: $P_c = 0.8$ and $P_m = 0.005$; the number of generations is 1000; the number of population is up-to-15000. The results are reported in Table 3. The first column h is the headway of BRT (frequency). The second column shows the number of BRT vehicles (No.Veh.). The third one indicates the schedule for each BRT vehicle: 0 means Normal; 1 is Zone; and 2 shows Express. For instance, 000 in the first line of the table indicates that all three vehicles is assigned to Normal scheduling. The forth column (ScheCost) gives the total cost calculated by GA. The

Table 3. Results obtained by GA for the normal hour

h (min)	No. Veh.	Schedule	ScheCost	IniCost	Save (%)	Time (m:ss)
20	3	0 0 0	322289	322289	0.00	0:00
15	4	0 0 0 1	295138	302723	2.51	0:01
12	5	0 0 0 1 1	279390	290997	3.99	0:02
10	6	0 0 0 0 1 1	267417	283189	5.57	0:15
6	10	2 0 0 0 2 1 1 1 1 1	234840	267636	12.25	1:54
5	12	2 0 0 0 0 2 1 1 2 1 1 1	226408	263778	14.17	2:10
4	15	2 0 0 0 1 0 2 1 1 2 1 2 1 1 1	227082	269672	15.79	2:25
3	20	2 0 0 0 0 0 0 1 0 2 2 1 2 1 1 1 1 2 2 1 1	231154	281469	17.88	2:14

fifth column (InCost) presents the total cost without the scheduling i.e. the total cost in the case that all vehicles chooses Normal one. The sixth column shows the percentage of saving cost when using scheduling. The last column is the computing time of GA.

Rush-Hour Scenario

We also test for the data in rush-hour (see Figure 5) and obtain the result in Table 4.

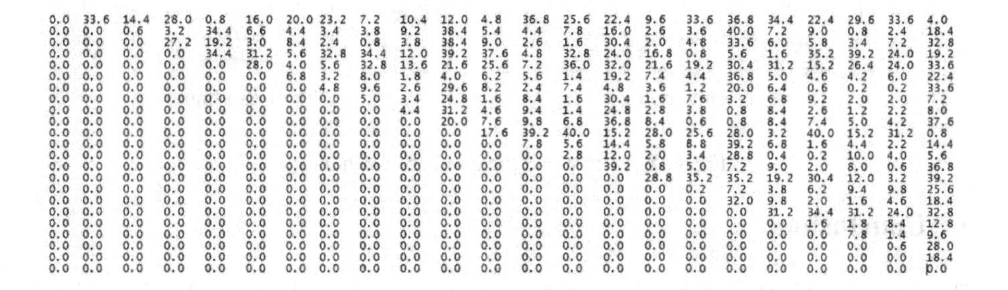

0.0	33.6	14.4	28.0	0.8	16.0	20.0	23.2	7.2	10.4	12.0	4.8	36.8	25.6	22.4	9.6	33.6	36.8	34.4	22.4	29.6	33.6	4.0
0.0	0.0	0.6	3.2	34.4	6.6	4.4	3.4	3.8	9.2	38.4	5.4	4.4	7.8	16.0	2.6	3.6	40.0	7.2	9.0	0.8	2.4	18.4
0.0	0.0	0.0	27.2	19.2	3.0	8.4	2.4	0.8	3.8	38.4	9.0	2.6	1.6	30.4	2.0	4.8	33.6	6.0	5.8	3.4	7.2	32.8
0.0	0.0	0.0	0.0	34.4	31.2	5.6	32.8	34.4	12.0	39.2	37.6	4.8	32.8	24.0	16.8	0.8	5.6	1.6	35.2	39.2	24.0	19.2
0.0	0.0	0.0	0.0	0.0	28.0	4.0	5.6	32.8	13.6	21.6	25.6	7.2	36.0	32.0	21.6	19.2	30.4	31.2	15.2	26.4	24.0	33.6
0.0	0.0	0.0	0.0	0.0	0.0	6.8	3.2	8.0	1.8	4.0	6.2	5.6	1.4	19.2	7.4	4.4	36.8	5.0	4.6	4.2	6.0	22.4
0.0	0.0	0.0	0.0	0.0	0.0	0.0	4.8	9.6	2.6	29.6	8.2	2.4	7.4	4.8	3.6	1.2	20.0	6.4	0.6	0.2	0.2	33.6
0.0	0.0	0.0	0.0	0.0	0.0	0.0	0.0	5.0	3.4	24.8	1.6	8.4	1.6	30.4	1.6	7.6	3.2	6.8	9.2	2.0	2.0	7.2
0.0	0.0	0.0	0.0	0.0	0.0	0.0	0.0	0.0	4.4	31.2	4.6	9.4	1.4	24.8	2.8	3.8	0.8	8.4	2.6	1.2	2.2	8.0
0.0	0.0	0.0	0.0	0.0	0.0	0.0	0.0	0.0	0.0	20.0	7.6	9.8	6.8	36.8	8.4	0.6	0.8	8.4	7.4	5.0	4.2	37.6
0.0	0.0	0.0	0.0	0.0	0.0	0.0	0.0	0.0	0.0	0.0	17.6	39.2	40.0	15.2	28.0	25.6	28.0	3.2	40.0	15.2	31.2	0.8
0.0	0.0	0.0	0.0	0.0	0.0	0.0	0.0	0.0	0.0	0.0	0.0	7.8	5.6	14.4	5.8	8.8	39.2	6.8	1.6	4.4	2.2	14.4
0.0	0.0	0.0	0.0	0.0	0.0	0.0	0.0	0.0	0.0	0.0	0.0	0.0	2.8	12.0	2.0	3.4	28.8	0.4	0.2	10.0	4.0	5.6
0.0	0.0	0.0	0.0	0.0	0.0	0.0	0.0	0.0	0.0	0.0	0.0	0.0	0.0	39.2	0.8	5.0	7.2	9.0	2.0	8.0	0.6	36.8
0.0	0.0	0.0	0.0	0.0	0.0	0.0	0.0	0.0	0.0	0.0	0.0	0.0	0.0	0.0	28.8	35.2	35.2	19.2	30.4	12.0	3.2	39.2
0.0	0.0	0.0	0.0	0.0	0.0	0.0	0.0	0.0	0.0	0.0	0.0	0.0	0.0	0.0	0.0	0.2	7.2	3.8	6.2	9.4	9.8	25.6
0.0	0.0	0.0	0.0	0.0	0.0	0.0	0.0	0.0	0.0	0.0	0.0	0.0	0.0	0.0	0.0	0.0	32.0	9.8	2.0	1.6	4.6	18.4
0.0	0.0	0.0	0.0	0.0	0.0	0.0	0.0	0.0	0.0	0.0	0.0	0.0	0.0	0.0	0.0	0.0	0.0	31.2	34.4	31.2	24.0	32.8
0.0	0.0	0.0	0.0	0.0	0.0	0.0	0.0	0.0	0.0	0.0	0.0	0.0	0.0	0.0	0.0	0.0	0.0	0.0	1.6	1.8	8.4	12.8
0.0	0.0	0.0	0.0	0.0	0.0	0.0	0.0	0.0	0.0	0.0	0.0	0.0	0.0	0.0	0.0	0.0	0.0	0.0	0.0	7.8	1.4	9.6
0.0	0.0	0.0	0.0	0.0	0.0	0.0	0.0	0.0	0.0	0.0	0.0	0.0	0.0	0.0	0.0	0.0	0.0	0.0	0.0	0.0	0.6	28.0
0.0	0.0	0.0	0.0	0.0	0.0	0.0	0.0	0.0	0.0	0.0	0.0	0.0	0.0	0.0	0.0	0.0	0.0	0.0	0.0	0.0	0.0	18.4
0.0	0.0	0.0	0.0	0.0	0.0	0.0	0.0	0.0	0.0	0.0	0.0	0.0	0.0	0.0	0.0	0.0	0.0	0.0	0.0	0.0	0.0	0.0

Fig. 5. Matrix of passenger arrival rates for the rush hour (people/min)

Some remarks on the results:

- Using scheduling combination, the cost of BRT system can be decreasing up-to-18%.
- In rush hour, the number of vehicle using zone scheduling/express scheduling is increasing.
- The computing time of the algorithm is trivial.
- From Table 3 and 4, the total cost can be visualized with respect to the number of vehicles (Figure 6). It shows that the optimal number of vehicle is 12 (headway h=5) for the normal hour. In rush-hour, the number of vehicle should be greater to get the optimal cost, namely, the vehicle number is 15 (headway h=4). This result is logical when remarking that one need more vehicles in rush hour.

Table 4. Results obtained by GA for the rush-hour

h (min)	No. Veh.	Schedule	ScheCost	IniCost	Save (%)	Time (m:ss)
20	3	0 0 1	1793877	1811648	0.98	0:00
15	4	0 0 0 1	1650732	1704609	3.16	0:01
12	5	0 0 0 1 1	1556245	1640398	5.13	0:02
10	6	1 0 0 1 1 1	1490410	1597601	6.71	0:14
6	10	2 0 0 0 2 1 2 2 2 2	1313327	1512068	13.14	1:30
5	12	2 0 0 0 2 1 1 2 2 2 2 2	1267757	1490761	14.96	2:11
4	15	2 0 0 0 0 0 2 2 1 2 2 2 2 2 2	1254554	1520776	17.51	3:15
3	20	2 1 0 0 0 2 0 1 2 1 2 2 1 1 1 2 2 2 2 2	1276884	1564966	18.41	3:07

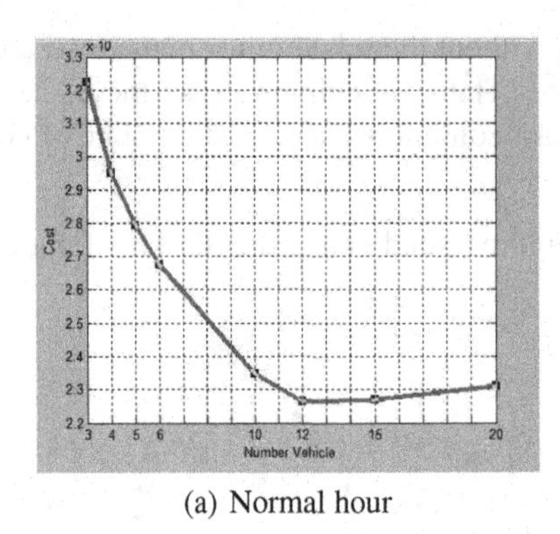

(a) Normal hour

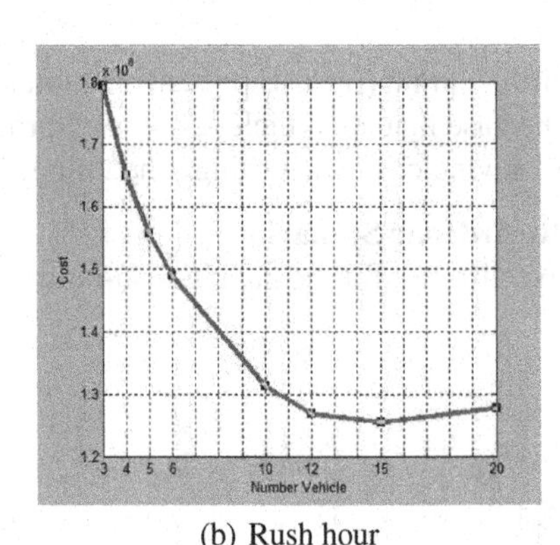

(b) Rush hour

Fig. 6. Total cost and the number of vehicles

5 Conclusions

This work consider the problem of determining an optimal headway and BRT scheduling combination. We propose a modified model that overcomes the limitation, namely, two BRT vehicles probably cross each other. The model is solved by the genetic algorithm. The case study of the BRT route in Hanoi is investigated. Experimental results show that total cost is reasonably decreasing after using scheduling combination strategies.

Note that the model is probably extended to the multi-objective optimization problem. An incompletely prioritized BRT system is also the objective for the future work.

Acknowledgements. This work is supported by Vietnam National Foundation for Science and Technology Development (NAFOSTED) under Grant Number 101.01-2013.10. The authors thank Dr. Pham Hong Quang, Center for Informatics and Computing, Vietnam Academy of Science and Technology for his useful comments.

References

1. Avishai, C.: Urban transit scheduling: framework, review and examples. Journal of Urban Planning and Development 128(4), 225–243 (2002)
2. Bai, Z.J., He, G.G., Zhao, S.Z.: Design and implementation of Tabu search algorithm for optimizing BRT Vehicles dispatch. Computer Engineering and Application 43(23), 229–232 (2007)
3. Dai, L.G., Liu, Z.D.: Research on the multi-objective assembled optimal model of departing interval on bus dispatch. Journal of Transportation Systems Engineering and Information Technology 7(4), 43–46 (2007)
4. Fan, Q.S., Pan, W.: Application research of genetic algorithm in intelligent transport systems scheduling of vehicle. Computer and Digital Engineering 35(5), 34–35 (2007)
5. Levinson, H., Zimmerman, S., Clinger, J., Rutherford, S., Smith, R.L., Cracknell, J., Soberman, R.: Bus Rapid Transit, Volume 1: Case Studies in Bus Rapid Transit, TCRP Report 90, Transportation Research Board, Washington, USA (2003)
6. Liang, S., He, Z., Sha, Z.: Bus Rapid Transit Scheduling Optimal Model Based on Genetic Algorithm. In: ICCTP 2011, pp. 1296–1305 (2011)
7. Miller, M.A., Yin, Y., Balvanyos, T., Avishai, C.: Framework for Bus Rapid Transit Development and Deployment Planning. Research report, California PATH, University of California Berkeley (2004)
8. Ren, C.X., Zhang, H., Fan, Y.Z.: Optimizing dispatching of public transit vehicles using genetic simulated annealing algorithm. Journal of System Simulation 17(9), 2075–2077 (2005)
9. Rickert, T.: Technical and Operational Challenges to Inclusive Bus Rapid Transit: A Guide for Practitioners, World Bank, Washington, USA (2010)
10. Schaefer, R.: Foundations of Global Genetic Optimization. SCI, vol. 74. Springer, Heidelberg (2007)
11. Shrivastava, P., Dhingra, S.L.: Development of coordinated schedules using genetic algorithms. Journal of Transportation Engineering 128(1), 89–96 (2002)
12. Sun, C., Zhou, W., Wang, Y.: Scheduling Combination and Headway Optimization of Bus Rapid Transit. Journal of Transportation Systems Engineering and Information Technology 8(5), 61–67 (2008)
13. Tong, G.: Application study of genetic algorithm on bus scheduling. Computer Engineering 31(13), 29–31 (2005)

Part II

Operational Research and Decision Making

Application of Recently Proposed Metaheuristics to the Sequence Dependent TSP

Samet Tonyali[1] and Ali Fuat Alkaya[2]

[1] Southern Illinois University, Department of Computer Science, Carbondale, IL 62901 USA
[2] Marmara University, Department of Computer Engineering, Istanbul, Turkey
samet.tonyali@siu.edu, falkaya@marmara.edu.tr

Abstract. The Sequence Dependent Traveling Salesman Problem (SDTSP) is a combinatorial optimization problem defined as a generalization of the TSP. It emerged during optimization of two kinds of commonly used placement machines for production of printed circuit boards. The difference between SDTSP and TSP is that the cost incurred by transition from one point to another is dependent not only the distance between these points but also subsequent k points. In this study, we applied Simulated Annealing (SA), Artificial Bee Colony (ABC) and Migrating Birds Optimization (MBO) to solve real-world and random SDTSP instances. The metaheuristics were tested with 10 neighbor functions. In our computational study, we conducted extensive computational experiments. Firstly, we obtained best parameter value combination for each metaheuristic. Secondly, we conducted experiments so as to determine best performing neighbor function for each metaheuristic. Computational experiments show that ***twoopt*** function can be considered as the most suitable function for all the three metaheuristics.

Keywords: metaheuristics, combinatorial optimization, sequence dependent traveling salesman problem.

1 Introduction

The Sequence Dependent Traveling Salesman Problem (SDTSP) is a recently introduced combinatorial optimization problem as a generalization of the Traveling Salesman Problem (TSP) [1]. TSP can be easily stated as follows: a salesman wants to visit n distinct cities and then returns home. He wants to determine the sequence of the travel so that the overall travelling time is minimized while visiting each city not more than once. Conceptually it seems simple, but obtaining an optimal solution requires a search in set with $n!$ feasible solutions. The difference between SDTSP and the TSP is that the cost incurred by transition from one vertex to another depends not only the distance between these points but also on the choice of subsequent k vertices due to an additional constraint added to the cost calculation function. Therefore, solution methods for the TSP are not guaranteed to generate optimum/near-optimum solutions for SDTSP instances.

H.A. Le Thi et al. (eds.), *Advanced Computational Methods for Knowledge Engineering*, Advances in Intelligent Systems and Computing 358, DOI: 10.1007/978-3-319-17996-4_8

In the SDTSP, the cities (points) have an additional property, called free time (ft), which affects the cost calculation (i.e., in this study we calculate cost in time units). For a given travel sequence, cost of travel between points i and j when point j is visited in pth position is calculated with the following equation.

$$C2(p, i, j) = \max\{C1(i, j), \max\{ft_p, ft_{p+1}, ft_{p+2}, \ldots, ft_{p+k-1}\}\} \quad (1)$$

where $C1(i, j)$ gives the travel cost from point i to point j, ft_{p+1} is the free time of the point visited in $(p+1)$st position, ft_{p+2} is the free time of the point visited in $(p+2)$nd position and so on. k represents the number of points in the travel sequence whose free time characteristic will be taken into account in the cost calculation. If k is zero, then $C2$ turns out to be $C1(i, j)$ and thus the problem becomes a classical TSP.

One can easily observe that the cost calculation of a tour can be done only after the tour is completely determined. The following example illustrates the difference between the classical TSP and the SDTSP. Table 1 provides the coordinates and free-time costs of the points. In this example, let the cost between any pair of points is affected by free time costs of subsequent four points (k=4). In order to calculate the $C2$ cost between two points, a and b, according to Eq. 1, we firstly calculate $C1$ (in Chebyshev metric for this small example). Then, by using the $C1$ cost and finding the maximum of free time values of following four points, the $C2$ cost is calculated.

Table 1. Data for a SDTSP example

Point	x	y	ft
A	2	7	2
B	4	6	2
C	6	7	2
D	8	5	2
E	7	5	2
F	4	5	2
G	8	1	2
H	4	1	2
J	1	5	2
K	2	5	4

If the problem had been handled as a classical TSP instance, the tour given in Figure 1 (Tour 1) would have been an optimal solution with 24 time units. However, this is a costly tour (34 time units) for the SDTSP. On the other hand, although the tour given in Figure 2 is more costly than an optimal solution for the classical TSP (25 time units) it will be the optimal solution for the SDTSP. Hence, it is not guaranteed that optimal solutions for a conventional TSP are also optimal for the SDTSP.

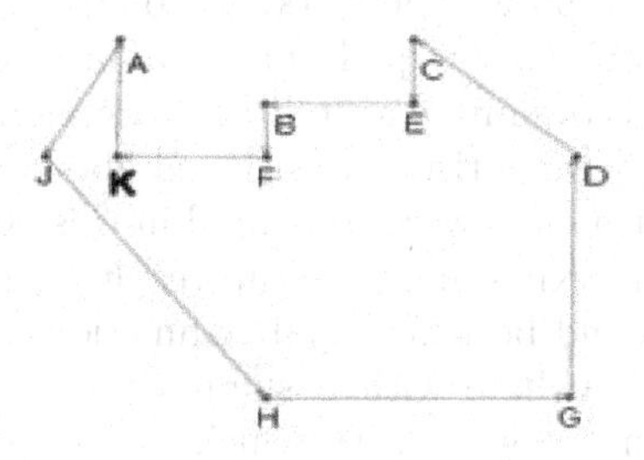

Fig. 1. Tour 1

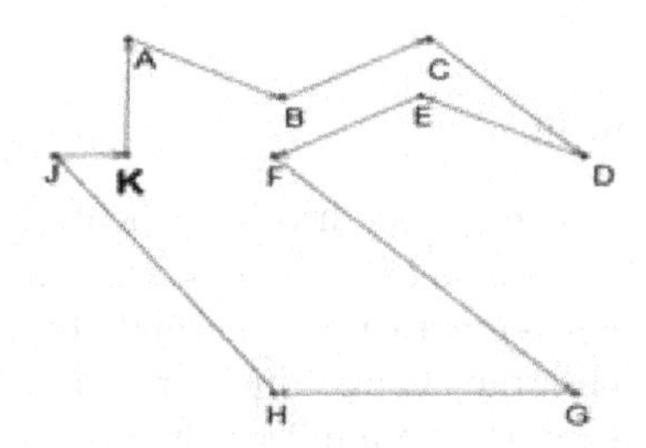

Fig. 2. Tour 2

The SDTSP emerged during the optimization of placement machines used in printed circuit board (PCB) manufacturing industry [2]. A placement machine is very expensive and therefore, the assembly lines are typically designed such that the placement machine is the limiting resource or bottleneck, which is the key issue for assembly line optimization [3,4]. Thus, for manufacturers to remain competitive in the growing PCB market, they must concentrate their efforts on improving the efficiency of their placement machines. Even small improvements in PCB assembly increase productivity and bring huge amounts of profits to the company.

It is agreed that one of common solution methods for combinatorial optimization problems is to utilize metaheuristics. Metaheuristics are nothing but the heuristic methods that can be applied for various problems with small modifications. Some of the well-known metaheuristics being used to solve combinatorial optimization problems are genetic algorithm, simulated annealing, ant colony optimization, tabu search and iterated local search. In this study, we applied three metaheuristics to find solutions to the SDTSP: simulated annealing (SA) [5], artificial bee colony algorithm (ABC) [6] and newly proposed migrating birds optimization (MBO) [7]. The reason why we have chosen ABC and MBO is that they are recently proposed and unexploited metaheuristics. Besides, all these metaheuristics rely on neighborhood search. Therefore, in this study, we also compared the efficacy and the efficiency of several neighbor functions on solution quality. In our computational work, we compared performances of these metaheuristics in terms of solution quality and running time on synthetic data.

The rest of this paper is organized as follows: In Section 2, we give algorithmic details about the three metaheuristics mentioned above. In Section 3, neighbor functions of which performances were compared in this study are introduced. In Section 4, we provide the experimental setup results regarding the best values of algorithm parameters and best-fit neighbor functions for each metaheuristic. In the last section, we conclude with a summary of the study and its main conclusions, and some future work with which we are planning to extend this study.

2 Background

2.1 Simulated Annealing

In metallurgy, annealing is a technique involving heating a solid to a high temperature (melting) and then controlled cooling it (freezing) very slowly to reach thermal equilibrium at each temperature. Thermal equilibrium is reached at a temperature value at which large changes in energy do not occur [8]. The main goal is to find a configuration that minimizes the energy of the system. In this analogy; system, configuration and energy correspond to a combinatorial optimization problem, solution generated for this problem and the cost of this proposed solution, respectively [5].

The algorithm starts by generating an initial solution (either randomly or heuristically constructed) and by initializing the so-called temperature parameter T. Then, at each iteration a solution s in $N(s)$ (the set of neighbor solutions of s) is randomly sampled by using a neighbor function (e.g., twoopt move, pair-wise exchange etc.) and it is accepted as new current solution depending on $f(s)$, $f(s\prime)$ (f function calculates cost of a solution) and T. $s\prime$ replaces s if $f(s\prime) < f(s)$ or, in case $f(s\prime) \geq f(s)$, with a probability, which is a function of T, and $f(s\prime) - f(s)$. The probability is generally computed following the Boltzmann distribution

$$\exp^{\frac{f(s')-f(s)}{T}} \tag{2}$$

Metropolis et al. [9] introduced a simple algorithm that simulates the annealing process in metallurgy. This algorithm was latterly named as SA by Kirkpatrick et al. [5]. We designed our SA originating from [10] as follows: There are two nested loops. An initial solution is generated randomly and assigned as the best solution before the loops start. Then, in the inner loop, current solution is perturbed by using a neighbor function and a new solution is generated. If the objective function value of this new solution is less than or equal to that of the current solution, then the new solution is assigned as the current solution for the next inner loop iteration. In addition, if the objective function value of the new solution is less than that of the best solution, then the best solution is replaced with the new solution. Otherwise, that is, if the objective function value of the new solution is greater than that of the current solution, then the new solution is accepted as the current solution with a probability. This probability value is calculated as above. The inner loop repeats this process R times. After inner loop finishes, this value is augmented by multiplying it by b where b is the expansion coefficient for R. Then, so-called temperature value T is updated by being divided by a, which is decrease ratio of temperature value. After these updates, next iteration for the outer loop starts again. The outer loop repeats this process until the number of solutions generated during iterations reaches the maximum number of solutions to be generated.

2.2 Artificial Bee Colony

The ABC algorithm is a nature-inspired and population based optimization algorithm simulating intelligent foraging behavior of a bee colony. In a bee colony, in order to maximize the amount of food collected, honey bees collaborate with each other. In that organization, there are three types of bees: employee bees, onlooker bees and scout bees. Employee bees are responsible for determining the quality of food source for onlooker bees. They dance for onlooker bees in order to exchange information specifying the quality of food source they visited. With the collective information from the employee bees, onlooker bees choose the food sources and try to find better ones. After a while, the food sources being exploited become exhausted. In such a case, associated employed bee abandons her food source and becomes a scout. Then, it starts to explore for a new food source randomly.

In this algorithm, all bees have the ability to keep their best solution. The steps of the ABC are as follows:

i Initialize the solutions randomly,
ii Send employee bees to find solutions
iii From the probability calculations of employee bees, send onlooker bees and find new better solutions
iv Send the scout bees to generate new solutions.
v Keep the best solution so far
vi Repeat ii-v until the stopping criteria are met.

ABC was proposed by Karaboga and Basturk where in their original study they used ABC algorithm for optimizing multimodal (a function which has two or more local optima) and multivariable functions [6]. Then, performance of ABC was compared with those of Genetic Algorithm (GA), Particle Swarm Optimization (PSO) algorithm and Particle Swarm Inspired Evolutionary Algorithm (PS-EA). ABC outperformed the other algorithms. The results showed that this new algorithm is competitive with commonly used algorithms for optimization problems in the literature.

2.3 Migrating Birds Optimization

The MBO algorithm is a newly proposed, population-based neighborhood search technique inspired from the V formation flight of the migrating birds which is proven to be an effective formation in energy minimization. In the analogy, initial solutions correspond to a flock of birds. Likewise the leader bird in the flock, a leader solution is chosen and the rest of the solutions is divided into two parts. Each solution generates a number of neighbor solutions. This number is a determiner value on exploration and it corresponds to the speed of the flock. The higher this value, the more detailed the flock explores its surroundings.

The algorithm starts with a number of initial solutions corresponding to birds in a V formation. Starting with the first solution (corresponding to the leader bird) and progressing on the lines towards the tales, each solution is tried to be improved by its neighbor solutions. If any of the neighbor solutions is better, the current solution is replaced by that one. There is also a benefit mechanism for the solutions (birds) from the solutions in front of them. Here we define the benefit mechanism as sharing the best unused neighbors with the solutions that follow. In other words, a solution evaluates a number of its own neighbors and a number of best neighbors of the previous solution and is replaced by the best of them. Once all solutions are improved (or tried to be improved) by neighbor solutions, this procedure is repeated a number of times (tours) after which the first solution becomes the last, and one of the second solutions becomes the first and another loop starts. The algorithm is terminated after a number of iterations.

This new metaheuristic was proposed by Duman et al. [7]. They applied it to solve quadratic assignment problem instances arising from printed circuit board assembly workshops. Its performance was compared with those of metaheuristics implemented and compared in two previous studies. These metaheuristics are simulated annealing, tabu search, genetic algorithm, scatter search, particle swarm optimization, differential evolution and guided evolutionary simulated annealing. In this comparison, the MBO outperformed the best performed metaheuristic (simulated annealing) in the previous studies by approximately three percent on the average. In addition, MBO was tested with some benchmark problem instances obtained from QAPLIB and in most of the instances it obtained the best known solutions. As a result of these tests, it is concluded that the MBO is a promising metaheuristic and it is a candidate to become one of the highly competitive metaheuristics. Duman and Elikucuk [12] applied MBO to solve fraud detection problem. They also proposed a new version of MBO where a different benefit mechanism is used. They tested the original MBO algorithm and its new version on real data and compared their performance with that of genetic algorithm hybridized with scatter search (GASS). Test results showed that the MBO algorithm and its new version performed significantly better than the GASS algorithm.

3 Neighbor Functions

Since a solution for an SDTSP instance answers the question of in which order vertices should be visited to minimize the total assembly time, it can be represented as a permutation of vertex indices (See Figure 3).

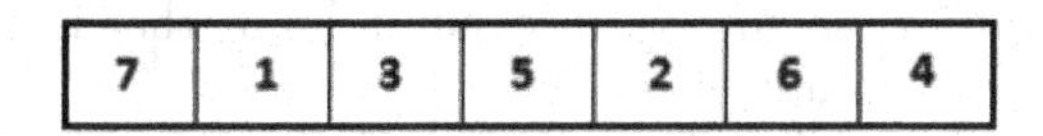

Fig. 3. Representation of a solution

In this study, we tested and used ten neighbor functions (Figure 4).

- **Adjacent Interchange:** One of the indices in the permutation is chosen randomly and this index is interchanged with its adjacent on the right. If the chosen index is in the last position, then it is interchanged with the index that is in the first position.
- **Random Interchange:** Two distinct indices in the permutation are chosen randomly and interchanged.
- **Move a Single Term:** One of the indices in the permutation is chosen randomly. Number of steps to be moved is chosen randomly from a predetermined set of candidate numbers, and then the index is moved rightwards by the number of steps further from the current position.
- **Move a Subsequence:** A subsequence of the permutation, whose length is determined by randomly choosing from a predetermined set of candidate numbers is designated, and then the subsequence is moved rightwards by the number, which is generated randomly in the range of 1 and the length of the permutation, further from the current position of the beginning index of the subsequence.
- **Reverse a Subsequence:** A subsequence of the permutation is determined as in the Move a Subsequence function, and then the subsequence is reversed.
- **Reverse and/or Move a Subsequence:** This function consists of two different functions: reverse and move a subsequence, and reverse or move a subsequence. A subsequence of the permutation is determined as in the Move a Subsequence function. Then, a random number which is in the range [0, 1) is generated. If the generated number is less than 0.5, then reversing and move a subsequence function is carried out. This function, firstly, reverses the subsequence and then moves the reversed subsequence rightwards by a random number of steps in the range of 1 and the length of the permutation further from the current position of the beginning index of the subsequence. If the generated number is equal or greater than 0.5, then the function reverses or moves the subsequence. A new random number in the range of [0, 1) is generated. If the generated number is less than 0.5, then the subsequence is reversed. Otherwise, it is moved rightwards as explained above.

- **Swap Subsequences:** The permutation is divided into two. One subsequence for each division is determined as explained in the Move a Subsequence function, and then these subsequences are swapped with each other.
- **Reverse and Swap Subsequences:** The subsequences are determined as in the Swap Subsequences function and they are reversed. Then, they are swapped with each other.
- **Twoopt:** This function is the same as Reverse a Subsequence function except that twoopt has no subsequence length limit, that is, the subsequence may be the permutation itself.
- **Random Remove and Reinsert:** One of the indices is chosen randomly and then it is inserted into a randomly chosen position.

4 Experimental Setup, Results and Discussion

To measure the performance of the metaheuristics we conducted two groups of tests:

1. Tests conducted to fine tune the parameters of the algorithms,
2. Tests conducted to determine the neighbor functions with which the algorithms can show their best performance,

The SDTSP originally emerged during the optimization of placement machines used in PCB manufacturing industry. For all tests, we used the specifications of a particular placement machine called ship shooters. That is, the board carrier speed is 280 mm/sec in both x and y directions and the placement machine transports the next head onto mount position in 0.15 sec., 0.19 sec., 0.24 sec. and 0.29 sec. for weight groups 1, 2, 3 and 4, respectively and k is 6.

The random data were generated via a random data generator developed for [2]. There are two kinds of model for generating the random data: structured and homogeneous model. In structured model, components in the same weight group are juxtaposed and this model resembles the real data. On the other hand, in homogeneous model, components are placed homogeneously on the board regardless of their weight category. We generated four types of PCB data each having 100, 200, 300 or 400 components. All random data include components from all weight groups, but mostly from group 1.

We compared the performances using two criteria:

1. **Total Assembly Time (TAT):** The time required for a chip mounter to complete assembly of a PCB. This gives us the objective function value for a problem instance.
2. **Run Time (RT):** The time required for a test case to be completed.

The tests are performed on a machine with Intel Xeon CPU E5 at 3.10 GHz. with 128 GB RAM using Windows 7 Professional OS. In the following subsections, each figure is an average of 10 runs.

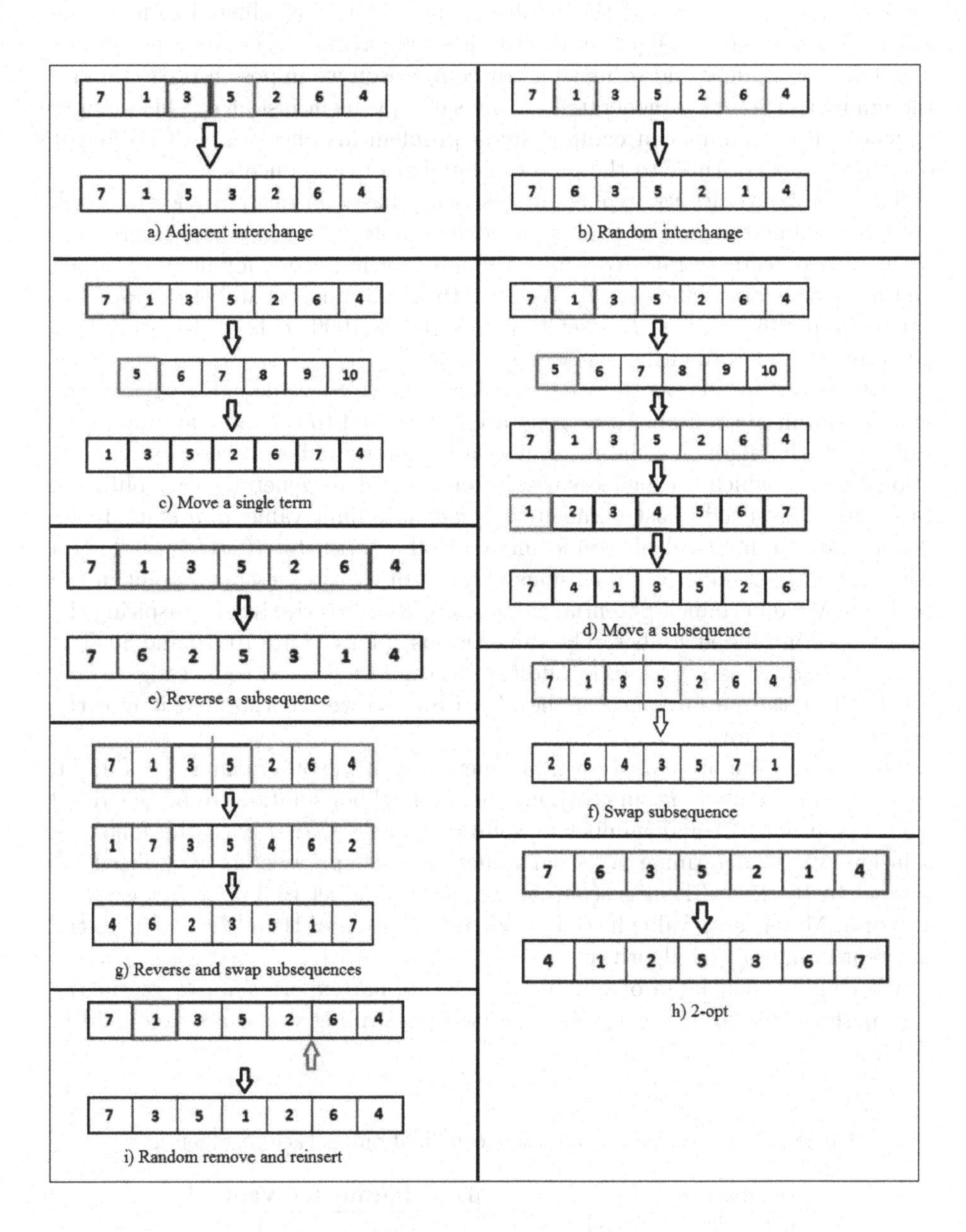

Fig. 4. Representation of a solution

4.1 Parameter Fine Tuning

In the following, we present the results of the first test conducted to fine tune the parameters of the algorithms. All these algorithms generate their initial solution(s) randomly and to make a fair comparison we impose a restriction on the number of solutions generated on the same problem instance. This number depends on the number of components a problem instance (or a PCB) and it one third of the of three to the power of number of components.

The metaheuristics we exploit in this study have various parameters which have to be fine-tuned. For SA, we need to fine-tune initial temperature (T), temperature decrease ratio (a), initial number of iterations for inner loop (R) and its expansion coefficient (b). We took these parameters and their potential values from [10]. That is, T is set to either 100 or 1000. R is set to either 5 or 20. a and b are set to either 1.1 or 1.5.

ABC algorithm inherently needs a colony size (CS) value. This value is the number of initial solutions to be generated. It is used to calculate an important value for ABC algorithm, limit. limit is maximum number of consecutive iterations during which an ineffectual solution is used to generate new solutions. In addition, we need a last parameter to calculate limit value; d. d is used as a denominator in limit calculation formula as follows: $limit = ((2xCSxD))/d$ [11] where D is the number of dimensions of a vector that represents a point in the problem. We determined potential values for CS arbitrarily but by inspiring the values used in related studies. The values we used for CS are 10, 20 and 30. The referenced study uses 3 for d. In addition to this, we wanted to investigate how 2 will affect the performance of the algorithm, so we incorporated it into the potential values for d.

MBO algorithm has many more parameters than the others: number of initial solutions (n), number of tours (m), number of neighbor solutions to be generated from a solution (p) and number of solutions to be shared with the following solution (x). To determine potential values for these parameters we utilized [7]: n is set to 13, 25 or 51. m is set to 5, 10 and 20. p is set to 3, 5 or 7. x is set to 1, 2 or 3. Moreover, p value has to be equal to or greater than $2*x+1$ due to the inherent design of the algorithm.

We conducted a large of number computational experiments to reveal the best performing parameter values. The best performing values are presented in Table 2.

Table 2. The best parameter value combination for each metaheuristic

Algorithm	Best Parameter Values
Artificial Bee Colony	CS=30, d=2
Migrating Birds Optimization	n=25, m=5, p=5, x=1
Simulated Annealing	T=100, R=5, a=1.5, b=1.5

4.2 Determining Best Performing Neighbor Functions

After selecting the best performing parameter value combinations for the algorithms we need to determine which neighbor function should be used for each algorithm. Hence, we carried out some computational experiments on synthetic data in order to determine best-fit functions. In Table 3, we present average runtimes and average total assembly times of the algorithms for each neighbor function.

To determine most eligible neighbor function for each algorithm, we compare the performances of the functions by using average of total assembly times obtained by the functions from all real PCB data. For all algorithms, adjacent interchange function shows the worst performance.

For ABC, ***twoopt*** function shows the best performance. The second- and the third-best functions are random remove and reinsert and reverse and/or move subsequence, respectively.

MBO also shows its best performance with ***twoopt*** function. The second- and the third-best functions are reverse and/or move subsequence and move subsequence functions, respectively.

For SA, ***randomremovereinsert*** function shows the best performance. It is slightly better than twoopt function. twoopt and reverse and/or move subsequence functions show the second- and the third-best performance, respectively.

Table 3. Neighboring function performance for each metaheuristic

Function	ABC		MBO		SA	
	RT	TAT	RT	TAT	RT	TAT
adjacentinterchange	115.97	34.45	144.83	31.81	113.03	26.43
movesingleterm	117.29	27.66	148.26	21.23	113.99	20.17
movesubsequence	116.66	24.90	146.50	17.06	114.50	16.90
randominterchange	116.36	25.75	146.00	17.88	113.42	17.05
randomremovereinsert	118.27	24.32	159.89	17.10	116.80	16.82
reverseandormovesubsequence	183.75	24.72	203.75	16.95	194.92	16.88
reverseandswapsubsequences	121.39	27.17	158.41	17.73	116.50	17.49
reversesubsequence	122.39	25.77	150.26	20.16	114.62	17.73
swapsubsequences	119.82	27.18	154.35	17.54	116.05	17.50
twoopt	***120.24***	***23.18***	***154.99***	***16.87***	***116.16***	***16.84***

5 Summary, Conclusions and Future Work

SDTSP is a newly defined combinatorial optimization problem which emerged during optimization of placement machines having been used PCB assembly. In this study, we determined the best parameter value combination and the best performing neighbor function for three metaheuristics by taking SDTSP into consideration. Firstly, best performing parameter value combination for each of the metaheuristics are found as a result of extensive computational experiments.

When the metaheuristics utilize their best parameter value combination it is observed that twoopt shows the best performance out of 10 neighbour functions. As an extension to this study, we plan to make a study that finds the effect of various values of k on the complexity of SDTSP. Besides, the performance of the metaheuristics will be compared with exact solutions on relatively small problem instances.

Acknowledgment. This work is supported by Marmara University Scientific Research Committee under the project ID FEN-A-150513-0172.

References

1. Alkaya, A.F., Duman, E.: A New Generalization of the Traveling Salesman Problem. Appl. Comput. Math-Bak 9(2), 162–175 (2010)
2. Alkaya, A.F., Duman, E.: An Application of the Sequence Dependent Traveling Salesman Problem in Printed Circuit Board Assembly. IEEE Transactions on Components, Packaging and Manufacturing Technology 3(6) (June 2013)
3. Csaszar, P., Tirpak, T.M., Nelson, P.C.: Optimization of a highspeed placement machine using tabu search algorithms. Ann. Oper. Res. 96(1–4), 125–147 (2000)
4. Tirpak, T.M.: Design to manufacturing information management for electronics assembly. Int. J. Flex Manuf. Syst. 12(2–3), 189–205 (2000)
5. Kirkpatrick, S., Gelatt, C.D., Vecchi, M.P.: Optimization by Simulated Annealing. Science 220(13), 4598 (1983), doi:10.1126/science.220.4598.671
6. Karaboga, D., Basturk, B.: A Powerful and Efficient Algorithm for Numerical Function Optimization: Artificial Bee Colony (ABC) Algorithm. Journal of Global Optimization 39(3), 171–459 (2007)
7. Duman, E., Uysal, M., Alkaya, A.F.: Migrating Birds Optimization: A New Metaheuristic Approach and its Performance on Quadratic Assignment Problem. Information Sciences 217, 65–77 (2012)
8. Malek, M., Guruswamy, M., Pandya, M., Owens, H.: Serial and Parallel Simulated Annealing and Tabu Search Algorithms for the Traveling Salesman Problem. Annals of Operations Research 21, 59–84 (1989)
9. Metropolis, N., Rosenbluth, A., Rosenbluth, M., Teller, A., Teller, E.: Equation of State Calculations by Fast Computing Machines. Journal of Chemical Physics 21(6), 1087–1092 (1953)
10. Duman, E., Or, I.: The Quadratic Assignment Problem in the Context of the Printed Circuit Board Assembly Process. Comput. Oper. Res. 34(1), 163–179 (2007)
11. Karaboga, D., Gorkemli, B.: A Combinatorial Artificial Bee Colony Algorithm for Traveling Salesman Problem. In: International Symposium on Innovations in Intelligent Systems and Applications (June 2011)
12. Duman, E., Elikucuk, I.: Solving Credit Card Fraud Detection Problem by the New Metaheuristics Migrating Birds Optimization. In: Rojas, I., Joya, G., Cabestany, J. (eds.) IWANN 2013, Part II. LNCS, vol. 7903, pp. 62–71. Springer, Heidelberg (2013)

Comparative Study of Extended Kalman Filter and Particle Filter for Attitude Estimation in Gyroless Low Earth Orbit Spacecraft

Nor Hazadura Hamzah[1], Sazali Yaacob[1], Hariharan Muthusamy[1], and Norhizam Hamzah[2]

[1] School of Mechatronic Engineering, Universiti Malaysia Perlis, Perlis, Malaysia
[2] Astronautic Technology Sdn. Bhd., Selangor, Malaysia
hazadura@gmail.com

Abstract. This paper presents attitude determination using a designed observer for gyroless Low Earth Orbit spacecraft using two different estimation approaches which are Extended Kalman Filter (EKF) and Particle Filter (PF) algorithms. This designed system contributes as an alternative or backup system during rate data absence resulted from the unexpected failure of existing gyros. The performance of EKF and PF as the estimator in the observer are compared in terms of statistical and computational aspects. The performance of the nonlinear observer to estimate the states is also verified using real flight data of Malaysian satellite.

Keywords: Nonlinear observer, Satellite attitude estimation, Extended Kalman Filter, Particle Filter.

1 Introduction

Satellite attitude is important to be determined in a satellite to be fed back to controller in accomplishing a specific satellite mission such as Earth observation, communication, scientific research and many other missions. However not all states are directly available may be due to malfunction sensor or as a way to obtain a substantial reduction of sensors which represents a cost reduction. In commonly practice of attitude determination system (ADS), the angular velocity and attitude information of a spacecraft are obtained respectively from measurement of rate sensor such as gyroscopes and also attitude sensor such as sun sensor, star sensor, or magnetometer. However, gyroscopes are generally expensive and are often prone to degradation or failure [1]. Therefore, as an alternative or backup system to circumvent the problem of gyroless measurement, an observer can be designed to provide the information of angular velocity by using only the measurement of Euler angles attitude.

Since decades, a great number of research works have been devoted to the problem of estimating the attitude of a spacecraft based on a sequence of noisy vector observations such as [2][3][4][5]. Different algorithms have been designed and implemented in satellite attitude estimation problem. Early applications relied mostly on the

H.A. Le Thi et al. (eds.), *Advanced Computational Methods for Knowledge Engineering*,
Advances in Intelligent Systems and Computing 358, DOI: 10.1007/978-3-319-17996-4_9

Kalman filter for attitude estimation. Kalman filter was the first applied algorithm for attitude estimation for the Apollo space program in 1960s. Due to limitation of Kalman filter which work optimal for linear system only, several famous new approaches have been implemented to deal with the nonlinearity in satellite attitude system including Extended Kalman Filter (EKF) [6][7][4], Unscented Kalman Filter (UKF) [8][9][10], Particle Filter (PF) [11][12][13], and predictive filtering [14][15]. EKF is an extended version of Kalman filter for nonlinear system whereby the nonlinear equation is approximated by linearized equation through Taylor series expansion. UKF, an alternative to the EKF uses a deterministic sampling technique known as the unscented transform to pick a minimal set of sample points called sigma points to propagate the non-linear functions. EKF and UKF approaches is restricted assume the noise in the system is Gaussian white noise process. While, PF is a nonlinear estimation algorithm that approximates the nonlinear function using a set of random samples without restricted to a specific noise distribution as EKF and UKF.

The organization of this paper proceeds as follows. Section 2 presents mathematical model of the derived nonlinear observer for satellite attitude estimation. Section 3 describes two estimation algorithms used in the observer system which are EKF and PF. Section 4 presents and discusses the performance of the observer system and Section 5 presents the paper's conclusions.

2 Nonlinear Mathematical Model of the Observer

A nonlinear observer is a nonlinear dynamic system that is used to estimate the unknown states from one or more noisy measurements. Mathematically, the nonlinear observer design is described as follows. Given the actual nonlinear system dynamics and measurement described by continuous-time model [16]

$$\dot{x} = f(x) + w \tag{1}$$

$$y = h(x) + v \tag{2}$$

Then, the observer is modelled as

$$\dot{\hat{x}} = f(\hat{x}) \tag{3}$$

$$\hat{y} = h(\hat{x}) \tag{4}$$

In Eqs. (1)-(4), $x \in R^n$ is the state vector and $y \in R^p$ is the output vector, w and v denote the noise or uncertainty vector in the process and measurement respectively. While $\hat{x}$ and $\hat{y}$ denote the corresponding estimates.

In this work, the system is designed to estimate the satellite's angular velocity $(\omega_x, \omega_y, \omega_z)$ by using Euler angles attitude $(\emptyset, \theta, \varphi)$ measurement only. Hence the state vector is $x = [\omega_x, \omega_y, \omega_z, \emptyset, \theta, \varphi]^T$, while the state equation is

$$\dot{x} = \left[\dot{\omega}_x, \dot{\omega}_y, \dot{\omega}_z, \dot{\emptyset}, \dot{\theta}, \dot{\varphi}\right]^T \tag{5}$$

with

$$\dot{\omega}_x = -\left(\frac{I_z-I_y}{I_x}\right)\omega_y\omega_z + 3\omega_0{}^2\frac{(I_z-I_y)}{I_x}s\emptyset c\emptyset c^2\theta \tag{6}$$

$$\dot{\omega}_y = -\left(\frac{I_x-I_z}{I_y}\right)\omega_x\omega_z + 3\omega_0{}^2\frac{(I_z-I_x)}{I_y}s\theta c\theta c\emptyset \tag{7}$$

$$\dot{\omega}_z = -\left(\frac{I_y-I_x}{I_z}\right)\omega_x\omega_y + 3\omega_0{}^2\frac{(I_x-I_y)}{I_z}s\emptyset c\theta s\theta \tag{8}$$

$$\dot{\emptyset} = [\omega_x + \omega_0 c\theta s\varphi\,] + s\emptyset t\theta\left[\omega_y + \omega_0(c\emptyset c\varphi + s\emptyset s\theta s\varphi)\right] + c\emptyset t\theta[\omega_z + \omega_0(-s\emptyset c\varphi + c\emptyset s\theta s\varphi)] \tag{9}$$

$$\dot{\theta} = c\emptyset\left[\omega_y + \omega_0(c\emptyset c\varphi + s\emptyset s\theta s\varphi)\right] - s\emptyset[\omega_z + \omega_0(-s\emptyset c\varphi + c\emptyset s\theta s\varphi)] \tag{10}$$

$$\dot{\varphi} = \frac{s\emptyset}{c\theta}\left[\omega_y + \omega_0(c\emptyset c\varphi + s\emptyset s\theta s\varphi)\right] + \frac{c\emptyset}{c\theta}[\omega_z + \omega_0(-s\emptyset c\varphi + c\emptyset s\theta s\varphi)] \tag{11}$$

where $I = diag[I_x, I_y, I_z]$, $\dot{\omega} = [\dot{\omega}_x, \dot{\omega}_y, \dot{\omega}_z]$, $\omega = [\omega_x, \omega_y, \omega_z]$, $T = [T_x, T_y, T_z]$ represent satellite's moment of inertia, angular acceleration, angular velocity and space environmental disturbances torque vectors respectively. While $\emptyset$ is rotational angle about X-axis (roll); θ is rotational angle about Y-axis (pitch); and φ is rotational angle about Z-axis (yaw). In the above equation *c, s* and *t* denote cosine, sine, and tangent functions, respectively. While, ω_0 is the orbital rate of the spacecraft.

While, the measurement equation of the designed observer is

$$y = h(x) = \begin{bmatrix}\emptyset \\ \theta \\ \varphi\end{bmatrix} \tag{12}$$

3 Nonlinear Estimation Algorithms

3.1 Extended Kalman Filter

In this work, EKF is used as one of the methods since it is widely used estimation algorithm in real practice of spacecraft community and theoretically attractive in the sense that it minimizes the variance of the estimation error. EKF algorithm is described as below. [17]

Let the continuous model in Eqs. (1) and (2) are transformed into the discrete-time model such that

$$x_k = f(x_{k-1}) + w_{k-1} \tag{13}$$

$$y_k = h(x_k) + v_k \tag{14}$$

Here the subscript of the variables denotes the time step, while w_{k-1} and v_k are restricted assumed as Gaussian distributed noises with mean zero and covariance R_w and R_v respectively such that $w_{k-1}{\sim}N(0, R_w)$ and $v_k{\sim}N(0, R_v)$. Then, the estimated state is obtained through the following step.

Step 1: Set the initial state estimate $\hat{x}_0 = \hat{x}_{0|0}$ and variance $P_0 = P_{0|0}$. (15)

Step 2: Repeat

(i) Prediction step (priori estimate)

- Jacobian of $f(x_{k-1})$: $$F_{k-1} = \left.\frac{\partial f}{\partial x}\right|_{\hat{x}_{k-1|k-1}} \quad (16)$$

- Predicted state estimate: $$\hat{x}_{k|k-1} = f(\hat{x}_{k-1|k-1}) \quad (17)$$

- Predicted covariance estimate: $$P_{k|k-1} = F_{k-1}P_{k-1|k-1}F_{k-1}^T + R_w \quad (18)$$

(ii) Update step (posteriori estimate)

- Jacobian of $h(x_k)$: $$H_k = \left.\frac{\partial h}{\partial x}\right|_{\hat{x}_{k|k-1}} \quad (19)$$

- Kalman gain: $$K_k = P_{k|k-1}H_k^T\left[H_kP_{k|k-1}H_k^T + R_v\right]^{-1} \quad (20)$$

- Updated state estimate: $$\hat{x}_{k|k} = \hat{x}_{k|k-1} + K_k\left[y_k - h(\hat{x}_{k|k-1})\right] \quad (21)$$

- Updated covariance estimate: $$P_{k|k} = [I - K_kH_k]P_{k|k-1} \quad (22)$$

3.2 Particle Filter

Particle Filter (PF) was developed in 1993 by Gordon et al [18] under the name of the 'bootstrap filter algorithm'. PF does not require any assumption about the state-space or the noise of the system to be Gaussian as restricted in conventional method EKF. The method approximates the posterior density using a set of particles. PF algorithm is described as below. [19]

Let the continuous-time model in Eqs. (1) and (2) are transformed into the discrete-time model as described in Eqs. (13) and (14) written again for convenience

$$x_k = f(x_{k-1}) + w_{k-1} \quad (23)$$

$$y_k = h(x_k) + v_k \quad (24)$$

Here the subscript of the variables denotes the time step, while w_{k-1} and v_k are process and measurement noises respectively with variance R_w and R_v. Then, the estimated state is obtained through the following step [19]:

Step 1: Set the number of particles N_s and set the initialization

- Initial state estimate: $$\hat{x}_0 = \hat{x}_{0|0} \quad (25)$$

- Initial particles: $$\chi_0^i = \hat{x}_{0|0} \text{ for } i = 1,2,\cdots,N_s \quad (26)$$

- Initial weight: $$w_0^i = \frac{1}{N_s} \quad \text{for } i = 1,2,\cdots,N_s \quad (27)$$

Step 2: Repeat

(i) Sequential importance sampling

- Draw particles: $$\chi_k^i \sim N\left(x_k; f_{k-1}(\chi_{k-1}^i), R_w\right) \text{ for } i = 1,2,\cdots,N_s \quad (28)$$

- Compute the weight for each particle: $w_k^i = w_{k-1}^i N\big(z_k; h_k(\chi_k^i), R_v\big)$ for $i = 1,2,\cdots,N_s$ (29)
- Calculate the total weight: $T = \sum_{i=1}^{N_s} w_k^i$ (30)
- Normalize the weight: $w_k^i = \frac{w_k^i}{T}$ for $i = 1,2,\cdots,N_s$ (31)

(ii) Resampling (To eliminate samples with low importance weights)

- Initialize cumulative sum of weight (CSW) : $c_1 = 0$ (32)
- Construct CSW: $c_i = c_{i-1} + w_k^i$ for $i = 2,3,\cdots,N_s$ (33)
- Start at the bottom of the CSW: $i = 1$ (34)
- Draw a starting point: $u_1 \sim U\left[0, \frac{1}{N_s}\right]$ (35)
- For $j = 1,2,\cdots,N_s$
  - Move along the CSW: $u_j = u_1 + \frac{1}{N_s}(j-1)$ (36)
  - Set if $u_j > c_i$, then update $i = i+1$ (37)
  - Assign particles: $\chi_k^j = \chi_k^i$ (38)
  - Assign weight: $w_k^j = \frac{1}{N_s}$ (39)

(iii) State estimation

- Compute estimated state: $\hat{x}_{k|k} = \sum_{i=1}^{N_s} \chi_k^i w_k^i$ (40)

4 Result and Discussion

The first study is to validate the capability of the designed observer to estimate the angular velocity by using measurement Euler angles attitude roll, pitch and yaw only. The accuracy of the estimated states is assessed by comparing the estimated states with the real states provided by sensor data of RazakSAT. RazakSAT is a Malaysian satellite which was launched into Low Earth Orbit near Equatorial in 2009. In the mission, the attitude was provided directly using sun sensor, one of the attitude sensor, while the angular velocity was provided by gyroscope sensor.

The performance of the PF and EKF as the algorithms in the designed observer to estimate the angular velocity without the gyroscopes is verified by comparing the estimated states with the real states provided by sensor data of RazakSAT. Figs. 1-3 show comparisons between the estimated angular velocities using PF and EKF algorithms compared to the real angular velocities measurement provided by gyroscope sensor in RazakSAT mission, respectively around X-axis, Y-axis, and Z-axis. From all the three figures, it is observed that the trendlines of the estimated states are still within 0.1 deg/s ranges which are suitable for moderate accuracy attitude determination.

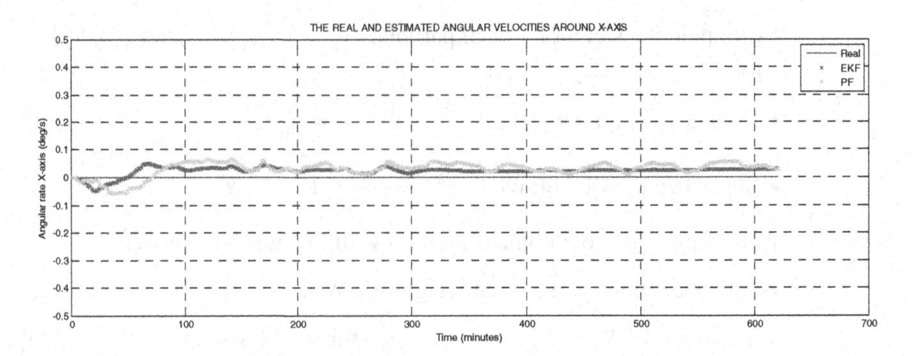

Fig. 1. Comparison between the estimated and the real angular velocity around X-axis

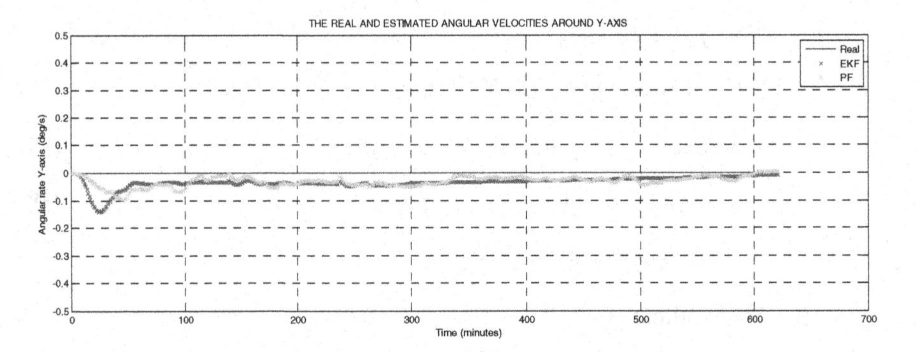

Fig. 2. Comparison between the estimated and the real angular velocity around Y-axis

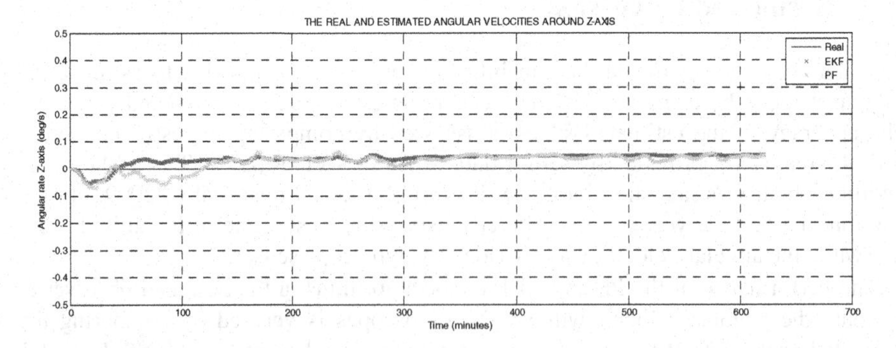

Fig. 3. Comparison between the estimated and the real angular velocity around Z-axis

The accuracy of the estimated states is validated using Root Mean Squared Error (RMSE) as tabulated in Table 1. By referring to the norm values in Table 1, it is concluded that PF provide slightly more accurate than EKF and could be a realistic option in practice when one considering statistical performance. This could be due to the non-Gaussian factor of the real data of RazakSAT as verified in previous work of the authors in [20].

Table 1. RMSE of the estimated angular velocity

Algorithm	RMSE (deg/s)			
	ω_x	ω_y	ω_z	Norm
EKF	0.0282	0.0393	0.0416	0.0638
PF	0.0339	0.0371	0.0387	0.0634

The second study is to investigate the effects of the process noise variance, R_w on both algorithms performance in terms of accuracy and computational aspects through simulation in MATLAB software. The process noise represents the uncertainty in the system model due to modeling errors and approximations. The accuracy and computational aspects are assessed respectively using norm of error and CPU time. Norm of error and CPU time generated for different values of the process noise variance, R_w are empirically tabulated in Table 2. Fig. 4 plots the norm of error as a function of the process noise variance, R_w for $0.001 \leq R_w \leq 0.01$. It is observed that for PF, the norm of error is drastically increased as the process noise variance R_w is increased. However, note that for EKF, there are insignificant changes of the norm of error values, which remains less than 0.002 as R_w is increased. Hence, it is concluded that EKF is more robust with respect to modeling errors and approximations. While the plot of the CPU time as a function of the process noise variance, R_w for $0.001 \leq R_w \leq 0.01$ is shown in Fig. 5. It is observed that CPU time for both algorithms do not increased as the variance R_w is increased. However, the CPU time taken by PF is about 28 times more than EKF. Hence, it is concluded that EKF is faster than PF to provide the estimation. This is because of more computational effort is required in PF to generate the particles to represent the density of the estimated state. Although EKF algorithm requires the Jacobian matrix computation at each iteration, the complexity does not contributes a large burden as generation of particles to represent the density in PF algorithm.

Table 2. Norm of error and CPU time for different values of process noise variance, R_w

Process noise variance, $\boldsymbol{R_w}$	Norm of error (deg/s)		CPU time (s)	
	EKF	PF	EKF	PF
0.001	0.0014	0.0017	0.067	1.379
0.002	0.0016	0.0023	0.053	1.617
0.003	0.0016	0.0026	0.050	1.440
0.004	0.0017	0.0031	0.062	1.561
0.005	0.0017	0.0044	0.040	1.444
0.006	0.0017	0.0045	0.049	1.542
0.007	0.0017	0.0048	0.059	1.463
0.008	0.0017	0.0060	0.053	1.493
0.009	0.0018	0.0066	0.052	1.561
0.010	0.0017	0.0085	0.053	1.480

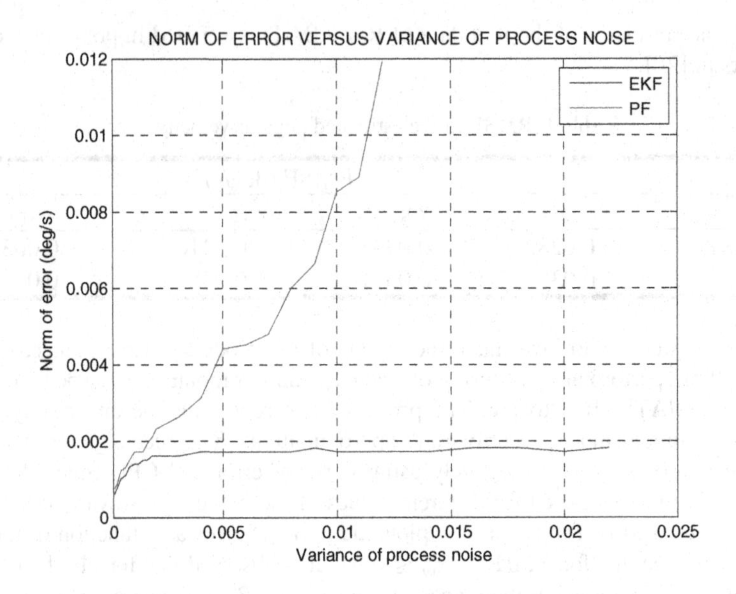

Fig. 4. The norm of error versus the variance of process noise, R_w

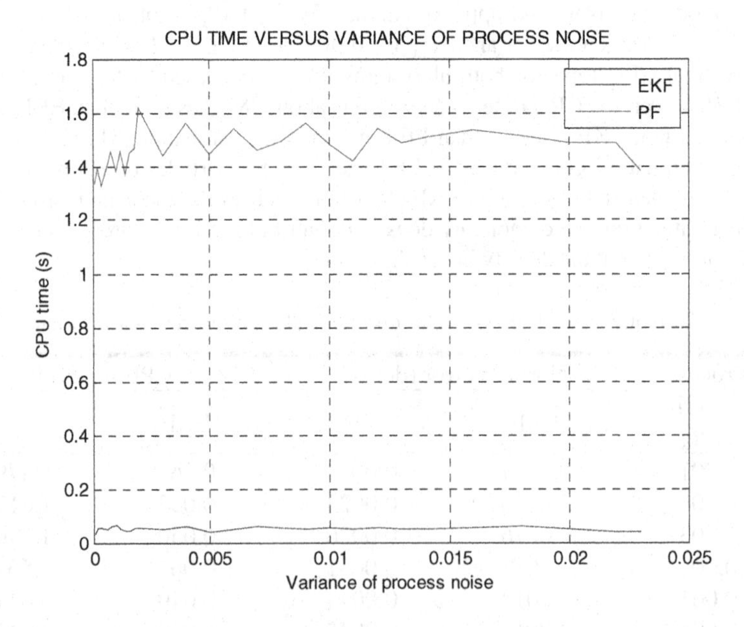

Fig. 5. The CPU time versus the variance of process noise, R_w

The third study is to investigate the effects of the measurement noise variance, R_v on both algorithms performance in terms of accuracy and computational aspects, also through simulation in MATLAB software. The measurement noise represents the uncertainty in the measurements accuracy. Smaller values of the measurement noise variance indicate more accurate of the measurements. Again, the accuracy and computational aspects are assessed respectively using norm of error and CPU time. Norm of error and CPU time generated for different values of the measurement noise variance, R_v are empirically tabulated in Table 3. Fig. 6 plots the norm of error as a function of the measurement noise variance, R_v for $0.001 \leq R_v \leq 4$. It is observed that for EKF, the norm of error is drastically increased as the measurement noise variance R_v is increased. However for PF, there are insignificant changes of the norm of error values, which remains less than 0.25 as R_v is increased. Hence, it is concluded that PF is more robust with respect to measurement uncertainty. While the plot of the CPU time as a function of the measurement noise variance, R_v for $0.001 \leq R_v \leq 4$ is shown in Fig. 7. It is observed that CPU time for both algorithms do not increased as the variance R_v is increased. As in the second analysis discussed before, the CPU time required by PF is about 28 times more than EKF also, which demonstrates that the EKF is faster than PF to provide the estimation. Hence, it is concluded that the size of process and measurement noise does not influenced the CPU time required to compute the estimation.

Table 3. Norm of error and CPU time for different values of measurement noise variance, R_v

Measurement noise variance, R_v	Norm of error (deg/s)		CPU time (s)	
	EKF	PF	EKF	PF
0.001	1.4102e-04	4.6745e-04	0.053	1.469
0.010	0.0156	0.0043	0.040	1.480
0.100	0.0300	0.0131	0.050	1.380
1.000	0.8965	0.0191	0.050	1.440
2.000	1.9548	0.0153	0.050	1.279
3.000	3.2486	0.0278	0.040	1.510
4.000	4.1421	0.0122	0.030	1.480

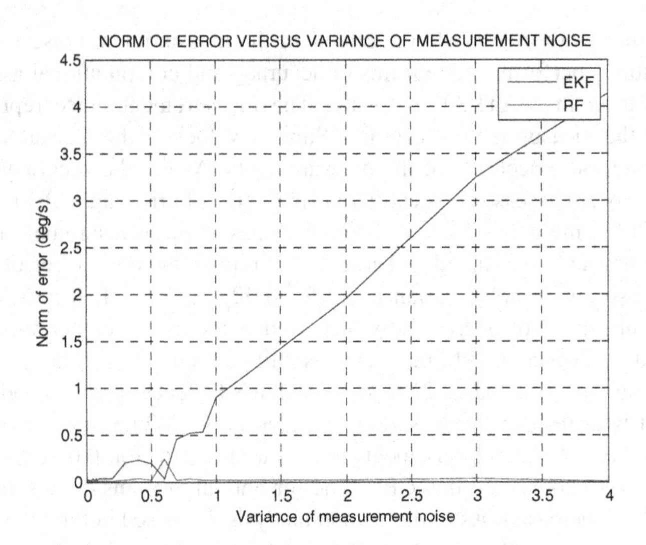

Fig. 6. The norm of error versus the variance of measurement noise, R_v

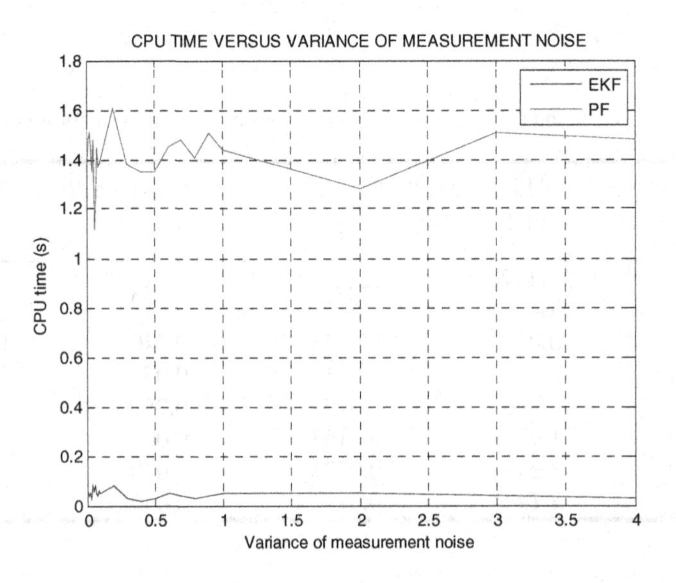

Fig. 7. The CPU time versus the variance of measurement noise, R_v

5 Conclusions

In this paper, a nonlinear observer to provide the angular velocity information in gyroless condition was studied using two different estimation approaches which are Extended Kalman Filter (EKF) and Particle Filter (PF) algorithms. The capability of the designed observer has been verified using real inflight attitude data of Malaysian satellite. The result shows that the designed observer is able to provide the information of angular velocity within 0.1 deg/s accuracy, which is suitable for moderate accuracy attitude determination such as during housekeeping and detumbling mode. Furthermore, the robustness of the algorithms with respect to the modeling uncertainty and measurement accuracy in this application are studied and compared in terms of statistical and computational aspects through MATLAB simulation. In terms of statistical aspect, the EKF shows its robustness against the modeling uncertainty, however not robust to the measurements uncertainty. Conversely, the PF shows its robustness against the measurements uncertainty, but not robust to the modeling uncertainty. While, in terms of computational aspect, the result shows that the EKF is faster than PF, with the computational time taken by PF is about 28 times more than EKF. This is due to more computational effort is required in PF to generate the particles to represent the density of the estimated states. Although EKF algorithm requires computing the Jacobian matrix at each iteration, the complexity does not contributes a large burden as generation of particles to represent the density in PF algorithm. Therefore for this application, the conventional method EKF is preferred to be used for efficient implementation as long as the method work fine and hence there is no need for further complications. However, the PF is strongly suggested during contingency condition of inaccurate or large uncertainty measurements such as due to faulty sensors.

References

1. Crassidis, J.L., Markley, F.L.: An MME-based attitude estimator using vector observations. NASA Conf. Publ., 137 (1995)
2. Christian, J.A., Lightsey, E.G.: The sequential optimal attitude recursion filter. Journal of Guidance, Control, and Dynamics, 1–20 (2010)
3. Emara-Shabaik, H.E.: Spacecraft spin axis attitude. IEEE Trans. Aerosp. Electron. Syst. 28(2), 529–534 (1992)
4. Appel, P.: Attitude estimation from magnetometer and earth-albedo-corrected coarse sun sensor measurements. Acta Astronaut 56(1–2), 115–126 (2005)
5. Liu, B., Chen, Z., Liu, X., Yang, F.: An efficient nonlinear filter for spacecraft attitude estimation. Int. J. Aerosp. Eng. 2014, 1–11 (2014)
6. Filipski, M.N., Varatharajoo, R.: Evaluation of a spacecraft attitude and rate estimation algorithm. Aircr. Eng. Aerosp. Technol. 82(3), 184–193 (2010)
7. Fadly, M., Sidek, O., Said, A., Djojodihardjo, H., Ain, A.: Deterministic and recursive approach in attitude determination for InnoSAT. TELKOMNIKA 9(3) (2011)
8. Cheon, Y.J., Kim, J.H.: Unscented filtering in a unit quaternion space for spacecraft attitude estimation. In: IEEE Int. Symp. Ind. Electron., pp. 66–71 (2007)
9. Vandyke, M.C., Schwartz, J.L., Hall, C.D.: Unscented Kalman filtering for spacecraft attitude state and parameter estimation. In: Proceedings of the AAS/AIAA Space Flight Mechanics Conference, pp. 1–13 (2004)
10. Bae, J.H., Kim, Y.-D.: Attitude estimation for satellite fault tolerant system using federated unscented Kalman filter. Int. J. Aeronaut. Sp. Sci. 11(2), 80–86 (2010)
11. Liu, Y., Jiang, X., Ma, G.: Marginalized particle filter for spacecraft attitude estimation from vector measurements. J. Control Theory Appl. 5(1), 60–66 (2007)

12. Yafei, Y., Jianguo, L.: Particle filtering for gyroless attitude/angular rate estimation algorithm. In: 2010 First Int. Conf. Pervasive Comput. Signal Process. Appl., pp. 1188–1191 (2010)
13. Cheng, Y., Crassidis, J.L.: Particle filtering for sequential spacecraft attitude estimation. In: AIAA Guidance, Navigation, and Control Conference and Exhibit (2000)
14. Crassidis, J.L., Markley, F.L., Lightsey, E.G., Ketchum, E.: Predictive attitude estimation using global positioning system signals. In: NASA Conference Publication, pp. 1–14 (1997)
15. Crassidis, J.L., Markley, F.L.: Predictive filtering for attitude estimation without rate sensors. J. Guid. Control. Dyn. 20(3), 522–527 (1997)
16. Aly, M.M., Abdel Fatah, H.A., Bahgat, A.: Nonlinear observers for spacecraft attitude estimation in case of yaw angle measurement absence. Int. J. Control. Autom. Syst. 8(5), 1018–1028 (2010)
17. Crassidis, J.L., Junkins, J.L.: Optimal Estimation of Dynamic Systems. CRC Press, Boca Raton (2004)
18. Gordon, N.J., Salmond, D.J., Smith, A.F.M.: Novel approach to nonlinear/non-Gaussian Bayesian state estimation. IEE Proc. F Radar Signal Process. 140(2), 107–113 (1993)
19. Ristic, B., Arulampalam, S.: Gordon. Artech House, Boston (2004)
20. Hamzah, H.N., Yaacob, S., Muthusamy, H., Hamzah, N.: Analysis of the residual between the model and the data using autocorrelation function for satellite attitude estimation. In: Proceedings - 2013 IEEE 9th International Colloquium on Signal Processing and its Applications, CSPA 2013, pp. 78–82 (2013)

Graph Coloring Tabu Search for Project Scheduling

Nicolas Zufferey

Geneva School of Economics and Management, GSEM - University of Geneva
Blvd du Pont-d'Arve 40, 1211 Geneva 4, Switzerland
n.zufferey@unige.ch

Abstract. Consider a project consisting of a set of n operations to be performed. Some pairs $\{j, j'\}$ of operations are *incompatible*, which can have two different meanings. On the one hand, it can be allowed to perform j and j' at common time periods. In such a case, incompatibility costs are encountered and penalized in the objective function. On the other hand, it can be strictly forbidden to perform j and j' concurrently. In such a case, the overall project duration has to be minimized. In this paper, three project scheduling problems (P_1), (P_2) and (P_3) are considered. It will be showed that tabu search relying on graph coloring models is a very competitive method for such problems. The overall approach is called *graph coloring tabu search* and denoted *GCTS*.

Keywords: Graph coloring, tabu search, project scheduling, combinatorial optimization, metaheuristics.

1 Introduction

Firstly, consider a project (P_1) consisting in n operations to be performed within k time periods, assuming that each operation has a duration of at most one time period. When an operation is assigned to a specific period, an assignment cost is encountered. In addition, for some pairs of operations, an incompatibility cost is encountered if they are performed at the same period. The goal is to assign a period to each operation while minimizing the costs. Secondly, consider problem (P_2), which is an extension of (P_1). For each operation, its duration is known as an integer number of time periods, and preemptions are allowed at integer points of time. The goal is to assign the required number of periods to each operation while minimizing the costs. Thirdly, consider problem (P_3), which consists in a set of operations to be performed, assuming the processing time of each operation is at most one time period. Precedence and incompatibility constraints between operations have to be satisfied. The goal is to assign a time period to each operation while minimizing the duration of the project.

Given a graph $G = (V, E)$ with vertex set V and edge set E, the k-coloring problem (k-GCP) consists in assigning an integer (called color) in $\{1, \ldots, k\}$ to every vertex such that two adjacent vertices have different colors. The *graph coloring problem* (GCP) consists in finding a k-coloring with the smallest possible

H.A. Le Thi et al. (eds.), *Advanced Computational Methods for Knowledge Engineering*,
Advances in Intelligent Systems and Computing 358, DOI: 10.1007/978-3-319-17996-4_10

value of k (the optimal value is denoted $\chi(G)$). Both problems are NP-hard [10] and many heuristics were proposed to solve them. It will be showed that problem (P_1) (resp. (P_2) and (P_3)) can be modeled as an extension of the *graph coloring problem* (resp. *graph multi-coloring problem*, *mixed graph coloring problem*).

Problems (P_1), (P_2) and (P_3) are new and there is no literature on it, except [3,22,30] on which this paper strongly relies. For a recent survey on graph coloring, the reader is referred to [21]. Relevant references for the multi-coloring problem with applications in scheduling are [5,9,12]. For more information on the mixed graph coloring problem and some applications in scheduling, the reader is referred to [1,8,13,26]. Graph coloring approaches for management and scheduling problems are given in [15,28,29]. The reader desiring a review on scheduling models and algorithms is referred to [24]. The reader interested in a general project management book with applications to planning and scheduling is referred to [17]. Finally, the reader interested in project scheduling is referred to [6,16,18,19].

The literature shows that tabu search has obtained competitive results for (P_1), (P_2) and (P_3). Let f be an objective function to minimize. Starting from an initial solution, a *local search* generates at each iteration a *neighbor* solution s' from a *current* solution s by performing a *move* m (i.e. a slight modification on s). In a *descent local search*, the best move is performed at each iteration and the process stops when the first local optimum is reached. To escape from a local optimum, in a tabu search, when a move m is performed to generate s' from s, then the reverse move is forbidden for *tab* (parameter) iterations. At each iteration, the best non tabu move is performed. The process is stopped for example when a time limit is reached. The reader interested in a recent book on metaheuristics is referred to [11], and to [27] for guidelines on an efficient design of metaheuristics according to various criteria.

In this work, it is showed that the project scheduling problems (P_1), (P_2) and (P_3) can be efficiently tackled with a tabu search metaheuristic relying on graph coloring models. The resulting overall approach is called *graph coloring tabu search* and denoted *GCTS*. In this paper, *GCTS* is adapted to (P_1) in Section 2 (relying on [30]), to (P_2) in Section 3 (relying on [3]), and to (P_3) in Section 4 (relying on [22]). The reader is referred to the above mentioned three references to have detailed information on the NP-hard state, the complexity issues, the literature review, the generation of the instances, the experimental conditions and the presentation of the results. For each problem, the main numerical results will highlight the efficiency of *GCTS*. A conclusion is provided in Section 5.

2 Problem (P_1)

2.1 Presentation of the Problem

Consider a project which consists of a set V of n operations to be performed. The project manager provides a target number k of time periods within which the project has to be performed. It is assumed that: each time period has the same duration (e.g. a working day); there is no precedence constraint between operations; each operation has a duration of at most one time period. Let $c(j, j') \geq 0$ denote an *incompatibility* cost between operations j and j', which is to be paid if j and j' are performed at the same time period. The incompatibility cost $c(j, j')$ represents for example that the same staff has to perform operations j and j', thus additional human resources must be hired in order to be able to perform both operations at the same period. In addition, for each operation j and each time period t, an *assignment* cost $a(j, t)$ has to be paid if j is performed at period t. $a(j, t)$ represents for example the cost of the staff and machines which have to perform operation j at period t. The goal is to assign a time period $t \in \{1, \ldots, k\}$ to each operation $j \in V$ while minimizing the total costs.

A solution using k periods can be generated by the use of a function $per : V \longrightarrow \{1, \ldots, k\}$. The value $per(j)$ of an operation j is the period assigned to j. With each period t can be associated a set C_t that contains the set of operations performed at period t. Thus, a solution s can be denoted $s = (C_1, \ldots, C_k)$, and the associated encountered costs are described in Equation (1). (P_1) consists in finding a solution with k periods which minimizes these costs.

$$f(s) = \sum_{t=1}^{k} \sum_{j \in C_t} a(j, t) + \sum_{j=1}^{n-1} \sum_{j' \in \{j+1, \ldots, n\} \cap C_{per(j)}} c(j, j') \qquad (1)$$

2.2 Graph Coloring Model Based on the *k*-GCP

Let $I(j)$ denote the set of operations j' such that $c(j, j') > 0$. From the input data of problem (P_1), an *incompatibility graph* $G = (V, E)$ can be built as follows. A vertex j is associated with each operation j, and an edge $[j, j']$ is drawn each time $j' \in I(j)$ (but not more than one edge between two vertices). A color t represents a time period t. Coloring G with k colors while minimizing the number of conflicting edges (which is exactly the k-GCP) is equivalent to assign a time period $t \in \{1, \ldots, k\}$ to each operation while minimizing the number of incompatibilities. (P_1) is actually an extension of the k-GCP, because the latter is a subcase of the former where $a(j, t) = 0$ $(\forall j, t)$, and $c(j, j') = 1$ $(\forall j$ with $j' \in I(j))$. From now on, the *project scheduling* terminology (e.g., operations, time periods) and the *graph coloring* terminology (e.g., vertices, colors) are indifferently used.

2.3 Tabu Search

An efficient approach for the k-GCP consists in giving a color to each vertex while minimizing the number of conflicts (a *conflict* occurs if two adjacent vertices have the same color). If this number reaches 0, a legal k-coloring is found. In such a context, a straightforward move is to change the color of a conflicting vertex [14]. For (P_1), the search space is the set of k-partitions of V and the objective function to minimize is the total cost f. A move consists in changing the period assigned to an operation. In order to avoid testing every possible move at each iteration, only the $q\%$ (parameter tuned to 40%) most costly operations are considered for a move at each iteration. If operation j moves from C_t to $C_{t'}$ when going from the current solution s to the neighbor solution s', it is forbidden to put j back in C_t during $tab(j, C_t)$ iterations as described in Equation (2), where $R(u, v)$ randomly returns an integer in interval $[u, v]$ (uniform distribution), and (u, v, α) are parameters tuned to $(10, 20, 15)$. The maximum is used to enforce $tab(j, C_t)$ to be positive. The last term of Equation (2) represents the improvement $Imp(s, s')$ of f when moving from s to s'. If s' is better than s, $Imp(s, s')$ is positive and the reverse of the performed move will be forbidden for a larger number of iterations than if $Imp(s, s') < 0$. In addition, if the diversity of the visited solutions is below a predetermined threshold, $tab(j, C_t)$ is augmented from that time (for all j and t), and if the diversity becomes above the threshold, the tabu durations are reduced from that time. Note that a diversification mechanism also favors moves which are unlikely to be performed.

$$tab(j, C_t) = \max\left\{1, R(u, v) + \alpha \cdot \frac{f(s) - f(s')}{f(s)}\right\} \quad (2)$$

2.4 Results

The stopping condition of each method is a time limit T of one hour on an Intel Pentium 4 (4.00 GHz, RAM 1024 Mo DDR2). The following methods *GR*, *DLS*, *GCTS* and *AMA* are compared on instances derived from the well-known graph coloring benchmark instances [21]. Note that *GR* and *DLS* are restarted as long as T is not reached, and the best encountered solution is returned.

- *GR*: a greedy constructive heuristic working as follows. Let J be the set of scheduled operations. Start with an empty solution s (i.e. $J = \emptyset$). Then, while s does not contain n operations, do: (1) randomly select an unscheduled operation j; (2) a time period $t \in \{1, \ldots, k\}$ is assigned to j such that the augmentation of the costs is as small as possible.
- *GCTS*: as described in Subsection 2.3.
- *DLS*: a descent local search derived from *GCTS* by setting $q = 100\%$ (i.e. considering all the possible moves at each iteration), without tabu tenures.
- *AMA*: an adaptive memory algorithm [25], where at each generation, an offspring solution s is built from a central memory M (containing 10 solutions), then s is improved with a tabu search procedure relying on *GCTS*, and finally s is used to update the content of M.

For a fixed value of k, Table 1 reports the average results (over 10 runs) obtained with the above methods. Let $f^{(GCTS)}$ be the average value of the solution returned by *GCTS* (rounded to the nearest integer) over the considered number of runs. $f^{(GR)}$, $f^{(DLS)}$ and $f^{(AMA)}$ are similarly defined. From left to right, the columns indicate: the instance name (incompatibility graph), its number n of operations, its density d (average number of edges between a pair of vertices), the considered value of k, $f^{(GCTS)}$, and the percentage gap between $f^{(GR)}$ (resp. $f^{(DLS)}$ and $f^{(AMA)}$) and $f^{(GCTS)}$. Average gaps are indicated in the last line. It can be observed that $f^{(GCTS)}$ clearly outperforms the other methods.

Table 1. Results on the (P_1) instances

Graph G	n	d	k	$f^{(GCTS)}$	GR	DLS	AMA
DSJC1000.1	1000	0.1	13	241601	57.23%	28.49%	5.42%
DSJC1000.5	1000	0.5	55	250977	33.30%	18.43%	-0.28%
DSJC1000.9	1000	0.9	149	166102	10.30%	11.35%	-4.98%
DSJC500.5	500	0.5	32	98102	55.03%	34.76%	3.50%
DSJC500.9	500	0.9	84	64224	43.69%	42.71%	2.69%
flat1000_50_0	1000	0.49	33	665449	24.97%	10.82%	-0.82%
flat1000_60_0	1000	0.49	40	462612	28.63%	14.16%	-1.18%
flat1000_76_0	1000	0.49	55	246157	32.15%	18.23%	-1.92%
flat300_28_0	300	0.48	19	62862	51.20%	29.19%	1.50%
le450_15c	450	0.16	10	149041	40.75%	20.45%	2.86%
le450_15d	450	0.17	10	146696	42.89%	22.49%	5.19%
le450_25c	450	0.17	17	72974	39.99%	27.11%	21.32%
le450_25d	450	0.17	17	70852	43.40%	29.40%	23.28%
Average					**38.73%**	**23.66%**	**4.35%**

3 Problem (P_2)

3.1 Presentation of the Problem

(P_2) is an extension of (P_1) where the duration of an operation j is not limited to one time period, but to p_j (integer) periods. Preemptions are allowed at integer time points. The goal is to assign p_j (not necessarily consecutive) periods to each operation j while minimizing assignment and incompatibility costs. The assignment cost $a(j,t)$ is defined as in Subsection 2.1. In addition, let $c^m(j,j') > 0$ (with $m \in \mathbb{N}^\star$) denote the *incompatibility* cost between incompatible operations j and j', which is to be paid if j and j' have m common time periods. From a practical standpoint, it is reasonable to assume that $c^{m+1}(j,j') \geq c^m(j,j')$ $(\forall m)$. For compatible operations j and j', $c^m_{j,j'} = 0$ $(\forall m)$.

In order to represent a solution s, with each time period $t \in \{1,\ldots,k\}$ is associated a set C_t containing the operations which are performed at period t. Each operation j has to belong to p_j sets of type C_t in order to be totally performed. Let $\delta^m_{j,j'} = 1$ if operations j and j' are performed within m common time periods, and 0 otherwise. Thus, a solution s can be denoted $s = (C_1,\ldots,C_k)$, and the associated objective function $f(s)$ to minimize is presented in Equation (3).

$$f(s) = \sum_t \sum_{j \in C_t} a(j,t) + \sum_{j<j',m} c^m(j,j') \cdot \delta^m_{j,j'} \tag{3}$$

3.2 Graph Coloring Model Based on the Multi-coloring Problem

In the k*-multi-coloring* problem, each vertex j has to receive a predefined number p_j of colors in $\{1, \ldots, k\}$ such that adjacent vertices have no common color. A *conflict* occurs if two adjacent vertices have at least one color in common. The *graph multi-coloring* problem consists in finding the smallest k for which a k-multi-coloring exists. Among the few existing methods for the multi-coloring problem, tabu search was shown to provide very competitive results [5,7].

A vertex represents an operation, a color is a time period, and the required number p_j of colors to assign to vertex j is the duration of operation j. In contrast with the k-multi-coloring problem, conflicts are allowed in (P_2), but lead to incompatibility costs. In addition, assignment costs are also considered (while they are all equal in the k-multi-coloring problem and can thus be ignored). Therefore, (P_2) is an extension of the k-multi-coloring problem.

3.3 Tabu Search

A feasible solution is any assignment of the correct number of colors to each vertex. The initial solution is a random assignment of p_j colors to each vertex j. A neighbor solution is produced by changing exactly one color on a vertex. Thus, a move (j, t, t') consists in replacing, for a single operation j, a time period t with another time period t'. Assume that move (j, t, t') has just been performed. The moves (j, t, t') and (j, t', t) are then forbidden for *tab* (parameter) iterations, where *tab* is tuned in interval $[1, 50]$ depending on the instance size n.

In order to avoid exhaustive search at each iteration (i.e. evaluating all the possible moves), it is proposed to control the size of the evaluated set of candidate moves by two sensitive parameters N and K, which respectively indicate the considered proportion of operations and time periods. It was observed that the larger is n, the smaller should be N.

3.4 Results

The stopping condition of all (meta)heuristics is a time limit of T seconds, where T depends on the number n of vertices of the graph. $T = 300$ for $n \leq 30$, 600 for $n = 50$, 1200 for $n = 100$, and 3600 for $n = 200$. The following methods *GR*, *DLS* and *GCTS* are compared on randomly generated instances. Note that *GR* and *DLS* are restarted as long as T is not reached, and the best encountered solution is returned.

- *GR*: a greedy constructive algorithm working as follows. At each step, fully color a vertex while minimizing the augmentation of the costs.
- *GCTS*: as described in Subsection 3.3.
- *DLS*: a descent local search derived from *GCTS* by setting $N = K = 100\%$ (an exhaustive search is performed at each iteration), without tabu tenures.

- *GLS*: a genetic local search algorithm, where at each generation, offspring solutions are built from a central memory M (containing 6 solutions), then these solutions are improved with a tabu search procedure relying on *GCTS*, and finally they are used to replace the solutions of M (except the best one).

The tests were executed on an Intel® Core^{TM} i7-2620M CPU @ 2.70GHz with 4GB of RAM (DDR3). For each instance is reported the very best objective function value $f^{\star}$ ever found by any algorithm during all the performed tests. Results on linear instances are reported in Table 2. The instance name is straightforward. For example, instance n10-d80-k14-p2-P5 has $n = 10$ operations, a density $d = 80\%$, $k = 14$ allowed time periods, and each operation j has a duration p_j in interval $[p, P] = [2, 5]$. On these linear instances, it was possible to use CPLEX 12.4 (during 4 hours, which is above T for any instance) to compute a lower (resp. upper) bound LB (resp. UB) on f. For each instance, the percentage gap of each method (with respect to $f^{\star}$) is given (averaged over 10 runs). The following observations can be made: (1) CPLEX can provide optimal solution for very small instances only (as $LB = UB$ only for the instances with $n = 10$); (2) *GR* and *DLS* performs very poorly, especially with larger n values; (3) *GCTS* and *GLS* have comparable performances, with a slight advantage to *GLS*, which highlights the benefit of the recombination operator when jointly used with *GCTS*.

Table 2. Results on the (P_2) linear instances

Instance	$f^{\star}$	LB	UB	GR	DLS	$GCTS$	GLS
n10-d50-k9-p2-P5	62.37	62.37	62.37	84.60%	0.00%	0.00%	0.00%
n10-d80-k14-p2-P5	44.66	44.66	44.66	181.40%	3.90%	0.20%	0.00%
n20-d50-k14-p2-P5	81.99	67.68	82.49	217.60%	2.90%	1.40%	0.50%
n20-d80-k17-p2-P5	181.12	95.2	199.17	138.90%	2.40%	1.30%	0.70%
n30-d50-k20-p2-P6	146.02	68.77	171.37	348.20%	14.40%	1.60%	1.90%
n30-d80-k30-p3-P6	438.28	73.53	640.02	166.30%	8.80%	0.60%	1.00%
n50-d20-k17-p2-P6	134.16	81.73	162.46	531.90%	29.30%	16.80%	3.50%
n50-d50-k22-p1-P4	85.94	38.73	183.59	1075.20%	43.20%	2.40%	3.00%
n100-d20-k20-p1-P5	170.84	85.18	449.86	1096.50%	66.30%	7.70%	7.50%
n200-d20-k25-p1-P5	883.17	100.91	∞	696.10%	65.30%	1.80%	3.70%
Average				**453.67%**	**23.65%**	**3.38%**	**2.18%**

4 Problem (P_3)

4.1 Presentation of the Problem

(P_3) is an extension of (P_1) where incompatibility costs are replaced with incompatibility constraints, assignment costs are ignored, precedence constraints have to be satisfied, and the total duration k of the project has to be minimized. An *incompatibility* constraint between operations j and j' is denoted by $[j, j']$.

For each operation j is given a set $P(j) \subset V$ of immediate predecessor operations. If $j' \in P(j)$, it means that operation j' has to be completely performed before j starts. Such a *precedence* constraint is denoted by (j', j). The goal is to assign a time period t to each operation j while minimizing the total duration of the project, and satisfying the incompatibility and precedence constraints.

Let $(P_3^{(k)})$ be the problem of searching for a feasible solution using k time periods. As explained in Subsection 2.1, such a solution can be generated by using a function *per*, and the notation $s = (C_1, \ldots, C_k)$ can be used. Problem (P_3) consists in finding a feasible solution s using k time periods with the smallest value of k. Starting with $k = n$, one can tackle (P_3) by solving a series of $(P_3^{(k)})$ with decreasing values of k, and the process stops when it is not possible to find a feasible solution with k time periods.

4.2 Graph Coloring Model Based on the Mixed GCP

A *mixed graph* $G = (V, E, A)$ is a graph with vertex set V, edge set E, and arc set A. By definition, an edge $[x, y]$ is not oriented and an arc (x, y) is an oriented edge (from x to y). In the $MGCP$ (mixed graph coloring problem), the goal is to assign a color to every vertex while using a minimum number of colors and satisfying the incompatibility constraints (i.e., two adjacent vertices must get different colors). In addition, for every arc (x, y), the precedence constraint $col(x) < col(y)$ has to be respected (where $col(x)$ is the color assigned to x). For (P_3), there is a *conflict* between vertices x and y if one of the following conditions is true: (1) $y \in I(x)$ (i.e. x and y are incompatible) and $col(x) = col(y)$ (incompatibility violation); (2) $y \in P(x)$ and $col(x) \leq col(y)$ (precedence violation). In both cases, x and y are *conflicting vertices*. In case (1), the conflict occurs on edge $[x, y]$, and in case (2), it occurs on arc (x, y).

From the input data of problem (P_3), one can construct a mixed graph $G = (V, E, A)$ as follows: vertex j represents operation j; if $j' \in I(j)$, then edge $[j, j']$ is drawn to represent an incompatibility (at most one edge between two vertices); if $j'' \in P(j)$, then an arc (j'', j) is drawn to represent a precedence constraint. In addition, a color t can be associated with each time period t. Coloring G with k colors while trying to minimize the number of conflicts is equivalent to assigning a time period $t \in \{1, \ldots, k\}$ to each operation while trying to minimize the number of violations of incompatibility and precedence constraints.

4.3 Tabu Search

GCTS is derived from the tabu search approach proposed for the k-GCP in [2]. The search space is the set of partial but legal solutions of $(P_3^{(k)})$, and the objective function f to minimize is the number of operations without an associated time period. Formally, any solution s can be denoted by $s = (C_1, \ldots, C_k; OUT)$, where C_t is the set of operations performed at time period t (without the occurrence of any conflict), and $|OUT|$ has to be minimized (all the vertices without a time period are in OUT). Note that if $|OUT| = 0$, it means that a feasible solution has been found with k periods, and the process is restarted with $k-1$ periods, and so on until no feasible solution is found. Then, the provided number of periods will be the last number for which a feasible solution has been found.

A move consists in assigning a time period t to an operation j belonging to OUT. If it creates conflicting operations (in C_t), their associated time period t is removed (i.e., such conflicting vertices are moved from C_t to OUT). When a time period t is assigned to an operation j, it is then tabu to remove t from j during tab (parameter) iterations. At each iteration, the best (according to function g defined below) neighbor solution s' of the current solution s is determined (ties are broken randomly), such that either s' is a non-tabu solution, or $f(s') < f^\star$, where $f^\star$ is the value of the best solution $s^\star$ encountered so far during the search. If operation j is removed from OUT when switching from s to s', it is forbidden to put j back into OUT during $tab(j) = R(u, v) + \gamma \cdot n_c$ iterations, where n_c is the number of conflicts in s, and function $R(u, v)$ is defined as in Subsection 2.3. Parameters u, v and γ are respectively tuned to 0, 9 and 0.6.

Let s be the current solution. Note that f may give the same value to several candidate neighbor solutions of s. At each iteration, in order to better discriminate the choice of a neighbor solution, another objective function g is used instead of f (thus, g is only used to evaluate candidate neighbor solutions). More precisely, a conflict can be due to an incompatibility constraint violation or to a precedence constraint violation. It was observed that it is better to give different weights to these two types of conflicts. Given a partial solution $s = (C_1, \dots, C_k; OUT)$, an operation $j \in OUT$ and a time period $t \in \{1, \dots, k\}$, two quantities are computed: (1) $A(j, t)$ is the set of incompatible operations which will be put in OUT if time period t is assigned to operation j; (2) $B(j, t)$ is the set of operations, involved in precedence constraint violations, which will be put in OUT if time period t is given to operation j. At each step of *GCTS*, the move which minimizes $g(j, t) = \alpha \cdot |A(j, t)| + \beta \cdot |B(j, t)|$ is performed, where α and β are parameters tuned to 4 and 1, respectively. With such an objective function g, it is very quick and accurate to evaluate a neighbor solution.

4.4 Results

The stopping condition of all (meta)heuristics is a time limit of $T = 3600$ seconds. The following methods *GR*, *GCTS* and *VNS* are compared on instances derived from the well-known graph coloring benchmark instances [21]. Note that *GR* is restarted as long as T is not reached, and the best encountered solution is returned.

- *GR*: a greedy constructive algorithm derived from [4] and working as follows. At each step, select a vertex j and assign to it the smallest possible color without creating any conflict. If it is not possible, put j in OUT.
- *GCTS*: as described in Subsection 4.3.
- *VNS*: a variable neighborhood search [23], using *GCTS* as intensification procedure.

Our algorithms were implemented in C++ and run on a computer with the following properties: Processor Intel Core2 Duo Processor E6700 (2.66GHz, 4MB Cache, 1066MHz FSB), RAM 2GB DDR2 667 ECC Dual Channel Memory (2x1GB). The results are presented in Table 3. The five first columns respectively indicate the following information: the name of the graph, the number n of vertices, the smallest number of colors $k^\star$ for which a legal $k^\star$-coloring was found by an algorithm or the chromatic number $\chi(G)$ if it is known, the edge density d, and the arc density $\hat{d}$. The last three columns respectively indicate the smallest number of colors for which a legal coloring was found by *GR*, *GCTS*, and *VNS*, with the number of successes among five runs in brackets. As expected, larger d and $\hat{d}$ values lead to a larger number of used colors. One can observe that *GCTS* outperforms both *GR* and *VNS*.

Table 3. Results on the (P_3) instances

Graph	n	$k^\star$	d	$\hat{d}$	*GR*	*GCTS*	*VNS*
DSJC250.1	250	8	0.1	0.005	9 (5)	8 (5)	8 (5)
				0.1	30 (5)	30 (5)	30 (5)
DSJC250.5	250	28	0.5	0.005	36 (5)	30 (1)	31 (4)
				0.01	39 (5)	35 (5)	38 (1)
DSJC250.9	250	72	0.9	0.005	89 (5)	78 (5)	82 (3)
				0.01	95 (5)	91 (2)	97 (1)
DSJR500.1	500	12	0.03	0.005	12 (5)	12 (5)	12 (5)
				0.1	19 (5)	19 (5)	19 (5)
DSJR500.1c	500	85	0.97	0.005	183 (5)	187 (1)	186 (1)
				0.01	279 (5)	285 (3)	285 (4)
DSJR500.5	500	122	0.47	0.005	137 (5)	132 (5)	138 (1)
				0.01	146 (5)	149 (1)	148 (2)
le450_15c	450	15	0.16	0.005	24 (5)	18 (5)	18 (2)
				0.01	25 (5)	21 (5)	22 (4)
le450_15d	450	15	0.17	0.005	24 (5)	18 (5)	18 (2)
				0.01	25 (5)	20 (1)	22 (1)
le450_25c	450	25	0.17	0.005	29 (5)	28 (5)	27 (1)
				0.01	30 (5)	29 (5)	29 (4)
le450_25d	450	25	0.17	0.005	29 (5)	28 (5)	28 (5)
				0.01	30 (5)	29 (5)	29 (4)
flat300_20_0	300	20	0.47	0.005	40 (5)	27 (3)	27 (1)
				0.01	42 (5)	32 (1)	40 (3)
flat300_26_0	300	26	0.48	0.005	41 (5)	34 (4)	35 (1)
				0.01	42 (5)	38 (5)	42 (2)
flat300_28_0	300	28	0.48	0.005	40 (5)	35 (5)	35 (2)
				0.01	44 (5)	40 (1)	45 (1)

5 Conclusion

In this work, *GCTS* is discussed (with a unified view), which is a tabu search approach relying on graph coloring models. It was showed that *GCTS* was efficiently adapted to three project scheduling problems. This success mainly relies on four aspects:

- the use of graph coloring models to represent the considered problem and its solutions;
- the use of an auxiliary objective function instead of the provided one (e.g., for problem (P_3), fix k and minimize OUT, instead of minimizing k directly);
- the use of moves focusing on the costly operations (if costs have to be minimized) or on the removal of conflicts (if constraint violations are penalized);
- an efficient management of the tabu durations (e.g., depending on the quality of the performed moves).

This paper contributes to build bridges between the graph coloring and the project scheduling communities. An avenue of research consists in reducing the dimension of the project graph before triggering the solution methods (e.g., [20]).

References

1. Al-Anzi, F.S., Sotskov, Y.N., Allahverdi, A., Andreev, G.V.: Using Mixed Graph Coloring to Minimize Total Completion Time in Job Shop Scheduling. Applied Mathematics and Computation 182(2), 1137–1148 (2006)
2. Bloechliger, I., Zufferey, N.: A graph coloring heuristic using partial solutions and a reactive tabu scheme. Computers & Operations Research 35, 960–975 (2008)
3. Bloechliger, I., Zufferey, N.: Multi-Coloring and Project-Scheduling with Incompatibility and Assignment Costs. Annals of Operations Research 211(1), 83–101 (2013)
4. Brélaz, D.: New Methods to Color Vertices of a Graph. Communications of the Association for Computing Machinery 22, 251–256 (1979)
5. Chiarandini, M., Stuetzle, T.: Stochastic local search algorithms for graph set T-colouring and frequency assignment. Constraints 12, 371–403 (2007)
6. Demeulemeester, E.L., Herroelen, W.S.: Project Scheduling: A Research Handbook. Kluwer Academic Publishers (2002)
7. Dorne, R., Hao, J.-K.: Meta-heuristics: Advances and trends in local search paradigms for optimization, chapter Tabu search for graph coloring, T-colorings and set T-colorings, pp. 77–92. Kluwer, Norwell (1998)
8. Furmańczyk, H., Kosowski, A., Żyliński, P.: Scheduling with precedence constraints: Mixed graph coloring in series-parallel graphs. In: Wyrzykowski, R., Dongarra, J., Karczewski, K., Wasniewski, J. (eds.) PPAM 2007. LNCS, vol. 4967, pp. 1001–1008. Springer, Heidelberg (2008)
9. Gandhi, R., Halldórsson, M.M., Kortsarz, G., Shachnai, H.: Improved bounds for sum multicoloring and scheduling dependent jobs with minsum criteria. In: Persiano, G., Solis-Oba, R. (eds.) WAOA 2004. LNCS, vol. 3351, pp. 68–82. Springer, Heidelberg (2005)
10. Garey, M., Johnson, D.S.: Computer and Intractability: a Guide to the Theory of NP-Completeness. Freeman, San Francisco (1979)
11. Gendreau, M., Potvin, J.-Y.: *Handbook of Metaheuristics*. International Series in Operations Research & Management Science, vol. 146. Springer, Heidelberg (2010)
12. Halldórsson, M.M., Kortsarz, G.: Multicoloring: Problems and techniques. In: Fiala, J., Koubek, V., Kratochvíl, J. (eds.) MFCS 2004. LNCS, vol. 3153, pp. 25–41. Springer, Heidelberg (2004)
13. Hansen, P., Kuplinsky, J., de Werra, D.: Mixed Graph Coloring. Mathematical Methods of Operations Research 45, 145–169 (1997)

14. Hertz, A., de Werra, D.: Using tabu search techniques for graph coloring. Computing 39, 345–351 (1987)
15. Hertz, A., Schindl, D., Zufferey, N.: A solution method for a car fleet management problem with maintenance constraints. Journal of Heuristics 15(5), 425–450 (2009)
16. Icmeli, O., Erenguc, S.S., Zappe, C.J.: Project scheduling problems: A survey. International Journal of Operations & Production Management 13(11), 80–91 (1993)
17. Kerzner, H.: Project Management: A Systems Approach to Planning, Scheduling, and Controlling. Wiley (2003)
18. Kolisch, R., Padman, R.: An integrated survey of deterministic project scheduling. Omega 29(3), 249–272 (2001)
19. Lancaster, J., Ozbayrak, M.: Evolutionary algorithms applied to project scheduling problems – a survey of the state-of-the-art. International Journal of Production Research 45(2), 425–450 (2007)
20. Luyet, L., Varone, S., Zufferey, N.: An Ant Algorithm for the Steiner Tree Problem in Graphs. In: Giacobini, M. (ed.) EvoWorkshops 2007. LNCS, vol. 4448, pp. 42–51. Springer, Heidelberg (2007)
21. Malaguti, E., Toth, P.: A survey on vertex coloring problems. International Transactions in Operational Research 17(1), 1–34 (2010)
22. Meuwly, F.-X., Ries, B., Zufferey, N.: Solution methods for a scheduling problem with incompatibility and precedence constraints. Algorithmic Operations Research 5(2), 75–85 (2010)
23. Mladenovic, N., Hansen, P.: Variable neighborhood search. Computers & Operations Research 24, 1097–1100 (1997)
24. Pinedo, M.: Scheduling: Theory, Algorithms, and Systemsmulti-coloring. Prentice Hall (2008)
25. Rochat, Y., Taillard, E.: Probabilistic diversification and intensification in local search for vehicle routing. Journal of Heuristics 1, 147–167 (1995)
26. Sotskov, Y.N., Dolgui, A., Werner, F.: Mixed Graph Coloring for Unit-Time Job-Shop Scheduling. International Journal of Mathematical Algorithms 2, 289–323 (2001)
27. Zufferey, N.: Metaheuristics: some Principles for an Efficient Design. Computer Technology and Applications 3(6), 446–462 (2012)
28. Zufferey, N.: Graph Coloring and Job Scheduling: from Models to Powerful Tabu Search Solution Methods. In: Proceedings of the 14th International Workshop on Project Management and Scheduling (PMS 2014), Munich, Germany, March 31 – April 2 (2014)
29. Zufferey, N., Amstutz, P., Giaccari, P.: Graph colouring approaches for a satellite range scheduling problem. Journal of Scheduling 11(4), 263–277 (2008)
30. Zufferey, N., Labarthe, O., Schindl, D.: Heuristics for a project management problem with incompatibility and assignment costs. Computational Optimization and Applications 51, 1231–1252 (2012)

Quality of the Approximation of Ruin Probabilities Regarding to Large Claims

Aicha Bareche, Mouloud Cherfaoui, and Djamil Aïssani

Research Unit LaMOS (Modeling and Optimization of Systems),
Faculty of Technology, University of Bejaia, 06000 Bejaia, Algeria
aicha_bareche@yahoo.fr,
mouloudcherfaoui2013@gmail.com,
lamos_bejaia@hotmail.com
http://www.lamos.org

Abstract. The aim of this work is to show, on the basis of numerical examples based on simulation results, how the strong stability bound on ruin probabilities established by Kalashnikov (2000) is affected regarding to different heavy-tailed distributions.

Keywords: Approximation, Risk model, Ruin probability, Strong stability, Large claim.

1 Introduction

In the actuarial literature, the evolution in time of the capital of an insurance company is often modeled by the process of reserve resulting from the difference between the premium-income and the pay-out process.

The probability of ruin is one of the basic characteristics of risk models and various authors investigate the problem of its evaluation (for example, see [1] and [11], Chapter 11). However, it cannot be found in an explicit form for many risk models. Furthermore, parameters governing these models are often unknown and one can only give some bounds for their values. In such a situation the question of stability becomes crucial.

Indeed, when using a stochastic model in insurance mathematics one has to consider this model as an approximation of the real insurance activities. The stochastic elements derived from these models represent an idealization of the real insurance phenomena under consideration. Hence the problem arises out of establishing the limits in which we can use our 'ideal' model. The practitioner has to know the accuracy of his recommendations, resulting from his investigations based on the ideal model [2]. Using approximations means here that we investigate 'ideal' models which are rather simple, but nevertheless close in some sense to the real (disturbed) model.

After introducing the problem of stability in insurance mathematics by Beirlant and Rachev [2], Kalashnikov [7] investigated the estimation of ruin probabilities in the univariate risk models, using the strong stability method, the reversed process notion and the supplementary variables technique.

H.A. Le Thi et al. (eds.), *Advanced Computational Methods for Knowledge Engineering*,
Advances in Intelligent Systems and Computing 358, DOI: 10.1007/978-3-319-17996-4_11

On the other hand, we often deal in insurance and finance with large claims that are described by heavy-tailed distributions (Pareto, Lognormal, Weibull, ...). It is worthy of notice the special importance of heavy-tailed distributions, which is increasing the last years because of occasional appearance of huge claims [4,9,5,6,12]. Indeed, the loss distribution in actuarial science and financial risk management is fundamental and of ultimate use. It describes the probability distribution of payment to the insured. In most situations losses are small, and extreme losses rarely occur. But the number and the size of the extreme losses can have a substantial influence on the profit of the company. Traditional methods in actuarial literature use parametric specifications to model loss distributions by a single parametric model or decide to analyze large and small losses separately. The most popular specifications are the lognormal, Weibull and Pareto distributions or a mixture of lognormal and Pareto distributions.

The aim of this work is to study, on the basis of numerical examples based on simulation results, the sensitivity of the strong stability bound on ruin probabilities established by Kalashnikov [7] regarding to the different heavy-tailed distributions mentioned above.

2 Strong Stability of a Univariate Classical Risk Model

2.1 Description of the Model

The classical risk process in the one-dimensional situation can be stated as

$$X(t) = u + ct - Z(t), \quad t \geq 0, \tag{1}$$

where $X(t)$ is the surplus of an insurance company at time $t \geq 0$, $u \geq 0$ the initial surplus, c the rate at which the premiums are received, and $Z(t)$ the aggregate of the claims between time 0 and t. $Z(t) = \sum_{i=1}^{N(t)} Z_i$, where $\{Z_i, i \geq 1\}$ is a sequence of iid random variables, representing the claim amounts of distribution function denoted by $F(x)$ and mean claim size denoted by μ, $\{N(t), t \geq 0\}$ being a Poisson process with parameter λ, representing the number of claims. The relative security loading θ is defined by $\theta = \frac{c-\lambda\mu}{\lambda\mu}$. We further assume that $c > \lambda\mu$, the expected payment per unit of time.

Ruin theory for the univariate risk process defined as (1) has been extensively discussed in the literature (for example, see [1] and [11], Chapter 11).

Let us denote the reversed process associated to the risk model by $\{V_n\}_{n\geq 0}$. The strong stability approach consists of identifying the ruin probability $\Psi_a(u)$ associated to the risk model governed by a vector parameter $a = (\lambda, \mu, c)$, with the stationary distribution of the reversed process $\{V_n\}_{n\geq 0}$ [7], i.e.

$$\Psi_a(u) = \lim_{n\to\infty} \mathbb{P}(V_n > u),$$

where u is the initial reserve.

2.2 Strong Stability of a Univariate Classical Risk Model

For a general framework on the strong stability method, the reader is referred to [8]. However, let us recall the following basic definition.

Definition 1. *[8] A Markov chain X with transition kernel P and invariant measure π is said to be v-strongly stable with respect to the norm $\|.\|_v$ ($\|\alpha\|_v = \int_0^\infty v(x)|\alpha|(dx)$, for a measure α), if $\|P\|_v < \infty$ and each stochastic kernel Q in some neighborhood $\{Q : \|Q - P\|_v < \epsilon\}$ has a unique invariant measure $\mu = \mu(Q)$ and $\|\pi - \mu\|_v \to 0$ as $\|Q - P\|_v \to 0$.*

More concrete, following the preceding definition, our approximation problem can be stated in the following way: if the input elements of the ideal and real models are 'close' to each other, then, can we estimate the deviation between the corresponding outputs? In other words, the stability theory in general renders the following picture: If we have as input characteristics the distribution function of the service times (claims distribution function for our risk model) and as output characteristics the stationary distribution of the waiting times (ruin probability for our risk model), the stability means that the convergence in $\mathcal{L}^1$ of the input characteristics implies the weak convergence of the output characteristics.

Let $a' = (\lambda', \mu', c')$ be the vector parameter governing another univariate risk model defined as above, its ruin probability and its reversed process being respectively $\Psi_{a'}(u)$ and $\{V'_n\}_{n\geq 0}$.

The following theorem determines the v-strong stability conditions of a univariate classical risk model. It also gives the estimates of the deviations between both transition operators and both ruin probabilities in the steady state.

Theorem 1. *[7] Consider a univariate classical risk model governed by a vector parameter a. Then, there exists $\varepsilon > 0$ such that the reversed process $\{V_n\}_{n\geq 0}$ (Markov chain) associated to this model is strongly stable with respect to the weight function $v(x) = e^{\epsilon x}$ $(\epsilon > 0)$, $x \in \mathbb{R}^+$.*

In addition, if $\mu(a, a') < (1 - \rho(\epsilon))^2$, then we obtain the margin between the transition operators P and P' of the Markov chains $\{V_n\}_{n\geq 0}$ and $\{V'_n\}_{n\geq 0}$:

$$\|P - P'\|_v \leq 2\mathbb{E}e^{\epsilon Z}|ln\frac{\lambda c'}{\lambda' c}| + \|F - F'\|_v,$$

where,

$$\mu(a, a') = 2\mathbb{E}e^{\epsilon Z}|ln\frac{\lambda c'}{\lambda' c}| + \|F - F'\|_v,$$
$$\rho(\epsilon) = \mathbb{E}(\exp\{\epsilon(Z_1 - c\theta_1)\}),$$
$$\|F - F'\|_v = \int_0^\infty v(u)|d(F - F')|(u) = \int_0^\infty v(u)|f - f'|(u)du.$$

Moreover, we have the deviation between the ruin probabilities:

$$\|\Psi_a - \Psi_{a'}\|_v \leq \frac{\mu(a, a')}{(1 - \rho(\epsilon))((1 - \rho(\epsilon))^2 - \mu(a, a'))} = \Gamma. \quad (2)$$

Remark 1. Without loss of generality, we relax some conditions by taking $\lambda' = \lambda$ and $c' = c$, then we have: $\mu(a, a') = \|F - F'\|_v = \int_0^\infty v(u)|f - f'|(u)du$. The perturbation may concern the mean claim size parameter (i.e. $\mu' = \mu + \varepsilon$) or the claim amounts distribution function F itself.

3 Simulation Based Study

We want to analyze the quality and the sensitivity of the bound defined as in formula (2) of Theorem 1 regarding to certain heavy-tailed distributions. To do so, we elaborated an algorithm which follows the following steps:

3.1 Algorithm

1) Introduce the parameters λ, μ, c of the ideal model, and λ', μ', c' of the perturbed (real) model.
2) Verify the positivity of the relative security loadings θ and θ' defined by: $\theta = \frac{c - \lambda\mu}{\lambda\mu}$ and $\theta' = \frac{c' - \lambda'\mu'}{\lambda'\mu'}$.
If yes, (*the ruin of the models is not sure*) go to step 3;
else return to step 1.
3) Initialize ϵ ($\epsilon > 0$) such that $0 < \rho(\epsilon) < 1$ and Γ be minimal.
4) Compute $\mu(a, a') = \int_0^\infty v(u)|f - f'|(u)du$, and test:
$\mu(a, a') < (1 - \rho(\epsilon))^2$.
If yes, (*we can deduce the strong stability inequality*) go to step 5;
else increment ϵ with step p, then return to step 4.
5) Compute the bound Γ on the deviation $\|\Psi_a - \Psi_{a'}\|_v$ such that:

$$\|\Psi_a - \Psi_{a'}\|_v \leq \frac{\mu(a, a')}{(1 - \rho(\epsilon))((1 - \rho(\epsilon))^2 - \mu(a, a'))} = \Gamma.$$

Using the above algorithm, we perform a comparative study (comparison of the resulting error on ruin probabilities) based on simulation results obtained with the following different distributions.

3.2 Simulated Distributions

In this section, we compare the following four distributions (Lognormal, Weibull, logistic, mixture (Lognormal-Pareto)). In order to well discuss and judge our results, we also use a benchmark distribution the exponential one (see Table 1).

1. The density of the Lognormal law

$$f(t/\alpha, \beta) = \frac{1}{t\beta\sqrt{2\pi}} e^{-\frac{(\log(t) - \alpha)^2}{2\beta^2}}, \quad t \geq 0. \tag{3}$$

2. The density of the Weibull law

$$f(t/\alpha, \beta) = \beta\alpha^{-\beta} t^{\beta - 1} e^{-(\frac{t}{\alpha})^\beta}, \ t \geq 0. \tag{4}$$

3. The density of the truncated logistic law

$$f(t) = \frac{2}{s} e^{\frac{t-\mu}{\sigma}} \left(1 + e^{\frac{t-\mu}{\sigma}}\right)^{-2}, \; t \geq \mu. \tag{5}$$

4. The density of the mixture (p Lognormal and $(1-p)$ Pareto) law

$$f(t) = p \left(\frac{1}{t\sigma\sqrt{2\pi}} e^{-\frac{(\log(t)-\mu)^2}{2\sigma^2}} \right) + (1-p) \left((t-c)^{-(\rho+1)} \rho\lambda^{\rho} \right), \; t \geq 0. \tag{6}$$

5. The density of the exponential law

$$f(t) = \frac{1}{\mu} e^{-t/\mu}, \; t \geq 0. \tag{7}$$

In general, these test distributions can be categorized as light (Weibull), medium (Lognormal) and heavy-tailed (Pareto) [3]. Another classification of heavy-tailed distributions can be found in [10], where the above distributions are defined to depend on their parameters, that is to say, they may be either in the class of heavy-tailed, light-tailed or medium-tailed distributions, and this according to their parameters.

Table 1. Different simulated distributions

Mean	Exp λ	LogNormal (a, b)	Weibull (a, b)	Logistic (μ, s)	Mixture: $p * LogN + (1-p)Pareto$ $(p, a, b, \alpha, \beta, c)$
2.00	2.00	(0.5816 , 0.4724)	(2.2397 , 3)	(1.0000 , 0.7213)	(0.7000 , 0.3051 , 0.4480 , 0 , 3.0000 , 2.1111)
2.10	2.10	(0.6398 , 0.4521)	(2.3517 , 3)	(1.1000 , 0.7213)	(0.7000 , 0.3051 , 0.4480 , 0 , 3.0000 , 2.3333)
2.20	2.20	(0.6945 , 0.4334)	(2.4637 , 3)	(1.2000 , 0.7213)	(0.7000 , 0.3051 , 0.4480 , 0 , 3.0000 , 2.5556)
2.30	2.30	(0.7463 , 0.4161)	(2.5756 , 3)	(1.3000 , 0.7213)	(0.7000 , 0.3051 , 0.4480 , 0 , 3.0000 , 2.7778)
2.40	2.40	(0.7954 , 0.4001)	(2.6876 , 3)	(1.4000 , 0.7213)	(0.7000 , 0.3051 , 0.4480 , 0 , 3.0000 , 3.0000)
2.50	2.50	(0.8421 , 0.3853)	(2.7996 , 3)	(1.5000 , 0.7213)	(0.7000 , 0.3051 , 0.4480 , 0 , 3.0000 , 3.2222)
2.60	2.60	(0.8865 , 0.3714)	(2.9116 , 3)	(1.6000 , 0.7213)	(0.7000 , 0.3051 , 0.4480 , 0 , 3.0000 , 3.4444)
2.70	2.70	(0.9290 , 0.3585)	(3.0236 , 3)	(1.7000 , 0.7213)	(0.7000 , 0.3051 , 0.4480 , 0 , 3.0000 , 3.6667)
2.80	2.80	(0.9696 , 0.3465)	(3.1356 , 3)	(1.8000 , 0.7213)	(0.7000 , 0.3051 , 0.4480 , 0 , 3.0000 , 3.8889)
2.90	2.90	(1.0085 , 0.3352)	(3.2476 , 3)	(1.9000 , 0.7213)	(0.7000 , 0.3051 , 0.4480 , 0 , 3.0000 , 4.1111)
3.00	3.00	(1.0459 , 0.3246)	(3.3595 , 3)	(2.0000 , 0.7213)	(0.7000 , 0.3051 , 0.4480 , 0 , 3.0000 , 4.3333)

3.3 Numerical and Graphical Results

This section is devoted to present the different numerical and graphical results obtained when studying the influence of heavy-tailed distributions on the stability of a risk model, by considering the distributions defined in the section above.

Table 2. Stability intervals regarding to different distributions

ϵ	Mean	Exp	Lognormal	Weibull	Logistic	Mixture
-0.5	2.00	[0.0002 , 0.3083]	]0 ,0.2955]	[0.0005 , 0.2685]	[0.0003 , 0.1933]	[0.0002 , 0.3002]
-0.4	2.10	[0.0002 , 0.3320]	]0 ,0.3726]	[0.0004 , 0.3424]	[0.0002 , 0.2807]	[0.0002 , 0.3238]
-0.3	2.20	[0.0001 , 0.3610]	]0 ,0.4645]	[0.0003 , 0.431]	[0.0002 , 0.3873]	[0.0002 , 0.3543]
-0.2	2.30	[0.0001 , 0.3999]	]0 ,0.5826]	[0.0003 , 0.547]	[0.0002 , 0.5264]	[0.0001 , 0.3965]
-0.1	2.40	[0.0001 , 0.4627]	]0 ,0.7565]	[0.0002 , 0.7306]	[0.0001 , 0.7357]	[0.0001 , 0.4657]
0.00	2.50	] 0 , ∞ [	]0 , ∞ [	] 0 , ∞ [	] 0 , ∞ [	] 0 , ∞ [
+0.1	2.60	[0.0001 , 0.6172]	] 0 , 0.7571]	[0.0002 , 0.7121]	[0.0001 , 0.7166]	[0.0001 , 0.5786]
+0.2	2.70	[0.0001 , 0.5295]	] 0 , 0.5590]	[0.0002 , 0.5261]	[0.0002 , 0.4921]	[0.0001 , 0.4722]
+0.3	2.80	[0.0001 , 0.4772]	] 0 , 0.4330]	[0.0003 , 0.4145]	[0.0002 , 0.3465]	[0.0002 , 0.4066]
+0.4	2.90	[0.0002 , 0.4398]	] 0 , 0.3399]	[0.0004 , 0.3352]	[0.0002 , 0.2397]	[0.0002 , 0.3591]
+0.5	3.00	[0.0002 , 0.4108]	] 0 , 0.2663]	[0.0004 , 0.2744]	[0.0003 , 0.1573]	[0.0002 , 0.3224]

Table 3. Stability bound Γ regarding to different distributions

ϵ	Mean	Exp	Lognormal	Weibull	Logistic	Mixture
-0.5	2.00	0.1954	1.0098	0.9509	2.0178	0.2286
-0.4	2.10	0.1463	0.6713	0.6224	1.1851	0.1823
-0.3	2.20	0.1032	0.4302	0.3943	0.6986	0.1361
-0.2	2.30	0.0649	0.2498	0.2273	0.3819	0.0903
-0.1	2.40	0.0307	0.1105	0.0999	0.1612	0.0449
0.00	2.50	0	0	0	0	0
+0.1	2.60	0.0267	0.1087	0.0957	0.1613	0.0427
+0.2	2.70	0.0534	0.2418	0.2064	0.3819	0.0853
+0.3	2.80	0.0801	0.4068	0.3357	0.6990	0.1277
+0.4	2.90	0.1067	0.6166	0.4883	1.1861	0.1695
+0.5	3.00	0.1333	0.8915	0.6704	2.0216	0.2106

3.4 Discussion of Results

Note, according to Table 2, that for all the distributions, the stability domain decreases with the increase of the perturbation ϵ. It is evident that a risk model tends to not be stable with a great perturbation. Note also the closure of the stability domains of the mixture distribution to those of the exponential one.

Notice also, following Table 3 and Figure 1, that the strong stability bound Γ increases with the increase of the perturbation ϵ. Even taking distributions having the same mean as the exponential one, one obtains bounds relatively far away from those of the exponential one. This can be explained by the influence of the weight of the tails of the different considered distributions. Comparing to the other distributions, we note that the strong stability bound for the mixture distribution is more closer to that of the exponential one. May be it is due to the special choice of the parameters of this distribution. That is to say, one may be able, in this case, to justify the approximation of the risk model with a general mixture claim distribution by another risk model governed by an exponential law.

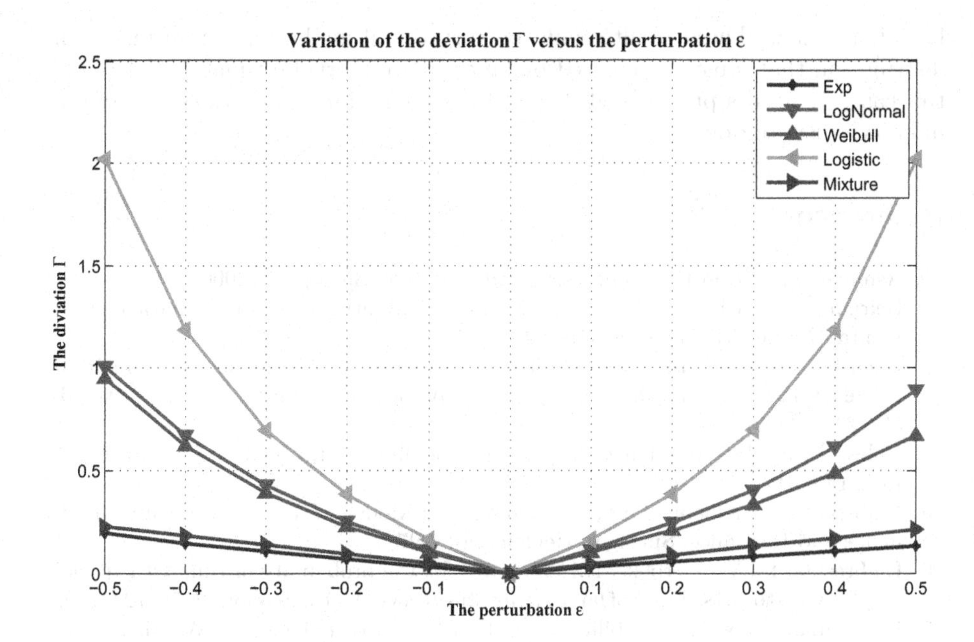

Fig. 1. Variation of the stability bound Γ regarding to different distributions

Note that in the literature, many authors pointed out the limits of the results of Kalashnikov [7] on the stability of risk models and the difficulty of applying them in case of large claims (heavy-tailed distributions). The present results show that in some situations, approximating the characteristics of a risk model with a general heavy-tailed distribution by a classical model is possible, that is to say, one may approach its characteristics by those of a model governed by an exponential distribution (see Tables 2 and 3 and Figure 1). This approximation is in connection not only with the weight of the tail but also with other criteria such as: the shape of the distribution, dispersion parameter, ...

4 Conclusion

We are interested, in this work, in the approximation of the ruin probability of a classical risk model by the strong stability method. We studied the impact of some large claims (heavy-tailed distributions) on the quality of this approximation. A comparative study based on numerical examples and simulation results, involving different heavy-tailed distributions, is performed.

The literature indicates that, in general, the results of Kalashnikov [7] on the stability of risk models, are not applicable for heavy-tailed distributions. The present results show that, in some situations, the approximation of the characteristics of a risk model with a heavy-tailed distribution by a classical model (with an exponential law) is possible. This approximation is linked not only with the weight of the tail but also with other criteria such as the shape of the distribution. These results could be very useful in the case of an unknown distribution that must be replaced by an estimate (kernel estimate). Indeed, in this case, we need a prior knowledge, at least approximately, of the shape of the unknown distribution.

References

1. Asmussen, S.: Ruin Probabilities. World Scientific, Singapore (2000)
2. Beirlant, J., Rachev, S.T.: The problems of stability in insurance mathematics. Insurance Math. Econom. 6, 179–188 (1987)
3. Buch-Larsen, T., Nielsen, J.P., Guillen, M., Bolancé, C.: Kernel density estimation for heavy-tailed distribution using the Champernowne transformation. Statistics 6, 503–518 (2005)
4. Coles, S.: An Introduction to Statistical Modelling of Extreme Values. Springer, Berlin (2001)
5. Embrechts, P., Klueppelberg, C., Mikosch, T.: Modelling Extremal Events for Finance and Insurance. Springer, Heidelberg (1997)
6. Embrechts, P., Veraverbeke, N.: Estimates for the probability of ruin with special emphasis on the possibility of large claims. Insurance: Math. Econom. 1, 55–72 (1982)
7. Kalashnikov, V.V.: The stability concept for stochastic risk models. Working Paper Nr 166, Laboratory of Actuarial Mathematics. University of Copenhagen (2000)
8. Kartashov, N.V.: Strong Stable Markov chains. TbiMC Scientific Publishers, VSPV, Utrecht (1996)
9. Konstantinidis, D.G.: Comparison of ruin probability estimates in the presence of heavy tails. Journal Mathem 93, 552–562 (1999)
10. Konstantinidis, D.G.: Risk models with extremal subexponentiality. Brazilian Journal of Probability and Statistics, Brazilian Statistical Association 21, 63–83 (2007)
11. Panjer, H.H., Willmot, G.E.: Insurance Risk Models. The Society of Actuaries (1992)
12. Tsitsiashvili, G., Konstantinides, D.G.: Supertails in risk theory. Far Eastern Mathem. J. 2, 68–76 (2001)

Part III

Machine Learning, Data Security, and Bioinformatics

An Improvement of Stability Based Method to Clustering

Minh Thuy Ta[1] and Hoai An Le Thi[2]

[1] University of Science and Technology of Hanoi (USTH)
18 Hoang Quoc Viet, Cau Giay, Hanoi, Vietnam
[2] Laboratory of Theoretical and Applied Computer Science (LITA)
UFR MIM, University of Lorraine, Ile du Saulcy, 57045 Metz, France
thuy.taminh@gmail.com,
hoai-an.le-thi@univ-lorraine.fr

Abstract. In recent years, the concept of clustering stability is widely used to determining the number of clusters in a given dataset. This paper proposes an improvement of stability methods based on bootstrap technique. This amelioration is achieved by combining the instability property with an evaluation criterion and using a DCA (Difference Convex Algorithm) based clustering algorithm. DCA is an innovative approach in nonconvex programming, which has been successfully applied to many (smooth or nonsmooth) large-scale nonconvex programs in various domains. Experimental results on both synthetic and real datasets are promising and demonstrate the effectiveness of our approach.

Keywords: Selection the number of clusters, k optimal, Clustering stability, Bootstrap, DC Programming, DCA.

1 Introduction

Clustering is a fundamental problem in unsupervised learning domain, which aims at dividing a dataset n points into k groups (clusters) containing similar data. Determining the optimal number of clusters (k) in a given dataset is a challenging issue in clustering task.

In the literature, there are several papers in this field. The first direction research is to perform clustering with the number of clusters belonging to an interval [*lower bound*, *upper bound*]. Then, one criterion evaluation is used to evaluate the results of clustering. The best value corresponds to the optimal number of clusters. Milligan et al. [25] used hierarchical clustering algorithm and evaluation via 30 criteria. Chiang and Boris [3], [4] introduced an *iK-Means* algorithm and tested via eight indexes. However, this approach only works well when the optimal number of clusters is small.

The second direction is based on genetic algorithm, which introduces new operations (crossover, mutation) or new encoding strategies ([8], [22], [23], [29]). These algorithms are usually center based or centroid based. The disadvantage of this direction is finding solutions in a high space, so the algorithms are slow.

H.A. Le Thi et al. (eds.), *Advanced Computational Methods for Knowledge Engineering*,
Advances in Intelligent Systems and Computing 358, DOI: 10.1007/978-3-319-17996-4_12

One other approach is based on statistical test. In 2000, Pelleg et Moore introduced X-Means algorithm [26]. The G-Means algorithm [6] was presented by Hamerly et Elkan in 2003. Both algorithms start with one cluster ($k = 1$). Then, the number of clusters will grow until the statistic contest is verified. G-Means algorithm assumes that each cluster adheres to a Gauss distribution and X-Means uses the Bayesian Information Criterion (BIC). But, G-Means depends on the choice of statistical significance level; while X-Means tends to overfit by choosing too many centers when the data is not strictly spherical [6]. Other methods such as: gap statistic (Tibshirani et al. [35]), jump statistic (Sugar et al. [30]) can be applied to this problem.

Another approach is based on the concept of clustering stability. The principle of this concept is: if we repeatedly draw samples from the data and apply a given clustering algorithm, a good one should produce clustering that does not vary much from one sample to another [37]. The researches study large methods to compute the stability scores and some different ways to generate the sample data. However, these methods do not work well in case the structure of data is asymmetric [36].

In this paper, we propose an improvement of stability methods, which combines the instability property [5] with CH criterion (Calinski and Harabasz [2]) for evaluating the number optimal of clusters. The numerical results of our experiments show that our propose algorithm works well even if the structure of data is asymmetric.

In addition, we also need a robust and effective clustering algorithm. In our study, we use a clustering algorithm, namely DCA–MSSC, has been shown to be an efficient and inexpensive algorithm ([17], [32], [33], [34]). This algorithm, introduced in 2007 by Le Thi et al. [17], based on DC programming and DCA, which were introduced in 1985 by Pham Dinh Tao and developed since 1994 by Le Thi Hoai An and Pham Dinh Tao. Now, they become classic and increasingly popular (see e.g. ([18], [19], [27], [28]) and the list of references in [11]). In recent years, DCA has been successfully applied to many fields of Applied Science, especially in Data Mining: in clustering ([9], [13], [14], ([15], [16]), in Feature Selection ([12], [20]),...

The remainder of the paper is organized as follows. The Section 2 introduces an algorithm based on bootstrap technique and DC Programming & DCA. The improvement algorithm is developed in Section 3. The computational results with large number of dimensions and large number of clusters are reported in Section 4. Finally, Section 5 concludes the paper.

2 The Bootstrap Technique

In this section, firstly, we will describe briefly an algorithm based on bootstrap technique for clustering. The detail of this algorithm can be found in Fang et Wang [5]. Secondly, we will introduce the DC Programming and DCA.

2.1 An Algorithm Based on Bootstrap for Clustering

The algorithm, namely **ABM** algorithm, is proposed in [5] and it is described as follows:

Step 1: Generate B sample-pairs (X, Y) from n points $\{x_1, x_2, ..., x_n\}$. Each sample data, consisting of n points, is random generated with replacement.

Step 2: Perform clustering on sample-pairs (X, Y) and construct two prototypes (ψ_X, ψ_Y) with fixed k.

Step 3: Affect ψ_X, ψ_Y on the original data, then calculate their empirical clustering distance:

$$d_{\hat{F}}(\psi_X, \psi_Y) = \frac{1}{n^2} \sum_{i=1}^{n} \sum_{j=1}^{n} | I\{\psi_X(x_i) = \psi_X(x_j)\} - I\{\psi_Y(x_i) = \psi_Y(x_j\} | . \quad (1)$$

Estimate the clustering instability by:

$$\hat{s}_B(\psi, k, n) = \frac{1}{B} \sum_{b=1}^{B} d_{\hat{F}}. \quad (2)$$

Step 4: The optimal number of clusters (namely k_{optimal}) can be estimated by (K is the maximum number of clusters):

$$\hat{s}_B(\psi, k_{\text{optimal}}, n) = \underset{2 \leqslant k \leqslant K}{\text{argmin}}\ \hat{s}_B(\psi, k, n). \quad (3)$$

In the literature, k-Means and hierarchical agglomerative clustering are two clustering algorithms widely used. Benhur et al. [1] uses the average-link hierarchical clustering, while Fang et Wang [5] uses k-Means for the distance-based datasets and spectral clustering for the non-distance-based datasets.

In our study, we will use DCA–MSSC. This algorithm has been shown to be more efficient than the classical k-Means ([17], [32], [33], [34]).

2.2 DC Programming and DCA

DC programming and DCA constitute the backbone of smooth/nonsmooth nonconvex programming and global optimization. They address the problem of minimizing a function $f(x) = g(x) - h(x)$ on $\mathbb{R}^d$, where $g(x)$ and $h(x)$ are lower semi-continuous proper convex functions. At each iteration, DCA constructs two sequences $\{x^k\}$ $\{y^k\}$, the candidates for optimal solutions of primal and dual programs.

$$\begin{cases} y^k \in \partial h(x^k) \\ x^{k+1} \in \arg\min\{g(x) - h(x^k) - \langle x - x^k, y^k \rangle\} \end{cases}$$

For a complete study of DC programming and DCA the reader is referred to [10], [19], [27], [28] and the references therein.

DC Algorithm for Solving Clustering Problem
The DCA for clustering problem MSSC (minimum sum of squares clustering) has been developed in [17]. We will give below a description of this algorithm:

Algorithm DCA–MSSC:
Initialization: Let $\epsilon > 0$ be given, $X^{(0)}$ be an initial point in $\mathbb{R}^{k \times d}, p := 0$;
Repeat:
Calculate $Y^{(p)} \in \partial H(X^{(p)})$ by: $Y^{(p)} = qX^{(p)} - B - \sum_{i=1}^{q} e_{j(i)}^{[k]}(X_{j(i)}^{(p)} - a^i)$
Calculate $X^{(p+1)}$ according to: $X^{(p+1)} := (B + Y^{(p)})/n$.
Set $p = p + 1$.
Until: convergence of $\{X^{(p)}\}$.
where $B \in \mathbb{R}^{k \times n}$, $B_\ell := \sum_{i=1}^{q} a^i, \ell = 1..k$; $e_j^{[k]}$ being the canonical basis of $\mathbb{R}^k$.

2.3 Experiments with Bootstrap Techinique and DCA

In this section, we study the efficiency of algorithm **ABM** using DCA–MSSC clustering algorithm in comparison with classical k-Means.

We perform experiments on 10 datasets. Firstly, two artificial datasets Synthetic150 and Synthetic500 (Figure 1) are generated by a Gauss distribution. Synthetic150 (resp. Synthetic500) contains 150 points randomly distributed to 3 separate clusters (resp. 500 points distributed to 5 clusters). Both datasets are symmetric. They are created by Data Generation [38] application. Secondly, four datasets *GD_000* $\rightarrow$ *GD_003* (Figure 2) are generated by Mix Sim [24] software that consist of 500 points and 5 clusters. They are asymmetric data. *GD_000* is well separate and *GD_001* $\rightarrow$ *GD_003* are explored from *GD_000* with others overlapping degree. The other four datasets are real datasets taken from UCI repository [21].

The information of test data is given in Table 1 (column 1 – 4), where n is the number of points, m: the number of dimensions and k: the real number of clusters. All algorithms have been implemented in the VisualC++2008, run on a PC Intel i5CPU650, 3.2 GHz of 4GB RAM.

The experiment results are presented in Table 1. From our results, we see that:

- **ABM** using DCA–MSSC finds correctly the real number of clusters on 5/10 datasets, while **ABM** using classical k-Means only does on 4/10 datasets.
- The structure of *GD_000* $\rightarrow$ *GD_003* and *Iris* datasets is asymmetric, which is bias into 2 groups (see Figure 2, 3). The datasets *GD_001*, *GD_002*, *GD_003* and *Iris* are also overlapping. Both algorithms can not found the real number of clusters in these cases. The results are same as the study of [30] and [5].
- The results are not improved even if we use different parameters: change the number of pair-sampling; sampling datasets are generated by sub sampling; or using Jaccard coefficient (as [1]) to compute the clustering distance. Moreover, the results is worse than using empirical clustering distance $d_{\hat{F}}$ (see Table 2).

Table 1. Datasets and the results of DCA vs. k-Means

				DCA		k-Means	
Data	n	m	k	k_optimal	Time(second)	k_optimal	Time(second)
Synthetic150	150	2	3	**3**	14.88	**3**	7.80
Synthetic500	500	2	5	**5**	26.89	**5**	17.46
GD_000	500	2	5	2	23.47	2	15.87
GD_001	500	2	5	2	28.75	2	17.16
GD_002	500	2	5	2	26.98	2	19.23
GD_003	500	2	5	2	31.39	2	20.60
Iris	150	4	3	2	13.91	2	8.36
Blood Tranfusion	748	4	2	**2**	51.35	4	34.05
Glass	214	9	6	**6**	20.39	**6**	9.99
Wine	178	13	3	**3**	19.35	**3**	9.14

Table 2. Results using criterion $d_{\hat{F}}$ vs. Jaccard coefficient

				Using $d_{\hat{F}}$		Jaccard coeff.	
Data	n	m	k	k_optimal	Time(s)	k_optimal	Time(s)
Synthetic150	150	2	3	**3**	14.88	**3**	13.08
Synthetic500	500	2	5	**5**	26.89	**5**	28.60
GD_000	500	2	5	2	23.47	2	20.69
GD_001	500	2	5	2	28.75	2	27.11
GD_002	500	2	5	2	26.98	2	28.27
GD_003	500	2	5	2	31.39	2	28.53
Iris	150	4	3	2	13.91	2	14.01
Blood Tranfusion	748	4	2	**2**	51.35	**2**	45.02
Glass	214	9	6	**6**	20.39	3	20.38
Wine	178	13	3	**3**	19.35	**3**	19.58

From the above results we see that the bootstrap technique does not work well in some cases whether we attempt to change the parameters. In the next section, we will propose a method that can overcome the errors of bootstrap technique.

3 An Improvement of Stability Based Method

We focus on the formula (3) and observe that there exists some values that is close to the value $\hat{s}_B(\psi, k_{\text{optimal}}, n)$ (see Table 3). Furthermore, Figure 3 of data *GD_002* with respect to $k = 2, 4, 5$ shows that the data can be divided into 2, 4 or 5 clusters, which is acceptable. So, we consider the k_{optimal} in these values.

First, we construct a neighbor values of $\hat{s}_B(\psi, k_{\text{optimal}}, n)$ by the values that are smaller or equal than $\hat{s}_B(\psi, k_{\text{optimal}}, n) + \Omega\%(\hat{s}_B(\psi, k_{\text{optimal}}, n) + 1)$ (same as [31]). The parameter Ω is a small value and the factor *1* takes into account to overcome $\hat{s}_B(\psi, k_{\text{optimal}}, n) = 0$.

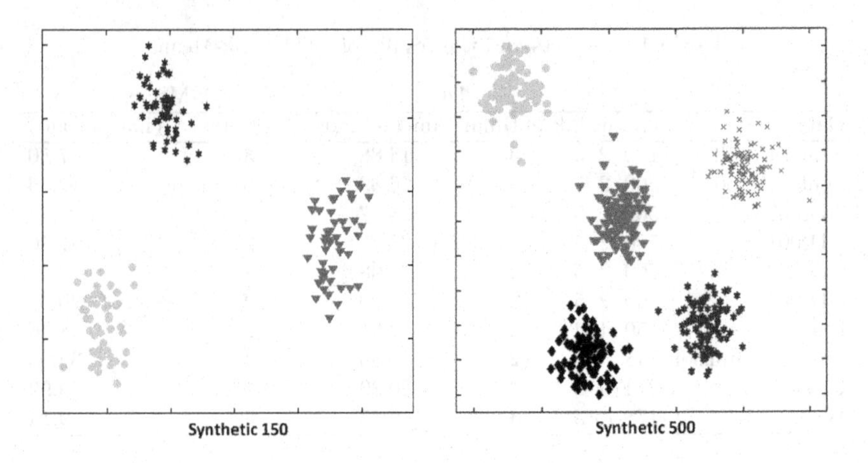

Fig. 1. Datasets Synthetic 150 & Synthetic 500

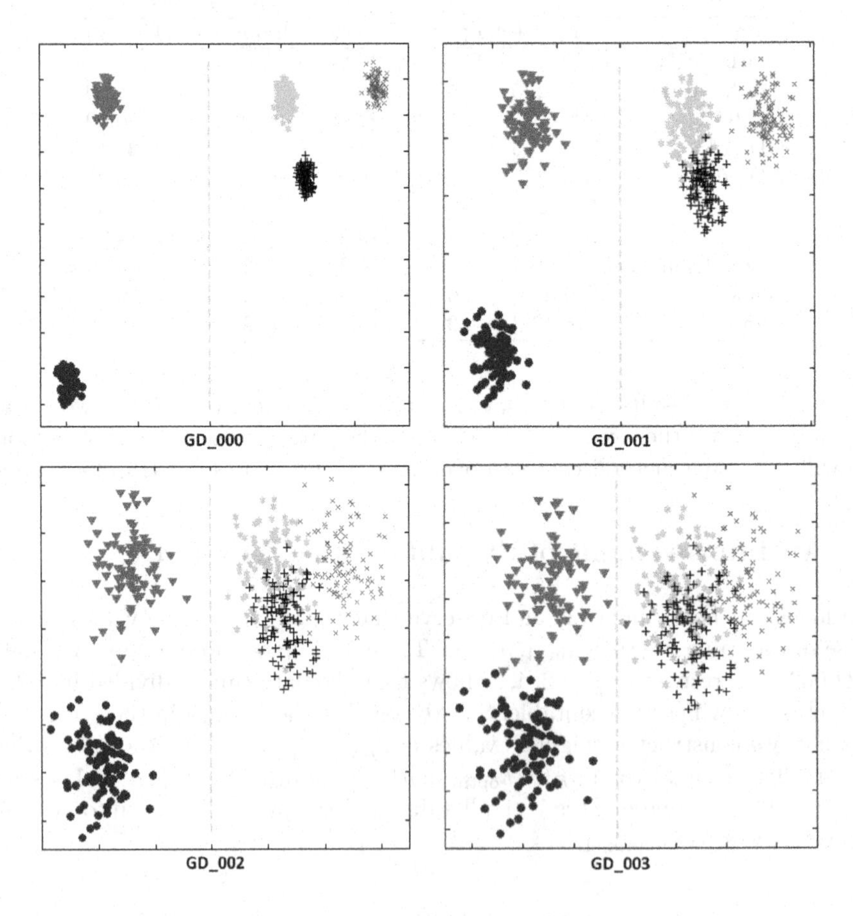

Fig. 2. Datasets $GD\_000 \rightarrow GD\_003$

Then, we calculate the CH(k) index [2] where k are taken from these values. The optimal number of clusters is chosen to correspond to the maximum of CH(k).

Table 3. The results of $\hat{s}_B$

	k							
Data	2	3	4	5	6	7	8	9
GD_000	**0**	0.1299	0.0676	**0.0395**	0.0579	0.0534	0.0646	0.0702
GD_001	**0**	0.0731	0.0750	**0.0527**	0.0640	0.0617	0.0741	0.0772
GD_002	**0**	0.0985	**0.0542**	**0.0597**	0.0810	0.0810	0.0838	0.0919
GD_003	**0.0017**	0.0807	**0.0478**	**0.0744**	0.0838	0.0856	0.0947	0.0984
Iris	**0.0057**	**0.0627**	0.0896	0.1017	0.0918	0.0978	0.0902	0.0827

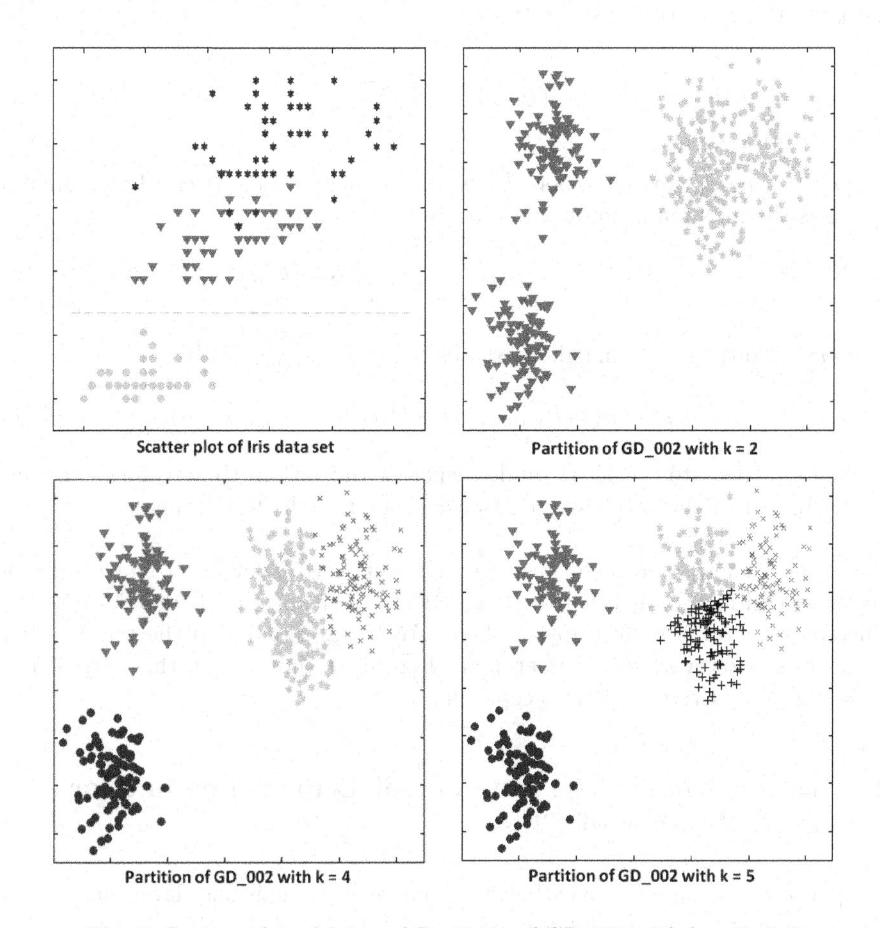

Fig. 3. Scatter plot of *Iris* and Partition of GD_002 with k = 2, 4, 5

Now, the improvement algorithm is described as follows (namely **IBM** algorithm):

The Improvement Algorithm – IBM

Step 1: Generate B sample-pairs (X, Y) from n points $\{x_1, x_2, ..., x_n\}$. Each sample data, consisting of n points, is random generated with replacement.

Step 2: Perform clustering on sample-pairs (X, Y) and construct two prototypes (ψ_X, ψ_Y) with fixed k.

Step 3: Affect ψ_X, ψ_Y on the original data, then calculate their empirical clustering distance:

$$d_{\hat{F}}(\psi_X, \psi_Y) = \frac{1}{n^2} \sum_{i=1}^{n} \sum_{j=1}^{n} \mid I\{\psi_X(x_i) = \psi_X(x_j)\} - I\{\psi_Y(x_i) = \psi_Y(x_j)\} \mid . \quad (4)$$

Estimate the clustering instability by:

$$\hat{s}_B(\psi, k, n) = \frac{1}{B} \sum_{b=1}^{B} d_{\hat{F}}. \quad (5)$$

Step 4: The optimal number of clusters (namely k_{optimal}) can be estimated by (K is the maximum number of clusters):

$$\hat{s}_B(\psi, k_{\text{optimal}}, n) = \underset{2 \leqslant k \leqslant K}{\text{argmin}} \hat{s}_B(\psi, k, n). \quad (6)$$

Step 5: Construct the neighbor values of $\hat{s}_B(\psi, k_{\text{optimal}}, n)$ by:

$$\hat{s}_B(\psi, k, n) \leq \hat{s}_B(\psi, k_{\text{optimal}}, n) + \Omega\%(\hat{s}_B(\psi, k_{\text{optimal}}, n) + 1). \quad (7)$$

Step 6: Calculate the CH(k) index where k takes from these values. The optimal number of clusters is chosen to correspond to the CH(k) maximize.

The improvement algorithm gives good results (see Table (4). The results of **IBM** algorithm are much better than **ABM** algorithm. In *GD_003* dataset, the clusters are too much overlapping, hence **IBM** can not found the real number of clusters. However, the scatter plot (Figure 4) shows that the partition is encouraging and the result is acceptable.

4 Testing with Large Number of Dimensions and/or Large Number of Clusters

We perform our algorithm with large number of dimensions, large number of objects and/or large number of clusters datasets.

The first experiment is tested with large number of objects and/or large number of clusters. The data are acquired from SIPU [40] (as Table 5).

Table 4. Multi-class datasets

		ABM-DCA		IBM-DCA	
Data	True class	*k*_optimal	Time(s)	*k*_optimal	Time(s)
Synthetic150	3	**3**	14.88	**3**	15.87
Synthetic500	5	**5**	26.89	**5**	27.02
GD_000	5	2	23.47	**5**	24.52
GD_001	5	2	28.75	**5**	29.38
GD_002	5	2	26.98	**5**	27.82
GD_003	5	2	31.39	4	33.06
Iris	3	2	13.91	**3**	14.10
Blood Tranfusion	2	**2**	51.35	**2**	53.54
Glass	6	**6**	20.39	**6**	21.13
Wine	3	**3**	19.35	**3**	20.45

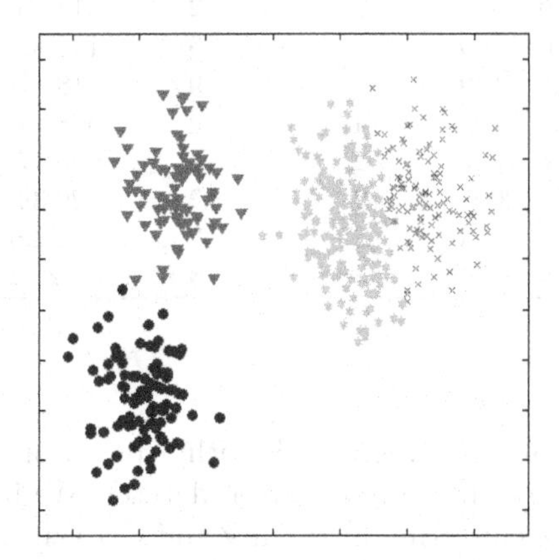

Fig. 4. Partition of GD_003 with k = 4

The second experiment is performed with large datases, which are medical datasets in Bio-medical Repository [7] and Nips Selection Challenge [39] (as Table 6).

In our experiments, we perform our algorithm by varying values of $\Omega = \{1\%, 2\%, 5\%\}$ and maximum number of clusters K is set to 10 (to 20 when datasets contain large number of objects or clusters).

From the results, we see that our approach is encouraging. We also obtain good results in case of high objects, high dimensions and high number of clusters.

Table 5. High objects and/or high clusters datasets

				ABM-DCA		IBM-DCA	
Data	n	m	True class	k_optimal	Time	k_optimal	Time
Dim32	1024	32	16	19	376.67	**16**	389.05
Dim64	1024	64	16	19	518.29	**16**	534.36
R15	600	2	15	19	135.85	**15**	139.41
Yeast	1484	8	10	14	245.31	11	249.09
S1	5000	2	15	19	1220.9	16	1244.44

Table 6. High dimensions datasets.

				ABM-DCA		IBM-DCA	
Data	n	m	True class	k_optimal	Time	k_optimal	Time
Madelon	600	500	2	**2**	514.76	**2**	515.43
Leukemia 2	34	7129	2	9	142.80	**2**	143.84
Leukemia 3	72	7129	3	9	518.17	**3**	531.01
Leukemia 4	72	7129	4	9	568.20	**4**	576.07
Embryonal Tumors	60	7129	2	8	600.47	**2**	610.64
Colon Tumor	62	2000	2	**2**	220.34	**2**	223.57
Arcene	100	10000	2	**2**	1109.96	**2**	1155.95
Mll	72	12582	3	8	1182.72	**3**	1209.33

5 Conclusion

This paper proposes an improvement algorithm based on bootstrap technique that finds the number of clusters in a given dataset. We investigate a new criterion, which combines clustering instability and CH value. We use also an efficient clustering algorithm DCA–MSSC instead of classical k-Means to improve our method. The numerical results on both synthetic and real datasets show the efficiency and the superiority of **IBM** algorithm (our approach) with respect to the standard algorithm **ABM**.

References

1. Ben-Hur, A., Elisseeff, A., Guyon, I.: A Stability Based Method for Discovering Structure in Clustered Data. In: Pacific Symposium on Biocomputing, vol. 7, pp. 6–17 (2002)
2. Calinski, T., Harabasz, J.: A dendrite method for cluster analysis. Communications in Statistics Simulation and Computation 3(1), 1–27 (1974)
3. Chiang, M.M., Mirkin, B.: Experiments for the Number of Clusters in K-Means. In: EPIA Workshops, pp. 395–405 (2007)
4. Chiang, M.M., Mirkin, B.: Intelligent Choice of the Number of Clusters in K-Means Clustering: An Experimental Study with Different Cluster Spreads. Journal Classification 27(1), 3–40 (2010)
5. Fang, Y., Wang, J.: Selection of the Number of Clusters via the Bootstrap Method. Computation Statistics and Data Analysis 56(3), 468–477 (2012)

6. Hamerly, G., Elkan, C.: Learning the K in K-Means. In: Neural Information Processing Systems. MIT Press (2003)
7. Jinyan, L., Huiqing, L.: Kent ridge bio-medical dataset repository (2002)m, http://datam.i2r.a-star.edu.sg/datasets/krbd/index.html (accessed on october 2014)
8. Kudova, P.: Clustering Genetic Algorithm. In: 18th International Workshop on DEXA, Regensburg, Germany (2007)
9. Minh, L.H., Thuy, T.M.: DC programming and DCA for solving Minimum Sum–of–Squares Clustering using weighted dissimilarity measures. Special Issue on Optimization and Machine Learning. Transaction on Computational Collective Intelligent XIII (2014)
10. Le Thi, H.A.: Contribution à l'optimisation non convexe et l'optimisation globale: Théorie, Algoritmes et Applications. HDR, Univesité. Rouen (1997)
11. Le Thi, H.A.: DC Programming and DCA, http://lita.sciences.univ-metz.fr/~lethi
12. Le Thi, H.A., Le Hoai, M., Van Nguyen, V.: A DC Programming approach for Feature Selection in Support Vector Machines learning. Journal of Advances in Data Analysis and Classification 2(3), 259–278 (2008)
13. Le Thi, H.A., Le Hoai, M., Pham Dinh, T.: Fuzzy clustering based on nonconvex optimisation approaches using difference of convex (DC) functions algorithms. Journal of Advances in Data Analysis and Classification 2, 1–20 (2007)
14. Le Thi, H.A., Le Hoai, M.: Optimization based DC programming and DCA for Hierarchical Clustering. European Journal of Operational Research 183, 1067–1085 (2006)
15. Le Thi, H.A., Le Hoai, M., Pham Dinh, T., Van Huynh, N.: Binary classification via spherical separator by DC programming and DCA. Journal of Global Optimization, 1–15 (2012)
16. Le Thi, H.A., Le Hoai, M., Pham Dinh, T., Van Huynh, N.: Block Clustering based on DC programming and DCA. Neural Computation 25(10) (2013)
17. Le Thi, H.A., Tayeb Belghiti, M., Pham Dinh, T.: A new efficient algorithm based on DC programming and DCA for clustering. Journal of Global Optimization 37(4), 593–608 (2007)
18. Le Thi, H.A., Pham Dinh, T.: DC programming: Theory, algorithms and applications. In: The State of the Proceedings of The First International Workshop on Global Constrained Optimization and Constraint Satisfaction (Cocos 2002), Valbonne-Sophia Antipolis, France (October 2002)
19. Le Thi, H.A., Pham Dinh, T.: The DC (Difference of Convex functions) Programming and DCA revisited with DC models of real world nonconvex optimization problems. Annals of Operations Research 46, 23–46 (2005)
20. Le Thi, H.A., Vo Xuan, T., Pham Dinh, T.: Feature Selection for linear SVMs under Uncertain Data: Robust optimization based on Difference of Convex functions Algorithms. Neural Networks 59, 36–50 (2014)
21. Lichman, M.: UCI Machine Learning Repository. University of California, School of Information and Computer Science, Irvine (2013), http://archive.ics.uci.edu/ml (accessed on October 2014)
22. Lu, Y., Lu, S., Fotouhi, F., Deng, Y., Susan, J.B.: Incremental genetic K-means algorithm and its application in gene expression data analysis. BMC Bioinformatics (2004)
23. Maulik, U., Bandyopadhyay, S.: Genetic algorithm-based clustering technique. Pattern Recognition 33(9), 1455–1465 (2000)

24. Melnykov, V., Chen, W.C., Maitra, R.: MixSim: An R Package for Simulating Data to Study Performance of Clustering Algorithms. Journal of Statistical Software 51(12), 1–25 (2012)
25. Milligan, G., Cooper, M.: An examination of procedures for determining the number of clusters in a dataset. Psychometrika 50(2), 159–179 (1985)
26. Pelleg, D., Moore, A.: X-means: Extending K-means with Efficient Estimation of the Number of Clusters. In: Pro. of the 17th International Conference on Machine Learning, pp. 727–734 (2000)
27. Pham Dinh, T., Le Thi, H.: Recent Advances in DC Programming and DCA. Transaction on Computational Collective Intelligence 8342, 1–37 (2014)
28. Pham Dinh, T., Le Thi, H.: Convex analysis approach to DC programming: theory, algorithms and applications. Acta Mathematica Vietnamica 1, 289–355 (1997)
29. Sharma, S., Rai, S.: Genetic K-Means Algorithm Implementation and Analysis. International Journal of Recent Technology and Engineering 1(2), 117–120 (2012)
30. Sugar, C.A., Gareth, J.M.: Finding the number of clusters in a dataset: An information theoretic approach. Journal of the American Statistical Association 33, 750–763 (2003)
31. Ta Minh Thuy: Techniques d'optimisation non convexe basée sur la programmation DC et DCA et méthodes evolutives pour la classification non supervisée. Ph.D thesis, University of Lorraine (2014), `http://docnum.univ-lorraine.fr/public/DDOC_T_2014_0099_TA.pdf` (accessed on January 2015)
32. Thuy, T.M., Le Thi, H.A., Boudjeloud-Assala, L.: An Efficient Clustering Method for Massive Dataset Based on DC Programming and DCA Approach. In: Lee, M., Hirose, A., Hou, Z.-G., Kil, R.M. (eds.) ICONIP 2013, Part II. LNCS, vol. 8227, pp. 538–545. Springer, Heidelberg (2013)
33. Ta, M.T., Le Thi, H.A., Boudjeloud-Assala, L.: Clustering Data Stream by a Subwindow Approach Using DCA. In: Perner, P. (ed.) MLDM 2012. LNCS, vol. 7376, pp. 279–292. Springer, Heidelberg (2012)
34. Thuy, T.M., Le An, T.H., Boudjeloud-Assala, L.: Clustering data streams over sliding windows by DCA. In: Nguyen, N.T., van Do, T., Thi, H.A. (eds.) ICCSAMA 2013. SCI, vol. 479, pp. 65–75. Springer, Heidelberg (2013)
35. Tibshirani, R., Walther, G., Hastie, T.: Estimating the number of clusters in a dataset via the Gap statistic. Journal of Royal Statistical Society, Series B 63, 411–423 (2000)
36. Ulrike von, L.: Clustering Stability: An Overview. Foundations and Trends in Machine Learning 2(3), 235–274 (2009)
37. Wang, J.: Consistent selection of the number of clusters via cross validation. Biometrika 97(4), 893–904 (2010)
38. `http://webdocs.cs.ualberta.ca/~yaling/Cluster/Php/data_gen.php` (accessed on (October 2014)
39. `http://www.nipsfsc.ecs.soton.ac.uk/datasets/` (accessed on October 2014)
40. `http://cs.joensuu.fi/sipu/datasets/` (accessed on October 2014)

A Method for Building a Labeled Named Entity Recognition Corpus Using Ontologies

Ngoc-Trinh Vu[1,2], Van-Hien Tran[1], Thi-Huyen-Trang Doan[1], Hoang-Quynh Le[1], and Mai-Vu Tran[1]

[1] Knowledge Technology Laboratory, University of Engineering and Technology, Vietnam National University Hanoi

[2] Vietnam Petroleum Institute, Vietnam National Oil and Gas Group

trinhvn@vpi.pvn.vn,
{hientv_55,trangdth_55,lhquynh,vutm}@vnu.edu.vn

Abstract. Building a labeled corpus which contains sufficient data and good coverage along with solving the problems of cost, effort and time is a popular research topic in natural language processing. The problem of constructing automatic or semi-automatic training data has become a matter of the research community. For this reason, we consider the problem of building a corpus in phenotype entity recognition problem, class-specific feature detectors from unlabeled data based on over 10260 unique terms (more than 15000 synonyms) describing human phenotypic features in the Human Phenotype Ontology (HPO) and about 9000 unique terms (about 24000 synonyms) of mouse abnormal phenotype descriptions in the Mammalian Phenotype Ontology. This corpus evaluated on three corpora: Khordad corpus, Phenominer 2012 and Phenominer 2013 corpora with Maximum Entropy and Beam Search method. The performance is good for three corpora, with F-scores of 31.71% and 35.77% for Phenominer 2012 corpus and Phenominer 2013 corpus; 78.36% for Khordad corpus.

Keywords: Named entity recognition, Phenotype, Machine learning, Biomedical ontology.

1 Introduction

Phenotype entity recognition is a sub-problem of biomedical information extraction, aiming to identify the phenotype entities. Despite the high performance, the supervised learning methods take a lot of time and efforts from domain experts to build a training corpus. Therefore, construction of a labeled corpus by the automatic method becomes a critical problem in biomedical natural language processing.

In many traditional approaches to machine learning, there are some researches using automatically generated training corpus from external domain ontologies e.g. the approach of [6] or [11]. Morgan et al.'s research built a model organism database to identify and normalize of gene entity based on FlyBase dictionary

H.A. Le Thi et al. (eds.), *Advanced Computational Methods for Knowledge Engineering*, Advances in Intelligent Systems and Computing 358, DOI: 10.1007/978-3-319-17996-4_13

[6]. They collected a large of related to abstracts and used Longest Matching method to annotate for gene entities in the abstracts. Soon after, Vlachos et al. also built by reproducing the experiments of [6] in bootstrapping a BioNER recognizer, it was based on creating training material automatically using existing domain resources and then training a supervised named entity recognition system [11]. Using an enlarged corpus and different toolkit, they applied this technique to the recognition of gene names in articles from the Drosophila literature. More recently, the notion of "silver standard" has also been introduced Rebholz-Schuhmann el al., referring to harmonization of automated system annotations [7].

In our study, we use the available large biomedical data resources to automatically build annotated phenotype entity recognition corpora, then create a new training corpus which is used for machine learning model. Finally, we describe the corpora which will be used to assess the quality of the training corpora based on the quality of machine learning models such as Phenominer 2012, Phenominer 2013, and Khordad's corpus (in section 2.1). Then we demonstrate how we apply Maximum Entropy method with Beam Search algorithm to evaluate performance of our corpus (in section 2.2). Then, (in section 3), we switch our focus to the methods and describe some shortcomings of the BioNER system built. We close the research with discussion of the results and pointers to conclusion (in section 4,5).

2 Phenotype Named Entity Recognition

Unlike genes or anatomic structures, phenotypes and their traits are complex concepts and do not constitute a homogeneous class of objects. Currently, there is no agreed definition of phenotype entity for using in the research community. In [9]'s research: a phenotype entity is defined as a (combination of) bodily features(s) of an organism determined by the interaction of its genetic make-up and environment. Collier et al.s works have described it in more detail: A phenotype entity is a mention of a bodily quality in an organism [1]. Some examples of phenotype entity are blue eyes, lack of kidney, absent ankle reflexes, no abnormality in his heart, etc.

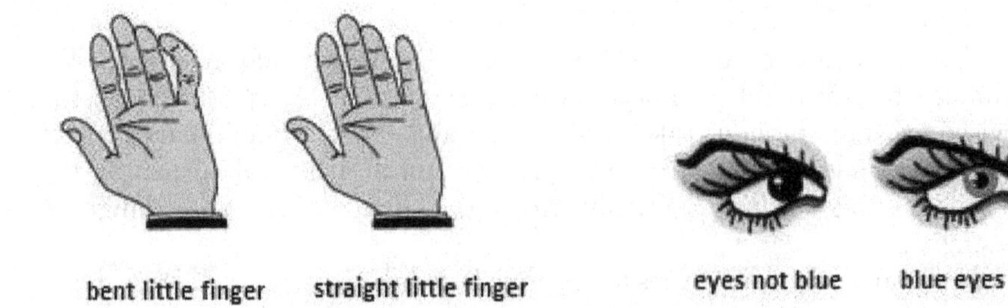

http://naturalsciences.sdsu.edu/ta/classes/lab2.4/TG.html)

The target of this study is to find out the PubMed's abstracts, using the phenotype entity in the available ontologies. To this end, firstly, we describe the ontologies and the databases which support to create the labeled corpus.

2.1 Phenotype Corpora

We aim to empirically build a corpus for phenotype entity recognition under the condition that the test and training data are relatively small and drawn from near domains. To do this, we used three corpora: (1) two Phenominer corpora about autoimmune diseases and cardiovascular disease in [3]'s work, (2) the corpus in [5]'s work, all of them are selected from Medline abstracts in PubMed that were cited by biocuration experts in the canonical database on heritable diseases, the Online Mendelian Inheritance of Man (OMIM) [4].

Phenominer Corpora. The Phenominer corpora contain Phenominer 2012 and Phenominer 2013. Phenominer 2012 corpus is a collection of 112 PubMed Central (PMC) abstracts chosen depending on 19 autoimmune diseases which were selected from OMIM, and from these records, citations were then chosen. These diseases include Type 1 diabetes, Grave's disease, Crohn's disease, autoimmune thyroid disease, multiple sclerosis and inflammatory arthritis. The total number of tokens in the corpus is 26,026 in which there were 472 phenotype entities (about 392 unique terms). Phenominer 2013 corpus includes 80 abstracts of Pubmed Central abstracts relate to cardiovascular diseases, contains 1211 phenotype entities (about 968 unique terms). Despite being small, all of the labeled entities in two corpora were carried out by the same highly experienced biomedical annotator who had annotated in the GENIA and BioNLP shared task corpus and event corpus annotation. The Brat tool supports recognising phenotype entities because of using the normal BIO labeling scheme(Begin In Out), where 'B' stands for the beginning of a concept, 'I' for inside a concept and 'O' for outside any concept, i.e: *between airway responsiveness* will be annotated as O B-PH I-PH, in which 'O' means outside a phenotype entity, 'B-PH' and 'I-PH' beginning of and inside a phenotype entity.

Khordad's Corpus. We use Khordad's corpus as a test corpus which is relevant to phenotypes from two available databases: PubMed (2009) and BioMedCentral (2004). All HPO phenotypes were searched for in these databases and every paper which contains at least three different phenotypes was added to the collection. The corpus is made from 100 papers and contains 2755 sentences with 4233 annotated phenotypes. It does not fully annotate all phenotype names. About 10 percent of the phenotype names are missed. But since we are currently lacking of annotated corpus for phenotype, the corpus is still a valuable choice. We will use this corpus for testing and analyzing our proposed model.

2.2 Maximum Entropy Model with Beam Search

Similar to [2], we also used an appropriate machine learning method called Maximum Entropy model with Beam Search. The use of this method is reasonable because it can train a large number of features and fast convergence. This assessment of the model is to evaluate the difference in possible minimum with the given information, it doesn't concern with the lack of information. Originally, Maximum Entropy model for labeled entity names uses the Viterbi algorithm, a dynamic programming technique to decode. However, recent researches use some approximate search algorithm such as Beam search. The benefit of using Beam Search is that it allows maximum use of entropy for easily labeling each decision but ignores the possibility of optimal label. The calculated complexity of Beam Search decoding is O(kT), compared with $O(N^T)$ for Viterbi decoder (T is the number of words, N is the number of labels). To implement Maximum Entropy with Beam Search, we used Java-based tool OpenNLP (`http://opennlp.apache.org/`) with the default parameters. To train phenotype entity recognition model, we use some features and external resources (dictionaries, ontologies), these are shown in the Table 1 and Table 2.

Table 1. The popular feature sets were used in the machine learning labeler. These were taken from a ±2 window around the focus word for parts of speech, orthography and surface word forms. POS tagging was done using the OpenNLP library with Maximum Entropy model and Genia Corpus + WSJ Corpus (F-score 98.4%), there are 44 Penn Treebank POS tags and all of them are used.

No.	Feature	Description
1	Lemma	The original of the token
2	GENIA POS tagger	Part of speech tag of the token
3	GENIA Chunk tagger	Phrase tag (the number of the token is larger than 1) such as noun phrase, phrasal verb,
4	GENIA named entity tagger	Output of the analysis of sentences in GENIA tagger..
5	Orthographic tag	Orthography of the token
6	Domain prefix	Prefix of the token
7	Domain suffix	Suffix of the token
8	Word length	Length of the word
9	In/Out parentheses	In parentheses will be tagged: Y, out parentheses will be tagged: N
10	Dictionary	Dictionary features

Table 2. Some external resources: dictionaries, biomedical ontologies and datasets

No.	Feature	Description
1	HPO	An ontology contains terms describing human phenotypic features
2	MP	An ontology has been applied to mouse phenotype descriptions. This ontology allows comparisons of data from diverse sources, can facilitate comparisons across mammalian species, assists in identifying appropriate experimental disease models, and aids in the discovery of candidate disease genes and molecular signaling pathways
3	PATO	An ontology of phenotypic qualities. This ontology can be used in conjunction with other ontologies such as GO or anatomical ontologies to refer to phenotypes
4	FMA	A domain ontology that represents a coherent body of explicit declarative knowledge about human anatomy. Its ontological framework can be applied and extended to all other species
5	MA	The mouse anatomy ontology was developed to provide standardized nomenclature for anatomical structures in the postnatal mouse
6	UMLS_DISEASE	The concepts of disease in UMLS
7	45CLUSTERS	45 cluster classes were derived by Richard Socher and Christopher Manning from PubMed
8	UMLS	A set of files and software that brings together many health and biomedical vocabularies and standards. It has three tools: Metathesaurus, semantic network and SPECIALIST Lexicon and Lexical Tools.

3 Building Annotated Corpora

3.1 Phenotype Knowledge Resources

Human Phenotype Ontology. Human Phenotype Ontology (HPO) aims to provide a standardized vocabulary of phenotypic abnormalities encountered in human diseases [8]. Terms in HPO describe a phenotypic abnormality, such as atrial septal defect. HPO was initially developed by using information from Online Mendelian Inheritance in Man (OMIM), which is a hugely important data resource in the field of human genetics and beyond. HPO is currently being developed using information from OMIM and the medical literature, contains approximately 10,000 terms. Over 50,000 annotations to hereditary diseases are available for download or can be browsed using the PhenExplorer. The HPO project encourages input from the medical and genetics community with regards to the ontology itself and to clinical annotations.

Mammalian Phenotype Ontology. Similarly to HPO, the Mammalian Phenotype Ontology (MP) is a standardized structured vocabulary [10]. The highest level terms describe physiological systems, survival, and behavior. The physiological systems branch into morphological and physiological phenotype terms at the next node level. This ontology helps to classify and organize phenotypic information related to the mouse and other mammalian species, MP ontology applied to mouse phenotype descriptions in the *Mouse Genome Informatics Database* (MGI, `http://www.informatics.jax.org`) and *Rat Genome Database* (RGD, `http://rgd.mcw.edu`), Online Mendelian Inheritance in Animals (OMIA, `http://omia.angis.org.au`).

MP has about 8800 unique terms (about 23700 synonyms) of mouse abnormal phenotype descriptions, it is maintained by OBO-Edit software to add new terms, synonyms and relationships.

3.2 Building Process

Firstly, we carried out to build a training corpus which identifies phenotype entities in humans. By combining the two relationships (the relationship between terms in HPO and documents from OMIM database extracted from the file *Phenotype_annotation.tab* and the relationship between each document of OMIM database and referenced Pubmed abstracts), we assembled relationships between each Pubmed abstract related phenotype entities in humans and HPO terms. Collecting all summaries in the above relationship list, depending on each abstract referenced to a separate list of HPO terms from the relationship file, we used a method named "*Noun Chunking*" to label the phenotype entities in each abstract. The *Noun Chunking* method found all nouns and noun phrases in each Pubmed abstract and matched them with the separate list which referenced some certain HPO phenotype terms to label. Finally, we obtained the corpus HPO_NC by this method.

We also built a training corpus which identifies phenotype entities in mammals. Firstly, we collected relationship between each Pubmed abstract related to terms in MP ontology from two statistics files: *MGI_GenoPheno.rpt* and *MGI_PhenoGenoMP.rpt*. Assembling Pubmed abstracts in the above relationship list, depending on each abstract referenced to a separate list of MP terms, we also used *Noun Chunking* to label phenotype entities in mammals for Pubmed abstracts. A training corpus MP_NC was created as a result of the above process.

At the next step, we joined the two sets HPO_NC and MP_NC to obtain the HPO_MP_NC set with large coverage of phenotype entities domain.

Table 3. Corpora statistics

	HPO_NC	MP_NC	HPO_MP_NC
Abstracts	18.021	4.035	22.056
Tokens	3.387.015	988.598	4.375.613
Phenotype entities	39.454	6.833	46.287
Unique phenotype entities	3.579	1.169	4.371

3.3 Error Analysis

The training corpora which were automatically generated still contain some errors, especially "**Missing case**" and "**Error case**", which appear in *Noun Chunking* method. For example, although the phrase noun "*Amyotrophic lateral sclerosis*" in the abstract ID: 9933298 was abbreviated as "*ALS*", some contexts appeared as "*ALS*" were still not recognized as a phenotype entity. Another example is that in the Pubmed abstract ID: 34999, the noun phrase "*hyperparathyroidism*" is a phenotype entity, but in other contexts, this concept had not been found.

Last example with "**Error case**", the noun phrase "*Severe combined immunodeficiency disease*" and "*Severe combined immunodeficiency*" from the Pubmed abstract ID: 18618 were identified as phenotype entities. However, in fact, each of them is a type of disease.

4 Result and Discussion

We have evaluated the effectiveness of automatically generated corpus using machine learning method (ME+BS) with 17 type features on three standard training corpora: Phenominer 2012, Phenominer 2013 and Khordad corpus. We also show the Table 4 as a result of the evaluation of the automatically generated training corpora on Phenominer 2012 and Phenominer 2013 and Khordad corpus.

Table 4. Evaluation results

Testing data	**Phenominer 2012**			**Phenominer 2013**			**Khordad corpus**		
Training data	P	R	F	P	R	F	P	R	F
HPO_NC	55.37	20.28	29.69	59.82	25.08	**35.34**	89.57	68.21	77.44
MP_NC	40.08	17.44	24.3	42.64	20.78	27.94	83.24	61.09	70.47
HPO_MP_NC	55.69	22.17	**31.71**	58.47	23.97	34	88.12	70.54	**78.36**

Through some experiments evaluating the effectiveness of the automatically generated corpora, the best F-score measures at **31.71%** in Phenominer 2012, **35.34%** in Phenominer 2013 and **78.36%** in Khordad's corpus. The results are not high due to some errors in the above corpora as well as the intersection of the domain of the automatically generated training corpora and the three evaluation corpora. However, a more important reason is the complexity of grammar in the two standard training corpora labeled by experts is higher than in the generated training corpora. We evaluated the average number of tokens per each phenotype entity over all the corpora in the Table 5.

From Table 5, we can see that the average number of tokens for each phenotype entity in Phenominer 2012 and Phenominer 2013 is approximately 3 token/entity whereas the number is 1.7 token/entity in the automatically generated training corpora. This issue affects the ability of identification in the sequence labeling model. It is a challenge for models using machine learning methods.

Table 5. The average number of tokens per phenotype entity over all the corpora

Corpora	The average number of tokens / phenotype entity
HPO_NC	1.710
MP_NC	1.778
HPO_MP_NC	1.761
Khordads corpus	1.688
Phenominer 2012	2.911
Phenominer 2013	3.204

The automatically generated training corpora achieved better results than on Khordad's corpus. The reason is the intersection between the domain of the automatically generated training corpora and the Khordad's corpus is quite large as well as the complexity of grammar in the Khordad's corpus is not too high. Table 4 shows that for Khordad's corpus F-score reached the best result at **78.36%** in HPO_MP_NC corpus, which is higher than in HPO_NC (F-score: 77.44%) and MP_NC (F-score: 70.47%). Therefore, the HPO_MP_NC corpus shows its wider coverage to help to increase the effectiveness of automatically generated training corpora.

5 Conclusion

In this work, we have presented a systematic research of how to build an automatic training corpus for phenotype entity recognition from various ontological resources and methods. We believe that it is the first study to evaluate such a rich set of features for the complex class of phenotypes. The corpus is evaluated using the recognition phenotype entity model called Maximum Entropy method with Beam Search algorithm. By this approach, we achieved the best micro-averaged F-score about **31.71%** on Phenominer 2012; **35.34%** on Phenominer 2013 and **78.36%** on Khordad's corpus.

In summary, our experiment brings overview of the effectiveness of the corpora generated by the automatic methods. Beside, labeled phenotype entity recognition corpus is important for the analysis of the molecular mechanism underlying diseases, and is also expected to play a key role in inferring gene function in complex heritable diseases. Therefore, in the near future, the collection of this corpus can be a useful resource for gene and disease domain. Our work in this direction will be reported in a future publication.

Acknowledgments. The authors gratefully acknowledge the many helpful comments from the anonymous reviewers of this paper.

References

1. Collier, N., Tran, M.-V., Le, H.-Q., Oellrich, A., Kawazoe, A., Hall-May, M., Rebholz-Schuhmann, D.: A hybrid approach to finding phenotype candidates in genetic texts. In: COLING, pp. 647–662 (2012)

2. Collier, N., Tran, M.-V., Le, H.-Q., Ha, Q.-T., Oellrich, A., Rebholz-Schuhmann, D.: Learning to recognize phenotype candidates in the auto-immune literature using svm re-ranking. PloS One 8(10), e72965 (2013)
3. Collier, N., Paster, F., Tran, M.-V.: The impact of near domain transfer on biomedical named entity recognition. In: Proceedings of the 5th International Workshop on Health Text Mining and Information Analysis (Louhi)@ EACL, pp. 11–20 (2014)
4. Hamosh, A., Scott, A.F., Amberger, J.S., Bocchini, C.A., McKusick, V.A.: Online mendelian inheritance in man (omim), a knowledgebase of human genes and genetic disorders. Nucleic Acids Research 33(suppl. 1), D514–D517 (2005)
5. Khordad, M., Mercer, R.E., Rogan, P.: Improving phenotype name recognition. In: Butz, C., Lingras, P. (eds.) Canadian AI 2011. LNCS, vol. 6657, pp. 246–257. Springer, Heidelberg (2011)
6. Morgan, A.A., Hirschman, L., Colosimo, M., Yeh, A.S., Colombe, J.B.: Gene name identification and normalization using a model organism database. Journal of Biomedical Informatics 37(6), 396–410 (2004)
7. Rebholz-Schuhmann, D., Yepes, A.J.J., Van Mulligen, E.M., Kang, N., Kors, J., Milward, D., Corbett, P., Buyko, E., Beisswanger, E., Hahn, U.: Calbc silver standard corpus. Journal of Bioinformatics and Computational Biology 8(01), 163–179 (2010)
8. Robinson, P.N., Köhler, S., Bauer, S., Seelow, D., Horn, D., Mundlos, S.: The human phenotype ontology: a tool for annotating and analyzing human hereditary disease. The American Journal of Human Genetics 83(5), 610–615 (2008)
9. Scheuermann, R.H., Ceusters, W., Smith, B.: Toward an ontological treatment of disease and diagnosis. Summit on Translational Bioinformatics 2009, 116 (2009)
10. Smith, C.L., Goldsmith, C.-A.W., Eppig, J.T.: The mammalian phenotype ontology as a tool for annotating, analyzing and comparing phenotypic information. Genome Biology 6(1), R7 (2004)
11. Vlachos, A.: Semi-supervised learning for biomedical information extraction. University of Cambridge, Computer Laboratory, Technical Report, UCAM-CL-TR-791 (2010)

[illegible] (polymorphisms) [illegible] using [illegible]
[illegible] numerical [illegible] Workshop on [illegible] (2014)
[illegible] with deep [illegible] J. Dent. Res. [illegible] (2013)
[illegible]
[illegible] and [illegible] (2007)
[illegible]
[illegible] (2008)
[illegible]
[illegible] (2012)
[illegible]
[illegible]

A New Method of Virus Detection Based on Maximum Entropy Model

Nhu Tuan Nguyen, Van Huong Pham, Ba Cuong Le, Duc Thuan Le, and Thi Hong Van Le

Academy of Cryptography Techniques of Government, Vietnam
nguyennhutuan@bcy.gov.vn,
{huongpv,cuonglb304,leducthuan255,hongvan.lt86}@gmail.com

Abstract. The paper presents a new method for detecting virus based on the Maximum Entropy Model. This method is also used to detect unknown viruses. Maximum Entropy is a machine learning method based on probability distribution and has been successfully applied in the classification problem. From the background knowledge, we improve and apply this model to solve the problem of virus detection. In the training phase, virus samples in the virus warehouse are extracted their features and trained to create the Entropy Model. In the detection process, the Entropy Model is used to recognize virus based on the corresponding features of a checked file.

Keywords: Virus detection, virus samples, virus database, Maximum Entropy Model, feature set, data set.

1 Introduction

The explosion in the use of computer and the internet has brought about a new avenue for computer virus to spread at an exponential rate. In 1992, the number of viruses was estimated from 1000 to 2300; 60,000 viruses in 2002; and approximately 100,000 viruses in 2008 [16]. This rapid change calls for an urgent improvement in virus detection techniques. Traditional methods often rely on virus signature recognition and sample comparison. The virus signatures, which are extracted from infected files, are stored in virus databases [3,4], [6,7]. In [16], the authors summarized some of the traditional virus detection methods, for example the string method, the wildcard method, the wrong pattern matching method, the skeleton detection method. In general, these traditional methods have some limitations:

- Large virus database
- Large searching space
- Unable to detect a virus that does not yet exist in the database.

H.A. Le Thi et al. (eds.), *Advanced Computational Methods for Knowledge Engineering*,
Advances in Intelligent Systems and Computing 358, DOI: 10.1007/978-3-319-17996-4_14

Machine learning method based on Maximum Entropy Model has been widely used in classification problems, especially in natural language processing [11,12] [14]. According to the study [13], the authors successfully applied this model in character recognition problems. In this paper, we propose a new method of virus detection based on Maximum Entropy Model. The rest of this paper is organized as follows. First, we represent the overall model of virus detection method based on Maximum Entropy Model in Section 2. Section 3 presents the detailed implementation steps. In Section 4, we show the experiment and evaluation. Finally, we show the conclusion and further works in Section 5.

2 Model of Virus Detection Systems

The virus detection problem has been studied and implemented for a long time ago. There are two approaches to the problem of detecting virus: the traditional approach and the approach based on machine learning. In the former, anti-virus systems are usually installed in the model shown in Fig. 1. These systems consist of two phases such as signature extraction of virus and virus detection. In extracting phase, the system analyzes the structure as well as the code of the virus pattern file to extract the signatures corresponding to each type of virus. Each virus has a set of signatures stored in the virus database [3], [7]. In other words, updating virus definition files is amending this database. In the other phase, the system extracts the signatures of a selected file, compares them with the signature set of each virus type in the virus database in order to draw conclusions and come up with final solutions.

As much as the machine learning approach is recent and prospective, it is hardly studied. The case studies following this approach is primarily based on neural networks [1,2] [5]. Based on promising results in [1], we proposed and implemented a new method for detecting viruses using Maximum Entropy Model. According to the overall model presented in Fig. 2, virus detection system is carried out in two phases: training and testing. During the training phase, a feature set of viruses extracted from a virus sample repository are stored and inputted into training module. The output of training phase is the best distribution model which is used in the testing phase. In this phase, each file which are need to test will be extracted a set of features. The training model and this set are transferred to detection module to find the probability corresponding to each output class. The file belongs to virus class, which has the highest probability, will be transferred to virus processing module.

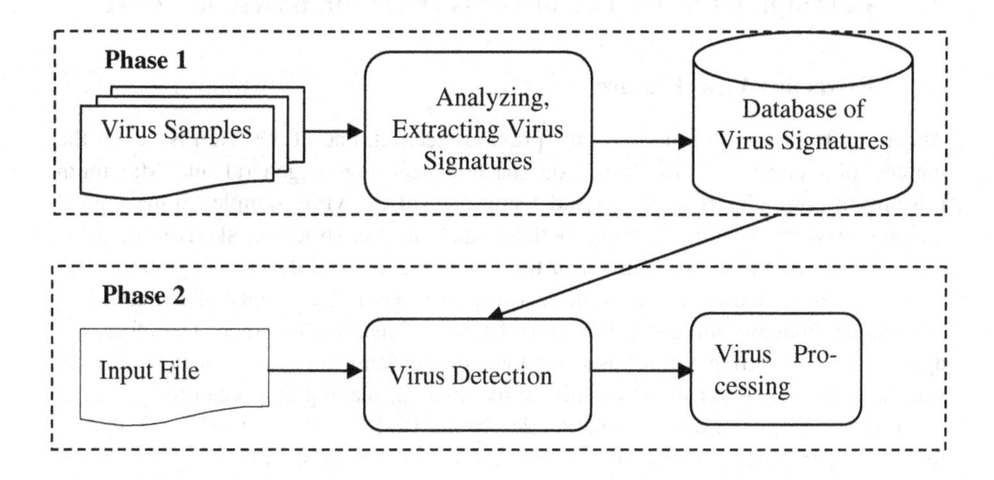

Fig. 1. The overall model of traditional virus detection method

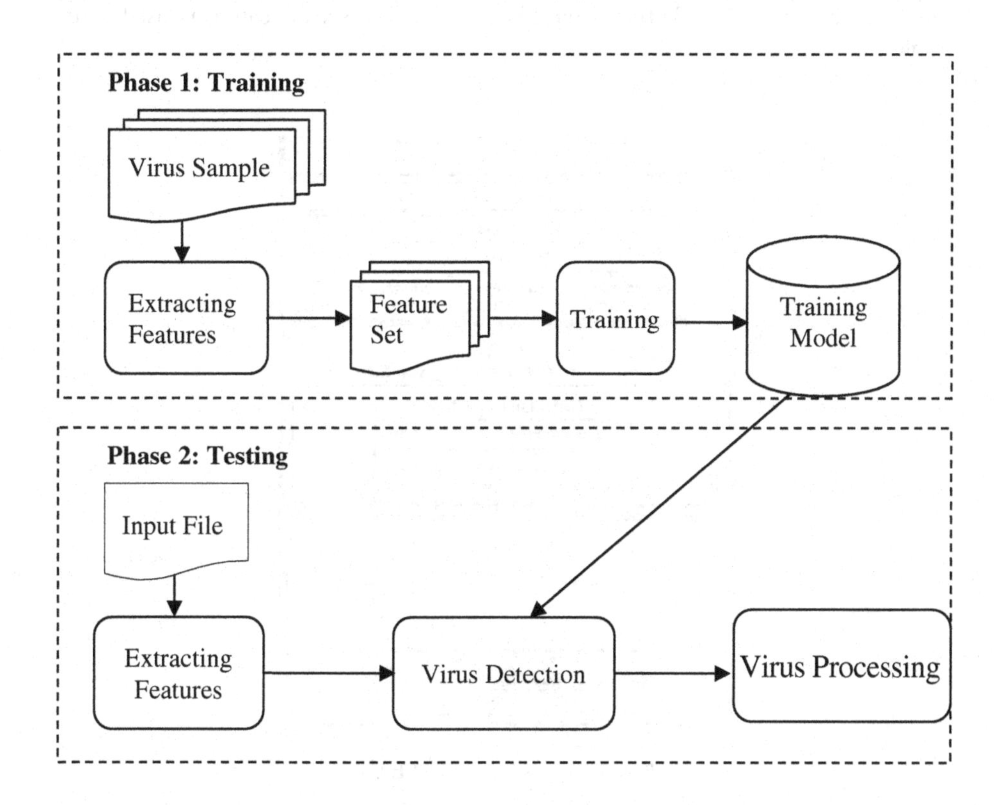

Fig. 2. The overall model of virus detection based on Maximum Entropy Model

3 Developing the Method of Virus Detection Based on MEM

3.1 Extracting Virus Feature

According to the overall model in the previous section, one of the main issues of the method of identifying virus based on machine learning in general and Maximum Entropy in particular is how to build feature set of the virus sample. Some typical features used in machine learning methods such as: file structure, skeleton of code, popular code fragment, the number of the function calls, and the set of symbols [16]. The main advantage of the machine learning methods is the variety of features in a knowledge database and a combination of types of features together. Therefore, the approach for detecting virus based on machine learning methods can be inherited, selected virus signatures in the traditional methods for making these features.

In this research, we use information describing PE file and a method for analyzing its structure in order to extract the feature set from the virus sample. Thus, the experiments in this research are only applied for the PE file of the windows operating system. We generate the feature set of each virus by combining the features of the PE structure in the research [1] and hash code of machine code in that file. The PE structure is described in Fig. 3. Based on this structure, the set of feature is used as in Table 1.

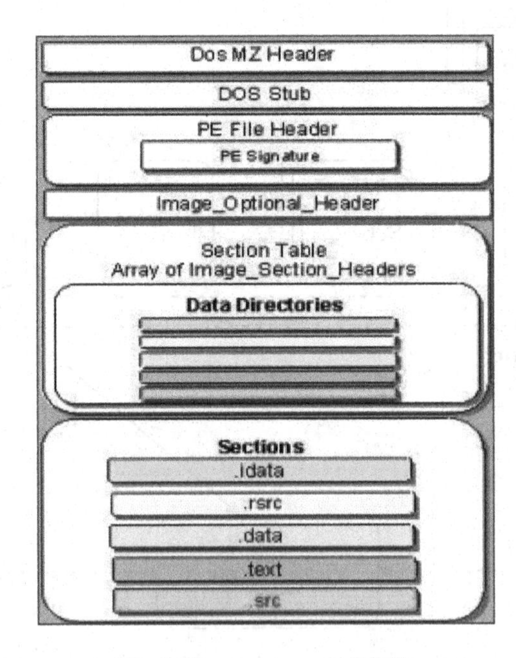

Fig. 3. The structure of a PE file

Table 1. The feature is based on the analysis of the structure

Order	Feature	Size (byte)	Address offset (PE 32 / PE 64)
1	MajorLinkerVersion	1	2 / 2
2	MinorLinkerVersion	1	3 / 3
3	SizeOfInitializedData	4	8 / 8
4	SizeOfUninitializedData	4	12 / 12
5	MajorOSVersion	2	40 / 40
6	MinorOSVersion	2	40 / 42
7	MajorImageVersion	2	44 / 44
8	MinorImageVersion	2	46 / 46
9	Checksum	4	64 / 64
10	DLLCharacteristics	2	70 / 70
11	HashCodeSegment		

3.2 Applying Maximum Entropy Model for Detecting Virus

In this section, we will construct a model for problem detection virus based on Maximum Entropy Model. Maximum Entropy Model, which is the probability distribution model, has been successfully applied for the problem of data classification, especially in natural language processing. A core principle of maximum entropy based on distribution model of each data set and a set of constraints that has to achieve the best possible balance [12]. The training data set used to generate the constraints of the model, which are basic factorial for estimating the probability distribution of each class. These constraints are represented by values of the features. From that, we calculate on the model to get a distribution for the maximum Entropy.

A specialty of the Maximum Entropy Model is shown by a function $f_i(d, c)$, in which d is observation data and c is a class. Maximum entropy allows limiting distribution model in order to obtain expected values for each specialty of data set. Therefore, the distribution probability of the data d for class c is $P(c|d)$ satisfying the Equation (1). During training phase, the distribution probability $P(d)$ is unknown and no need to concern, so we just use the sample data as a condition for distributing data according to Equation (2) .

$$\frac{1}{|D|}\sum_{d\in D} f_i(d,c(d)) = \sum_{d} P(d) \sum_{c} P(c|d) f_i(d,c) \tag{1}$$

$$\frac{1}{|D|}\sum_{d\in D} f_i(d,c(d)) = \frac{1}{|D|}\sum_{d}\sum_{c} P(c|d) f_i(d,c) \tag{2}$$

Maximum Entropy Model provides a simple mechanism to combine features in different contexts in order to estimate the probability of some classes appearing in these contexts. The main idea of the maximum Entropy method is to find a model with probability distribution satisfying all constraints between observation data and a set of classes. According to the principle of maximum entropy, the distribution has to satisfy the observation data and maximize the conditional entropy measure. Maximum Entropy Model is the probability distribution as in Equations (3) and (4) where $P*$ is the optimal probability distribution.

$$H(p) \equiv -\sum_{c,d} \tilde{p}(c)p(c|d)\log(p(d|c)) \tag{3}$$

$$P^* = \underset{p \in C}{\operatorname{argmax}}(H(p)) \tag{4}$$

In Maximum Entropy Model, each specialty is shown by a function with only one input value by true or false. The constraint set will be established from these specialties. A constraint is a condition from the data that the model must satisfy. Each specialty f_i is assigned with a weight λ_i. Meanwhile, the problem of classification is given to the problem of estimating the conditional probability. By applying Lagrange, we can prove that probability fraction shall be in Equation (5) and (6) [12,13].

$$p(c|\ d) = \frac{1}{Z(d)} \times e^{\sum_i \lambda_i f_i(d,c)} \tag{5}$$

$$Z(d) = \sum_{c} e^{\sum_i \lambda_i f_i(d,c)} \tag{6}$$

where, $Z(d)$ is the expression normalized to ensure conditions $\Sigma p\ (c \mid d) = 1$ and λ_i is the weight corresponded with the ith specialty.

According to the Equations (5) and (6), in order to calculate the probability $p(c|d)$, we need to determine the weight λ_i. Each Entropy model has an important set of weights $\lambda = \{\lambda_i \mid 1 \le i \le |D|\}$. Maximum Entropy Model is the model with the best set of weights. The objective of the training phase is to find the best set of weights based on the set of classes and specialties. The objective of the test period is the distribution probability of the object to be checked with the set of classes based on the set of weights created during the training phase. From this distribution probability, we find the class for the object with the largest probability.

Based on initial Maximum Entropy Model, we modified for applying to the problem of virus detection. In essence, the problem of detecting virus is a type of classification problem. We describe this problem as a mapping ε as in Equation (7).

$$\varepsilon = F \rightarrow V \tag{7}$$

where,

- F is a set of files to be checked
- $V = \{v_i \mid i = 0..N\}$ is a set of labels: v_0 corresponding with files without virus, v_i with $i = 1..N$ is classes, corresponding with each type of virus.

Let $S = \{s \mid s$ is a specialty$\}$ is a set of specialties selected from the sample virus. For each pair (s, v), constraints f_i (s, v) is defined as in Equation (8).

$$f_i(s,v) = \begin{cases} 1 \text{ if } s \text{ is an attribute of } v \\ 0 \text{ if } s \text{ is not an attribute of } v \end{cases} \quad (8)$$

3.3 Training Phase

During the training phase, we select features s of each virus file from the virus samples database. After that, we need to calculate a set of weights λ based on the set of features S and the set of classes V. Recently, there are a number of algorithms, which are commonly used to figure out λ such as GIS, IIS, L-BFGS [12]. In this study, we describe the data structure and implement algorithm GIS as follows:

1. Calculate $E(1)i = \sum_{s \in S,\, v \in V} p(v|s) \times f_i(s,v)$
2. Initiate $\lambda_i^{(1)}$ with $i = 1..|S| \times |V|$ a certain value
3. Repeat until convergence or exceeding standard allowed

For each i

3.1 Calculate $E^{(n)} = \sum_{s \in S,\, v \in V} p^{(n)}(v \mid s) f_i(s, v)$

Where, $p^{(n)}(v \mid s) = \dfrac{e^{\sum_{s \in S,\, v \in V} \lambda_i f_i(s,\, v)}}{Z(S)}$

3.2 Update $\lambda_i^{(n+1)} = \lambda_i^{(n)} + \frac{1}{C}(\log \frac{E_i^{(1)}}{E_i^{(n)}})$ where, $C = \sum_{s \in S,\, v \in V} f_i(s,v)$

3.4 Phase of Detection Virus

In order to determine whether a file is a virus or not, we use the set of weights λ in training phase. That file should be selected a set of features. From this set with the weights λ, the probability distribution is calculated using the Equations (5) and (6). Based on this probability distribution, the file is assigned to the class with the largest probability. This class can be a specific virus or not.

4 Experiment

4.1 Program and Experimental Data

In this experiment, we develop the experimental program based on *ShapEntropy* library with Visual Studio.Net 2008 and C# language. Our program includes three major modules such as feature extraction, training and detection. In the feature extraction module, we implement the extraction technique mentioned in the previous section. Fig. 4 shows interface of feature extraction module and Fig. 6 illustrates a part of virus feature. Interface of training module and detection module is shows in Fig. 5 and Fig. 7. In order to do the experiment, we use three data sets described in Table 2 and Table 3. We use virus samples in [15] to build our data sets. The data set is built under *10-fold* cross validation.

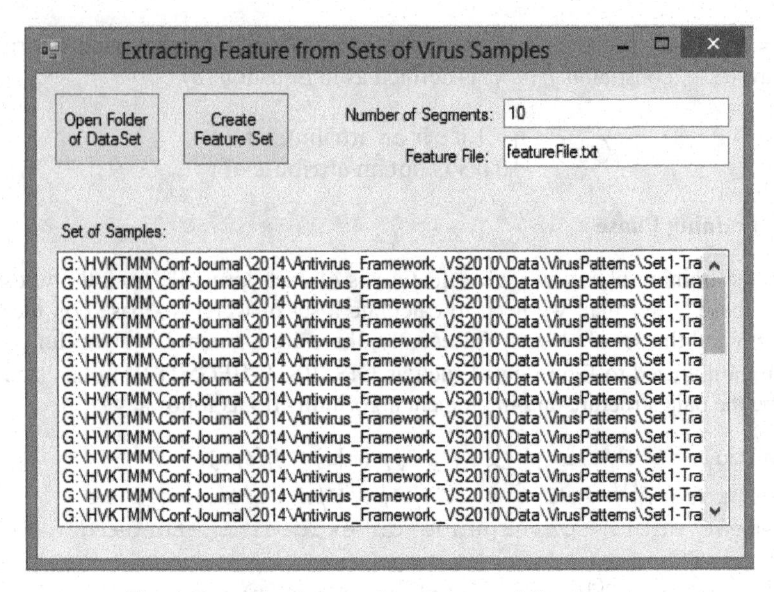

Fig. 4. Extracting features from the sets of virus samples

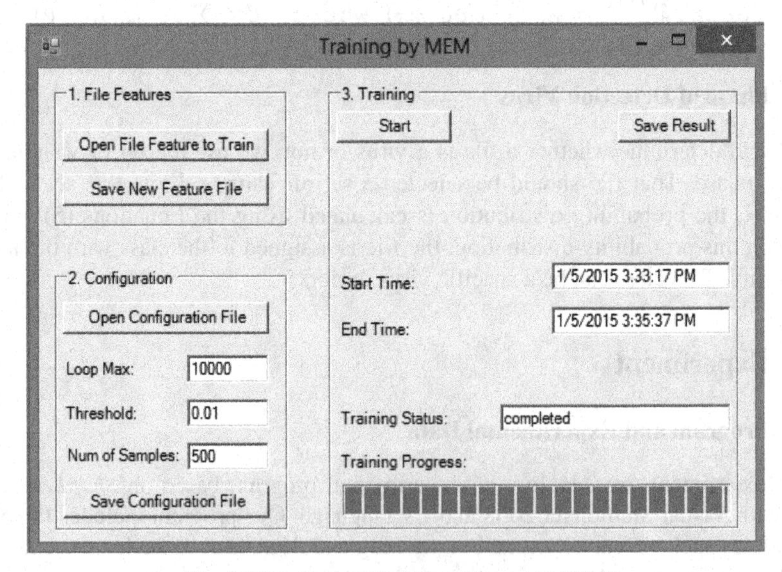

Fig. 5. Training by Maximum Entropy Model

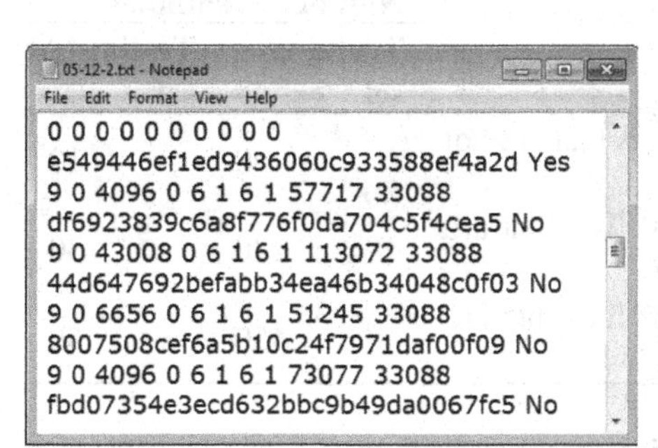

Fig. 6. A part of virus feature

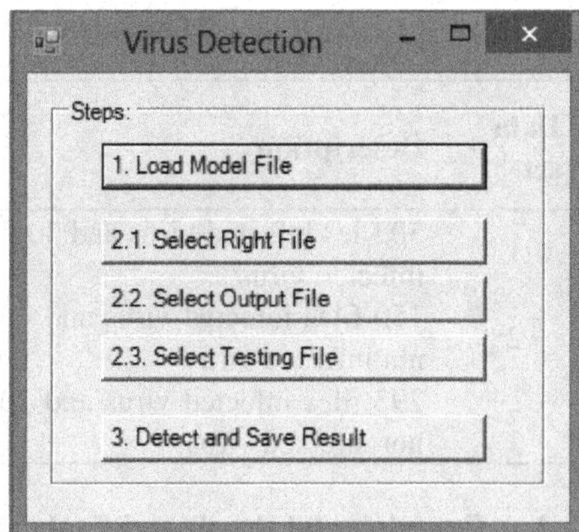

Fig. 7. Virus detection module

Table 2. Some of virus sample in the experiment

Order	Virus name	Virus samples
1	Elsahes.A	VirusShare_268988aa1df82ab073f527b5b6c8bff7
		VirusShare_5c4806b5859b35a3df03763e9c7ecbf6
2	Minaps.A	VirusShare_1328eaceb140a3863951d18661b097af
		VirusShare_c99fa835350aa9e2427ce69323b061a9
		VirusShare_e476e4a24f8b4ff4c8a0b260aa35fc9f
3	Sharat.gen!A	VirusShare_0b680e7bd5c0501d5dd73164122a7faf
		VirusShare_605c1dc91a5c85024160ce78dfac842d
		VirusShare_eef80511aa490b2168ed4c9fa5eafef0
4	Small.XR	VirusShare_4192479b055b2b21cb7e6c803b765d34
		VirusShare_d22863c5e6f098a4b52688b021beef0a
		VirusShare_f1e5d9bf7705b4dc5be0b8a90b73a863
5	Orsam!rts	VirusShare_225e33508861984dd2a774760bfdfc52
		VirusShare_b3848edbabfbce246a9faf5466e743bf
6	Sisproc	VirusShare_0ca6e2ad69826c8e3287fc8576112814
		VirusShare_23059de2797774bbdd9b21f979aaec51
		VirusShare_6fbf667e82c1477c4ce635b57b83bfa0
7	Noobot.A	VirusShare_0908d8b3e459551039bade50930e4c1b
		VirusShare_2bd02b41817d227058522cca40acd390
		VirusShare_6808ec6dbb23f0fa7637c108f44c5c80
		VirusShare_95f25d3afc5370f5d9fd8e65c17d3599

Table 3. Description of the data sets

Data set	Description	Number of samples	
		Training set	Testing set
1	50 files infected virus and 50 files that are not infected virus	90	10
2	150 files infected virus and 150 files that are not infected virus	270	30
3	293 files infected virus and 207 files that are not infected virus	450	50

4.2 Experimental Result and Evaluation

After doing the experiment under the program and data sets above, we summary experimental result in Table 4. Based on this result, we evaluate the application of Maximum Entropy Model in the problem of virus detection as follows: size of training model is small, training time is long but it is not important, accurate rate is high, detection time is small. Moreover, this method is better than traditional methods because it can detect unknown virus (new virus) [1].

Table 4. Summary of experimental result

Statistical information	Data set 1	Data set 2	Data set 3
Size (MB)	9.7	22.5	34.9
Training time (Minute)	0.8	1.5	2.3
Average of detection time (Second)	20	23	27
Size of training model (KB)	15.1	23.3	36.5
Accurate rate in case virus trained (%)	100	100	100
Accurate rate in case untrained virus (%)	36.2	47.5	54.8

5 Conclusion and Future Work

Detecting virus based on Maximum Entropy Model is a prospect method and high ability of application. This method has some advantages as follows: can present many complex problems by feature functions; allow to associate different types of feature; has high accurate rate; has a good ability of classification with large number of classes. In addition, it can detect unknown virus. The main contribution of the paper is that we have proposed a new method of virus detection based on Maximum Entropy Model. In addition, practically, we have built the virus detection program and data set to deploy and test this method. Based on the research result of the paper, we continue doing further researches such as applying MEM in intrusion detection system, applying MEM in malware recognition, associating MEM with other machine leaning algorithm to detect virus, malware.

References

[1] Shah, S., Jani, H., Shetty, S., Bhowmick, K.: Virus Detection using Artificial Neural Networks. International Journal of Computer Applications 84(5), 17–23 (2013)
[2] Stopel, D., Boger, Z., Moskovitch, R., Shahar, Y., Elovici, Y.: Improving Worm Detection with Artificial Neural Networks through Feature Selection and Temporal Analysis Techniques. Transactions on Engineering, Computing and Technology 15, 202–208 (2006)
[3] Griffin, K., Schneider, S., Hu, X., Chiueh, T.-C.: Automatic Generation of String Signatures for Malware Detection. In: Kirda, E., Jha, S., Balzarotti, D. (eds.) RAID 2009. LNCS, vol. 5758, pp. 101–120. Springer, Heidelberg (2009)
[4] Kephart, J., Arnold, W.: Automatic Extraction of Computer Virus Signatures. In: Proceedings of the Virus Bulletin International Conference, Abingdon, England, pp. 179–194 (1994)
[5] Assaleh, T.A., Cercone, N., Keselj, V., Sweidan, R.: Detection of New Malicious Code Using N-grams Signatures. In: PST, pp. 193–196 (2004)
[6] Chaumette, S., Ly, O., Tabary, R.: Automated extraction of polymorphic virus signatures using abstract interpretation. In: NSS, pp. 41–48. IEEE (2011)
[7] John, A.: Computer Viruses and Malware, pp. 27–45. Springer (2006) ISBN 978-0-387-30236-2
[8] Serazzi, G., Zanero, S.: Computer Virus Propagation Models. In: Calzarossa, M.C., Gelenbe, E. (eds.) MASCOTS 2003. LNCS, vol. 2965, pp. 26–50. Springer, Heidelberg (2004)
[9] Jussi, P.: Digital Contagions: A Media Archaeology of Computer Viruses, pp. 50–70. Peter Lang, New York (2007) ISBN 978-0-8204-8837-0
[10] Yu, Z., et al.: A Novel Immune Based Approach For Detection of Windows PE Virus. In: Proceeding of ADMA, Chengdu, China, pp. 250–260 (2008)
[11] Ratnaparkhi, R.: A simple introduction to maximum entropy models for natural language processing. Technical Report 97-08, Institute for Research in Cognitive Science, University of Pennsylvania, pp. 1–18 (1997)
[12] Guiasu, S., Shenitzer, C.: The principle of maximum entropy. The Mathematical Intelligencer 7, 42–48 (1985)
[13] Huong, P.V., et al.: Some Approaches to Nôm Optical Character Recognition. VNU, Hanoi. J. of Science, Natural Sciences and Technology 24(3S), 90–99 (2008)
[14] Chistopher, D.M., Schutze, H.: Foundations of Statistical Natural Language Processing, pp. 5–25. MIT Press (1999)
[15] Database of virus patterns, http://samples.virussign.com/samples/, http://virusshare.com/
[16] Daoud, E.A., Jebril, I., Zaquaibeh, B.: Computer virus strategies and detection methods. Int. J. Open problems Compt. Math. 1(2), 123–130 (2008)

A Parallel Algorithm for Frequent Subgraph Mining

Bay Vo[1,2], Dang Nguyen[1,2], and Thanh-Long Nguyen[3]

[1] Division of Data Science, Ton Duc Thang University, Ho Chi Minh City, Vietnam
[2] Faculty of Information Technology, Ton Duc Thang University, Ho Chi Minh City, Vietnam
[3] Center for Information Technology, Ho Chi Minh City of Food Industry, Ho Chi Minh City, Vietnam
{vodinhbay,nguyenphamhaidang}@tdt.edu.vn, longnt@cntp.edu.vn

Abstract. Graph mining has practical applications in many areas such as molecular substructure explorer, web link analysis, fraud detection, outlier detection, chemical molecules, and social networks. Frequent subgraph mining is an important topic of graph mining. The mining process is to find all frequent subgraphs over a collection of graphs. Numerous algorithms for mining frequent subgraphs have been proposed; most of them, however, used sequential strategies which are not scalable on large datasets. In this paper, we propose a parallel algorithm to overcome this weakness. Firstly, the multi-core processor architecture is introduced; the way to apply it to data mining is also discussed. Secondly, we present the gSpan algorithm as the basic framework of our algorithm. Finally, we develop an efficient algorithm for mining frequent subgraphs relied on parallel computing. The performance and scalability of the proposed algorithm is illustrated through extensive experiments on two datasets, chemical and compound.

Keywords: data mining, frequent subgraph mining, parallel computing, multi-core processor.

1 Introduction

Graph mining is a well-known topic in machine learning and data mining. There are several applications of graph mining such as molecular substructure exploration [1], web link analysis [2], outlier detection [3], chemical molecules [4], and social networks [5]. Frequent subgraph mining (FSM) is an essential part of graph mining. The goal of FSM is to discover all frequent subgraphs in a given graph dataset. A subgraph is called frequent if its occurrence is above a user-specified threshold. Numerous algorithms for mining frequent subgraphs have been proposed in recent years [6-10]. Nevertheless, most of them are sequential algorithms, causing that they require much effort and time to mine large datasets. Along with the development of modern hardware, multi-core processors became a mainstream when Intel and AMD introduced their commercial multi-core chips in 2008 [11], which allows parallel computing to be even more easy and feasible. Therefore, this study aims to propose

H.A. Le Thi et al. (eds.), *Advanced Computational Methods for Knowledge Engineering*,
Advances in Intelligent Systems and Computing 358, DOI: 10.1007/978-3-319-17996-4_15

an efficient strategy for parallel frequent subgraph mining on multi-core processor computers. Firstly, we introduce the multi-core processor architecture and its applications in data mining. Secondly, we present gSpan (graph-based Substructure pattern mining) which explores frequent substructure without candidate generation [6] as the basic framework of our proposed algorithm. Although FFSM [7] and Gaston [1] are also efficient algorithms in substructure pattern mining and their performances are slightly better than gSpan, we decided to adapt gSpan to the parallel version because of its good parallel and scale up properties. Finally, we develop a parallel strategy for gSpan based on the parallelism model in .NET Framework 4.0.

The rest of this paper is as follows. Section 2 reviews some sequential and parallel algorithms for mining frequent subgraphs in the literature. Section 3 introduces the multi-core processor architecture and benefits of parallel computing in data mining. Sections 4 and 5 describe gSpan and the proposed parallel algorithm. Section 6 presents experiments to show the performance of our algorithm. Conclusions and future work are discussed in Section 7.

2 Related Work

The problem of mining frequent subgraphs was firstly introduced in [10]. The authors proposed AGM, an algorithm which shares similar characteristic with the Apriori-based frequent itemset mining [12]. The Apriori property was also used in other algorithms for FSM such as FSG [8], the path-join algorithm [13], and AGM-Hash (an adaption of AGM) [14]. These algorithms inherit two weaknesses from Apriori: (1) joining two *k*-frequent subgraphs to generate (*k*+1)-subgraph candidates; (2) checking the frequency of these candidates separately. In order to avoid the overheads occurred in Apriori-based algorithms, several algorithms without candidate generation have been developed. Examples include gSpan [6], FFSM [7], Gaston [1], gRed [9], and G-Tries [15]. These algorithms adopt the concept of pattern growth mentioned in [16], which intends to expand patterns from a single pattern directly. While Apriori-based approach must use the breath-first search (BFS) strategy because of its level wise candidate generation, the pattern growth approach can use both BFS and depth-first search (DFS).

However, all FSM algorithms mentioned above are implemented by sequential strategies, which causes that they may require much effort and time to mine massive datasets. Along with the development of modern hardware, multi-core CPUs, GPUs, and Map/Reduce become potential and feasible tools for parallel computing [17-19]. Some parallel algorithms have been developed for FSM. For example, Cook and his colleagues proposed a parallel approach for graph-based knowledge discovery on the multi-processor systems [20]. In 2005, Buehrer et al. [21] developed a parallel algorithm for graph mining on the shared memory architecture. Recently, Kessl et al. [22] used CUDA to mine graph-based substructure patterns on GPUs. In addition, some studies have tried to parallelize FSM algorithms in the Map/Reduce paradigm [23]. It can be seen that applying parallelism to FSM is an emerging trend.

3 Multi-core Processor Architecture

A multi-core processor (shown in Figure 1) is a single computing component with two or more independent central processing units (cores) in the same physical package. Compared to a computer cluster (shown in Figure 2) or a SMP (Symmetric Multi-processor) system (shown in Figure 3), the multi-core processor architecture has many desirable properties, for example each core has direct and equal access to all the system's memory and the multi-core chip also allows higher performance at lower energy and cost. Parallel mining on multi-core processor computers has been widely adopted in many research fields such as frequent itemset mining [17], class association rule mining [18], and correlated pattern mining [11].

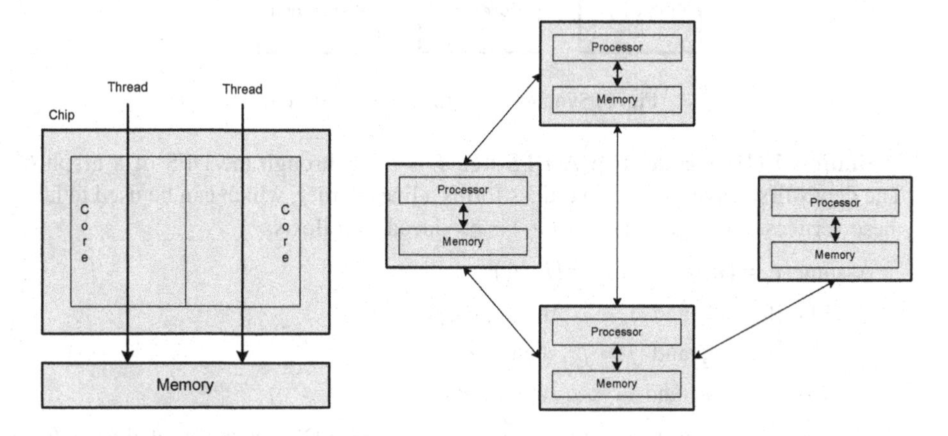

Fig. 1. Multi-core processor: one chip, two cores, two threads[1]

Fig. 2. Computer cluster[2]

4 gSpan Algorithm

In this section, we briefly summarize the gSpan algorithm because it forms the basic framework of our proposed parallel algorithm.

gSpan uses the depth-first search (DFS) strategy to traverse its search tree. To generate a child node, gSpan extends the parent node by adding one new edge. Each node in the tree is assigned a unique DFS-code and this code is used to determine the isomorphism of subgraphs. The idea of gSpan is to label a subgraph by a DFS-code and create children DFS-codes from the right-most path of DFS tree. If a subgraph has a minimal DFS-code, it is added to the result and used to find the next subgraph. This process is recursively executed until the DFS-code of subgraph is non-minimal.

---

[1] Source: `http://software.intel.com/en-us/articles/multi-core-processor-architecture-explained`

[2] Source: `http://en.wikipedia.org/wiki/Distributed_computing`

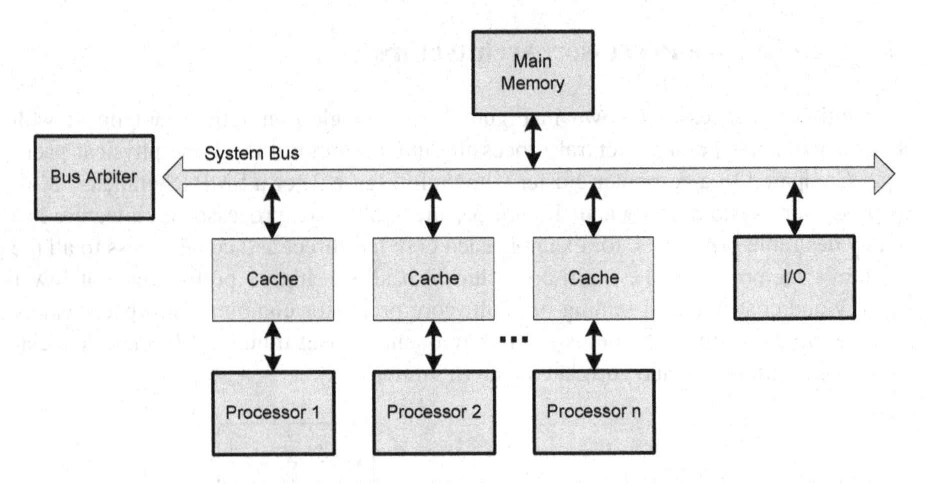

Fig. 3. Symmetric Multi-processor system[3]

Definition 1 (DFS code) [6]: A DFS tree *T* is built through the DFS of a graph *G*. The depth-first traverse of the vertices forms a linear order, which can be used to label these vertices. An edge sequence (e_i) is produced as follows.

Assume $e_1 = (i_1, j_1)$ and $e_2 = (i_2, j_2)$:

(1) if $i_1 = i_2$ and $j_1 < j_2$, then $e_1 < e_2$

(2) if $i_1 < i_2$ and $j_1 = j_2$, then $e_1 < e_2$

(3) if $e_1 < e_2$ and $e_2 < e_3$, then $e_1 < e_3$

The sequence e_i in which $i = 0,...,|E|-1$ is called a DFS-code, denoted as *code*(*G*, *T*).

Definition 2 (DFS Lexicographic Order) [6]: Suppose $Z = \{code(G,T) \mid T \text{is a DFS tree of } G\}$, *Z* is a set which contains all DFS-codes for all graphs. DFS Lexicographic Order is a linear order of DFS-codes defined as follows.

If $\alpha = code(G_\alpha, T_\alpha) = (a_0, a_1, ..., a_m)$ and $\beta = code(G_\beta, T_\beta) = (b_0, b_1, ..., b_m)$ with $\alpha, \beta \in Z$, then $\alpha \le \beta$ if one of two conditions is true:

(1) $\exists t, 0 \le t \le \min(m,n), a_k = b_k$ for $k < t, a_t < b_t$

(2) $a_k = b_k$ for $0 \le k \le m$ and $n \ge m$

Definition 3 (Minimal DFS-code) [6]: Given a graph *G* and *Z*(*G*). Regarding DFS Lexicographic Order, the minimal one, *min*(*Z*(*G*)), is called Minimal DFS-code of *G*.

Theorem 1 [6]: Given two graphs *G* and *G'*, *G* is isomorphic to *G'* if and only if min(*G*) = min(*G'*).

---

[3] Source: `http://en.wikipedia.org/wiki/Symmetric_multiprocessing`

Definition 4 (Right-most path extension rules) [6]: Given a DFS-code *s* and an edge *e*, in either of the following two cases, $s \cup e$ is called the right-most path extension:

(1) *e* connects the right-most vertex and the vertices on the right-most path in the DFS tree

(2) *e* connects the right-most path in the DFS tree and a new vertex

Based on four definitions and Theorem 1, the sequential version of gSpan is represented in Figure 4.

Input: Graph dataset *D* and minimum support threshold *minSup*
Output: All frequent subgraphs in *D*
Procedure:
GraphSet_Projection(*D*, *S*, *minSup*)
1. sort labels of vertices and edges in *D* by their frequency;
2. remove infrequent vertices and edges;
3. re-label the remaining vertices and edges in descending frequency;
4. *S1* = all frequent 1-edge graphs in *D*;
5. sort *S1* in DFS lexicographic order;
6. $S = S \cup S1$;
7. for each edge $e \in S1$ do
8. initialize *s* with *e*, set $s.D = \{g \mid \forall g \in D, e \in E(g)\}$; (only graph *id* is stored)
9. SubGraph_Mining(*D*, *S*, *s*);
10. $D = D \cup D \setminus e$;
11. if $|D| < minSup$ then
12. break;

SubGraph_Mining(*D*, *S*, *s*)
13. if $s \neq \min(s)$ then
14. return;
15. $S = S \cup \{s\}$;
16. generate all *s*' potential children with one edge growth;
17. enumerate(*s*);
18. for each *c*, *c* is *s*' child do
19. if support(*c*) ≥ *minSup* then
20. *s* = *c*;
21. SubGraph_Mining(*D*, *S*, *s*);

Fig. 4. gSpan algorithm [6]

5 Proposed Algorithms

In this section, we introduce our parallel version for gSpan. We adopt key features of gSpan such as isomorphism test and children subgraph generation in our algorithm. A graph can be considered as an object so that we can find a graph of (*k*+1)-edge from a graph of *k*-edge.

5.1 Parallel Mining Frequent Subgraphs with Independent Branch Strategy

In this strategy (called PMFS-IB), we distribute each branch of the DFS tree to a single task which mines assigned branch independently from other tasks. The pseudo code of PMFS-IB is represented in Figure 5.

PMFS-IB explores all frequent subgraphs in the same way as gSpan does, except that PMFS-IB mines each branch of the DFS tree in parallel. Firstly, PMFS-IB finds all frequent 1-edge subgraphs in the dataset (Lines 1-6). For each 1-edge subgraph, PMFS-IB creates a new task t_i (Line 9) and calls the EXPAND-SubGraph procedure inside that task with three parameters: dataset *D*, a single 1-edge subgraph *s*, and the set of frequent subgraphs returned by the task (*FS*). Procedure EXPAND-SubGraph is recursively called inside a task to find all frequent subgraphs (Lines 17-25). Finally, after all tasks are completed, their results are collected to form the full set of frequent subgraphs (Lines 12-14). However, the Independent Branch strategy has a significant disadvantage that it is not able to shrink the graph dataset after all descendants of a 1-edge subgraph have been searched. While the original gSpan algorithm removes edges and vertices to project the whole dataset to a smaller one at each iteration (Figure 4, Line 10), PMFS-IB cannot shrink the dataset because it mines all branches in parallel which requires the whole dataset as an input parameter (Figure 5, Line 11).

Input: Graph dataset *D* and minimum support threshold *minSup*
Output: All frequent subgraphs in *D*
Procedure:
FIND-SubGraph(*D*, *S*, *minSup*)
1. sort labels of vertices and edges in *D* by their frequency;
2. remove infrequent vertices and edges;
3. re-label the remaining vertices and edges in descending frequency;
4. *S1* = all frequent 1-edge subgraphs in *D*;
5. sort *S1* in DFS lexicographic order;
6. $S = S \cup S1$;
7. for each edge $e \in S1$ do
8. initialize *s* with *e*, set $s.D = \{g \mid \forall g \in D, e \in E(g)\}$; (only graph *id* is stored)
9. Task t_i = new Task(() => {
10. $FS = \varnothing$; *// list of frequent subgraphs returned by this task*
11. EXPAND-SubGraph(*D*, *s*, *FS*)});
12. for each task in the list of created tasks do
13. collect the set of frequent subgraphs (*FS*) returned by each task;
14. $S = S \cup FS$;
15. if $|D| < minSup$ then
16. break;

EXPAND-SubGraph(*D*, *s*, *FS*)
17. if $s \neq \min(s)$ then
18. return;

19. $FS = FS \cup \{s\}$;
20. generate all *s*' potential children with one edge growth;
21. enumerate(*s*);
22. for each *c*, *c* is *s*' child do
23. if support(*c*) $\geq$ *minSup* then
24. *s* = *c*;
25. EXPAND-SubGraph(*D*, *s*, *FS*);

Fig. 5. PMFS-IB algorithm

5.2 Example

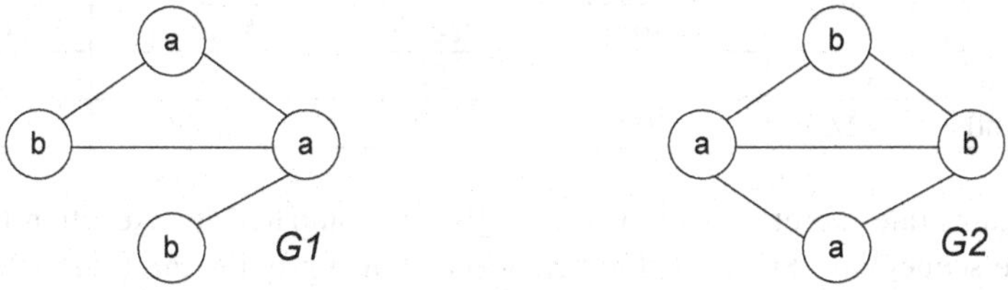

Fig. 6. Example of a graph dataset

We apply PMFS-IB to a sample dataset shown in Figure 6 with *minSup* = 100% to illustrate its process. The DFS tree constructed by PMFS-IB is shown in Figure 7. It can be seen that PMFS-IB creates four tasks *t1*, *t2*, *t3*, and *t4* (represented by solid blocks) to parallel mine four branches "a-a", "a-b", "b-a", and "b-b" independently (Figure 7).

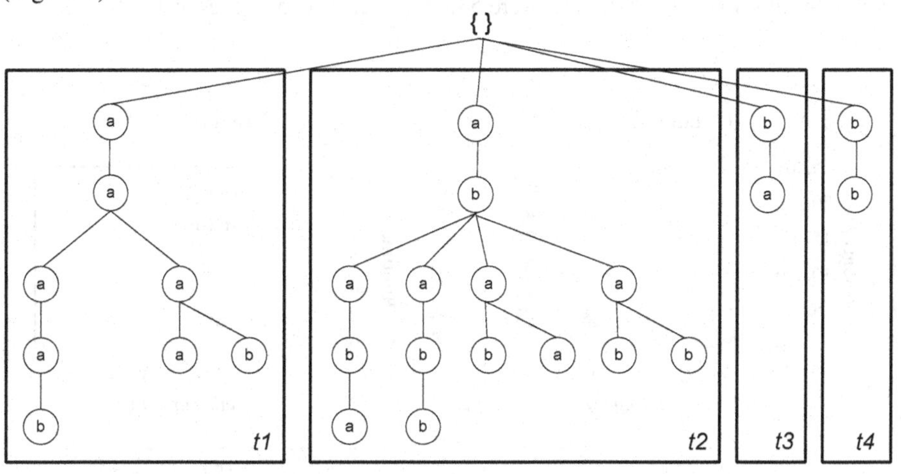

Fig. 7. DFS tree constructed by PMFS-IB for the sample dataset in Fig. 6

6 Experiments

All experiments were conducted on a computer with an Intel Core i7-2600 CPU at 3.4 GHz and 4 GB of RAM, which runs Windows 7 Enterprise (64-bit) SP1. The processor has 4 cores and an 8 MB L3-cache; it also supports Hyper-threading. The experimental datasets were obtained from website http://www.cs.ucsb.edu/~xyan/software/. The algorithms were coded in C# by using Microsoft Visual Studio .NET 2013. Characteristics of the experimental datasets are described in Table 1. The table shows the number of graphs, the average graph size, and the largest graph size in the dataset.

Table 1. Characteristics of the experimental datasets

Dataset	#graphs	Average size		Largest size	
		#nodes	#edges	#nodes	#edges
Chemical	340	27	28	214	214
Compound	422	40	42	189	196

To demonstrate the efficiency of PMFS-IB, we compared its execution time with that of the sequential gSpan [6]. Figures 8(a) and 9(a) provide information about the number of frequent subgraphs which is found in Chemical and Compound respectively while Figures 8(b) and 9(b) compare the runtimes of two algorithms.

The results show that PMFS-IB outperforms gSpan in all experiments. This is because the parallel algorithm can utilize the power of the multi-core processor architecture. When the values of *minSup* are low, the mining times of gSpan dramatically rise while the figures for PMFS-IB are smaller. For example, consider dataset Compound with *minSup* = 7%. The runtime of gSpan was 112.927(s) while that of PMFS-IB was 101.430(s). Similarly, consider the Chemical dataset with *minSup* = 3%. The execution times of gSpan and PMFS-IB were 53.037(s) and 46.657(s) respectively.

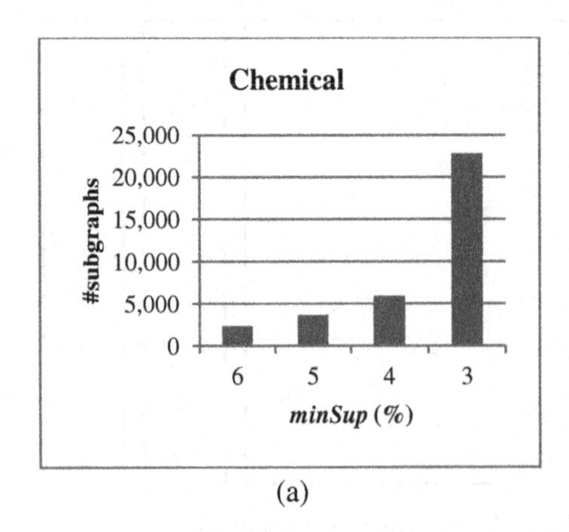

(a)

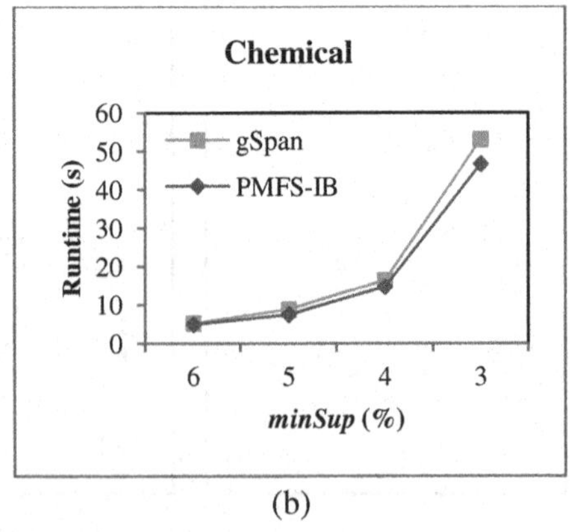

(b)

Fig. 8. Runtimes of gSpan and PMFS-IB for the Chemical dataset

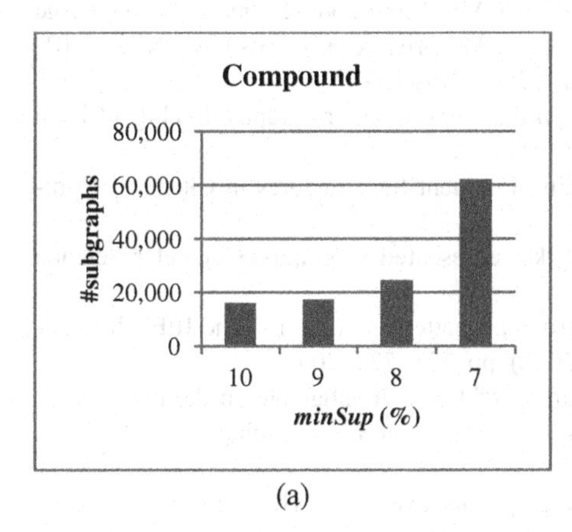

(a)

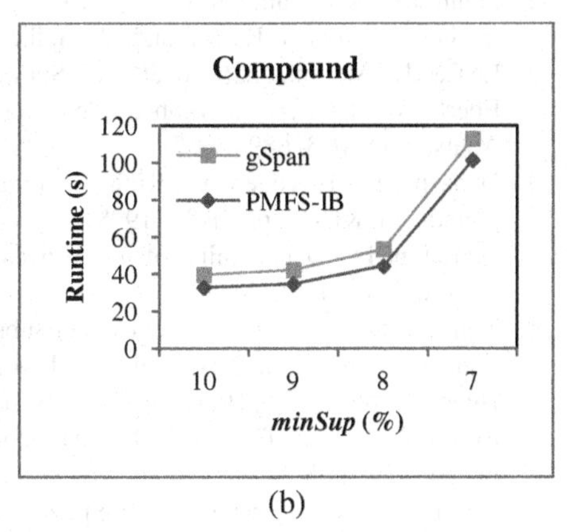

(b)

Fig. 9. Runtimes of gSpan and PMFS-IB for the Compound dataset

7 Conclusions and Future Work

This paper proposes a strategy for parallel mining frequent subgraphs. The basic idea of the proposed algorithm is to adapt gSpan (an efficient algorithm for frequent subgraph mining) to a parallel version based on the parallelism model in .NET Framework 4.0. The parallel feature is implemented on the multi-core processor architecture which does not require an extra cost for synchronization among processors like a computer cluster.

To validate the efficiency of the proposed algorithm, experiments were conducted on two popular datasets, namely Chemical and Compound. The experimental results show that the proposed method is superior to the sequential algorithm gSpan. However, the proposed method has a weakness that it is not able to shrink the dataset. This issue will be further investigated and solved. In addition, when the minimum support value is very low, the memory consumption is high, which may cause the memory leakage. We will study the solutions for reducing the memory consumption. We also expand our work to closed subgraph and maximal subgraph mining in the future.

Acknowledgments. This research was funded by Vietnam National Foundation for Science and Technology Development (NAFOSTED) under grant number 102.01-2012.17.

References

1. Nijssen, S., Kok, J.: Frequent graph mining and its application to molecular databases. In: The IEEE International Conference on Systems, Man and Cybernetics (SMC 2004), pp. 4571–4577 (2004)

2. Punin, J.R., Krishnamoorthy, M.S., Zaki, M.J.: LOGML: Log markup language for web usage mining. In: Kohavi, R., Masand, B., Spiliopoulou, M., Srivastava, J. (eds.) WebKDD 2001. LNCS (LNAI), vol. 2356, pp. 88–112. Springer, Heidelberg (2002)
3. Eberle, W., Holder, L.: Anomaly detection in data represented as graphs. Intelligent Data Analysis 11, 663–689 (2007)
4. Dehaspe, L., Toivonen, H., King, R.: Finding Frequent Substructures in Chemical Compounds. In: KDD, pp. 30–36 (1998)
5. Nettleton, D.: Data mining of social networks represented as graphs. Computer Science Review 7, 1–34 (2013)
6. Yan, X., Han, J.: gspan: Graph-based substructure pattern mining. In: The IEEE International Conference on Data Mining (ICDM 2002), pp. 721–724 (2002)
7. Huan, J., Wang, W., Prins, J.: Efficient mining of frequent subgraphs in the presence of isomorphism. In: The IEEE International Conference on Data Mining (ICDM 2003), pp. 549–552 (2003)
8. Kuramochi, M., Karypis, G.: Frequent subgraph discovery. In: The IEEE International Conference on Data Mining (ICDM 2001), pp. 313-320. (2001)
9. Gago Alonso, A., Medina Pagola, J.E., Carrasco-Ochoa, J.A., Martínez-Trinidad, J.F.: Mining frequent connected subgraphs reducing the number of candidates. In: Daelemans, W., Goethals, B., Morik, K. (eds.) ECML PKDD 2008, Part I. LNCS (LNAI), vol. 5211, pp. 365–376. Springer, Heidelberg (2008)
10. Inokuchi, A., Washio, T., Motoda, H.: An apriori-based algorithm for mining frequent substructures from graph data. In: Zighed, D.A., Komorowski, J., Żytkow, J.M. (eds.) PKDD 2000. LNCS (LNAI), vol. 1910, pp. 13–23. Springer, Heidelberg (2000)
11. Casali, A., Ernst, C.: Extracting Correlated Patterns on Multicore Architectures. In: Cuzzocrea, A., Kittl, C., Simos, D.E., Weippl, E., Xu, L. (eds.) CD-ARES 2013. LNCS, vol. 8127, pp. 118–133. Springer, Heidelberg (2013)
12. Agrawal, R., Srikant, R.: Fast Algorithms for Mining Association Rules in Large Databases. In: The 20th International Conference on Very Large Data Bases, pp. 487–499. Morgan Kaufmann Publishers Inc. (1994)
13. Vanetik, N., Gudes, E., Shimony, S.: Computing frequent graph patterns from semistructured data. In: The IEEE International Conference on Data Mining (ICDM 2002), pp. 458–465. IEEE (2002)
14. Nguyen, P.C., Washio, T., Ohara, K., Motoda, H.: Using a hash-based method for apriori-based graph mining. In: Boulicaut, J.-F., Esposito, F., Giannotti, F., Pedreschi, D. (eds.) PKDD 2004. LNCS (LNAI), vol. 3202, pp. 349–361. Springer, Heidelberg (2004)
15. Ribeiro, P., Silva, F.: G-Tries: a data structure for storing and finding subgraphs. Data Mining and Knowledge Discovery 28, 337–377 (2014)
16. Han, J., Pei, J., Yin, Y.: Mining frequent patterns without candidate generation. In: ACM SIGMOD Record, pp. 1–12. ACM (2000)
17. Schlegel, B., Karnagel, T., Kiefer, T., Lehner, W.: Scalable frequent itemset mining on many-core processors. In: The 9th International Workshop on Data Management on New Hardware, Article No. 3. ACM (2013)
18. Nguyen, D., Vo, B., Le, B.: Efficient Strategies for Parallel Mining Class Association Rules. Expert Systems with Applications 41, 4716–4729 (2014)
19. Zhang, F., Zhang, Y., Bakos, J.D.: Accelerating frequent itemset mining on graphics processing units. The Journal of Supercomputing 66, 94–117 (2013)
20. Cook, D., Holder, L., Galal, G., Maglothin, R.: Approaches to parallel graph-based knowledge discovery. Journal of Parallel and Distributed Computing 61, 427–446 (2001)

21. Buehrer, G., Parthasarathy, S., Nguyen, A., Kim, D., Chen, Y.-K., Dubey, P.: Parallel Graph Mining on Shared Memory Architectures. Technical report, Columbus, OH, USA (2005)
22. Kessl, R., Talukder, N., Anchuri, P., Zaki, M.: Parallel Graph Mining with GPUs. In: The 3rd International Workshop on Big Data, Streams and Heterogeneous Source Mining: Algorithms, Systems, Programming Models and Applications, pp. 1–16 (2014)
23. Lin, W., Xiao, X., Ghinita, G.: Large-scale frequent subgraph mining in MapReduce. In: The IEEE 30th International Conference on Data Engineering (ICDE 2014), pp. 844–855. IEEE (2014)

Combining Random Sub Space Algorithm and Support Vector Machines Classifier for Arabic Opinions Analysis

Amel Ziani[1], Nabiha Azizi[2], and Yamina Tlili Guiyassa[1]

[1] Computer Depatment, Lri Laboratory: Computer Research Laboratory, University of Annaba, Algeria
[2] Labged Laboratory: Electronic Documents Control Laboratory, Badji Mokhatr University of Annaba, P.O. Box 12, Annaba, Algeria
z_amel1911@live.fr, {nabiha111,guiyam}@yahoo.fr

Abstract. In this paper, an Arabic Opinion Analysis system is proposed. These sorts of applications produce data with a large number of features, while the number of samples is limited. The large number of features compared to the number of samples causes over-training when proper measures are not taken. In order to overcome this problem, we introduce a new approach based on Random sub space (RSS) algorithm integrating Support vector machine (SVM) learner as individual classifiers to offer an operational system able to identify opinions presented in reader's comments found in Arabic newspapers blogs. The main steps of this study is based primarily on corpus construction, Statistical features extraction and then classifying opinion by the hybrid approach RSS-SVM. Experiments results based on 800 comments collected from Algerian newspapers are very encouraging; however, an automatic natural language processing must be added to enhance primitives' vector.

Keywords: RSS (Random Sub Space), SVM (Support Vector Machine), Arabic opinion mining.

1 Introduction

Other's opinions can be crucial when it's time to make a decision or choose among multiple options. When those choices involve valuable resources (for example, spending time and money to buy products or services) people often rely on their peers' past experiences. Until recently, the main sources of information were friends and specialized magazine or websites. Opinion Mining (OM), also known as Sentiment Analysis (SA) is a challenging task that combines data mining and Natural Language Processing (NLP) techniques in order to computationally treat subjectivity in textual documents. This new area of research is becoming more and more important mainly due to the growth of social media where users continually generate contents on the web in the form of comments, opinions, emotions, etc. Most of built resources and systems are intended to English or other Indo-European languages. Despite the fact that Chinese, Arabic and Spanish are currently among the top ten

H.A. Le Thi et al. (eds.), *Advanced Computational Methods for Knowledge Engineering*,
Advances in Intelligent Systems and Computing 358, DOI: 10.1007/978-3-319-17996-4_16

languages most used on the Internet according to the Internet World State rank1, there are very few resources for managing sentiments or opinions in these languages.

However, people increasingly comment on their experiences, opinions, and points of views not only in English but in many other languages. A published study by Semiocast, which is a French based company, has revealed that Arabic was the fastest growing language on Twitter in 2011, and was the 6th most used language on Twitter in 2012[1]. The same study revealed that the number of twitter users in Saudi Arabia almost doubled (grew by 93%) in the span of 6 months in 2012 and that Riyadah is now the 10th most active city on Twitter. A breakdown of Facebook users by country, places Egypt with 11,804,060 users as the Arabic speaking country with the largest number of FB users, and ranks it at 20among all countries of the world. Saudi Arabia follows with a user base of 5,760,900 and an overall rank of 33[2].

The opinion mining consists of several tasks; it is useful to implement the intended applications [3]:

1. Detecting the presence or absence of Opinion;
2. Classification of opinion's axiology (positive, negative, neutral);
3. Classification of opinion's intensity;
4. Opinion's object Identification (on which the opinion relates);
5. Identifying the source of the opinion (which expresses the opinion).

This work is part of a national research project. It is primarily concerned with the task of extracting relevant characteristics and classification of the opinion's axiology in Arabic language from readers' comments regarding the national press. The approach is based on novel paradigm of supervised machine learning techniques.

In this study, an important extracted statistical characteristics set (such as the number of positive words, strongly positive, emotionality... etc.) are defined and analyzed in section 3. We noticed that the flow characteristics used may be incompatible and probably redundant information.

In such growing number of domains the data collected has a large number of features. This poses a challenge to classical pattern recognition techniques, since the number of samples often is still limited with respect to the feature size. In order to reduce the dimensionality of the feature space, the selection of informative features becomes an essential step towards the classification task. The large number of features compared to the number of samples causes over-training when proper measures are not taken. In order to overcome this problem, we introduce a new multivariate approach for feature selection based on the Random Subspace Method (RSM) proposed by Ho [4]. Ho introduced the RSM to avoid overfiting on the training set while preserving the maximum accuracy when training decision tree classifiers. However, SVM classifiers have showed its efficiency in opinion mining as learning algorithm. We propose a hybrid approach incorporating individual SVMs classifiers in RSM algorithm.

The exploitation of the different results produced by these classifiers using one of the methods of classifiers combination, usually results in a high rate of generalization [5]. Although the classifier is less efficient with the input data, the knowledge of behavior provides some useful information about the true class during the merger.

The final results are a set of classifiers ensuring a level of complementarity that will be combined to generate the final decision. To achieve this objective, two methods of fusion are used: Majority Voting and Weighted Voting.

This work is organized as follows: in section two we present the main concepts of "Random Sub Space". The main steps of our approach are detailed in Section three, beginning with the acquisition of the comments; to the classification with SVMs. Also, in the future, we plan to extend our work to be able to classify the Arabic opinions for the recommendation systems.

2 Related Works

Previous work has proposed methods for identifying subjective text that expresses opinion and distinguishing it from objective text that presents factual information ([6]; [7]; [8]; [9]).

Subjective text may express positive, negative, or neutral opinion. Previous work addressed the problem of identifying the polarity of subjective text ([7]; [10]; [11]). Many of the proposed methods for text polarity identification depend on the availability of polarity lexicons (i.e. lists of positive and negative words). Several approaches have been devised for building such lexicons ([12]; [13]; [14]; [15]). Other research efforts focused on identifying the holders and the targets of opinion ([16]; [17]; [18]).

Opinion mining and sentiment analysis techniques have been used in various applications. One example of such applications is identifying perspectives ([19]; [20]). For example, in [20], the authors experiment with several supervised and statistical models to capture how perspectives are expressed at the document and the sentence levels. In [21] they proposed a method for extracting perspectives from political texts. They used their method to estimate the policy positions of political parties in Britain and Ireland, on both economic and social policy dimensions. In [22]they present an unsupervised opinion analysis method for debate side classification. They mine the web to teach associations that are indicative of opinion stances in debates and combine this knowledge with discourse information.

3 Random Sub Space

The random sub space method introduced by Ho in 1998 represents a distinct aggregating method that has a fundamentally different philosophy from the other aggregating methods. The random subspace method originated from the stochastic discriminant analysis, which uses weak classifiers that are not necessarily very accurate but generalize well as base learner and achieves good prediction accuracy by aggregating many different such individual classifiers. The random subspace method generates multiple individual classifiers by projecting the original feature space into different subspaces. The projection from the high dimensional feature space into the low dimensional space can avoid the problems caused by high dimensionality. And all the information in the original data is maintained by aggregating many individual classifiers based on different subspaces. However, the original random subspace method did not emphasize the importance of selecting a base learner that should generalize well. Also, all individual classifiers have equal weights so the existence of uninformative subspaces would still deteriorate the performance of the aggregated classifier [23].

4 System Architecture

In this section, we will explain our proposed system architecture for Arabic opinion mining and classification (Fig. 1). When we started to design our system architecture, we realized that this architecture must meet the following requirements:

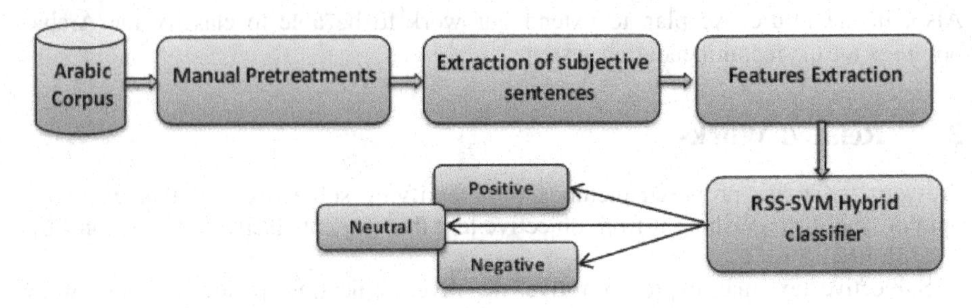

Fig. 1. The proposed process for opinions classification

4.1 The Arabic Corpus Construction and Manual Pre-treatment

The Arabic language has become a popular area of research in IR (Information Retrieval), but it presents serious challenges due to its richness, complex morphology, and syntactic flexibility [24]. Although in the last decade, a lot of works have been carried out to make the task of Natural Language Process of Arabic easy, the systems to analyze Arabic automatically are not as easily available as they are for other languages such as English [25]. In addition, available digital resources such as, for example, the corpora for Arabic NER are still limited although efforts are being made to remedy this [26]. Based on our national project, we have constructed an Arabic corpus that contains almost 800 reviews: which have been prepared from a collection of articles that were manually obtained from various sources on the web as: reader comments blogs from Algerian newspaper.

Table 1. Examples of reviews

Polarity	The commentaries in English	The commentaries in Arabic
Positive	Congratulation for us all, you made the good choice. The journalist Leila bouzidi is really competent and she has a professional personality allowing her to really manage her career, god help her.	مبروك علينا وعليكم لقد أحسنتم الاختيار. الصحفية القديرة ليلى بوزيدي حقا هي متمكنة وذات شخصية مهنية متحكمة في إدارة مهنتها أعانها الله ووفقها.
Negative	Honestly, I didn't like the writer's style, and also his point of view towards this topic and I hated his mockeries for the art.	صراحة لم أحب أبدا أسلوب الكاتب ولا رأيه في الموضوع وكرهت استهزاءه بالفن.
Neutral	No, it is his personal point of view.	كلا انه رأيه الشخصي.

4.2 Features Extraction

Feature engineering is an extremely basic and essential task for Sentiment Analysis. Converting a piece of text to a feature vector is the basic step in any data driven approach to Sentiment Analysis. It is important to convert a piece of text into a feature vector, so as to process text in a much efficient manner. In text domain, effective feature selection is a must in order to make the learning task effective and accurate. But in text classification, it's a group of statistical features.

In a previous work [27], we have analyzed different statistical features and after an empirical study, the selected features for Arabic polarity detection can be summarized as follow:

Table 2. The selected statistical features

Sentence	Number of sentences
Positive Words	Number of positive words
Negative Words	Number of negative words
Neutral Words	Number of neutral words
Polarity	Sum of polarity words
Positive polarity	Average of positive polarity words
Negative polarity	Average of negative polarity words
Neutral polarity	Average of neutral polarity words
Predicates	Number of predicates
	Average of predicates
Adverbs	Number of adverbs
	Average of adverbs
Adjectives	Number of adjectives
	Average of adjectives
Emotionalism	The emotionalism of the document
Addressage	The addressage of the document
Reflexivity	The reflexivity of the document

Some have noticed that the large vector of features can affect the performance of the SVM classifier, to outperform this limit and at the same time taking into account information issue from all features; our idea is use RSS to generate different small features vectors. Each novel SVM will be responsible to learn all comments by each vector. More details are explained in the next section.

4.3 RSS-SVM Hybrid Classifier

The Sentiment Polarity Classification is a binary classification task where an opinionated document is labeled with an overall positive or negative sentiment. Sentiment Polarity Classification can also be termed as a binary decision task. The input to the Sentiment Classifier can be opinionated or sometime not. When a news article is given as an input, analyzing and classifying it as a good or bad news is considered to be a text categorization task as in [5]. Furthermore, this piece of information can be good or bad news, but not necessarily subjective (i.e., without expressing the view of the author).

It is mentioned above, that the use of all features with limited dataset like Arabic corpus for opinion analysis causes over fitting of used classifier. To avoid this, the best solution is to combine the Random sub space technique with the SVM classifier; we have analyzed in this work the behavior of classical RSS, the SVM classifier trained by the all features and the proposed approach replacing decision tree individual classifier which is used in classical form of RSS with several individual SVM classifiers. The number of this pool of classifiers depends on "p" parameter of RSS which represents the size of sub set of selected features generating a vector $V_k$at each iteration. We must precise that the used kernel in individual SVM classifier is Gaussian kernel, chosen after some empirical test.

The main code of RSS-SVMs can be outlined as follow:

1. For each k=1,2,..K
   (a) Select a p dimensional random subspace, $X^k{}_p$, from initial features vector X.
   (b) Project the probe and gallery set onto the subspace $X^k{}_p$.
   (c) Construct the support vector machine classifier, $SVM^k{}_p$,
2. Aggregate the generated $SVM^k{}_p$ using voting and weighted voting combination techniques.

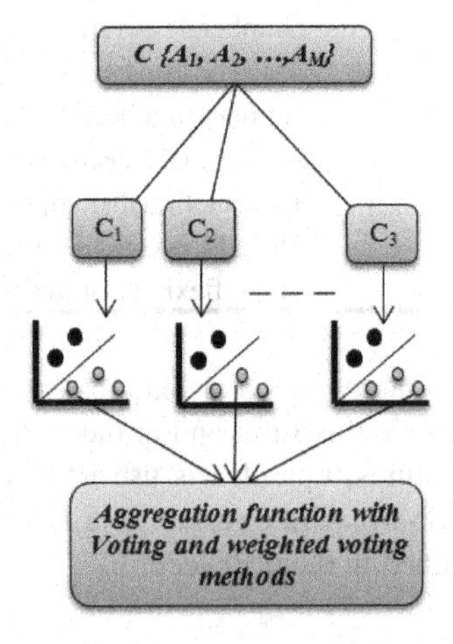

Fig. 2. The Hybrid system RSS-SVM classifier

5 Analysis of Results

In order to evaluate our approach, we applied widely used measurements such as Precision, Recall, F measure, and Accuracy [28].

The terms TruePositives (TP counts the positive opinions correctly classified), FalsePositives (FP counts the positive opinions incorrectly classified), FalseNegatives (FN counts the negative opinions incorrectly classified), TrueNegatives (TN counts the negative opinions correctly classified), TrueNeutral (TNe counts the neutral opinions correctly classified), FalseNeutral (FNe counts the neutral opinions incorrectly classified), were gathered to calculate the previous measures, according to the following formulas:

$$Precision = \frac{TP}{TP+FP} \tag{1}$$

$$Recall = \frac{TP}{TP+FN+FNe} \tag{2}$$

$$Fmeasure = 2 * \frac{Precision.Recall}{Precision+Recall} \tag{3}$$

$$Accuracy = \frac{TP+TN+TNe}{TP+FP+TN+FN+TNe+FNe} \tag{4}$$

To highlight the importance, and to clarify the effect of the random sub space in the Arabic opinion classification results, we carried out three experiments to calculate the previous measures. The first experiment applied on an SVM classifier that uses the Gaussian kernel, the second experiment uses a classic random sub space method with the decision tree, while in the last one a the random sub space algorithm is used to minimize the features vector for the same SVM classifier. Tables 3, 4 and 5 show the results of the experiments.

Table 3. The results of the first experiment (SVM classifier with the Gaussian kernel)

	Precision	Recall	Fmeasure	Accuracy
SVM Classifier	49.22	54.32	61.11	38.56

Table 4. Classical RSSclassifier obtained results

Combination function	RSS	Precision	Recall	Fmeasure	Accuracy
Voting	P=2	33.12	29.23	25.34	41.00
	P=3	30.12	28.21	27.51	24.36
	P=4	32.21	20.22	25.01	19.99
	P=5	23.00	21.00	25.01	20.39
Weighted Voting	P=2	23.22	39.33	29.37	31.10
	P=3	32.22	31.29	34.21	25.16
	P=4	37.23	28.23	19.31	20.59
	P=5	30.02	20.53	22.01	21.39

Table 5. Hybrid RSS-SVM classifier obtained results

Combination function	RSS	Precision	Recall	Fmeasure	Accuracy
Voting	P=2	65.22	64.32	67.11	66.56
	P=3	69.21	62.01	65.22	55.99
	P=4	87.21	85.12	86.20	91.00
	P=5	79.12	74.22	68.56	69.33
Weighted Voting	P=2	55.22	74.42	77.33	76.57
	P=3	79.22	72.11	68.12	66.99
	P=4	97.32	95.32	96.10	98.00
	P=5	89.02	64.32	69.12	79.13

From the above results, we can distinguish the advantages of combing the two methods RSS and SVM. From the last table, the RSS-SVM classifier with the vector size "P" equal to 4 and with the use of weighted voting techniques as aggregation function, our system generate the best results with 98.00.

6 Conclusion

In the last decade, the Arabic Language has become a popular area of research in IR (Information Retrieval) in general and in opinion mining in particular. Unfortunately, working with Arabic adds more difficulties than the languages that derive from Latin, because it implies the solving of different types of problems such as the short vowels, al-hamzah, prefixes, suffixes, etc. In this work, we have tried to combine both the random sub space method and support vector machine classifier in order to avoid over fitting creating by the used of all features and beneficiate from proven SVM classifier performances. Although obtained results with individual SVM classifier were promising, we have shown that the hybrid classifier and the combination techniques improved on the performances achieved individually. These results encourage us to continue working along this line. Thus, we could enrich feature vector with another sort of morphological primitives using natural language processing. We must also study the choice of "p" selected feature at each iteration for RSS algorithm. This study must take into account the correlation criteria between features in each sub vector.

References

1. Semiocast, Geolocation analysis of Twitter account sand tweets by Semiocast (2012), http://semiocast.com/publications/2012_07_30_Twitter_reaches_half_a_billion_accounts_140m_in_the_US
2. "Facebook Statistics by Country" (2012), http://www.socialbakers.com/facebook-statistics/
3. Kim, S.M., Hovy, E.: Determining the sentiment of opinions. In: Proceedings of the 20th International Conference on Computational Linguistics, COLING 2004. Association for Computational Linguistics, Morristown (2004)

4. Ho, T.K.: The Random Subspace Method for Constructing Decision Forests. IEEE Transactions on Pattern Analysis and Machine Intelligence 20(8), 832–844 (1998)
5. Turney, P.: Thumbs up or thumbs down? Semantic orientation applied to unsupervised classification of reviews. In: Proceedings of the Association for Computational Linguistics, ACL (2002)
6. Wiebe, J.: Learning subjective adjectives from corpora. In: Proceedings of the Seventeenth National Conference on Artificial Intelligence and Twelfth Conference on Innovative Applications of Artificial Intelligence, pp. 735–740 (2000)
7. Hatzivassiloglou, V., Wiebe, J.: Effects of adjective orientation and gradability on sentence subjectivity. In: COLING (2000)
8. Banea, C., Mihalcea, R., Wiebe, J.: A bootstrapping method for building subjectivity lexicons for languages with scarce resources. In: LREC 2008 (2008)
9. Riloff, E., Wiebe, J.: Learning extraction patterns for subjective expressions. In: EMNLP 2003, pp. 105–112 (2003)
10. Hassan, A., Qazvinian, V., Radev, V.: What's with the attitude?: identifying sentences with attitude in online discussions. In: Proceedings of the 2010 Conference on Empirical Methods in Natural Language Processing, pp. 1245–1255 (2010)
11. Riloff, E., Patwardhan, S., Wiebe, J.: Feature sub sumption for opinion analysis. In: Proceedings of the 2006 Conference on Empirical Methods in Natural Language Processing, pp. 440–448 (2006)
12. Turney, P., Littman, M.: Measuring praise and criticism: Inference of semantic orientation from association. ACM Transactions on Information Systems 21, 315–346 (2003)
13. Kanayama, H., Nasukawa, T.: Fully automatic lexicon expansion for domain oriented sentiment analysis. In: EMNLP 2006, pp. 355–363 (2006)
14. Takamura, H., Inui, T., Okumura, M.: Extracting semantic orientations of words using spin model. In: ACL 2005, pp. 133–140 (2005)
15. Hassan, A., Radev, D.: Identifying text polarity using random walks. In: ACL 2010 (2010)
16. Zhai, Z., Liu, B., Xu, H., Jia, P.: Grouping product features using semi-supervised learning with soft-constraints. In: Proceedings of the23rd International Conference on Computational Linguistics, pp. 1272–1280 (2010)
17. Popescu, A., Etzioni, O.: Extracting product features and opinions from reviews. In: Natural Language Processing and Text Mining, pp. 9–28. Springer (2007)
18. Bethard, B., Yu, H., Thornton, A., Hatzivassiloglou, V., Jurafsky, D.: Automatic extraction of opinion propositions and their holders. In: 2004 AAAI Spring Symposium on Exploring Attitude and Affect in Text, p 2224 (2004)
19. Grefenstette, G., Qu, Y., Shanahan, J., Evans, D.A.: Coupling niche browsers and affect analysis for an opinion mining application. Proceedings of RIAO 4, 186–194 (2004)
20. Lin, W., Wilson, T., Wiebe, J., Hauptmann, A.: Which side are you on?: identifying perspectives at the document and sentence levels. In: Proceedings of the Tenth Conference on Computational Natural Language Learning, pp. 109–116 (2006)
21. Laver, M., Benoit, K., Garry, J.: Extracting policy positions from political texts using words as data. American Political Science Review 97(02), 311–331 (2003)
22. Somasundaran, S., Wiebe, J.: Recognizing stances in online debates. In: Proceedings of the Joint Conference of the 47th Annual Meeting of the ACL and the 4th International Joint Conference on Natural Language Processing of the AFNLP, Suntec, Singapore, pp. 226–234 (August 2009)
23. Li, X., Zhao, H.: Weighted random subspace method for high dimensional data classification. The National Institutes of Health, PMC (2011)

24. Attia, M.: Handling Arabic morphological and syntactic ambiguities within the LFG framework with a view to machine translation, PhD Dissertation, University of Manchester (2008)
25. Sawalha, M., Atwell, E.: Comparative evaluation of Arabic language morphological analyzers and stemmers. In: Proceedings of COLING 2008 22nd International Conference on Computational Linguistics (2008)
26. Farghaly, A., Shaalan, K.: Arabic natural language processing: Challenges and solutions. ACM Transactions on Asian Language Information Processing, 8(4), Article 14, (2009).
27. Ziani, A., Azizi, N., Tlili, G.Y.: Détection de polarité d'opinions dans les forums en langue arabe par combinaison des SVMs. In: TALN-RÉCITAL 2013, Juin 17-21, Les Sables d'Olonne (2013)
28. Yang, Y.: An evaluation of statistical approaches to text categorization. Journal of Information Retrieval 1(1/2), 67–88 (1999)

Efficient Privacy Preserving Data Audit in Cloud

Hai-Van Dang, Thai-Son Tran, Duc-Than Nguyen,
Thach V. Bui, and Dinh-Thuc Nguyen

Faculty of Information Technology, University of Science, VNU-HCM, Vietnam
{dhvan,ttson,bvthach,ndthuc}@fit.hcmus.edu.vn,
thannguyen.hcmus@gmail.com

Abstract. With the development of database-as-a-service (DaS), data in cloud is more interesting for researchers in both academia and commercial societies. Despite DaS's convenience, there exist many considerable problems which concern end users about data loss and malicious deletion. In order to avoid these cases, users can rely on data auditing, which means verifying the existence of data stored in cloud without any malicious changes. Data owner can perform data auditing by itself or hire a third-party auditor. Until now, there are two challenges of data auditing as the computation cost in case of self auditing and data privacy preservation in case of hiding an auditor. In this paper, we propose a solution for auditing by a third-party auditor to verify data integrity with efficient computation and data privacy preservation. Our solution is built upon cryptographic hash function and Chinese Theorem Remainder with the advantage in efficient computation in all three sides including data owner, cloud server, and auditor. In addition, the privacy preservation can be guaranteed by proving the third-party auditor learns nothing about user's data during auditing process.

Keywords: Data Audit, Database-as-a-service (DaS), Cryptographic hash function, Chinese Remainder Theorem.

1 Introduction

Cloud computing and data service are two fast developing applications now. And the fields of those researches are also reinforced by requirements of the booming of smart phone and smart devices such as home appliances and wearable devices. It is trivial to see that two of the most requirements are data storage service and data privacy preservation. In those two requirements, data audit is the most important research because its result can give a direct solution of two those above requirements. In data storage service or database as a service, server of the third party is often assumed to be semi-trusted, i.e. it is believed not to use or reveal users data without permission. However, it may not reveal unexpected data loss or corruption to users, or delete some rarely accessed data sections. In order to avoid such risks, users must trust and rely on system of data auditing. The data auditing means verifying the existence of data stored in servers of cloud storage provider without any malicious deletion or modification. Data owner can perform data auditing by itself or hire an auditor. Here two challenges of data auditing are the computation cost in case of self-auditing and data privacy preservation in case of hiring an auditor. Since a decision made by a third party auditor will be not affected by

H.A. Le Thi et al. (eds.), *Advanced Computational Methods for Knowledge Engineering*,
Advances in Intelligent Systems and Computing 358, DOI: 10.1007/978-3-319-17996-4_17

different interest or benefit of both data owner and cloud server, the solution of hiring an auditor seems to be fair for the two sides.

Previous works of auditing methods by third party auditors consist of private verification and public verification. In detail, private verification scheme by Juels and Kaliski [4] encrypts data file and then embeds random value blocks (called sentinels) randomly. It is believed that if the server deletes a substantial portion of the file, a number of sentinels are also deleted and the auditor can discover this. This approach focus on only static file. The scheme by Shacham and Waters [5] bases on Message Authentication Code. They split erasure code of a file into blocks and each block into sectors, and create authenticators for each block based on Message Authentication Code (MAC). Every time the auditor wants to verify data, it sends a set of indices of blocks and the server returns corresponding data blocks along with their aggregated authenticator. The auditor afterwards applies MAC to verify integrity of blocks using its secret key shared by the data owner. This scheme gains advantage of fast computation but its downside is the server's response length and disclosure about verified data to the auditor.

The first public verification scheme by Ateniese et al. [1] was based on public key cryptosystem RSA. With this scheme, data are protected against the auditor and the server's response length is fixed by parameters of RSA cryptosystem. The scheme's downside is computation cost due to modular exponentiation operations. Another public verification scheme by Shacham and Waters [5] applies bilinear map [2] to aggregate authenticators of data blocks into one in order to reduce the server's response length. Later, a scheme by Wang et al. [6] applies pseudorandom generator to prevent the auditor from recovering data content from the message sent by the server; therefore, it provides data privacy-preservation against the auditor. The scheme's extension for batch auditing is straightforward and other schemes can be extended in the same manner.

In summary, auditing methods by third party auditors seem to focus on one or more problems among: reduction the server's response length, reduction computation cost in the server and the auditor, reduction storage overhead in the server and the auditor and data privacy preservation against the auditor, static data file or dynamic data file support. Among them, our motivation consists of computation cost reduction, efficient storage, data privacy preservation for data audit by an auditor and dynamic data file or database support.

In this paper, we propose solution for auditing by an auditor that uses secret key shared between data owner and auditor based on cryptographic hash function and Chinese Remainder Theorem. Thanks to fixed size of output of hash function, extra storage overhead for auditing is reduced. Shared secret key and Chinese Remainder Theorem provide the auditor a way to verify data in privacy preservation manner. In addition, the proposed approach gains efficient computation thanks to its simple calculations.

2 The Three-Parties Auditing Model

2.1 System Model

The model consists of three factors:

- Cloud server (CS): server that provides cloud database-as-a-service (DaS). The server is semi-trusted. It is believed not to use or reveal users data without permission. However, the server may not reveal unexpected data loss or corruption. Besides, the server may possibly decide to delete some data sections that have been rarely accessed by users.
- Data owner (DO): user who use DaS to keep their data. User wants to detect if any data has been lost, deleted or modified maliciously so it hires a third-party to audit data. However, the user does not want its data to be revealed to the third-party.
- Third-party auditor (TPA): a third-party that is hired by data owner to audit its data without knowing them. Data owner may send some information to TPA. When TPA wants to audit data in cloud server, it sends query to cloud server. Cloud server then returns corresponding metadata or proof back so that TPA can rely on to audit if data has been kept intact or not.

The design criteria of the above 3-parties model is a trade-off of three following goals:

- Low storage overhead. The additional storage used for auditing should be as small as possible on both TPA and CS.
- Low communication. The communication cost required by auditing process should be as low as possible.
- Low computational complexity. The computation for TPA should be low.

3 Proposed Method

3.1 Description

In this section, we give detailed exposition of the proposed auditing method and an example to illustrate it. In addition, we discuss about parameter selection and purpose of steps in the method.

Our proposed auditing method consists of three phases as described in the figure below.

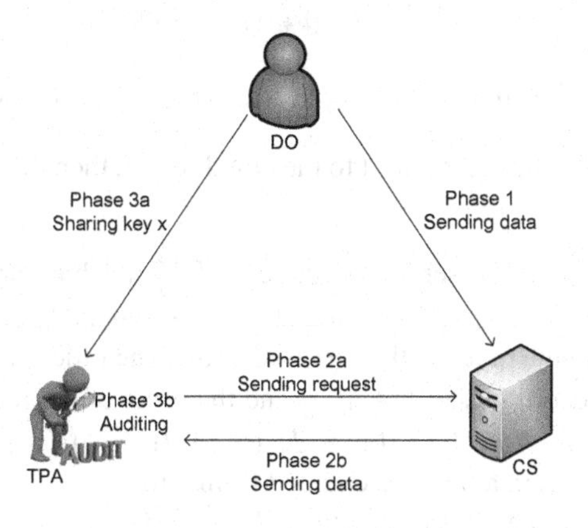

Fig. 1. The auditing process

Some conventions will be necessary for our description:

- x is shared secret key between the data owner and the auditor. Supposed that only data owner can upload data to server. the auditor cannot thanks to other access control policies of the server which we will not cover in this paper.
- $h : \{0,1\}^* \times \{0,1\}^* \rightarrow \{0,1\}^n$ is a keyed hash function that returns a value of n bits (n is a system parameter). $\{0,1\}^*$ is a binary string.
- $P : \{0,1\}^m \times \mathbb{N} \rightarrow \{0,1\}^m$ is a one-way permutation (this may be a uncompression hash function). $\mathbb{N}$ is the set of natural numbers and $\{0,1\}^m$ is a binary string of m bits.
- $Q : \{0,1\}^* \times \{0,1\}^* \rightarrow \{0,1\}^{n+1}$ is a prime generator that returns a $(n+1)$-bit prime.

Assuming that index of record in database is counted from 1.

Phase 1: The data owner (DO) sends data to the cloud server.

Each time the data owner sends a block of d data items. Assuming that the order of blocks is significant and counted from 0. Let us denote d records of the kth block by $\{R_{kd+1}, \ldots, R_{(k+1)d}\}$. The data owner will

(1) Compute $S_k = \bigoplus_{i=1}^{d} P(R_{kd+i}, kd+i)$ where $\oplus$ is bitwise XOR operation.
(2) Compute $\overline{R_{kd+i}} = S_k \oplus P(R_{kd+i}, kd+i)\ (\forall i = 1, \ldots, d) = \bigoplus_{u=1, u\neq i}^{d} P(R_{kd+u}, kd+u)$.
(3) Compute $f_{kd+i} = h(\overline{R_{kd+i}}, x)\ (\forall i = 1, \ldots, d)$.
(4) Generate primes $q_i = Q(kd+i, x), (\forall i = 1, \ldots, d)$.
(5) Solve the system of congruences for the solution X_k

$$\begin{cases} X_k \equiv f_{kd+1} \pmod{q_1} \\ X_k \equiv f_{kd+2} \pmod{q_2} \\ \vdots \\ X_k \equiv f_{(k+1)d} \pmod{q_d} \end{cases} \tag{1}$$

(6) Send the kth block and its metadata, $\{\{R_{kd+i}\}_{i=1}^{d}, X_k\}$, to the server.

Phase 2: The auditor sends its request to the cloud server, then the server sends back its proof

(1) The auditor sends to the server its request $\{I_j\}_{j=1}^{\ell}$ which are indices of records $R_{I_1}, \cdots, R_{I_\ell}, \forall j = 1, \cdots, \ell$. Given that $\{I_j = k_j d + i_j\}_{j=1}^{\ell}, 1 \leq i_j \leq d$.
(2) For each index in the request, the server identifies the order of block containing the corresponding record $\{k_j = \lfloor \frac{I_j}{d} \rfloor\}_{j=1}^{\ell}$ and the order of the record in the identified block $\{i_j \equiv I_j \pmod{d}\}_{j=1}^{\ell}$. If $i_j \equiv I_j \pmod{d} = 0$ then assign $i_j = d$ (because the order or records in a block is counted from 1 to d)
(3) For each index I_j, the server computes $\overline{R_{I_j+1}} = \bigoplus_{u=1, u\neq i_j+1}^{d} P(R_{k_j d+u}, k_j d + u)$. If $i_j + 1 = d + 1$ then assign $i_j + 1 = 1$.
(4) The server returns $\{\overline{R_{I_j+1}}, X_{k_j}\}_{j=1}^{\ell}$ to the auditor.

Phase 3: The auditor verifies if the queried records are intact or not.

(1) For each index I_j of its request (given that $I_j = k_j d + i_j, 1 \leq i_j \leq d$), the auditor generates a prime $q_{i_j+1} = Q(I_j + 1, x)$ $(j = 1, \ldots, \ell)$.
(2) For each index I_j, the auditor compares whether

$$X_{k_j} \pmod{q_{i_j+1}} = h(\overline{R_{I_j+1}}, x)$$

is true or false. If true, all $d-1$ records of the block that contains R_{I_j}, except R_{I_j+1}, are ensured to be intact. If false, at least one record among those $d-1$ records has been being modified.

Discussion

Parameter Selection: For privacy preserving security, we select the size of block $d = 3$ (see section 4.2).
Purpose of Permutation Step: The permutation function P in our scheme helps to ensure to create different metadata $\overline{R}$ for the two same records R at different indices.
Purpose of the Hash Function h: It ensures to prevent finding back the record value R from the metadata $\overline{R}$.
Selection of the Number of Requested Records: According to A. Giuseppe et al. [3], if the server deletes t records and t is a fraction of n (the number of all records), for e.g. $t = 1\% \times n$, the auditor will detect this after challenging $c = 460$ records for auditing with probability of 95%. In our scheme, with a requested record, the auditor can verify $d-1$ records in a block. Therefore, with the same $c = 460$, the probability of successful detection is higher than 95%.

Next, an example will show how the audit method works.

Example 1. Without loss of generality, assuming that we will audit the first block (block of order $k = 0$) containing $d = 3$ records, $R_1 = 23, R_2 = 17, R_3 = 29$. Given secret key $x = 10$ and supposed that the three functions are:
$h(\overline{R_j}, x) = R_j \pmod{Q(j, x)}, \forall j = 1, .., 3$
$P(R_j, j) = R_j, \forall j = 1, .., 3$
$Q(1, x) = q_1 = 13, Q(2, x) = q_2 = 11, Q(3, x) = q_3 = 7.$
Phase 1:

(1) $S_0 = P(R_1) \oplus P(R_2) \oplus P(R_3) = R_1 \oplus R_2 \oplus R_3 = 23 \oplus 17 \oplus 29 = 10111_2 \oplus 10001_2 \oplus 11101_2 = 11011_2.$
(2) $\overline{R_1} = S_0 \oplus P(R_1) = S_0 \oplus R_1 = 11011_2 \oplus 10111_2 = 1100_2 = 12$
$\overline{R_2} = S_0 \oplus P(R_2) = S_0 \oplus R_2 = 11011_2 \oplus 10001_2 = 1010_2 = 10$
$\overline{R_3} = S_0 \oplus P(R_3) = S_0 \oplus R_3 = 11011_2 \oplus 11101_2 = 0110_2 = 6$
(3) $f_1 = h(\overline{R_1}, x) = \overline{R_1} \pmod{q_1} = 1100_2 = 12,$
$f_2 = h(\overline{R_2}, x) = \overline{R_2} \pmod{q_2} = 1010_2 = 10,$
$f_3 = h(\overline{R_3}, x) = \overline{R_3} \pmod{q_3} = 0110_2 = 6$
(4) Solve system of congruences

$$\begin{cases} X_0 \equiv f_1 \pmod{q_1} \\ X_0 \equiv f_2 \pmod{q_2} \\ X_0 \equiv f_3 \pmod{q_3} \end{cases} \Leftrightarrow \begin{cases} X_0 \equiv 12 \pmod{13} \\ X_0 \equiv 10 \pmod{11} \\ X_0 \equiv 6 \pmod{7} \end{cases}$$

We have $X_0 = 1000$

(5) Send $\{R_1 = 23, R_2 = 17, R_3 = 29, X_0 = 1000\}$ to the server.

Phase 2:
Assuming that R_2, a record in database, is modified to $R'_2 = 23 = 10111_2$. The server then computes $\overline{R'_3} = R_1 \oplus R'_2 = 10111_2 \oplus 10111_2 = 0$ and returns $\{\overline{R_3} = 0, X_0 = 1000\}$ to the auditor.
Phase 3:
The auditor can detect this change. Indeed, we have
With $X_0 = 1000, q_3 = 7, X_0 \pmod{q_3} = 6 \neq 0 = \overline{R'_3} \pmod{q_3}$.
It means all records the block containing the record R_2, except the record R_3, has been ensured to be intact.

3.2 Correctness

In order to prove correctness of the proposed audit method, at first we need to prove uniqueness property of solution of the system of congruences in Equation 1.

Theorem 1. *Let X be the solution of system of congruences that established in Phase 1 of proposed scheme (Equation 1):*

$$\begin{cases} X \equiv f_1 \pmod{q_1} \\ \vdots \\ X \equiv f_d \pmod{q_d} \end{cases}$$

then

(i) The system 1 has unique solution X in $\mathbb{Z}_{q_1 \times \cdots \times q_n}$ and
(ii) $X \pmod{q_i} = f_i (\forall i = 1, \ldots, d)$ and
(iii) $X \pmod{q_i} \neq g$ given that $g \neq f_i (\forall i = 1, \ldots, d)$.

Proof. (i) In Equation 1, $f_{kd+i} < q_i, \forall i = 1, \ldots, d$ because f_{kd+i} is a n-bit output of the keyed hash function h while q_i is $n+1$-prime output of prime generator Q. The primes $q_1, \ldots, q_d$ are different because they are output of the same generator Q with different input. This is a variation of the system in Chinese Remainder Theorem because $q_1, \ldots, q_n$ are distinguished primes. Therefore, it is correct due to Chinese Remainder Theorem.
(i) and (ii) It is clear.

Theorem 2 (Correctness). *If the server passes phase 3 of the proposed audit method, the corresponding blocks containing audited data must be kept intact in the server.*

Proof. For each index I_j of the request (given that $I_j = k_j d + i_j, 1 \leq i_j \leq d$), if the corresponding block of record R_{I_j} is kept intact, the server returns to the auditor $\{\overline{R_{I_j+1}}, X_{k_j}\}$ and X_{k_j} be the solution of the congruence system:

$$\begin{cases} X_{k_j} \equiv h\left(\overline{R_{k_j d+1}}, x\right) \pmod{q_1} \\ \vdots \\ X_{k_j} \equiv h\left(\overline{R_{k_j d+d}}, x\right) \pmod{q_d} \end{cases}$$

At the auditor side, by Theorem 1 (ii), it has $X_{k_j} \pmod{q_{i_j+1}} = h(\overline{R_{I_j+1}}, x)$, and the server passes phase 3.

If there is any record $R_{I_u} (I_u \neq I_j + 1)$ in this block that has been changed, then computed value $\overline{R_{I_j+1}}$ will be different from the computed value from the original data. This leads to change of the value $h(\overline{R_{I_j} + 1}, x)$. Meanwhile, X_{k_j} is computed by the data owner based on the original data. Therefore, at the auditor side, $X_{k_j} \pmod{q_{i_j+1}} \neq h(\overline{R_{I_j+1}}, x)$. This is correct due to the Theorem 1 (iii).

4 Analysis

We will need some conventions in Table 1.

Table 1. Notations

Notation	Meaning
N	The number of records of database.
d	The number of records of a block.
R,$\|R\|$	R is a record. $\|R\|$ is the number of bits of the record R.
$\overline{R}$,$\|\overline{R}\|$	Combination of all records in the block except R. The combination is computed by bitwise XOR operation (same as step (2) in phase 1). It is easy to see that $\|\overline{R}\| = \|R\|$.
n	The number of bits of hash value created by the function $h : \{0, 1, \}^* \times \{0, 1\}^* \rightarrow \{0, 1\}^n$
X	Metadata computed from a data block as in step (5) in phase (1). It is the solution of the system on d congruence equations in Equation 1 where each prime has $n + 1$ bits. Therefore the number of bits of X is about $d \times (n + 1)$.
$\|Int\|$	The number of bits of an integer
Xor^d	Take bitwise XOR operation of d records
$Prime_n^d$	Generating d primes of n-bit
$Perm_n^d$	Permute bits of d values of n-bit
$Hash_n^d$	Hash d values into d values of n-bit
$Solve_n^d$	Solve the solution for the system of d equations of congruences modulo n bit
$Modulo_n^\ell$	Compute ℓ modulo n-bit
$Compare_n^\ell$	Compare ℓ pairs of values of n-bit

4.1 Storage Overhead, Communication Cost and Computation Complexity

Due to the limit of pages, proofs for the following results are put in Appendix.

Theorem 3. *The storage overhead at the server side is* $\mathcal{O}(N \times (|R| + n))$.

Theorem 4. *1. The communication cost when the data owner sends a data block to the server is* $\mathcal{O}(d \times (|R| + n))$.
2. *The communication cost is* $\mathcal{O}(\ell \times |Int|)$ *from the auditor to the server and* $\mathcal{O}(\ell \times (|R| + d \times n))$ *from the server to the auditor when the auditor requests to audit* ℓ *records.*

Theorem 5. *1. The computation complexity at the data owner side is* $Xor^d + Perm_n^d + Hash_n^d + Prime_n^d + Solve_n^d$ *when the data owner sends one data block to the server.*

2. *The computation complexity at the auditor side is* $Prime_n^\ell + Modulo_n^\ell + Hash_n^\ell + Compare_n^\ell$ *when the auditor requests to audit* ℓ *records.*

4.2 Security

We concern the list of attack scenarios below:

1. Privacy preserving attack: The auditor recovers the content of record R from the proof $\overline{R}, X_k$ sent by the server
2. Replace attack: The server has deleted a record R. When the auditor requests to audit on the record R, the server selects another valid and uncorrupted record R' and its metadata to return to the auditor.
3. Replay attack: The server has deleted a record R_j. When the auditor requests to audit on the record R_j,the server generates proof from previous proof without querying the actual data.
   The proof of a record R consists of $\overline{R_{j+1}}, X$ where the size of $\overline{R_{j+1}}$ is the same as the size of R. In our context, the server is semi-trusted. Data loss is caused by unexpected corruption or by server's deletion to save storage. Therefore, the fact of deleting R_j but keeping its proof $\overline{R_{j+1}}$ does not make sense. We skip this kind of attack.
4. Forge attack: The server has deleted a record R_j. When the auditor requests to audit on the record R_j,the server forges its proof without querying the actual data.
   From details of phase 1 of the scheme, it is seen that the server needs to know the actual data and secret key x to create metadata. Therefore, we consider the secret key attack instead.
5. Secret key attack: The server finds the secret key shared between data owner and auditor.

Privacy-Preserving Attack. We have the following lemma.

Lemma 1. *Given* $a \oplus b, a \oplus c, b \oplus c$*, it cannot recover* a, b *or* c*.*

Proof. It can be proved by the following table

Table 2. Proof of lemma 1

ID	Given value	Equal expression
(1)	$a \oplus b$	$(2) \oplus (3)$
(2)	$a \oplus c$	$(1) \oplus (3)$
(3)	$b \oplus c$	$(1) \oplus (2)$
(4)	0	$(1) \oplus (2) \oplus (3)$

We have $(1) = a \oplus b = (a \oplus b) \oplus (a \oplus c) = (2) \oplus (3)$. Similarly, $(2) = (1) \oplus (3)$, $(3) = (1) \oplus (2)$. Therefore, from $(1), (2), (3)$, there is no way to find out a, b, c.

Based on the lemma 1, we will prove security for the proposed auditing method if the number of records in a block is $d = 3$.

Theorem 6. *From the server response* $\{\overline{R_{I_j+1}}, X_{k_j}\}_{j=1}^{\ell}$, *no information of records* $\{R_{I_j}\}_{j=1}^{\ell}$ *is leaked to the auditor if choosing* $d = 3$.

Proof. For each requested index I_j, given $I_j = k_j d + i_j (1 \leq i_j \leq d)$ and X_{k_j} is the solution of the system of congruences

$$\begin{cases} X_{k_j} \equiv h(\overline{R_{k_j d+1}}, x) \pmod{q_1} \\ \vdots \\ X_{k_j} \equiv h(\overline{R_{k_j d+d}}, x) \pmod{q_d} \end{cases}$$

where x is secret key, $q_i = Q(k_j d + i, x)$ is a prime, $\overline{R_{k_j d+i}} = \bigoplus_{u=1, u\neq i}^{d} R_{k_j d+u}$ ($\forall i = 1, \ldots, d$).

With $d = 3$, $\overline{R_{k_j d+1}} = R_{k_j d+2} \oplus R_{k_j d+3}$, $\overline{R_{k_j d+2}} = R_{k_j d+1} \oplus R_{k_j d+3}$, $\overline{R_{k_j d+3}} = R_{k_j d+1} \oplus R_{k_j d+2}$. According to Lemma 1, it cannot recover any record $R_{k_j d+1}, R_{k_j d+2}, R_{k_j d+3}$ in the block k_j from $\overline{R_{k_j d+1}}, \overline{R_{k_j d+2}}, \overline{R_{k_j d+3}}$.

Replace Attack Assuming that the auditor requests on the record R_I but the server returns proof $\overline{R_{J+1}}, X'$ for another record R_J. When the auditor verifies, it computes a prime $q = Q(I, x)$ and compares $X' \pmod{q} = h(\overline{R_{J+1}}, x)$. Actually $X' \pmod{q'} = h(\overline{R_{J+1}}, x)$ where $q' = Q(J, x)$ and $q \neq q'$. Therefore, $X' \pmod{q} \neq h(\overline{R_{J+1}}, x)$ and the server cannot pass the auditor's verifying.

Secret Key Attack

Theorem 7. *Supposed that the cryptographic hash function h is secure. From data sent from the data owner,* $\{\overline{R_{kd+i}}\}_{i=1}^{d}, X_k$, *no information of the secret key* x *is leaked to the server.*

Proof. X_k is the solution of the system of congruences in Equation 1. Knowing X_k but $q_1, \ldots, q_d$, the server cannot compute $f_{kd+1}, \ldots, f_{kd+d}$. Note that even the server knows $f_{kd+1}, \ldots, f_{kd+d}$, it cannot recover the secret key x because $f_{kd+i} = h(\overline{R_{kd+i}}, x)$ where h is supposed to be secure.

4.3 Comparison to the State-of-Art Schemes

Compared to the two methods of private verification and public verification in the state-of-art [5], our proposed scheme has more efficient computation and storage overhead but heavier communication cost. Our efficient computation is thanks to simpler calculations of hashing and modulo compared to exponential and pairing [5]. Moreover, our storage overhead is proportional to the fixed size of hash function output instead of size of records as in [5], therefore it is more efficient. However, our proposed scheme has heavier communication since it can detect the exact block that is not intact and preserve data privacy against the third party auditor. Details of comparison is in the Appendix.

In addition, both state-of-art schemes and the proposed scheme can be applied for dynamic data file. This proposed scheme can be also applied for database. Basic operations (deletion, insertion, modification) are described in Appendix.

5 Conclusion

In this paper, we proposed an efficient and secure method of data audit. We then proved that our data auditing is privacy preserving based on cryptographic hash function and CRT, which can work online in cloud sufficiently. This can detect malicious deletion or modification with low computation cost. However, it will make an offset of communication cost to identify exactly the modified block while keeping data privacy. We believe that this is the future work of big data area with promising results.

References

1. Ateniese, G., Burns, R., Curtmola, R., Herring, J., Khan, O., Kissner, L., Peterson, Z., Song, D.: Remote data checking using provable data possession. ACM Transactions on Information and System Security (TISSEC) 14(1), 12 (2011)
2. Boneh, D., Lynn, B., Shacham, H.: Short signatures from the weil pairing. In: Boyd, C. (ed.) ASIACRYPT 2001. LNCS, vol. 2248, pp. 514–532. Springer, Heidelberg (2001)
3. Giuseppe, A., Randal, B., Reza, C., et al.: Provable data possession at untrusted stores. Proceedings of CCS 10, 598–609 (2007)
4. Juels, A., Kaliski Jr., B.S.: Pors: Proofs of retrievability for large files. In: Proceedings of the 14th ACM Conference on Computer and Communications Security, pp. 584–597. ACM (2007)
5. Shacham, H., Waters, B.: Compact proofs of retrievability. In: Pieprzyk, J. (ed.) ASIACRYPT 2008. LNCS, vol. 5350, pp. 90–107. Springer, Heidelberg (2008)
6. Wang, C., Wang, Q., Ren, K., Lou, W.: Privacy-preserving public auditing for data storage security in cloud computing. In: 2010 Proceedings IEEE INFOCOM, pp. 1–9. IEEE (2010)

A Appendix

A.1 Proof for Storage Overhead, Communication Cost and Computation Complexity

Theorem 8. *The storage overhead at the server side is $\mathcal{O}(N \times (|R| + n))$.*

Proof. Following phase 1 of the proposed auditing method in section 3.1, for each data block, a metadata is created and sent from the data owner. The number of blocks for the whole database is $\frac{N}{d}$. The size of a metadata is $d \times (n+1)$. Therefore, storage overhead at the server side is computed $N \times |R| + \frac{N}{d} \times d \times (n+1) = N \times (|R| + n + 1)$ or $\mathcal{O}(N \times (|R| + n))$

Theorem 9. *1. The communication cost when the data owner sends a data block to the server is $\mathcal{O}(d \times (|R| + n))$.*

2. *The communication cost is $\mathcal{O}(\ell \times |Int|)$ from the auditor to the server and $\mathcal{O}(\ell \times (|R| + d \times n))$ from the server to the auditor when the auditor requests to audit ℓ records.*

Proof. 1. Each time the owner sends a block of d records and its metadata to the server. Each record has size of $|R|$ bits and each metadata has size of $d \times (n+1)$ bits. Total cost of the data block and its metadata is

$$\text{Cost} = d \times |R| + d \times (n+1) = d \times (|R| + n + 1) \text{ or } \mathcal{O}(d \times (|R| + n))$$

2. When the auditor needs to audit data in the server, it sends a set of ℓ indices $\{I_j\}_{j=1}^{\ell}$ to the server. Each index is an integer. Therefore, total cost is $\mathcal{O}(\ell \times |Int|)$. After receiving request $\{I_j\}_{j=1}^{\ell}$ from the auditor, the server sends back $\{\overline{R_{I_j+1}}, X_{k_j}\}_{j=1}^{\ell}$ where $|\overline{R_{I_j+1}}| = |R|$ bits and X_{k_j} has size of $d \times (n+1)$ bits. Therefore, communication cost is

$$\text{Cost} = \ell \times (2|R| + d \times (n+1)) \text{ or } \mathcal{O}(\ell \times (|R| + d \times n)).$$

A.2 Comparison with the State-of-Art Schemes

Denote that

- N is the number of records
- $|R|$ is size of a record
- n is size of output of hash function
- d is the number of record of a block
- $Mult^d$ is d multiplications
- $Rand^d$ is d random generating operations
- $Hash^d$ is d hashing operations
- Xor^d is d Xor operations
- $Prime^d$ is d prime generation operations
- $Solve^d$ is solving a system of d congruences

Comparison about computation cost, communication cost and storage overhead are presented in Table 3.

A.3 Database Operations

The three basic operations for database consists of data insertion, deletion and modification. Here we describe details of each operation so that our auditing method can be applied for database.

- Insertion: Each time the data owner sends a block of d records and the server always inserts a block at the end of the database. In case of less than d records, the data owner can wait until there are enough or create a few fake constant records.

Table 3. Comparison between the proposed scheme and the state-of-art schemes

	Private verification [5]	Pubic verification [5]	The proposed scheme
Keys	Secret key	A pair of public and private key	Secret shared key
Who can audit data?	Data owner	Any third party	Third party who is shared key with data owner
Aggregated proof	Yes	Yes	No
Computation cost to send 1 block (at data owner)	$Mult^d + Rand^d$	$Hash^d + Exp^d + Mult^d$	$Xor^d + Hash^d + Prime^d + Solve^d$
Computation cost of auditing ℓ records (at auditor)	$Mult^\ell$	$Hash^\ell + Exp^\ell + Mult^\ell + Pairing^2$	$Prime^\ell + Modulo^\ell + Hash^\ell$
Storage cost (at server)	$2N \times (\|R\|)$	$2N \times (\|R\|)$	$N \times (\|R\| + n)$
Communication cost (from server to auditor) when auditing ℓ records	$\|R\|d$	$\|R\|d$	$\ell(\|R\| + dn)$

- Deletion: If the data owner wants to delete a record R_i, he/she sends query to the server. The server firstly identifies the block containing R_i. Assuming that there are D blocks in the database and the identified block is the kth block. If the last block has full of d records, he/she permutes the kth block and D^{th} block, then sends the new last block to data owner for deleting record R_i and updating X_D before sending back to server. Otherwise, he/she sends the kth and D^{th} blocks to data owner. The data owner then merges data of the two blocks so that the kth block is full of d records. Data owner computes X_k and X_D. Finally he/she resends the updated k^{th} block with X_k and D^{th} block with X_D to server.
- Modification: If the data owner wants to modify a record R_i into R'_i, he/she sends query to the server. The server firstly identifies which block contains R_i. Assuming that it is k^{th} block. It returns the kth block to the data owner. The data owner modifies R_i and recalculates X_k before sending back to server.

Incremental Mining Class Association Rules Using Diffsets

Loan T.T. Nguyen[1,2] and Ngoc Thanh Nguyen[2]

[1] Division of Knowledge and System Engineering for ICT, Faculty of Information Technology, Ton Duc Thang University, Ho Chi Minh City, Vietnam
[2] Department of Information Systems, Faculty of Computer Science and Management, Wroclaw University of Technology, Poland
nguyenthithuyloan@tdt.edu.vn, nthithuyloan@gmail.com, Ngoc-Thanh.Nguyen@pwr.edu.pl

Abstract. Class association rule (CAR) mining is to find rules that their right hand sides contain class labels. Some recent studies have proposed algorithms for mining CARs; however, they use the batch process to solve the problem. In the previous work, we proposed a method for mining CARs from incremental datasets. This method uses MECR-tree to store itemsets and rules are easy to generate based on this tree. Pre-large concept is used to reduce the number of rescan datasets and Obidsets (set of object identifiers) are used to fast compute the support of itemsets. CAR-Incre algorithm has been proposed. However, when the number of inserted records is large, this method still consumes much time to compute the intersection of Obidsets on dense datasets. In this paper, we modify CAR-Incre algorithm by using Diffsets (difference between Obidsets) instead of Obidsets. CAR-Incre is adjusted to fit Diffsets. We verify the improved algorithm by experiments.

Keywords: data mining, class association rule, incremental dataset, Obidset, Diffset.

1 Introduction

The first method for mining CARs was proposed by Liu et al. in 1998 [8]. It is based on Apriori to mine CARs. After that, CMAR, an FP-tree-based approach, was proposed by Li et al. in 2001 [6] to reduce the number of scans dataset and improve the accuracy of classifiers. Some efficient methods have been proposed after that such as ECR-CARM [13], MAC [1], and CAR-Miner [10, 12]. These methods only focus on solving the problem of CAR mining based on the batch process approach. However, real datasets typically change due to insertion/deletion. Algorithms for effectively mining CARs from incremental datasets are thus required. A naïve method is to re-run the CAR mining algorithm on the updated dataset. The original dataset is often very large whereas the inserted portion is often small. This approach is not effective because the entire dataset must be re-scanned. In addition, previous mining results cannot be reused. An efficient algorithm for update CARs when some records are inserted to the original dataset, named CAR-Incre, has been proposed [11]. CAR-Incre only updates some nodes in MECR-tree if the number of inserted records is smaller than or equal to a safety threshold. In this approach, Obidsets are used to

H.A. Le Thi et al. (eds.), *Advanced Computational Methods for Knowledge Engineering*, Advances in Intelligent Systems and Computing 358, DOI: 10.1007/978-3-319-17996-4_18

fast compute the support of itemsets. The experimental results show a good performance of CAR-Incre compared to CAR-Miner [12]. When the number of inserted records on a dense dataset is large, the required time for computing the intersection of Obidsets is large as well. This paper presents an improved version of CAR-Incre using Diffsets. Firstly, CAR-Miner-Diff is used to construct MECR-tree from the original dataset. When records are inserted into the original dataset, we update MECR-tree by computing information of changed nodes using Diffsets.

The rest of the paper is organized as follows: Section 2 presents related works related to CAR mining, (class) association rule mining from incremental datasets. Section 3 presents the proposed algorithm. An example to illustrate the proposed algorithm is also presented in this section. Section 4 presents the experimental results. Conclusions and future work are described in Section 5.

2 Related Work

2.1 Basic Concepts

Let $D = \{((A_1, a_{i_1}), (A_2, a_{i_2}), \ldots, (A_n, a_{i_n}), (C, c_i)) : i = 1, .., |D|\}$ be a training dataset, where A_1, A_2, …, A_n are attributes and $a_{i_1}, a_{i_2}, \ldots, a_{i_n}$ are their values. Attribute C is the class attribute for which different values c_1, c_2, …, c_k ($k \leq |D|$) represent classes in D. Without inconsistency, we can use the symbol C to denote the set of classes, that is we can write $C = \{c_1, c_2, \ldots, c_k\}$. One of attributes among A_1, A_2, …, A_n is used to identify the records in D, often denoted by the symbol OID. Let $A = \{A_1, A_2, \ldots, A_n\} \backslash \{OID\}$. Attributes from A are often called description attributes.

For example, let D be the training dataset shown in Table 1, which contains 8 records (|D| = 8), where A = {A, B, C}, class is the decision attribute, and C = {y, n}, i.e., there are 2 classes.

Table 1. An example of training dataset

OID	*A*	*B*	*C*	*class*
1	*a1*	*b1*	*c1*	*y*
2	*a1*	*b1*	*c2*	*n*
3	*a1*	*b2*	*c2*	*n*
4	*a1*	*b3*	*c3*	*y*
5	*a2*	*b3*	*c1*	*n*
6	*a1*	*b3*	*c3*	*y*
7	*a2*	*b1*	*c3*	*y*
8	*a2*	*b2*	*c2*	*n*

Definition 1: An itemset is a set of pairs (a, v) where a is an attribute from A, and v is its value. Any attribute from A can appear at most once in the itemset.

From Table 1, we have many itemsets, such as {(B, b3)}, {(A, a1), (B, b1)}, etc.

Definition 2: A class association rule r is in the form of X→ c, where X is an itemset and c is a class.

Definition 3: The support of an itemset X, denoted by Supp(X), is the number of records in D containing X.

Definition 4: The support of a class association rule r = X→ c_i, denoted by Supp(r), is the number of records in D containing X and (C, c_i).

Definition 5: The confidence of a class association rule r = X→ c_i, denoted by Conf(r), is defined as:

$$\text{Conf(r)} = \frac{Supp(r)}{Supp(X)}.$$

For Example: consider rule r: (A, a1) → y with X = (A, a1) and c_i = y. We have Supp(X) = 3, Supp(r) = 2, and Conf(r) = $\frac{Supp(r)}{Supp(X)} = \frac{2}{3}$.

2.2 Mining Class Association Rules

There are some methods to mine class association rules (CARs) in static datasets. They are divided into three main approaches: 1). Method based on Apriori [8]; 2). Method based on FP-Tree approach [6]; 3) Methods based on equivalence class rule tree [10, 12, 13].

The first method to mine CARs is CBA[8]. It has two parts: First, the authors are based on Apriori to mine all CARs. Second, after mining all CARs, they build a classifier using heuristics. The second method for mining CARs is CMAR [6]. It also includes two parts: mine all CARs based on FP-tree and use database coverage to build the classifier. In 2008, Vo and Le proposed a method to mine CARs named ECR-CARM [13]. They proposed a tree structure to store all frequent itemsets and rules are easily generated based on this tree. Besides, Obidset was used to fast compute the support of itemsets. Then they proposed a theorem to prune some rules with confidence 100%. In 2013, CAR-Miner was proposed [10]. In this approach, we used MECR-tree (Modification of Equivalence class rule tree) to store all frequent itemsets and generate rules. Two theorems were proposed to eliminate some nodes on the tree.

2.3 Mining Association Rules from Incremental Datasets

In 1996, Cheung et al proposed Fast Update algorithm (FUP) [2]. The authors divide itemsets in the original dataset and inserted dataset into two categories: frequent and infrequent. So there are four cases that need to be considered:

1). Itemset is frequent on both original dataset and inserted dataset: It must be frequent on the updated dataset. This case does not need to be considered to know whether it is frequent.

2). Itemset is frequent on original dataset but infrequent on inserted dataset: It may be frequent or infrequent on the updated dataset. However, only its new support count in the inserted dataset needs to be considered.

3). Itemset is infrequent on the original dataset but frequent on the inserted dataset: It may be frequent or infrequent on the updated dataset. In this case it must be rescanned in the updated dataset to determine whether this itemset is frequent.

4). Itemset is infrequent on both original and inserted datasets: It must be infrequent on the updated dataset. This case does not need to be considered the same as case 1.

We must rescan the updated dataset in case 3 so it requires a lot of time and effort if the dataset is too large. In order to minimize the number of scans of the original dataset and the number of generated itemsets from the new data, Hong et al. [3] proposed the concept of pre-large itemsets. A pre-large itemset is an infrequent itemset, but its support is larger than or equal to a lower support threshold. In the concept of pre-large itemsets, two minimum support thresholds are used. The first is the upper minimum support S_U (which is also the minimum support threshold) and the second is the lower minimum support S_L. With these two minimum support thresholds, an itemset is placed into one of three categories: frequent, pre-large, or infrequent. Thus, there are 9 cases when considering an itemset in two datasets (original and inserted):

1). Itemset is frequent on both original and inserted datasets: It must be frequent on the updated dataset. This case does not need to be considered.
2). Itemset is frequent on the original dataset but pre-large on the inserted dataset: It may be frequent or pre-large on the updated dataset. This is easily determined by using the information of this itemset.
3). Itemset is frequent on the original dataset but infrequent on the inserted dataset: It may be frequent, pre-large, or infrequent on the updated dataset. This is easily determined by using the information of this itemset.
4). Itemset is pre-large on the original dataset but frequent on the inserted dataset: It may be frequent or pre-large on the updated dataset. This is easily determined by using the information of this itemset.
5). Itemset is pre-large on both original and inserted datasets: It must be pre-large on the updated dataset. This case does not need to be considered.
6). Itemset is pre-large on the original dataset but infrequent on the inserted dataset: It may be pre-large or infrequent on the updated dataset. This is easily determined by using the information of this itemset.
7). Itemset is infrequent on the original dataset but frequent on the inserted dataset: We must re-scan the original dataset to compute the support of this itemset.
8). Itemset is infrequent on the original dataset but pre-large on the inserted dataset: It may be pre-large or infrequent. This is easily updated by using the information of this itemset.
9). Itemset is infrequent in both original dataset and inserted dataset: It must be infrequent on the updated dataset. This case does not need to be considered.

To reduce the number of re-scans on the original dataset, the authors proposed the following safe threshold formula f (i.e., if the number of added records does not exceed the threshold, then the original dataset does not need to be considered):

$$f = \left\lfloor \frac{(S_U - S_L) \times |D|}{1 - S_U} \right\rfloor$$

where |D| is the number of records in the original dataset.

In 2009, Lin et al. [7] proposed the Pre-FUFP algorithm for mining frequent itemsets in a dataset by combining the FP-tree and the pre-large concept. They proposed an algorithm that updates the FP-tree when a new dataset is inserted using the *f* safety threshold. After the FP-tree is updated, the FP-Growth algorithm is applied to mine frequent itemsets in the whole FP-tree (created from the original dataset and inserted dataset). The updated FP-tree contains the entire resulting dataset, so this method does not reuse information of previously mined frequent itemsets and thus frequent itemsets must be re-mined from the FP-tree. Some effective methods for mining itemsets in incremental datasets based on the pre-large concept have been proposed, such as methods based on Trie [5] and the IT-tree [4], a method for fast updating frequent itemsets based on a lattice [14].

2.4 Mining Class Association Rules from Incremental Datasets

In 2014, we proposed a method to mine all CARs in incremental dataset in the case of records inserted [11]. We used pre-large concept to reduce the number of re-scans on the dataset. Besides, a theorem is developed to eliminate some nodes that do not generate rules on the tree. This approach used intersection between Obidsets to compute the support of itemsets.

3 A Method for Updating CARs in Incremental Dataset Using Diffsets

In this section, we present an improved version of CAR-Incre [12] using Diffset instead of using Obidset. The modified algorithm is presented in Figure 1.

3.1 CAR-Incre-Diff Algorithm

Input: The MECR-tree built from original dataset D in which L_r is the root node, inserted dataset D', two minimum support thresholds S_U and S_L, and *minConf*
Output: Class association rules that satisfy S_U and *minConf* from $D + D'$

CAR-Incre-Diff( )
Begin
1. if $|D| = 0$ then
   Begin
2. Call the procedure **Modified-CAR-Miner** [11] to mine CARs in D' using S_L.
3. Compute $f = \left\lfloor \frac{(S_U - S_L) \times |D'|}{1 - S_U} \right\rfloor$
   Endif
4. else if $|D'| > f$ then
   Begin
5. Call the procedure **Modified-CAR-Miner** to mine CARs in $D + D'$ using S_L.
6. Compute $f = \left\lfloor \frac{(S_U - S_L) \times (|D| + |D'|)}{1 - S_U} \right\rfloor$
   Endif
7. Else
   Begin
8. Clear the Obidset of each node in the MECR-tree

Fig. 1. Algorithm for updating the MECR-tree on an incremental dataset

```
9.              Update information of nodes in the first level of Lr including Obidset,
                count, and pos and mark them
10.             Call the procedure UDATE-TREE to update the MECR-tree
11.             Call the procedure GENERATE-RULE [11] to generate CAR from the
                MECR-tree
12.             f = f - |D'|
            End
13.     D = D + D'
End.
UPDATE-TREE(Lr)
Begin
14.     for all li ∈ Lr.children do
          Begin
15.       if li is not marked then
16.           TRAVERSE-TREE-TO-CHECK(li);
17.       else
            Begin
18.           if li.count[li.pos] < SL × (|D| + |D'|) then
19.               DELETE-TREE(li);
20.           else
21.               for all lj ∈ Lr.children, with j > i do
22.                   if li.att ≠ lj.att and lj is marked then
                      Begin
23.                       Let O be a direct child node of li and lj;
24.                       if O has existed in the tree then
                            Begin
25.                         if Lr = null then
26.                              O.Obidset = li.Obidset \ lj.Obidset;
27.                         else
28.                              O.Obidset = lj.Obidset \ li.Obidset;
29.                         O.total = li.total - |O.Obidset|;
30.                         if O.total > 0 then
                              Begin
31.                             Update O.count based on O.Obidset;
32.                             O.pos = argmax_{i∈[1,k]} {O.count_i};
33.                             if O.count[O.pos] ≥ SL × (|D| + |D'|) then
34.                                 Mark O;
                              Endif
                            Endif
                      Endif
35.           UPDATE-TREE(li);
            Endelse
        Endfor
End.
```

Fig. 1. *(continued)*

Figure 1 shows the algorithm for updating the MECR-tree when dataset *D'* is inserted. Firstly, the algorithm checks whether the MECR-tree was created by considering the number of records in the original dataset. If the number of records is 0 (line 1), it means that the tree was not created yet. **Modified-CAR-Miner** is called to create the MECR-tree for dataset *D'* (line 2) and the *f* safety threshold is computed using Eq. (1) (line 3). If the number of records in dataset *D'* is larger than the *f* safety threshold, then the algorithm calls **Modified-CAR-Miner** to generate rules in the entire dataset *D* + *D'* (lines 4 and 5), and then computes the *f* safety threshold based on the integrated dataset *D* + *D'* (line 6). If the number of records in dataset *D'* is not larger than *f*, then the algorithm simply updates the MECR-tree as follows. First, all Obidsets of nodes on the tree (line 8) are deleted to ensure that the algorithm works on the inserted dataset only. The information of nodes in the first level of the MECR-tree whose itemsets are contained in the inserted dataset is marked and these nodes are marked (line 9). Second, the **UPDATE-TREE** procedure with root node L_r is called to update the information of nodes on the tree (line 10). Third, the procedure **GENERATE-RULES** with L_r is called to generate rules whose supports and confidences satisfy S_U and *minConf* (line 11). The *f* safety threshold is reduced to $f - |D'|$ (line 12). Finally, the original dataset is supplemented by *D'* (line 13).

Procedure UPDATE-TREE, TRAVERSE-TREE-TO-CHECK, DELETE-TREE, and GENERATE-RULES are the same [11].

3.2 An Illustrative Example

Assume that the dataset in Table 1 is the original dataset and the inserted dataset has one record, as shown in Table 2.

Table 2. Inserted dataset

OID	*A*	*B*	*C*	*class*
9	*a1*	*b1*	*c2*	*y*

With $S_U = 25\%$ and $S_L = 12.5\%$, the process of creating and updating the MECR-tree is illustrated step by step as follows.

The *f* safety threshold is computed as follows:

$$f = \left\lfloor \frac{(0.25 - 0.125) \times 8}{1 - 0.25} \right\rfloor = 1$$

Consider the inserted dataset. Because the number of records is 1, $|D'| = 1 \leq f = 1$, the algorithm updates the information of nodes in the tree without re-scanning original dataset *D*.

The process for updating the MECR-tree is as follows. First, the first level of the ME CR-tree is updated. The results are shown in Figure 2.

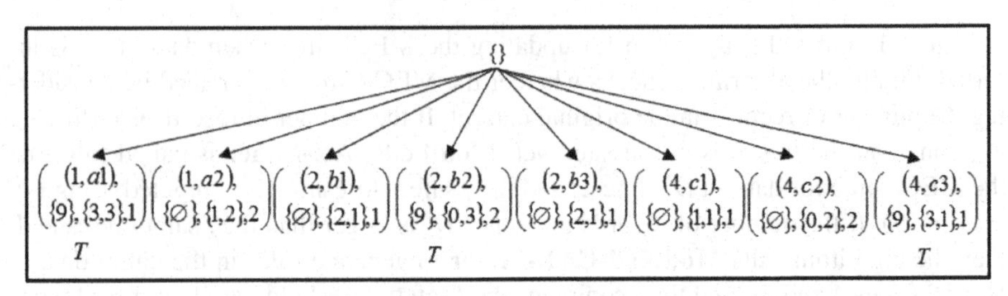

Fig. 2. Results of updating level 1 of MECR-tree

The new record (record 9) contains items (*A*, a1), (*B*, b2), and (*C*, c3), so three nodes in the MECR-tree are changed (marked by *T* in Figure 2).

- Consider node $l_i = \begin{pmatrix} (1, a1), \\ \{9\}, \{3,3\}, 1 \end{pmatrix}$. Because it has been changed, it needs to be checked with its following nodes (only changed nodes) in the same level to update information:
  - With node $l_j = \begin{pmatrix} (2, b2), \\ \{9\}, \{0,3\}, 2 \end{pmatrix}$, the node created from these two nodes (after update) is $\begin{pmatrix} (3, a1b2), \\ \{\varnothing\}, \{0,2\}, 2 \end{pmatrix}$. This node has *count*[*pos*] = $2 \geq S_L \times (8+1) = 1.125$, so it is marked as a changed node.
  - With node $l_j = \begin{pmatrix} (4, c3), \\ \{9\}, \{3,1\}, 1 \end{pmatrix}$, the node created from these two nodes (after update) is $\begin{pmatrix} (5, a1c3), \\ \{\varnothing\}, \{2,1\}, 1 \end{pmatrix}$. This node has *count*[*pos*] = $2 \geq S_L \times (8+1) = 1.125$, so it is marked as a changed node.
- Results obtained after considering node $\begin{pmatrix} (1, a1), \\ \{9\}, \{3,3\}, 1 \end{pmatrix}$ are shown in Figure 3.

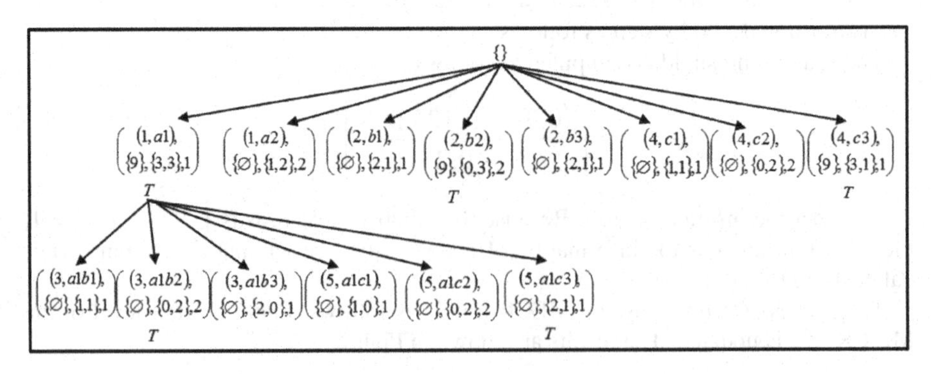

Fig. 3. Results obtained after considering node $\begin{pmatrix} (1, a1), \\ \{9\}, \{3,3\}, 1 \end{pmatrix}$

- After considering node $\begin{pmatrix} (1, a1), \\ \{9\}, \{3,3\}, 1 \end{pmatrix}$, the algorithm is called recursively to update all its child nodes.
  - Consider node $\begin{pmatrix} (3, a1b1), \\ \{9\}, \{2,1\}, 1 \end{pmatrix}$. Because $count[pos]$ =1 < S_L × (8+1), node $\begin{pmatrix} (3, a1b1), \\ \{9\}, \{2,1\}, 1 \end{pmatrix}$ is deleted. All its child nodes are also deleted because their supports are smaller than $S_L \times (8+1)$.
  - Consider node $\begin{pmatrix} (3, a1b2), \\ \{\varnothing\}, \{0,2\}, 2 \end{pmatrix}$. Because $count[pos] \geq S_L \times (8+1)$, it will check with node $\begin{pmatrix} (5, a1c3), \\ \{\varnothing\}, \{2,1\}, 1 \end{pmatrix}$. The node created from these two nodes (after update) is $\begin{pmatrix} (7, a1b2c3), \\ \{\varnothing\}, \{0,1\}, 1 \end{pmatrix}$. This node is deleted because $count[pos] = 1 < S_L \times (8+1)$. All its child nodes are deleted.
  - Similarly, nodes $\begin{pmatrix} (5, a1c1), \\ \{\varnothing\}, \{1,0\}, 1 \end{pmatrix}$ is also deleted.
- Consider node $\begin{pmatrix} (1, a2), \\ \{\varnothing\}, \{1,2\}, 2 \end{pmatrix}$. This node is not deleted.

The MECR-tree after all updates is shown in Figure 4.

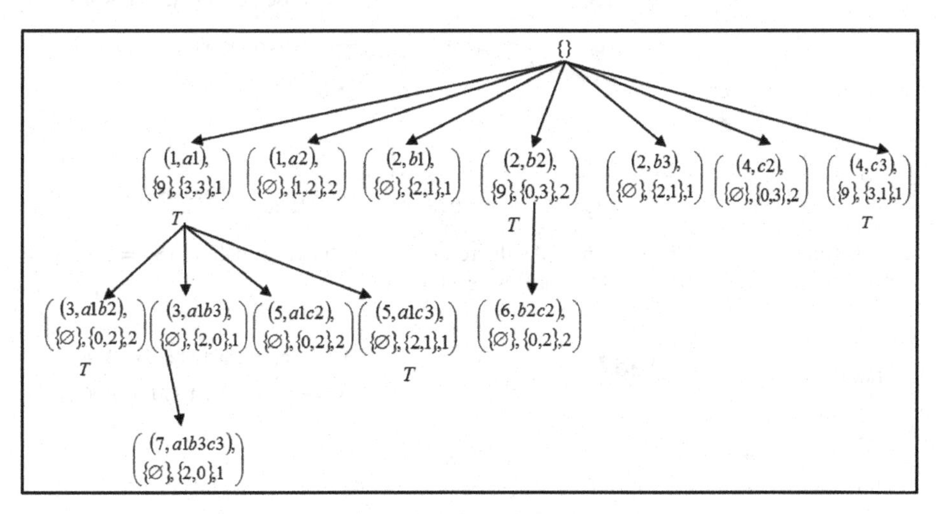

Fig. 4. Updated MECR-tree

Note that after the MECR-tree is updated, the *f* safety threshold is decreased by 1 $\Rightarrow f = 0$, which means that if a new dataset is inserted, then the algorithm re-builds MECR-tree for the original and inserted datasets. $D = D + D'$ includes nine records.

4 Experiments

The experimental results were tested in the datasets obtained from the UCI Machine Learning Repository (http://mlearn.ics.uci.edu). Table 3 shows the characteristics of the experimental datasets.

Table 3. The characteristics of the experimental datasets

Dataset	#attrs	#classes	#distinct values	#Objs
German	21	2	1,077	1,000
Led7	8	10	24	3,200
Chess	37	2	75	3,196
Connect	43	3	129	67,557
Poker-hand	11	10	95	1,000,000

The algorithms used in the experiments were coded on a personal computer with C#2012, Windows 8.1, i5-4200U, 1.60 GHz, 4 GB RAM, and 128 GB main memory. Experiments were made to compare the execution time between CAR-Incre (Nguyen and Nguyen, 2014) and CAR-Incre-Diff. The results are shown from Figure 5 to Figure 9.

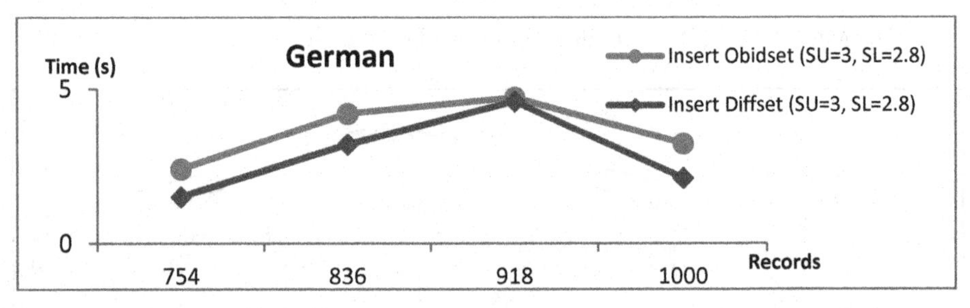

Fig. 5. Runtimes of CAR-Incre and CAR-Incre-Diff for the German dataset (S_U = 3%, S_L = 2.8%) for each inserted dataset (82 records for each insert)

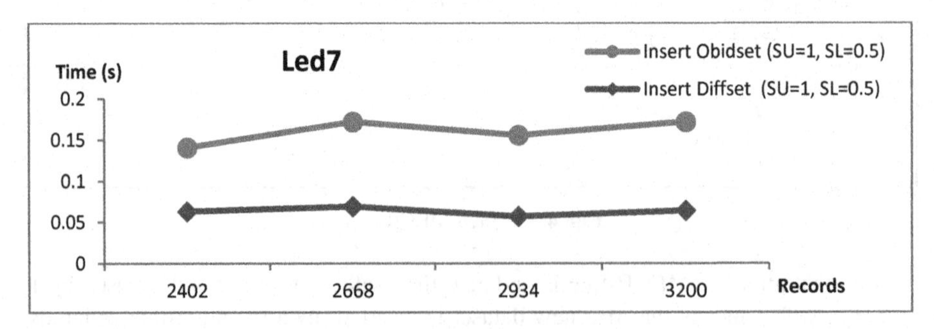

Fig. 6. Runtimes of CAR-Incre and CAR-Incre-Diff for the Led7 dataset (S_U = 1%, S_L = 0.5%) for each inserted dataset (266 records for each insert)

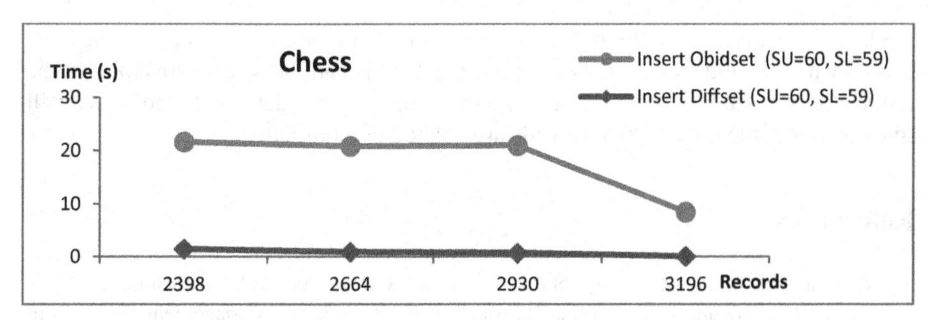

Fig. 7. Runtimes of CAR-Incre and CAR-Incre-Diff for the Chess dataset (S_U = 60%, S_L = 59%) for each inserted dataset (266 records for each insert)

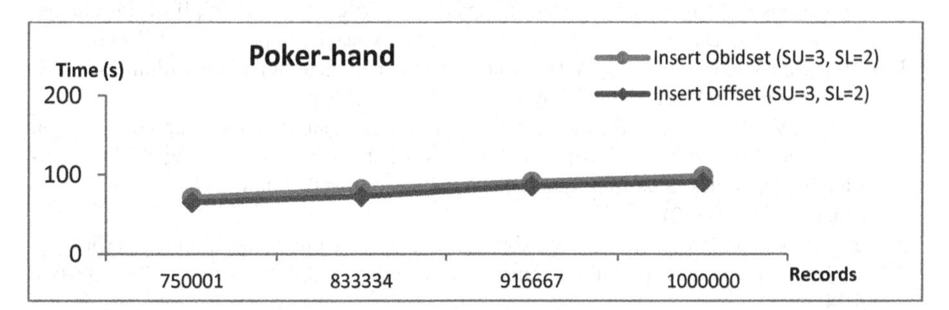

Fig. 8. Runtimes of CAR-Incre and CAR-Incre-Diff for the Poker-hand dataset (S_U = 3%, S_L = 2%) for each inserted dataset (83333 records for each insert)

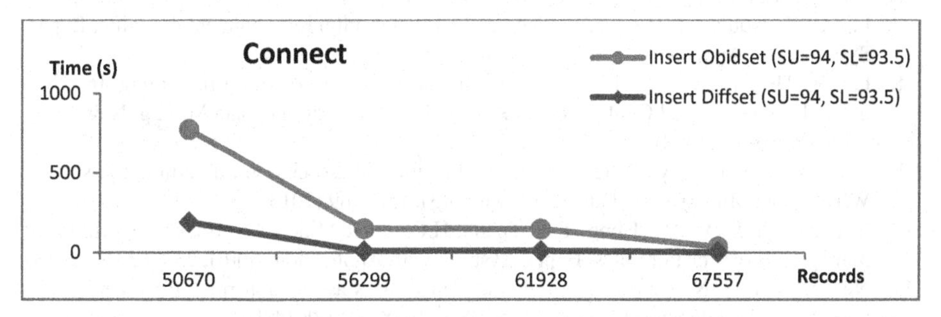

Fig. 9. Runtimes of CAR-Incre and CAR-Incre-Diff for the Connect dataset (S_U = 94%, S_L = 93.5%) for each inserted dataset (5629 records for each insert)

The experimental results from Figures 5 to 9 show that CAR-Incre-Diff is more efficient than CAR-Incre in most cases, especially on dense datasets.

5 Conclusions and Future Work

This paper proposed a method for mining all CARs in incremental dataset with records inserted. We use a Diffset to compute the support of itemsets. Based on Diffset, we can speed up the runtime and save memory usage.

One of weaknesses of the proposed method is that it needs to rescan the updated dataset when the number of inserted records is larger than the safety threshold. In the future, we will study how to avoid rescanning the original dataset. Besides, we will study how to mine CARs when records are deleted or modified.

References

1. Abdelhamid, N., Ayesh, A., Thabtah, F., Ahmadi, S., Hadi, W.: MAC: A Multiclass Associative Classification Algorithm. Information & Knowledge Management 11(2), 1–10 (2012)
2. Cheung, D.W., Han, J., Ng, V.T., Wong, C.Y.: Maintenance of discovered association rules in large databases: An incremental updating approach. In: Proc. of the Twelfth IEEE International Conference on Data Engineering, New Orleans, Louisiana, USA, pp. 106–114 (1996)
3. Hong, T.P., Wang, C.Y., Tao, Y.H.: A new incremental data mining algorithm using pre-large itemsets. Intelligent Data Analysis 5(2), 111–129 (2001)
4. Le, T.P., Vo, B., Hong, T.P., Le, B.: An efficient incremental mining approach based on IT-tree. In: Proc. of the 2012 IEEE International Conference on Computing & Communication Technologies, Research, Innovation, and Vision for the Future, Ho Chi Minh, Viet Nam, pp. 57–61 (2012)
5. Le, T.P., Vo, B., Hong, T.P., Le, B., Hwang, D.: Improving incremental mining efficiency by Trie structure and pre-large itemsets. Computing and Informatics 33(3), 609–632 (2014)
6. Li, W., Han, J., Pei, J.: CMAR: Accurate and efficient classification based on multiple class-association rules. In: Proc. of the 1st IEEE International Conference on Data Mining, San Jose, California, USA, pp. 369–376 (2001)
7. Lin, C.W., Hong, T.P., Lu, W.H.: The Pre-FUFP algorithm for incremental mining. Expert Systems with Applications 36(5), 9498–9505 (2009)
8. Liu, B., Hsu, W., Ma, Y.: Integrating classification and association rule mining. In: Proc. of the 4th International Conference on Knowledge Discovery and Data Mining, New York, USA, pp. 80–86 (1998)
9. Nath, B., Bhattacharyya, D.K., Ghosh, A.: Incremental association rule mining: A survey. WIREs Data Mining Knowledge Discovery 3(3), 157–169 (2013)
10. Nguyen, L.T.T., Vo, B., Hong, T.P., Thanh, H.C.: CAR-Miner: An efficient algorithm for mining class-association rules. Expert Systems with Applications 40(6), 2305–2311 (2013)
11. Nguyen, L.T.T., Nguyen, N.T.: Updating mined class association rules for record insertion. Applied Intelligence (2014), doi:10.1007/s10489-014-0614-1
12. Nguyen, L.T.T., Nguyen, N.T.: An improved algorithm for mining class association rules using the difference of Obidsets. Expert Systems with Applications (2015), doi:10.1016/j.eswa.2015.01.002
13. Vo, B., Le, B.: A novel classification algorithm based on association rules mining. In: Richards, D., Kang, B.-H. (eds.) PKAW 2008. LNCS, vol. 5465, pp. 61–75. Springer, Heidelberg (2009)
14. Vo, B., Le, T., Hong, T.P., Le, B.: An effective approach for maintenance of pre-large-based frequent-itemset lattice in incremental mining. Applied Intelligence 41(3), 759–775 (2014)

Mathematical Morphology on Soft Sets for Application to Metabolic Networks

Mehmet Ali Balcı and Ömer Akgüller

Mugla Sitki Kocman University,
Faculty of Science, Department of Mathematics, Mugla, 48000, Turkey
{mehmetalibalci,oakguller}@mu.edu.tr

Abstract. In this paper, we introduce mathematical morphological operators such as dilation and erosion on soft sets. Rather than the common approach which is based on expressing soft set analogues of the classical theory, we use the classical theory to morphologically analyse a soft set. The main goal of our study is to apply this morphological concepts to metabolic networks to derive an algebraic picture of chemical organizations. For this purpose we first introduce various types of complete lattices on soft sets which represent metabolic networks, then study morphological operations respect to corresponding lattice structure. We also give the concept of duality on soft sets.

1 Introduction

Soft set theory, introduced by Molodtsov in [4], is a mathematical tool for dealing with uncertainty of real world problems that involve data which are not always crisp. It differs from the theories of same kind like fuzzy sets theory, vague sets theory, and rough sets theory by the adequacy of the parametrization. Basically, a soft set over an initial universe set U is a pair (F, A), where F is a mapping $F : A \subset E \to P(U)$, E is a set of parameters, and $P(U)$ is the power set of the universe set. One may conclude that a soft set over U is a parameterized family of subsets of the universe U. Mathematically, a soft set is just a set-valued function mapping elements of a "parameter space" to subsets of a universe; alternatively, it can also be seen as a relation between these sets. Similar settings have been studied in a multitude of other fields. A striking example is Formal Concept Analysis [2]. In Formal Concept Analysis, the point of departure is a binary relation R between objects Y and properties X. From such a relation, one can of course construct set-valued functions, for example a function mapping a property x to the subset of objects y such that $R(x, y)$, i.e., those objects having property x; formally, this corresponds exactly to the soft set representation. However, the other way around, one can also map an object y to the subset of properties it has, i.e., to the subset of x such that $R(x, y)$. Formal Concept Analysis is looking at these mappings simultaneously and constructing so-called formal concepts from this. For the sake of simplicity, we prefer the analysis of soft sets which involves the more simple structure. It can be also seen that a soft

H.A. Le Thi et al. (eds.), *Advanced Computational Methods for Knowledge Engineering*,
Advances in Intelligent Systems and Computing 358, DOI: 10.1007/978-3-319-17996-4_19

set is not a set but set systems. By the arise of the theory its algebraic [9, 35] and topological [34] properties, its relation with other theories [7, 31, 33], and also implicational feature of the theory [8, 29, 30] have been studied intensively. We refer [25] to the interested readers for soft set theoretical analogues of the basic set operations. Throughout this paper we will denote the sets of soft sets over the universe U as $S(U)$, and the image of a subset of the parameter set under the mapping F as $F(A)$.

Mathematical Morphology is a well-founded and widely used non-linear theory for information processing. It has started by analysing set of points with certain kinds of set theoretical operations like Minkowski sum and Minkowski substraction [19, 23], then with the development of information systems, it was necessary to generalize set theoretical notions to complete lattices [3, 6, 10, 17]. Beside the general fashion to use mathematical morphology in image and signal processing, it also finds application areas in topology optimization [32], spatial reasoning [15, 16], preference modeling and logics [13, 14]. Extending mathematical morphology to soft set theory will increase the modeling and reasoning in all these areas. This extension can be performed by defining complete lattices as the underlying structure of soft sets, and then lead us the definitions of algebraic dilations and erosions.

Organization and functioning of metabolic processes is a subject of biochemistry that still has lots of research attention by the new opportunities to study the mechanisms and interactions that govern metabolic processes. Algebraic approaches to chemical organizations in terms of set-valued set-functions that encapsulate the production rules of the individual reactions lead us to mathematically define generalized topological spaces on the set of chemical species. From the mathematical point of view, these approaches mostly depend on the kinetic models and topological network analysis, with their own benefits [1, 5, 11, 18, 24, 26–28]. Topological network analysis does not have to suppose any knowledge of kinetic parameters, therefore is a useful mathematical tool for less well characterized organisms. It is also applicable to complex systems such as chemical reaction networks. Therefore it allows us to study topological properties with less computational complexities.

In this study, we consider morphological operations on soft sets which are set systems with various application benefits to metabolic networks. In Section 2, we introduce certain kinds of complete lattices that can be constructed on a soft set. Rather than straightforward ones, we define the complete lattices involving information about both universe and parameter set. In Section 3, we study the basic morphological operators like dilation and erosion on soft sets. And by the trivial decompositions of soft sets, we give the idea of structuring element for a soft set. In Section 4, we present the soft set model and the mathematical morphological operations of the KEGG [22] representation of the cysteine metabolism network in P.fluorescens PfO-1. For the sake of clarifying the method, we consider trivial lattices in this model. These lattices can be extended to the other ones and different structuring elements can be obtained. In

Section 5, we introduce the concept of the dual of a soft set and study this idea with respect to morphological dilations.

2 Lattice Structures on Soft Sets

A classical lattice structure on the universe set may be defined directly by $\mathcal{L}_U = (P(U), \subseteq)$ where $P(U)$ is the power set of the universe set. Since this lattice do not involve the complete information about the soft set, it may not be useful. To take the parameter sets into the account we may define more useful lattice structure as follows.

Definition 1. *Let (F, A) be a soft set and $x \in U$. Degree of a point is the number of parameters assigned by F to x denoted by $d(x)$. If $H(x) = \{a \in A \subset E \mid x \in F(a)\}$, then $d(x) = s(H(x))$.*

Definition 2. *Let (F, A) be a soft set and $U' \subseteq U$. If for all $(x, y) \in U' \times U$ $F(H(x) \cap H(y)) \subseteq U'$, then we say U' is a closed universe set. We denote the family of closed sets by $\mathcal{C}(U)$.*

Remark 1. It's straightforward to see that $\emptyset \in \mathcal{C}(U)$.

Proposition 1. *The structure $(\mathcal{C}, \subseteq)$ is a complete lattice with for all $(V', V'') \in \mathcal{C}(U) \times \mathcal{C}(U)$*

$$V' \wedge V'' = \cup\{V''' \in \mathcal{C}(U) \mid V''' \subseteq V' \cap V''\}$$

is the infimum and

$$V' \vee V'' = \cap\{V''' \in \mathcal{C}(U) \mid V' \cup V'' \subseteq V'''\}$$

is the supremum.

A complete lattice on the power set of the parameter set A can be defined directly by $\mathcal{L} = (P(A), \subseteq)$. One may apply classical lattice results here directly. However, the next lattice definitions are noteful since they also involve information about the universe set.

Definition 3. *Let (F_1, A_1) and (F_2, A_2) be two soft sets over $U_1 \subseteq U$ and $U_2 \subseteq U$ respectively. A partial order $\preceq$ can be defined as*

$$(F_1, A_1) \preceq (F_2, A_2) \iff U_1 \subseteq U_2,\ A_1 \subseteq A_2.$$

Proposition 2. *$(S(U), \preceq)$ is a complete lattice. It's infimum is*

$$(F_1, A_1) \wedge (F_2, A_2) = (U_1 \cap U_2, A_1 \cap A_2)$$

and for any family $\bigwedge_i (F_i, A_i) = (\bigcap_i U_i, \bigcap_i A_i)$. It's supremum is $(F_1, A_1) \vee (F_2, A_2) = (U_1 \cup U_2, A_1 \cup A_2)$ and for any family $\bigvee_i (F_i, A_i) = (\bigcup_i U_i, \bigcup_i A_i)$.

Definition 4. *Let (F, A) be a soft set over the universe U. (F, B) is an induced soft subset of (F, A) if $B = \{b_i \in U \mid b_i = F(a_i), a_i \in A\}$ for $i \in \{1, 2, \ldots\}$.*

Definition 5. *Let (F, A) be a soft set over the universe U and (F, A_1) and (F, A_2) are induced soft subsets of (F, A). A partial order $\preceq_{in}$ can be defined as*

$$(F, A_1) \preceq_{in} (F, A_2) \iff A_1 \subseteq \{F(a) \cap U \mid a \in A_2\}.$$

Proposition 3. *$(S(U), \preceq_{in})$ is a complete lattice with the infimum*

$$(F, A_1) \wedge_{in} (F, A_2) = (F, \{F(a_1) \cap F(A_2), F(a_2) \cap F(A_1) \mid a_1 \in A_1, a_2 \in A_2\})$$

and the supremum

$$(F, A_1) \vee_{in} (F, A_2) = (F, A_1 \cup A_2).$$

The smallest element of the lattice is the null soft set.

Definition 6. *Let I be the set of isomorphism classes of soft sets, and . A partial order $\preceq_f$ on I can be defined as*

$$(F_1, A_1) \preceq_f (F_2, A_2) \iff (F_1, A_1) \text{ is isomorphic to an induced soft subset of } (F_2, A_2) \text{ by } f$$

3 Morphological Operators on Soft Sets

Let $(\mathcal{L}, \preceq)$ and $(\mathcal{L}', \preceq')$ be two complete lattices.

Definition 7. *An operator $\delta : \mathcal{L} \to \mathcal{L}'$ is a dilation on a soft set (F_i, A_i) for $i \in \{1, 2, \ldots\}$ if*

$$\forall (F_i, A_i) \in \mathcal{L}, \ \delta(\vee_i (F_i, A_i)) = \vee'_i \delta(F_i, A_i)$$

where $\vee$ and $\vee'$ denote the supremums to $\preceq$ and $\preceq'$,respectively. Similarly, an operator $\epsilon : \mathcal{L}' \to \mathcal{L}$ is a erosion on a soft set (F_j, A_j) for $j \in \{1, 2, \ldots\}$ if

$$\forall (F_j, A_j) \in \mathcal{L}', \ \epsilon(\wedge_j (F_j, A_j)) = \wedge'_j \delta(F_j, A_j)$$

where $\wedge$ and $\wedge'$ denote the infimums to $\preceq$ and $\preceq'$,respectively.

In mathematical morphology, an essential part of the dilation and erosion operations is the structuring element which allows us to define arbitrary neighborhood structures. Structuring element can be interpreted as a binary relation between two elements. Therefore, it is possible to extend this idea to any lattice. Generally, the structuring element of x is defined as $B_x = \delta(\{x\})$, where x can be element of the universe set or the parameter set. Hence, it is directly depended on the definition of the corresponding dilation. Just before clarify this idea with an example, we would like to introduce canonical decomposition of a soft set with respect to considered lattice.

For the lattice $(P(U), \subseteq)$, trivial decomposition of each subset of the universe set U is $U = \bigcup_{x \in U} \{x\}$ and its corresponding morphological dilation is $\bigcup_{x \in U} B_x = \bigcup_{x \in U} \delta(\{x\})$. In the same sense, for the lattice $(P(E), \subseteq)$, trivial decomposition of each subset of the parameter set E is $E = \bigcup_{a \in E}\{a\}$ and its corresponding morphological dilation is $\bigcup_{a \in E} B_a = \bigcup_{a \in E} \delta(\{a\})$. Let us now consider a complete lattice structure defined on a soft set (F, A) as Definition 3. For A, a natural decomposition is $A = \bigvee_{a \in A}\{a\}$. For the decomposition of U, we need a reliable one with the assignment of the parameters. Hence, for $U^{\emptyset} = \{x \in U \mid x \notin F(A)\}$,

$$(F, A) = \left(\bigvee_{a \in A} (F(a), \{a\}) \right) \vee \left(\bigvee_{x \in U^{\emptyset}} (\{x\}, \emptyset) \right)$$

is the canonical decomposition of a soft set. The other types of canonical decompositions respect to given lattice structure on the soft set can be obtained by following the same procedure.

4 Mathematical Morphology on Metabolical Networks

Directed graphs are main mathematical tools to represent metabolic networks [12, 20, 21]. In this representation, metabolites are nodes and an arc corresponds to the state of being a reaction between any two metabolites. Since a common reaction have more than one substrate and/or product, and any two metabolites may be involved by more than one reaction, directed graph representation of a metabolic network gets being handicapped. To handle this problem, we propose soft set representation of such metabolic networks.

Let us consider the standard representation of the cysteine metabolism network in *P.fluorescens* PfO-1 in Figure 1. One reads the soft set representation of this network as:

$F_{out}(R00568) = \{C00010, C00024, C00065, C00979\}$ $F_{out}(R00590) = \{C02218\}$
$F_{out}(R01874) = \{C00014, C00022, C00283\}$ $F_{out}(R03132) = \{C00033, C05824\}$
$F_{out}(R04859) = \{C00033, C00097, C00343, C00094\}$ $F_{out}(R03650) = \{C00013, C03125\}$
$F_{out}(R02433) = \{C00026, C00302, C00506, C05528\}$ $F_{out}(R00896) = \{C00010, C00097\}$
$F_{out}(R02619) = \{C00302, C05527\}$ $F_{out}(R04861) = \{C00029, C05532\}$
$F_{out}(R03105) = \{C00094, C00957\}$ $F_{out}(R00897) = \{C00033, C00097\}$
$F_{out}(Null) = \{C00022\}$

$F_{in}(R00568) = \{C00010, C00024, C00065, C00979\}$ $F_{in}(R00590) = \{C00065\}$
$F_{in}(R01874) = \{C00097\}$ $F_{in}(R03132) = \{C00320, C00979\}$
$F_{in}(R04859) = \{C00320, C00342, C00979\}$ $F_{in}(R03650) = \{C00097, C01639\}$
$F_{in}(R02433) = \{C00026, C00302, C00506, C05528\}$ $F_{in}(R00896) = \{C00302, C00957\}$
$F_{in}(R02619) = \{C00026, C00606\}$ $F_{in}(R04861) = \{C05527\}$
$F_{in}(R03105) = \{C05529\}$ $F_{in}(R00897) = \{C00283, C00979\}$
$F_{in}(Null) = \{C05529\}$

with F_{out} and F_{in} parameter mappings that defined on the set of biochemical reactions R to the universe set of metabolites M. More precisely, F_{out} and F_{in} maps the corresponding biochemical reaction to the produced and substrate metabolites, respectively. Moreover, by adding a null reaction to the R, this mapping becomes well-defined. The reader also shall note that the compounds such as

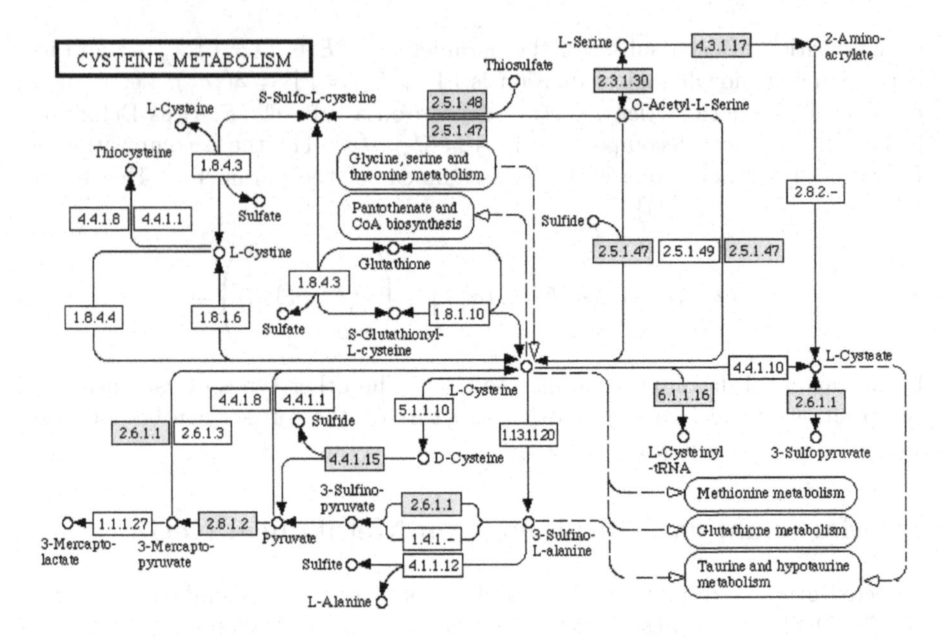

Fig. 1. A standard representation of the cysteine metabolism network in P. fluorescens PfO-1 from KEGG [22].

ATP, AMP, and H_2O are not considered in the soft sets (F_{in}, R) and (F_{out}, R). We also use the corresponding abbreviations of the reactions and compounds respect to KEGG database [22] for the improve readability.

Let us now consider the complete lattice $\mathcal{L} = (P(M), \subseteq)$. Now consider structuring elements of dilation on biochemical reaction set:

$$\forall a \in R,\ B_a = \delta(\{a\}) = \{a' \in M \mid F_{out}(a) \cap F_{in}(a') \neq \emptyset\}$$

and then applying these morphological dilations, one may obtain following results:

$$B_{R00568} = \{R00568, R00590, R00897\} \quad B_{R00896} = \{R00568, R00590, R00897\}$$
$$B_{R04859} = \{R01874, R03650\} \quad B_{R02433} = \{R00896, R02433\}$$
$$B_{R00897} = \{R01874, R03650\} \quad B_{R01874} = \{R00897\}$$
$$B_{R03105} = \{R00896\}$$

Another dilation can be defined on reactions to the universe of metabolites as

$$\forall a \in R,\ B'_a = \cup\{F_{in}(a') \mid F_{out}(a) \cap F_{in}(a') \neq \emptyset\},$$

and then the dilation reads

$$B'_{R00896} = \{C00010, C00024, C00065, C00097, C00979, C01639\}$$
$$B'_{R00568} = \{C00010, C00024, C00065, C00213, C00979\}$$
$$B'_{R02433} = \{C00026, C00302, C00506, C00957, C05528\}$$
$$B'_{R01874} = \{C00283, C00979\}$$
$$B'_{R03105} = \{C00302, C00957\}$$
$$B'_{R00897} = \{C00097, C01639\}.$$

5 Morphological Dilation Dualities

Definition 8. *Let (F, A) be a soft set with non-empty parameter and universe set. The dual of the soft set (F, A) is defined with $F^* : A^* \subset U \to E$, where U is the universe set and E is the parameter set of (F, A) and denoted by (F^*, A^*).*

Remark 2. One may conclude by the definition that the dual of two isomorphic soft sets are also isomorphic to each other.

Now, let $\delta : U \to P(U)$ be a mapping and its dual be a mapping $\delta^* : U \to P(U)$ defined as $\delta^*(\{x\}) = \{y \in U \mid x \in \delta(\{y\})\}$. It also follows that $\delta^{**} : U \to P(U)$, $\delta^{**}(\{x\}) = \{y \in U \mid x \in \delta^*(\{y\})\}$.

Theorem 1. *Let (F, A) be a soft set over the non-empty universe set and $A \neq \emptyset$. $\forall V \in P(U)$*

1. $\delta^*(V) = \bigcup_{x \in V} = \{y \in U \mid V \cap \delta(\{y\}) \neq \emptyset\}$ *iff* δ^* *is a dilation.*
2. $\bigcup_{x \in V} \delta^*(\{x\}) = U$ *implies that* $V \subseteq \bigcup_{V \cap \delta^*(\{y\}) \neq \emptyset} \delta^*(\{y\})$.
3. *On the universe set* U, $\delta^{**} = \delta$.

Proof. 1. *If δ^* is a dilation, then it is straightforward by the definition. Now let us consider $y \in \delta^*(V)$ for $V \in P(U)$, then there exists $x \in V$ such that $y \in \delta^*(x)$ iff $x \in \delta(\{y\})$. Therefore, $y \in \{z \in U \mid V \cap \delta(\{z\}) \neq \emptyset\}$. Now, let $V \in P(U)$ and $y \in \{z \in U \mid V \cap \delta(\{z\}) \neq \emptyset\}$, then there exists $x \in V$ such that $x \in \delta(\{y\})$ iff $y \in \delta^*(\{x\})$. This implies $y \in \bigcup_{x \in V} \delta^*(\{y\})$. Conversely, if the equalities hold, since δ^* commutes with the supremum, δ^* is a dilation.*
2. *Straightforward*
3. *Let $z \in \delta^{**}(\{x\})$. This implies that $x \in \delta^*(\{z\})$, hence $z \in \delta(\{x\})$. Similarly, $z \in \delta(\{x\})$ implies that $x \in \delta^*(\{z\})$, therefore $z \in \delta^{**}(\{x\})$.*

One of the significant results arise with the choose of parameter map as the structuring element. Let $\delta : P(U) \to P(U)$ be a dilation. Then, for $x \in U$, $(F, A)_\delta = (\delta(\{x\}), A)$ is a soft set. We can also define a dilation to any soft set (F, A) by considering the structuring element as $\delta : V \to P(A)$, $\delta(\{x\}) = \{a \in A \mid x \in F(a)\}$. These ideas lead us to following theorem:

Theorem 2. *Let $\delta : U \to P(U)$ be a mapping and (F, A) be a soft set with the bijective parameter map. Then, (F, A) is isomorphic to $(F, A)_\delta$ iff (F^*, A^*) is isomorphic to $(F, A^*)_{\delta^*}$.*

Proof. Let (F, A) and $(F, A)_\delta$ be two isomorphic soft sets. By the definition of dilation, if $x \neq y$ then $\delta(\{x\}) \neq \delta(\{y\})$, that is $\delta(\{x\}) = \delta(\{y\})$ implies that $x = y$. Therefore, δ is injective on U.

Moreover, by the isomorphism there exists a bijection $f : (F, A) \to (F, A)_\delta$ such that $a \in A$ iff for $x \in U$, $f(a) = \delta(\{x\}) \in A_\delta$. Notice that $A_\delta = \{\delta(\{x\}), x \in U\}$.

Let (F^*, A^*) be a dual of (F, A). (F^*, A^*) is a soft set over the universe U^*. By the definition of the duality, U^* is isomorphic to A and A^* is isomorphic to U. Hence, (F, A) is isomorphic to $(F, A)_\delta$ iff (F^*, A^*) is isomorphic to $(F^*, A^*)_\delta$.

Now it is sufficient to show that $(F^*, A^*)_\delta$ *is isomorphic to* $(F, A^*)_{\delta^*}$. *Here,* $(F^*, A^*)_\delta$ *is the soft set with* U^*_δ *is isomorphic to* A_δ *and* A^*_δ *is isomorphic to* $H^*(x) = \{u \in A^* \subset U \mid x \in F^*(u)\}$.

Let $g : \{\delta(\{y\}), y \in U\} \to U$ *defined by* $g(\delta(\{y\})) = y$. g *is well defined, since* $\delta(\{x\}) = \delta(\{y\}) \implies g(\delta(\{x\})) = g(\delta(\{y\}))$ *for* $x = y$.

It can be clearly seen that g *is surjective; and by the equality of the cardinalities of* $\{\delta(\{y\}), y \in U\}$ *and* U, g *is injective.*

Now, $H^*(x) \in A^*_\delta$ *iff* $H^*(x) = \{\delta(\{u_i\}), x \in \delta(\{u_i\})\} \in A^*_\delta$ *iff* $g(H^*(x)) = \{g(\delta(\{u_i\})), i \in \{1, 2, \ldots, n\}\} = \{u_1, u_2, \ldots, u_n\} = \delta^*(\{x\})$; *since* $x \in \delta(u_i)$ *iff* $u_i \in \delta^*(\{x\})$. *Therefore,* $(F^*, A^*)_\delta$ *is isomorphic to* $(F, A^*)_{\delta^*}$.

6 Conclusions

In this paper, we introduced mathematical morphology on soft sets. To obtain the morphological operations such as dilation and erosion, we studied some introductive complete lattices on soft sets. While constructing these lattices, we consider the cases where information about parameters and universe is preserved. By the help of complete lattices, we were able to define morphological operations and the structuring element. A new duality definition for the soft set theory was also given to show the relevance of relationship between mathematical morphology and soft sets.

The morphological interpretation of metabolic networks led us to derive an algebraic picture of chemical organizations. By the discontinuous nature of the metabolic networks, we need to state this new approach to define concepts such as connectedness, similarity, and continuity of change. For the representation of the cysteine metabolism network in P. fluorescens PfO-1, it is possible to obtain such concepts by the help of mathematical structuring element. We may also conclude from our analysis that the reactions $R00590, R03132, R03650, R04861$, and $Null$ do not have any dilations. Different structuring elements also lead us to obtain different morphological characteristic of the network. For instance, by new structuring element

$$\forall a \in R,\ B^k_a = \delta(\{a\}) = \{a' \in R \mid |F_{out}(a) \cap F_{in}(a')| \geq k\},$$

where $|,|$ denotes the maximum cardinality, it can be seen that only $R00586$ and $R02433$ have the stronger connectedness for $k \geq 2$. Different kinds of structuring elements which may be more suitable for the model under consideration can be obtained by choosing different lattices on soft sets.

Authors' Contributions

Mehmet Ali Balcı and Ömer Akgüller worked together in the derivation of the mathematical results. Both authors read and approved the final manuscript.

References

1. Barabási, A.L., Oltvai, Z.N.: Network biology: Understanding the cell's functional organization. Nature Reviews Genetics 5, 101–113 (2004)
2. Ganter, B., Stumme, G., Wille, R. (eds.): Formal Concept Analysis. LNCS (LNAI), vol. 3626. Springer, Heidelberg (2005)
3. Ronse, C.: Why mathematical morphology needs complete lattices. Signal Processing 21(2), 129–154 (1990)
4. Molodtsov, D.: Soft set theory – First results. Computers and Mathematics with Applications 37(4-5), 19–31 (1999)
5. Fell, D.A., Wagner, A.: The small world of metabolism. Nature Biotechnology 18, 1121–1122 (2000)
6. Aptoula, E., Lefèvre, S.: A comparative study on multivariate mathematical morphology. Pattern Recognition 40(11), 2914–2929 (2007)
7. Feng, F., Li, C., Davvaz, B., Ali, M.I.: Soft sets combined with fuzzy sets and rough sets: A tentative approach. Soft Computing 14(9), 899–911 (2010)
8. Feng, F., Jun, Y.B., Liu, X., Li, L.: An adjustable approach to fuzzy soft set based decision making. Journal of Computational and Applied Mathematics 234(1), 10–20 (2010)
9. Aktas, H., Cagman, N.: Soft sets and soft groups. Information Sciences 177(13), 2726–2735 (2007)
10. Heijmans, H.J.A.M., Ronse, C.: The algebraic basis of mathematical morphology 1. Dilations and erosions. Computer Vision, Graphics and Image Processing 50(3), 245–295 (1990)
11. Jeong, H., Tombor, B., Albert, R., Oltvai, Z.N., Baraási, A.L.: The largescale organization of metabolic networks. Nature 407, 651–654 (2000)
12. Ma, H., Zeng, A.-P.: Reconstruction of metabolic networks from genome data and analysis of their global structure for various organisms. Bioinformatics 19, 270–277 (2003)
13. Bloch, I., Pino-Pérez, R., Uzcategui, C.: A Unified Treatment of Knowledge Dynamics. In: International Conference on the Principles of Knowledge Representation and Reasoning, KR 2004, Canada, pp. 329–337 (2004)
14. Bloch, I., Pino-Pérez, R., Uzcategui, C.: Mediation in the Framework of Morphologic. In: European Conference on Artificial Intelligence, ECAI 2006, Riva del Garda, Italy, pp. 190–194 (2006)
15. Bloch, I.: Spatial Reasoning under Imprecision using Fuzzy Set Theory, Formal Logics and Mathematical Morphology. International Journal of Approximate Reasoning 41, 77–95 (2006)
16. Bloch, I., Heijmans, H., Ronse, C.: Mathematical Morphology. In: Aiello, M., Pratt-Hartman, I., van Benthem, J. (eds.) Handbook of Spatial Logics, ch. 13, pp. 857–947. Springer (2007)
17. Bloch, I., Bretto, A.: Mathematical Morphology on hypergraphs: Preliminary definitions and results. In: Debled-Rennesson, I., Domenjoud, E., Kerautret, B., Even, P. (eds.) DGCI 2011. LNCS, vol. 6607, pp. 429–440. Springer, Heidelberg (2011)
18. Morgan, J.A., Rhodes, D.: Mathematical modeling of plant metabolic pathways. Metab. Eng. 4, 80–89 (2002)
19. Serra, J.: Introduction to Mathematical Morphology. Computer Vision, Graphics, and Image Processing 35(3), 283–305 (1986)
20. Arita, M.: Metabolic reconstruction using shortest paths. Simulation Pract. Theory 8, 109–125 (2000)

21. Arita, M.: Graph modeling of metabolism. J. Jpn. Soc. Artif. Intell. (JSAI) 15, 703–710 (2000)
22. Kanehisa, M., Goto, S., Hattori, M., Aoki-Kinoshita, K.F., Itoh, M., Kawashima, S., Katayama, T., Araki, M., Hirakawa, M.: From genomics to chemical genomics: new developments in KEGG. Nucleic Acids Res. 34, D354–D357 (2006)
23. Maragos, P., Schafer, R.W.: Morphological Filters – Part I: Their Set-Theoretic Analysis and Relations to Linear Shift - Invariant Filters. IEEE Transactions on Acoustics, Speech, and Signal Processing ASSP-35(8), 1170–1184 (1987)
24. Mulquiney, P.J., Kuchel, P.W.: Modelling Metabolism with Mathematica. CRC Press, London (2003)
25. Maji, P.K., Biswas, R., Roy, A.R.: Soft Set Theory. Computers and Mathematics with Applications 45(4-5), 555–562 (2003)
26. Albert, R., Barabási, A.L.: Statistical mechanics of complex networks. Rev. Mod. Phys. 74, 47–97 (2002)
27. Heinrich, R., Rapoport, S.M., Rapoport, T.A.: Metabolic regulation andmathematical models. Prog. Biophys. Molec. Biol. 32, 1–82 (1977)
28. Rios Estepa, R., Lange, B.M.: Experimental andmathematical approaches tomodeling plant metabolic networks. Phytochemistry 68(16-18), 2351–2374 (2007)
29. Kalayatkankal, S.J., Suresh Singh, G.: A fuzzy soft flood alarm model. Mathematics and Computers in Simulation 80(5), 887–893 (2010)
30. Herewan, T., Deris, M.M.: A soft set approach for association rules mining. Knowledge-Based Systems 24(1), 186–195 (2011)
31. Xu, W., Ma, J., Wang, S., Hao, G.: Vague soft sets and their peoperties. Computers and Mathematics with Applications 59(2), 787–794 (2010)
32. Zhang, W., Zhong, W., Guo, X.: An explicit length scale control approach in SIMP-based topology optimization. Computer Methods in Applied Mechanics and Engineering 282, 71–86 (2014)
33. Yang, X., Ling, T.Y., Yang, J., Li, Y., Yu, D.: Combination of interval-valued fuzzy set and soft set. Computers and Mathematics with Applications 58(3), 521–527 (2009)
34. Ge, X., Li, Z., Ge, Y.: Topological spaces and soft sets. Journal of Computational Analysis and Applications 13(5), 881–885 (2011)
35. Jun, Y.B.: Soft BCK/BCI-algebras. Computers and Mathematics with Applications 56(10), 2621–2628 (2008)

Molecular Screening of Azurin-Like Anticancer Bacteriocins from Human Gut Microflora Using Bioinformatics

Van Duy Nguyen[1] and Ha Hung Chuong Nguyen[2,3]

[1] Institute of Biotechnology and Environment, Nha Trang University, Nha Trang, Khanh Hoa, Vietnam
duy.1981@yahoo.com, duynv@ntu.edu.vn
[2] Department for Management of Science and Technology Development, Ton Duc Thang University, Ho Chi Minh City, Vietnam
[3] Faculty of Applied Sciences, Ton Duc Thang University, Ho Chi Minh City, Vietnam
nguyenhahungchuong@tdt.edu.vn

Abstract. There has been renewed interest in the development of new cancer therapeutic drugs based on the use of live bacteria and their purified products. Bacteriocins are antimicrobial peptides produced by bacteria to inhibit the growth of closely related bacterial strains, and sometimes against a wide spectrum of species. As one of few known anticancer bacteriocins, Azurin can specifically penetrate human cancer cells and exerts cytostatic and apoptotic effects. We hypothesized that pathogenic and commensal bacteria with long term residence in human body can produce Azurin-like bacteriocins as a weapon against the invasion of cancers. Putative bacteriocins were screened from complete genomes of 66 dominant bacteria species in human gut microbiota using Bagel database and subsequently characterized by subjecting them as functional annotation algorithms using Prot fun2.2, Kihara Bioinformatics and BaCelLo with Azurin as control. The prediction of functional characterization including cell envelope activity, enzymatic activity, immune response, the number of protein phosphorylation sites and localization within the cell revealed that 14 bacteriocins possessed functional properties very similar to those of Azurin. This is first study on genomic-scale screening of anticancer bacteriocins from all dominant bacteria species in human gut microbiota.

Keywords: anticancer drug, azurin, bacteriocin, bioinformatics, human gut microbiome.

1 Introduction

Cancer is a complex disease involving dis-regulation of mammalian cell differentiation and growth [1]. At present there is no conceivable way current drugs can prevent cancer relapse once the cancer is in remission. Cancer is often treated by

H.A. Le Thi et al. (eds.), *Advanced Computational Methods for Knowledge Engineering*,
Advances in Intelligent Systems and Computing 358, DOI: 10.1007/978-3-319-17996-4_20

the surgical removal of the tumors followed by radiation therapy to kill surrounding hidden cancer cells and long term chemotherapy to prevent any residual tumor cells to proliferate. Among treatment options, chemotherapy can cause the most devastating side effects on the growth of normal cells and lead to the rapid resistance to drugs developed by the cancer cells using alternate pathways for growth, or use efflux pumps to pump out drugs [2, 3]. Therefore, new therapies are urgently needed.

Recently, there has been renewed interest in the development of new therapeutic anticancer modalities based on the use of live bacteria and their purified products. In late 1890s, William B. Coley first observed regression in tumors in bacteria-infected patients, which led to the discovery of anticancer activity in bacteria. The so-called Coley's toxins were used against different types of cancer, but despite anti-tumor activity, a few patients developed systemic infections and eventually died [4]. Nowadays, the problems with systemic infections after bacterial delivery are being overcome either by using engineered attenuated bacteria with low infection capabilities or bacterial products which are capable of targeting and specifically killing tumor cells. In the latter case, products derived from them have been engineered to target specific receptors overexpressed in cancer cells or enter preferentially in cancer cells when compared to normal tissues. These products include bacterial toxins, proteins, peptides and enzymes [5].

A number of bacterial proteins and peptides have recently been described to exert an anticancer activity at pre-clinical level toward diverse types of cancer cells [6]. Bacteriocins are ribosomally synthesized antimicrobial peptides or proteins produced by bacteria to inhibit the growth of similar or closely related bacterial strains (narrow spectrum), and sometimes against a wide spectrum of species. They have been looking for a positive health benefit to the host including human, livestock, aquaculture animals and some plants [7]. Bacteriocins promises to be effective therapeutic agent and its biochemical properties have been studied; their antineoplastic capability has also identified after its discovery in the late 1970s by using crude bacteriocin preparation toxic to mammalian cells [8]. Common bacteriocins like pyocin, colicin, pediocin, and microcin have been shown to possess inhibitory properties against different neoplastic line cells [1].

Azurin is an important bacteriocin, a member of the cupredoxin family of redox proteins, secreted by the bacterium Pseudomonas aeruginosa, which can preferentially penetrate human cancer cells and exerts cytostatic and apoptotic effects with no apparent activity on normal cells [9-11]. Azurin can directly interact and stabilize the tumor suppressor p53 [9]. The azurin domain responsible for its specific entry in cancer cells was demonstrated that it spans residues 50–77 (termed p28) and adopts an amphipathic alphahelical conformation [12]. Cell penetration is not accompanied by membrane disruption, which could cause cell death. Preclinical evaluation of pharmacokinetics, metabolism, and toxicity of azurin-p28 was evaluated [13], establishing it as non-immunogenic and nontoxic in mice and non-human primates. Moreover, the protein–protein interactions between azurin and p53 have recently been analyzed by bioinformatics and atomic force microscopy [14-16].

Interestingly, azurin has not only anticancer activity but it also strongly inhibits host cell invasion by the AIDS virus HIV-1, the malarial parasite Plasmodium falciparum [17] and the toxoplasmosis-causing parasite Toxoplasma gondii [18, 19]. Thus azurin is believed to be a weapon used by P. aeruginosa to keep invaders of the human body for long term residence without harming or exerting any toxicity to the host [19]. This also suggests that azurin maybe specific for tumors in the organs where P. aeruginosa normally resides during infection. In fact, Neisseria meningitides produces an azurin-like protein called Laz (lipidated azurin) with a 127 amino acid moiety with 56% amino acid sequence identity to P. aeruginosa azurin. Several US patents have been issued to cover the use of azurin and Laz in cancer therapies [20], and azurin has shown significant activity, as well as enhancement of the activity of other drugs, in oral squamous carcinoma cells [21].

The very important question is whether azurin is the only bacteriocin produced by P. aeruginosa as an anticancer weapon or whether there are other bacteriocins, produced by other bacteria with the ability to cause chronic infections and have long term residence in human bodies, as well as to defend the body from invaders such as cancers, viruses and parasites. It is thus interesting to note that azurin is not the only anticancer bacteriocins produced by human microflora. In fact, their antineoplastic properties have been inadequately revealed in the late 1970s by using crude bacteriocin preparation toxic to mammalian cells. Nowadays, purified bacteriocins are available and have shown inhibitory properties toward diverse neoplastic line cells. Pyocin, colicin, pediocin, and microcin are among bacteriocins reported to present such activity [8, 22].

Although bacteriocins have been found in many major lineages of Bacteria and some members of the Archaea, many new bacteriocins with new characteristics and origins are still awaiting discovery. By now bacteriocins have mainly been derived from lactic acid bacteria with mostly fermented food origins. Besides, colicins from E. coli were used as model Gram-negative bacteriocins. Only few basic researches were carried out on non-colicin bacteriocins of human origins and bacteriocins with killing activity against eukaryotic and human cells.

Here we hypothesized that bacteria from human microflora, especially pathogenic and commensal bacteria, with long term residence in human body can produce azurin-like bacteriocins as a weapon to protect their habitat from cancers. In this study, we screened putative bacteriocins from complete genomes of 66 dominant bacteria species in human gut microbiota and subsequently characterized by subjecting them as functional annotation algorithms with azurin as control. This is first study on genomic-scale screening of anticancer bacteriocins from all dominant bacteria species in human gut microbiota.

2 Methods

2.1 Selection of the Human Gut Microbiome

Using metagenomic variation analysis, Schloissnig et al. identified 66 dominant species among 101 prevalent gut microbial species [23]. These 66 species were selected for bacteriocin screening in this work. Their genomes in FASTA format were downloaded from the Genome database of NCBI (http://www.ncbi.nlm.nih.gov/genome/).

2.2 Identification of Probable Bacteriocins from Human Gut Microbiome

The selected complete genomes of 66 species were scanned using the BAGEL web server (http://bagel.molgenrug.nl) in order to identify putative genes encoding bacteriocins. BAGEL is a web-based bacteriocin mining tool which uses single or multiple DNA nucleotide sequences in FASTA format as input [24]. The output of BAGEL includes graphical and tabular results in html format, a log file as well as the nucleotide sequences and protein sequences of the identified areas of interest (AOI). The protein sequences of hypothetical bacteriocins were retrieved from the output and then were saved as FASTA format.

2.3 Screening of Potentially Anticancer Bacteriocins

The ProtFun 2.2 server (http://www.cbs.dtu.dk/services/ProtFun-2.2/) and Kihara Bioinformatics (http://kiharalab.org/pfp/) were used to predict the function of protein from sequence. The method queries a large number of other feature prediction servers to obtain information on various post-translational and localizational aspects of the protein [25, 26]. For each input sequence, the server predicts cellular role, enzyme class (if any), and selected Gene Ontology category. The output also includes the scores which represent the probability and the odds that the sequence belongs to that class/category. Firstly, azurin sequence (Uniprot entry: P00282) was submitted to the server. Then, all the selected bacteriocin protein sequences from the previous step were subjected to the ProtFun 2.2 server. After thorough functional characterization, the properties of the analyzed bacteriocins were compared to that of azurin and only those hypothetical bacteriocins which seemed to posse properties very similar to azurin were considered as potentially anticancer bacteriocins.

In addition, sub-cellular localization tools such as Balanced Sub-cellular Localization Predictor BaCelLo (http://gpcr.biocomp.unibo.it/bacello/pred.htm) were utilized to identify the putative location (secretory way, cytoplasm, nucleus, mitochondrion and chloroplast) of the bacteriocins. Finally, multiple sequence alignment of shortlisted bacteriocins with azurin was carried out by ClustalW2 (http://www.ebi.ac.uk/) to determine identities.

3 Results

3.1 Identification of Probable Bacteriocins from Human Gut Microbiome

Among 66 species, there are 26 of them that BAGEL identifies putative bacteriocins. The total number of predicted bacteriocin is 81. Among the 26 species, Clostridium nexile DSM 1787 has the largest number of putative bacteriocins which is 25. The length of bacteriocin sequences varies from 30 to 518 amino acids.

The number of bacteriocin predicted as small peptides is 68 and that of bacteriocin predicted as large proteins is 13. Most of the small peptide bacteriocins (50/68) are predicted to belong to the class Sactipeptides, a type of ribosomally assembled and posttranslational modified natural product peptides [27]. They share a common feature that they have an intramolecular thioether bond between the sulfur atom of a cysteine residue with the alpha-carbon of an acceptor amino acid. They are the new emerging class of ribosomally assembled peptides that their biological function has not yet been fully understood [27]. Meanwhile, most of the large protein bacteriocins (11/13) belong to the class Zoocin A, a penicillin-binding protein and presumably a D-alanyl endopeptidase [28].

3.2 Screening of Potentially Anticancer Bacteriocins

Azurin is an important bacteriocin that can penetrate human cancer cells and exerts cytostatic and apoptotic effects with no apparent activity on normal cells [9-11]. The results from ProtFun 2.2 and Kihara Bioinformatics servers of azurin include (i) 59.2% chance being part of cell envelope; (ii) 81.8% chance of not being an enzyme; and (iii) 50.0% chance of eliciting an immune response. Laz, an azurin-like protein, share two first properties of azurin but its Gene Ontology category is classified belonging to stress response. Among 81 putative bacteriocin sequences submitted to the ProtFun 2.2 server, 14 bacteriocins were chosen which possessed at least two properties similar with azurin or laz (Table 1).

Table 1. List of 14 bacteriocins which possessed at least two properties similar with azurin or laz

No.	Protein ID	Species/ Strain	Sequence length	Functional category	Odds	Enzyme activity	Odds	Gene Ontology (#)	Odds	P(*)	Location
1	AOI1 orf013	*Bacteroides pectinophilus* ATCC 43243	515	Cell Envelope	7.008	Enzyme (Ligase)	2.194	Im	2.119	51	Cytoplasm
2	AOI1 orf028	*Dorea longicatena* DSM 13814	49	Translation	6.849	Nonenzyme	1.131	Im	1.969	0	Cytoplasm
3	AOI1 orf012	*Eubacterium ventriosum* ATCC 27560	212	Cell Envelope	6.77	Enzyme (Isomerase)	1.693	St	2.998	13	Secretory
4	AOI1 orf019	*Ruminococcus torques* ATCC 27756	360	Cell Envelope	8.429	Enzyme (Ligase)	1.58	Im	2.31	19	Secretory
5	AOI2 smallORF1	*Clostridium nexile* DSM 1787	41	Transport and binding	1.826	Nonenzyme	1.122	Im	3.856	1	Secretory
6	AOI1 orf034	*Clostridium nexile* DSM 1787	75	Amino acid biosynthesis	11.014	Nonenzyme	1.039	Im	1.325	2	Cytoplasm
7	AOI1 orf036	*Clostridium nexile* DSM 1787	33	Translation	4.624	Nonenzyme	1.09	St	4.483	4	Cytoplasm
8	AOI1 orf041	*Clostridium nexile* DSM 1787	73	Energy metabolism	3.313	Nonenzyme	1.016	Im	1.085	4	Secretory
9	AOI1 smallORF20	*Clostridium nexile* DSM 1787	35	Regulatory functions	1.593	Nonenzyme	1.01	St	2.982	3	Secretory
10	AOI1 orf021	*Clostridium nexile* DSM 1787	357	Cell Envelope	6.821	Enzyme	1.768	Im	1.624	14	Secretory
11	AOI1 orf059	*Clostridium nexile* DSM 1787	33	Translation	6.246	Nonenzyme	1.022	St	1.565	1	Secretory
12	AOI1 smallORF42	*Bacteroides* sp. 2.1.16	57	Translation	1.635	Nonenzyme	1.002	Im	4.345	5	Secretory
13	AOI1 orf022	*Bacteroides vulgatus* PC510	82	Translation	2.823	Nonenzyme	1.137	Im	3.435	6	Secretory
14	AOI1 orf007	*Clostridium hathewayi* DSM 13479	77	Cell Envelope	2.065	Nonenzyme	1.104	Gr	1.631	3	Secretory
15	**Azurin**	*Pseudomonas aeruginosa*	148	Cell Envelope	9.71	Nonenzyme	1.147	Im	5.877	6	Secretory
16	**Laz**	*Neisseria meningitides*	183	Cell Envelope	9.987	Nonenzyme	1.134	St	2.438	8	Secretory

(#): Im: Immune response, St: Stress response, Gr: Growth factor; P(*): The number of phosphorylation sites

In addition, while azurin and laz have 6-8 protein phosphorylation sites, all our bacteriocins have at least 1 and at most 51 sites (Table 1). Only one exception of no phosphorylation residue comes from the putative bacteriocin encoded by AOI1 orf028 in Dorea longicatena DSM 13814. As predicting the number of protein phosphorylation sites is crucial in any research pursuit attempting to identifying proteins which can be potential anti neoplastic agents, the results showed an anticancer potential of our bacteriocins. Besides, the results from BaCelLo server showed 10 out of these 14 bacteriocins are secretory proteins, which are the same as azurin and laz (Table 1).

Finally, multiple sequence alignment of our bacteriocins with azurin showed highly conserved motifs. For example, the azurin-like bacteriocin encoded by AOI1 orf021 in Clostridium nexile DSM 1787, called Cnaz, was illustrated in Fig. 1. Interestingly, an azurin p28-like region was also found in Cnaz.

4 Discussions

In this study we identified 14 novel putative bacteriocins which possess functional properties very similar to azurin and laz. Protfun results predicted that azurin and laz have cell envelope activity whereas BaCelLo tool indicated them as secretory proteins. Most of our bacteriocins expressed the same properties, revealing a great chance of possessing cancer cell attack activity similar to that of azurin. The more important functional characterization, which was to predict the propensity of the bacteriocins, was enzymatic activity. The low propensity of azurin to have enzymatic activity indicates that it has little chance of interacting with any other substrates or alters the normal cellular kinetics. This is consistent with the special feature of azurin which targets human cancer cells without exerting activity on normal cells [9-11]. Like azurin, our bacteriocins showed immune responses which have very low chances of Drug-induced adverse reactions of type B which comprise idiosyncratic and immune-mediated side effects [29]. These results signify that our selected bacteriocins have great chance of having anticancer activity similar to that of azurin.

Interestingly, our bacteriocins are found in many pathogenic bacteria such as Clostridium nexile, Clostridium hathewayi (reclassified as Hungatella hathewayi), Bacteroides pectinophilus, Bacteroides vulgates and Bacteroides sp. 2.1.16, which fit well our hypothesis. A molecular genetic analysis of rDNA amplicons generated directly from a human faecal sample showed that more than 90% of the flora could be assigned to three major phylogenetic lineages (the Bacteroides, Clostridium coccoides and Clostridium leptum groups) [30]. Thus our bacteriocins were produced by the most dominant bacterial species in human gut.

In addition, these are unrelated bacteria with anticancer activity. Actually, the results are in agreement with cases of azurin-producing Pseudomonas aeruginosa and laz-producing Neisseria meningitides. Another example, SSL10, a superantigen-like protein from Staphylococcus aureus, inhibits the CXCl12-induced migration of leukemic Jurkat cell line and carcinoma cell line Hela [31]. SSL5, another superatingen-like protein from S. aureus, has a role in preventing adhesion of leukemic cells to endothelial cells and platelets. Interactions between tumor cells and endothelial cells are important for tumor progression since it mediates processes such as angiogenesis and metastasis formation [32]. Also, the actin assembly-inducing

```
Cnaz                MLKKKVKKYLLISGISFAIGTLGIIFVSVLIEEVVRAIAGEEANKQITQSDLEGLPAWIT 60
azurin              MLRKLA-----------AVSLLSLLSAPLLAAECSVDIQGNDQ-MQFNTNAIT-----VD 43
                    **:* .           *:. *.:: ..:*  *    * *::   *:. . :      :

Cnaz                VEMVQAAIDMMNETGYPASVVLGQMILEAGADGSELANPPYYNCLGQKAHCYKENGTVVM 120
azurin              KSCKQFTVNLSHPGNLPKNVMGHNWVLSTAADMQGVVTDG--MASGLDKDYLKPDDSRVI 101
                     .  * :::: :  . * .*:  : :*.:.** . :..     . * . .  * ::  *:

Cnaz                RTEEAWGTVTAEFSTFANYVDCMLAWGNKFTRQPYVDNVTACKRDPVTGHYDADSFITAL 180
azurin              AHTKLIGSGEKDSVTFD------------------------------------------- 118
                       :  *:   :  **

Cnaz                WKSGYATDPAYVSKVIAVMKSRNLYRFNYMTSADLENGLGEIGTGMFTHPCPGMTYQSSY 240
azurin              ---------------VSKLKEGEQY--------------------MFFCTFPGHS----- 138
                                   :: :*. : *                    **   . ** :

Cnaz                FGEIREFETGGHKGNDYAAPAGTPTLAAADGTVTVAGWSDSAGNWVVIDHGNGLTTKYMH 300
azurin              --------------------------ALMKGTLTLK------------------------ 148
                                              *  .**:*:

Cnaz                HSRLLVKTGDTVKKGQQIGEVGSTGQSTGNHLHFQVEENGVPVNPDKYLKGEGNERE 357
azurin              ---------------------------------------------------------
```

Fig. 1. Sequence alignment between azurin and Cnaz (*Clostridium nexile* azurin-like bacteriocin) in which an azurin p28-like region indicated in bold

protein (ActA), which plays an important role in pathogenesis of L. monocytogenes, has widely been used in immunotherapy to facilitate CD8-mediated immune responses [33]. Romidepsin (FK228) is a naturally occurring bicyclic dipeptide isolated from Chromobacterium violaceum which acts as histone deacetylase inhibitor. Histone deacetylases are implicated in leukemia development and progression and therefore are important therapeutic targets in this malignancy [34]. Spiruchostatin B (SP-B) is a structurally related peptide isolated from a culture broth of Pseudomonas sp. which displays the same activity toward cancer cells [35]. Finally, Pep27anal2 is an analog of Pep27, a peptide from Streptococcus pneumoniae where it initiates a program of cell death by signal transduction mechanisms. Peptide engineering was used to increase its anticancer activity at lower concentrations [36, 37].

There are some more recent examples of anticancer bacteriocins beside azurin. Nisin, a 34-amino acid polycyclic bacteriocin from Lactococcus lactis and commonly used food preservative, is not toxic to animals, is safe for human consumption, and was approved for human use by the WHO in 1969 and by the FDA in 1988. Current findings support the use of nisin as a novel potential therapeutic for treating head and neck squamous cell carcinoma (HNSCC), as it induces preferential apoptosis, cell cycle arrest, and reduces cell proliferation in HNSCC cells, compared with primary keratinocytes [38].

Using bioinformatics approach, the entire genome of a human commensal bacterium Lactobacillus salavarius was scanned for putative bacteriocins and potentially anticancer bacteriocins were screened through structure prediction and docking studies against the common cancer targets p53, Rb1 and AR with Azurin as control. The results have revealed that Lsl_0510 possessed highest binding affinity towards the all the three receptors [39]. However, Lactobacillus salavarius is not dominant species in human gut. In this work, for the first time, a genome-scale screening of anticancer bacteriocins from all 66 dominant bacteria species in human gut microbiota were performed to identify 14 candidate peptide drugs, thus further structure modelling and docking studies of these bacteriocins against cancer targets was required to make them to ideal candidates for future cancer therapeutics.

In order to develop a cancer therapeutic drug, at least four important properties should be considered: i) non-toxicity for long term use, ii) inhibiting and killing any preformed tumor cells, iii) preventing oncogenic transformation of normal cells to cancer cells, and iv) taken orally and not through intravenous injections. The three former properties are shared by p28 and probably by azurin, although azurin's lack of toxicity and side effects in humans has not yet been assessed. For the fourth feature, p28 is now given intravenously but future technological advances might overcome this problem. For example, a protein such as the 128 amino acid long azurin can be chemically synthesized at a modest cost. Also, advanced techniques can make novel peptide drugs more stable to stomach acids, easily absorbed through gut tract, stable in serum and less immunogenic, all keeping the cancer away.

In conclusion, using bioinformatics approach at least 14 novel putative bacteriocins from human gut pathogenic and commensal bacteria were found to possess functional properties very similar to those of Azurin, with anticancer activities. The results herald a new era of drug development and contributing to better human health. If the pathogenic and commensal bacteria with long term residence in human body produce these proteins to defend their habitat from invaders such as cancers and other deadly diseases, this can lead us to identify the novel anticancer drugs from human microflora. The discovery of these drugs has just been started.

Acknowledgments. The research was funded by Vietnam National Foundation for Science and Technology Development (NAFOSTED) under grant number 106-YS.04-2014.40 to VDN.

References

1. Chakrabarty, A.M.: Microbial pathogenicity: a new approach to drug development. Adv. Exp. Med. Biol. 808, 41–49 (2014)
2. Morrissey, D., O'Sullivan, G., Tangney, M.: Tumour Targeting with Systemically Administered Bacteria. CGT 10, 3–14 (2010)
3. Avner, B.S., Fialho, A.M., Chakrabarty, A.M.: Overcoming drug resistance in multi-drug resistant cancers and microorganisms: a conceptual framework. Bioengineered 3, 262–270 (2012)
4. Chakrabarty, A.M.: Microorganisms and cancer: quest for a therapy. J. Bacteriol. 185, 2683–2686 (2003)
5. Bernardes, N., Seruca, R., Chakrabarty, A.M., Fialho, A.M.: Microbial-based therapy of cancer: current progress and future prospects. Bioeng. Bugs 1, 178–190 (2010)
6. Mahfouz, M., Hashimoto, W., Das Gupta, T.K., Chakrabarty, A.M.: Bacterial proteins and CpG-rich extrachromosomal DNA in potential cancer therapy. Plasmid. 57, 4–17 (2007)

7. Riley, M.: Bacteriocins, biology, ecology, and evolution. In: Encyclopedia of Microbiology, pp. 32–44 (2009)
8. Cornut, G., Fortin, C., Soulières, D.: Antineoplastic properties of bacteriocins: revisiting potential active agents. Am. J. Clin. Oncol. 31, 399–404 (2008)
9. Punj, V., Das Gupta, T.K., Chakrabarty, A.M.: Bacterial cupredoxin azurin and its interactions with the tumor suppressor protein p53. Biochem. Biophys. Res. Commun. 312, 109–114 (2003)
10. Yamada, T., Goto, M., Punj, V., Zaborina, O., Chen, M.L., Kimbara, K., Majumdar, D., Cunningham, E., Das Gupta, T.K., Chakrabarty, A.M.: Bacterial redox protein azurin, tumor suppressor protein p53, and regression of cancer. Proc. Natl. Acad. Sci. USA 99, 14098–14103 (2002)
11. Yamada, T., Hiraoka, Y., Ikehata, M., Kimbara, K., Avner, B.S., Das Gupta, T.K., Chakrabarty, A.M.: Apoptosis or growth arrest: Modulation of tumor suppressor p53's specificity by bacterial redox protein azurin. Proc. Natl. Acad. Sci. USA 101, 4770–4775 (2004)
12. Yamada, T., Fialho, A.M., Punj, V., Bratescu, L., Das Gupta, T.K., Chakrabarty, A.M.: Internalization of bacterial redox protein azurin in mammalian cells: entry domain and specificity. Cell. Microbiol. 7, 1418–1431 (2005)
13. Jia, L., Gorman, G.S., Coward, L.U., Noker, P.E., McCormick, D., Horn, T.L., Harder, J.B., Muzzio, M., Prabhakar, B., Ganesh, B., Das Gupta, T.K., Beattie, C.W.: Preclinical pharmacokinetics, metabolism, and toxicity of azurin-p28 (NSC745104) a peptide inhibitor of p53 ubiquitination. Cancer Chemother. Pharmacol. 68, 513–524 (2011)
14. De Grandis, V., Bizzarri, A.R., Cannistraro, S.: Docking study and free energy simulation of the complex between p53 DNA-binding domain and azurin. Journal of Molecular Recognition: JMR 20, 215–226 (2007)
15. Taranta, M., Bizzarri, A.R., Cannistraro, S.: Probing the interaction between p53 and the bacterial protein azurin by single molecule force spectroscopy. Journal of Molecular Recognition: JMR 21, 63–70 (2008)
16. Taranta, M., Bizzarri, A.R., Cannistraro, S.: Modeling the interaction between the N-terminal domain of the tumor suppressor p53 and azurin. Journal of Molecular Recognition: JMR 22, 215–222 (2009)
17. Chaudhari, A., Fialho, A.M., Ratner, D., Gupta, P., Hong, C.S., Kahali, S., Yamada, T., Haldar, K., Murphy, S., Cho, W., Chauhan, V.S., Das Gupta, T.K., Chakrabarty, A.M.: Azurin, Plasmodium falciparum malaria and HIV/AIDS: inhibition of parasitic and viral growth by Azurin. Cell Cycle 5, 1642–1648 (2006)
18. Naguleswaran, A., Fialho, A.M., Chaudhari, A., Hong, C.S., Chakrabarty, A.M., Sullivan, W.J.: Azurin-like protein blocks invasion of Toxoplasma gondii through potential interactions with parasite surface antigen SAG1. Antimicrob. Agents Chemother. 52, 402–408 (2008)
19. Fialho, A.M., Chakrabarty, A.M.: Promiscuous Anticancer Drugs from Pathogenic Bacteria: Rational Versus Intelligent Drug Design. In: Microbial Approaches and Biotechnological Tools, pp. 179–198. John Wiley & Sons, Inc., Hoboken (2010)
20. Fialho, A.M., Bernardes, N., Chakrabarty, A.M.: Recent patents on live bacteria and their products as potential anticancer agents. Recent Pat. Anticancer Drug Discov. 7, 31–55 (2012)
21. Cho, J.-H., Lee, M.-H., Cho, Y.-J., Park, B.-S., Kim, S., Kim, G.-C.: The bacterial protein azurin enhances sensitivity of oral squamous carcinoma cells to anticancer drugs. Yonsei Med. J. 52, 773–778 (2011)

22. Hetz, C., Bono, M.R., Barros, L.F., Lagos, R.: Microcin E492, a channel-forming bacteriocin from Klebsiella pneumoniae, induces apoptosis in some human cell lines. Proc. Natl. Acad. Sci. USA 99, 2696–2701 (2002)
23. Schloissnig, S., Arumugam, M., Sunagawa, S., Mitreva, M., Tap, J., Zhu, A., Waller, A., Mende, D.R., Kultima, J.R., Martin, J., Kota, K., Sunyaev, S.R., Weinstock, G.M., Bork, P.: Genomic variation landscape of the human gut microbiome. Nature 493, 45–50 (2013)
24. van Heel, A.J., de Jong, A., Montalbán-López, M., Kok, J., Kuipers, O.P.: BAGEL3: automated identification of genes encoding bacteriocins and (non-)bactericidal posttranslationally modified peptides. Nucleic Acids Res. 41, W448 (2013)
25. Jensen, L.J., Gupta, R., Blom, N., Devos, D., Tamames, J., Kesmir, C., Nielsen, H., Stærfeldt, H.H., Rapacki, K., Workman, C., Andersen, C.A.F., Knudsen, S., Krogh, A., Valencia, A., Brunak, S.: Prediction of Human Protein Function from Post-translational Modifications and Localization Features. J. Mol. Biol. 319, 1257–1265 (2002)
26. Jensen, L.J., Gupta, R., Stærfeldt, H.H., Brunak, S.: Prediction of human protein function according to Gene Ontology categories
27. Flühe, L., Marahiel, M.A.: Radical S-adenosylmethionine enzyme catalyzed thioether bond formation in sactipeptide biosynthesis. Curr. Opin. Chem. Biol. 17, 605–612 (2013)
28. Heath, L.S., Heath, H.E., LeBlanc, P.A., Smithberg, S.R., Dufour, M., Simmonds, R.S., Sloan, G.L.: The streptococcolytic enzyme zoocin A is a penicillin-binding protein. FEMS Microbiol. Lett. 236, 205–211 (2004)
29. Naisbitt, D.J., Gordon, S.F., Pirmohamed, M., Park, B.K.: Immunological principles of adverse drug reactions: the initiation and propagation of immune responses elicited by drug treatment. Drug Saf. 23, 483–507 (2000)
30. Suau, A., Bonnet, R., Sutren, M., Godon, J.J., Gibson, G.R., Collins, M.D., Doré, J.: Direct analysis of genes encoding 16S rRNA from complex communities reveals many novel molecular species within the human gut. Appl. Environ. Microbiol. 65, 4799–4807 (1999)
31. Walenkamp, A.M.E., Boer, I.G.J., Bestebroer, J., Rozeveld, D., Timmer-Bosscha, H., Hemrika, W., van Strijp, J.A.G., de Haas, C.J.C.: Staphylococcal superantigen-like 10 inhibits CXCL12-induced human tumor cell migration. Neoplasia 11, 333–344 (2009)
32. Walenkamp, A.M.E., Bestebroer, J., Boer, I.G.J., Kruizinga, R., Verheul, H.M., van Strijp, J.A.G., de Haas, C.J.C.: Staphylococcal SSL5 binding to human leukemia cells inhibits cell adhesion to endothelial cells and platelets. Cell. Oncol. 32, 1–10 (2010)
33. Wood, L.M., Pan, Z.-K., Shahabi, V., Paterson, Y.: Listeria-derived ActA is an effective adjuvant for primary and metastatic tumor immunotherapy. Cancer Immunol. Immunother. 59, 1049–1058 (2010)
34. Vinodhkumar, R., Song, Y.-S., Devaki, T.: Romidepsin (depsipeptide) induced cell cycle arrest, apoptosis and histone hyperacetylation in lung carcinoma cells (A549) are associated with increase in p21 and hypophosphorylated retinoblastoma proteins expression. Biomed. Pharmacother. 62, 85–93 (2008)
35. Kanno, S.-I., Maeda, N., Tomizawa, A., Yomogida, S., Katoh, T., Ishikawa, M.: Involvement of p21waf1/cip1 expression in the cytotoxicity of the potent histone deacetylase inhibitor spiruchostatin B towards susceptible NALM-6 human B cell leukemia cells. Int. J. Oncol. 40, 1391–1396 (2012)
36. Lee, C.-H., Wu, C.-L., Shiau, A.-L.: Systemic administration of attenuated Salmonella choleraesuis carrying thrombospondin-1 gene leads to tumor-specific transgene expression, delayed tumor growth and prolonged survival in the murine melanoma model. Cancer Gene. Ther. 12, 175–184 (2005)

37. Lee, D.G., Hahm, K.-S., Park, Y., Kim, H.-Y., Lee, W., Lim, S.-C., Seo, Y.-K., Choi, C.-H.: Functional and structural characteristics of anticancer peptide Pep27 analogues. Cancer Cell Int. 5, 21 (2005)
38. Joo, N.E., Ritchie, K., Kamarajan, P., Miao, D., Kapila, Y.L.: Nisin, an apoptogenic bacteriocin and food preservative, attenuates HNSCC tumorigenesis via CHAC1. Cancer Med 1, 295–305 (2012)
39. Shaikh, F., Abhinand, P., Ragunath, P.: Identification & Characterization of lactobacillus salavarius bacteriocins and its relevance in cancer therapeutics. Bioinformation 8, 589–594 (2012)

Non-linear Classification of Massive Datasets with a Parallel Algorithm of Local Support Vector Machines

Thanh-Nghi Do

College of Information Technology, Can Tho University
No 1, Ly Tu Trong Street, Ninh Kieu District, Can Tho City, Viet Nam
dtnghi@cit.ctu.edu.vn

Abstract. We propose a new parallel algorithm of local support vector machines, called kSVM for the effectively non-linear classification of large datasets. The learning strategy of kSVM uses *k*means algorithm to partition the data into *k* clusters, followed which it constructs a non-linear SVM in each cluster to classify the data locally in the parallel way on multi-core computers. The kSVM algorithm is faster than the standard SVM in the non-linear classification of large datasets while maintaining the classification correstness. The numerical test results on 4 datasets from UCI repository and 3 benchmarks of handwritten letters recognition showed that our proposal is efficient compared to the standard SVM.

Keywords: Support vector machines, local support vector machines, large-scale non-linear classification.

1 Introduction

In recent years, the SVM algorithm proposed by [1] and kernel-based methods have shown practical relevance for classification, regression and novelty detection. Successful applications of SVMs have been reported for various fields like face identification, text categorization and bioinformatics [2]. They become increasingly popular data analysis tools. In spite of the prominent properties of SVM, they are not favorable to deal with the challenge of large datasets. SVM solutions are obtained from quadratic programming (QP), so that the computational cost of a SVM approach is at least square of the number of training datapoints and the memory requirement making SVM impractical. There is a need to scale up learning algorithms to handle massive datasets on personal computers (PCs).

Our investigation aims at proposing a new parallel algorithm of local SVM, called kSVM for the effectively non-linear classification of large datasets. Instead of building a global SVM model, as done by the classical algorithm is very difficult to deal with large datasets, the kSVM algorithm is to construct an ensemble of local ones that are easily trained by the standard SVM algorithms. The kSVM algorithm performs the training task with two steps. The first one is to use *k*means algorithm [3] to partition the data into *k* clusters, and then the second one is to learn a non-linear SVM in each cluster to classify the data locally in the parallel way on multi-core computers. The numerical test results on 4 datasets from UCI repository [4] and 3 benchmarks of handwritten letters recognition [5], MNIST [6], [7] showed that our proposal is efficient compared

H.A. Le Thi et al. (eds.), *Advanced Computational Methods for Knowledge Engineering*,
Advances in Intelligent Systems and Computing 358, DOI: 10.1007/978-3-319-17996-4_21

to the standard SVM in terms of training time and accuracy. The kSVM algorithm is faster than the standard SVM in the non-linear classification of large datasets while maintaining the high classification accuracy.

The paper is organized as follows. Section 2 briefly introduces the SVM algorithm. Section 3 presents our proposed parallel algorithm of local SVM for the non-linear classification of large datasets. Section 4 shows the experimental results. Section 5 discusses about related works. We then conclude in section 6.

2 Support Vector Machines

Let us consider a linear binary classification task, as depicted in Figure 1, with m datapoints x_i $(i = 1, \ldots, m)$ in the n-dimensional input space R^n, having corresponding labels $y_i = \pm 1$. For this problem, the SVM algorithms [1] try to find the best separating plane (denoted by the normal vector $w \in R^n$ and the scalar $b \in R$), i.e. furthest from both class $+1$ and class -1. It can simply maximize the distance or the margin between the supporting planes for each class ($x.w - b = +1$ for class $+1$, $x.w - b = -1$ for class -1). The margin between these supporting planes is $2/\|w\|$ (where $\|w\|$ is the 2-norm of the vector w). Any point x_i falling on the wrong side of its supporting plane is considered to be an error, denoted by z_i ($z_i \geq 0$). Therefore, SVM has to simultaneously maximize the margin and minimize the error. The standard SVMs pursue these goals with the quadratic programming of (1).

$$
\begin{aligned}
& min_{\alpha}(1/2)\sum_{i=1}^{m}\sum_{j=1}^{m} y_i y_j \alpha_i \alpha_j K\langle x_i, x_j\rangle - \sum_{i=1}^{m}\alpha_i \\
& s.t. \begin{cases} \sum_{i=1}^{m} y_i \alpha_i = 0 \\ 0 \leq \alpha_i \leq C \quad \forall i = 1,2,...,m \end{cases}
\end{aligned}
\tag{1}
$$

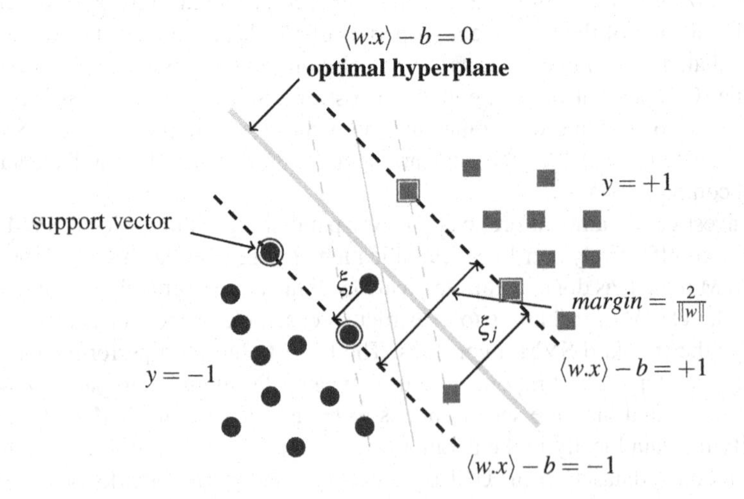

Fig. 1. Linear separation of the datapoints into two classes

where C is a positive constant used to tune the margin and the error and a linear kernel function $K\langle x_i, x_j\rangle = \langle x_i \cdot x_j\rangle$.

The support vectors (for which $\alpha_i > 0$) are given by the solution of the quadratic program (1), and then, the separating surface and the scalar b are determined by the support vectors. The classification of a new data point x based on the SVM model is as follows:

$$predict(x, SVMmodel) = sign(\sum_{i=1}^{\#SV} y_i \alpha_i K\langle x, x_i\rangle - b) \quad (2)$$

Variations on SVM algorithms use different classification functions [8]. No algorithmic changes are required from the usual kernel function $K\langle x_i, x_j\rangle$ as a linear inner product, $K\langle x_i, x_j\rangle = \langle x_i \cdot x_j\rangle$ other than the modification of the kernel function evaluation. We can get different support vector classification models. There are two other popular non-linear kernel functions as follows:

- a polynomial function of degree d: $K\langle x_i, x_j\rangle = (\langle x_i \cdot x_j\rangle + 1)^d$
- a RBF (Radial Basis Function): $K\langle x_i, x_j\rangle = e^{-\gamma \|x_i - x_j\|^2}$

SVMs are accurate models with practical relevance for classification, regression and novelty detection. Successful applications of SVMs have been reported for such varied fields including facial recognition, text categorization and bioinformatics [2].

3 Parallel Algorithm of Local Support Vector Machines

The study in [9] illustrated that the computational cost requirements of the SVM solutions in (1) are at least $O(m^2)$ (where m is the number of training datapoints), making standard SVM intractable for large datasets. Learning a global SVM model on the full massive dataset is challenge due to the very high computational cost and the very large memory requirement.

Learning SVM Models

Our proposed kSVM algorithm uses *k*means algorithm [3] to partition the full dataset into k clusters, and then it is easily to learn a non-linear SVM in each cluster to classify the data locally. Figure 2 shows the comparison between a global SVM model (left part) and 3 local SVM models (right part), using a non-linear RBF kernel function with $\gamma = 10$ and a positive constant $C = 10^6$.

Let now examine the complexity of building k local SVM models with the kSVM algorithm. The full dataset with m individuals is partitioned into k balanced clusters (the cluster size is about $\frac{m}{k}$). Therefore, the complexity of k local SVM models is $O(k(\frac{m}{k})^2) = O(\frac{m^2}{k})$. This complexity analysis illustrates that learning k local SVM models in the kSVM algorithm [1] is faster than building a global SVM model (the complexity is at least $O(m^2)$).

[1] It must be noted that the complexity of the kSVM approach does not include the *k*means clustering used to partition the full dataset. But this step requires insignificant time compared with the quadratic programming solution.

It must be remarked that the parameter k is used in the kSVM to give a trade-off between the generalization capacity and the computational cost. In [10], [11], [12], Vapnik points out the trade-off between the capacity of the local learning system and the number of available individuals. In the context of k local SVM models, this point can be understood as follows:

- If k is large then the kSVM algorithm reduces significant training time (the complexity of kSVM is $O(\frac{m^2}{k})$). And then, the size of a cluster is small; The locality is extremely with a very low capacity.
- If k is small then the kSVM algorithm reduces insignificant training time. However, the size of a cluster is large; It improves the capacity.

It leads to set k so that the cluster size is a large enough (e.g. 200 proposed by [11]).

Furthermore, the kSVM learns independently k local models from k clusters. This is a nice property for parallel learning. The parallel kSVM does take into account the benefits of high performance computing, e.g. multi-core computers or grids. The simplest development of the parallel kSVM algorithm is based on the shared memory multiprocessing programming model OpenMP [13] on multi-core computers. The parallel training of kSVM is described in algorithm 1.

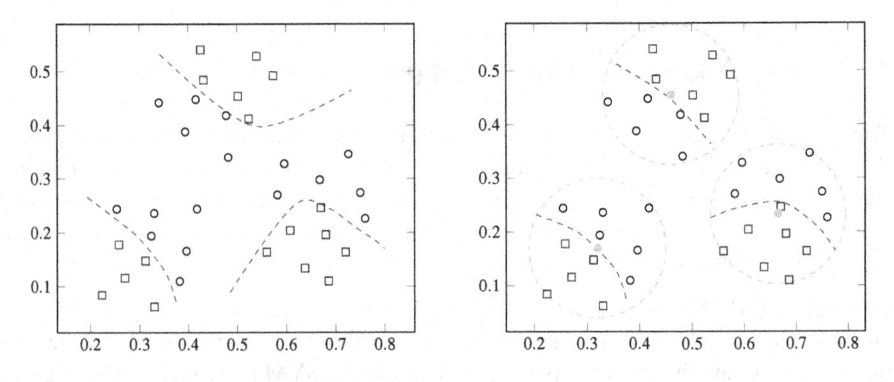

Fig. 2. Global SVM model (left part) versus local SVM models (right part)

Prediction of a New Individual Using Local SVM Models

The $kSVM - model = \{(c_1, lsvm_1), (c_2, lsvm_2), \ldots, (c_k, lsvm_k)\}$ is used to predict the class of a new individual x as follows. The first step is to find the closest cluster based on the distance between x and the cluster centers:

$$c_{NN} = \arg\min_{c} distance(x, c) \tag{3}$$

And then, the class of x is predicted by the local SVM model $lsvm_{NN}$ (corresponding to c_{NN}):

$$predict(x, kSVMmodel) = predict(x, lsvm_{NN}) \tag{4}$$

Algorithm 1. Local support vector machines algorithm

```
  input :
          training dataset D
          number of local models k
          hyper-parameter of RBF kernel function γ
          positive constant for tuning margin and errors of SVMs C
  output:
          k local support vector machines models
1 begin
2   /*kmeans performs the data clustering on dataset D;*/
3   creating k clusters denoted by D_1, D_2, ..., D_k and
4   their corresponding centers c_1, c_2, ..., c_k
5   #pragma omp parallel for
6   for i ← 1 to k do
7     /*learning local support vector machine model from D_i;*/
8     lsvm_i = svm(D_i, γ, C)
9   end
10  return kSVM − model = {(c_1, lsvm_1), (c_2, lsvm_2), ..., (c_k, lsvm_k)}
11 end
```

4 Evaluation

We are interested in the performance of the new parallel algorithm of local SVM (denoted by kSVM) for data classification. We have implemented kSVM in C/C++, OpenMP [13], using the highly efficient standard library SVM, LibSVM [14]. Our evaluation of the classification performance is reported in terms of correctness and training time. We are interested in the comparison obtained by our proposed kSVM with LibSVM.

All experiments are run on machine Linux Fedora 20, Intel(R) Core i7-4790 CPU, 3.6 GHz, 4 cores and 32 GB main memory.

Experiments are conducted with the 4 datasets collected from UCI repository [4] and the 3 benchmarks of handwritten letters recognition, including USPS [5], MNIST [6], a new benchmark for handwritten character recognition [7]. Table 1 presents the description of datasets. The evaluation protocols are illustrated in the last column of table 1. Datasets are already divided in training set (Trn) and testing set (Tst). We used the training data to build the SVM models. Then, we classified the testing set using the resulting models.

We propose to use RBF kernel type in kSVM and SVM models because it is general and efficient [15]. We also tried to tune the hyper-parameter γ of RBF kernel (RBF kernel of two individuals x_i, x_j, $K[i,j] = exp(-\gamma\|x_i - x_j\|^2)$) and the cost C (a trade-off between the margin size and the errors) to obtain a good accuracy. Furthermore, our kSVM uses the parameter k local models (number of clusters). We propose to set k so that each cluster has about 1000 individuals. The idea gives a trade-off between the generalization capacity [12] and the computational cost. Table 2 presents the hyper-parameters of kSVM and SVM in the classification.

Table 1. Description of datasets

ID	Dataset	Individuals	Attributes	Classes	Evaluation protocol
1	Opt. Rec. of Handwritten Digits	5620	64	10	3832 Trn - 1797 Tst
2	Letter	20000	16	26	13334 Trn - 6666 Tst
3	Isolet	7797	617	26	6238 Trn - 1559 Tst
4	USPS Handwritten Digit	9298	256	10	7291 Trn - 2007 Tst
5	A New Benchmark for Hand. Char. Rec.	40133	3136	36	36000 Trn - 4133 Tst
6	MNIST	70000	784	10	60000 Trn - 10000 Tst
7	Forest Cover Types	581012	54	7	400000 Trn - 181012 Tst

Table 2. Hyper-parameters of kSVM and SVM

ID	Datasets	γ	C	k
1	Opt. Rec. of Handwritten Digits	0.0001	100000	10
2	Letter	0.0001	100000	30
3	Isolet	0.0001	100000	10
4	USPS Handwritten Digit	0.0001	100000	10
5	A New Benchmark for Hand. Char. Rec.	0.001	100000	50
6	MNIST	0.05	100000	100
7	Forest Cover Types	0.0001	100000	500

The classification results of LibSVM and kSVM on the 7 datasets are given in table 3 and figure 3, figure 4. As it was expected, our kSVM algorithm outperforms LibSVM in terms of training time. In terms of test correctness, our kSVM achieves very competitive performances compared to LibSVM.

Table 3. Classification results in terms of accuracy (%) and training time (s)

ID	Datasets	Classification accuracy(%)		Training time(s)	
		LibSVM	kSVM	LibSVM	kSVM
1	Opt. Rec. of Handwritten Digits	98.33	97.05	0.58	0.21
2	Letter	97.40	96.14	2.87	0.5
3	Isolet	96.47	95.44	8.37	2.94
4	USPS Handwritten Digit	96.86	95.86	5.88	3.82
5	A New Benchmark for Hand. Char. Rec.	95.14	92.98	107.07	35.7
6	MNIST	98.37	98.11	1531.06	45.50
7	Forest Cover Types	NA	97.06	NA	223.7

With 5 first small datasets, the improvement of kSVM is not significant. With large datasets, kSVM achieves a significant speed-up in learning. For MNIST dataset, kSVM is 33.64 times faster than LibSVM. Typically, Forest cover type dataset is well-known as a difficult dataset for non-linear SVM [16], [17]; LibSVM ran for 23 days without any result. kSVM performed this non-linear classification in 223.7 seconds with 97.06% accuracy.

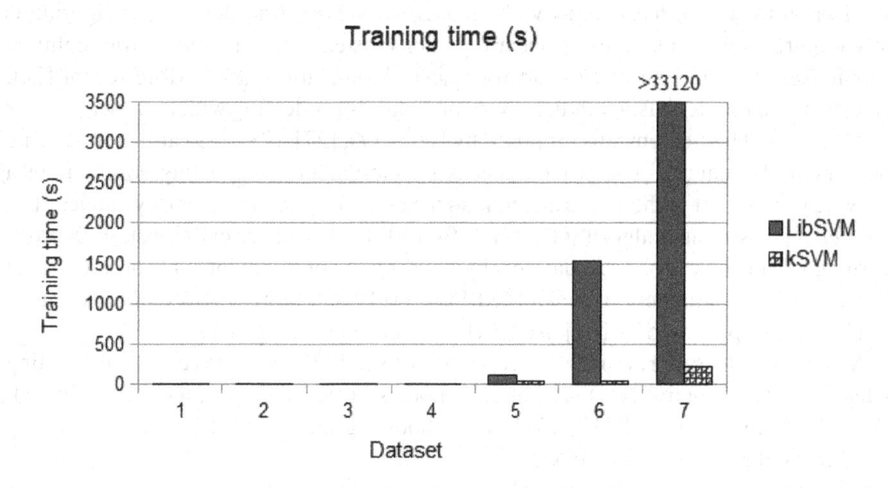

Fig. 3. Comparison of training time

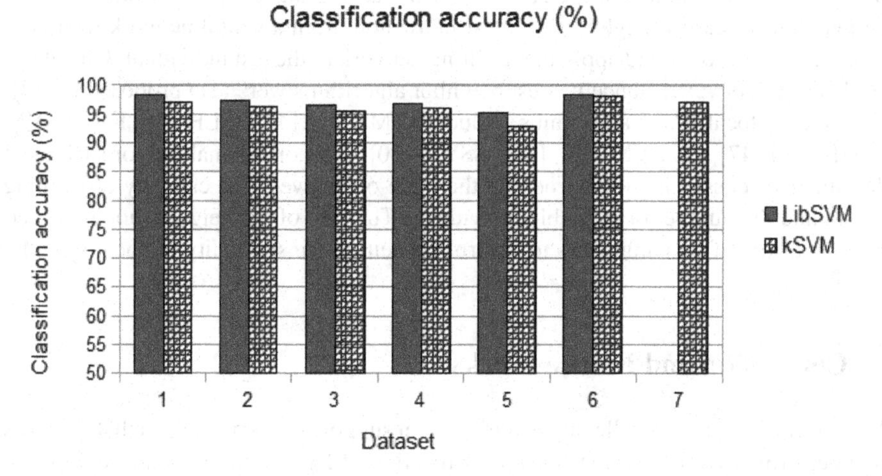

Fig. 4. Comparison of accuracy

5 Discussion on Related Works

Our proposal is in some aspects related to large-scale SVM learning algorithms. The improvements of SVM training on very large datasets include effective heuristic methods in the decomposition of the original quadratic programming into series of small problems [18], [19], [9] and [14].

Mangasarian and his colleagues proposed to modify SVM problems to obtain new formulas, including Lagrangian SVM [20], proximal SVM [21], Newton SVM [22]. The Least Squares SVM proposed by Suykens and Vandewalle [23] changes standard

SVM optimization to lead the new efficient SVM solver. And then, these algorithms only require solving a system of linear equations instead of a quadratic programming. This makes training time very short for linear classification tasks. More recent [24], [25] proposed the stochastic gradient descent methods for dealing with large scale linear SVM solvers. Their extensions proposed by [26], [17], [27], [28], [29] aim at improving memory performance for massive datasets by incrementally updating solutions in a growing training set without needing to load the entire dataset into memory at once. The parallel and distributed algorithms [27], [29], [30] for the linear classification improve learning performance for large datasets by dividing the problem into sub-problems that execute on large numbers of networked PCs, grid computing, multi-core computers. Parallel SVMs proposed by [31] use GPU to speed-up training tasks.

Active SVM learning algorithms proposed by [32], [33], [34], [35] choose interesting datapoint subsets (active sets) to construct models, instead of using the whole dataset. SVM algorithms [36], [37], [17], [38] use the boosting strategy [39], [40] for classifying very large datasets on standard PCs.

Our proposal of local SVM models is also related to local learning algorithms. The first paper of [41] proposed to use the expectation-maximization algorithm [42] for partitioning the training set into k clusters; for each cluster, a neural network is learnt to classify the individuals in the cluster. Local learning algorithms of Bottou & Vapnik [11] find k nearest neighbors of a test individual; train a neural network with only these k neighborhoods and apply the resulting network to the test individual. k-local hyperplane and convex distance nearest neighbor algorithms were also proposed in [43]. More recent local SVM algorithms include SVM-kNN [44], ALH [45],FaLK-SVM [46], LSVM [47], LL-SVM [48], [49], CSVM [50]. A theorical analysis for such local algorithms discussed in [10] introduces the trade-off between the capacity of learning system and the number of available individuals. The size of the neighborhoods is used as an additional free parameters to control capacity against locality of local learning algorithms.

6 Conclusion and Future Works

We presented the new parallel algorithm of local support vector machines that achieves high performances for the non-linear classification of large datasets. The training task of kSVM is to partition the data into k clusters and then it constructs a non-linear SVM in each cluster to classify the data locally in the parallel way. The numerical test results on 4 datasets from UCI repository and 3 benchmarks of handwritten letters recognition showed that our proposal is efficient in terms of training time and accuracy compared to the standard SVM. An example of its effectiveness is given with the non-linear classification of Forest Cover Types dataset (having 400000 individuals, 54 attributes) into 7 classes in 223.7 seconds and 97.06% accuracy.

In the near future, we intend to provide more empirical test on large benchmarks and comparisons with other algorithms. A promising avenue for future research aims at improving the classification accuracy of kSVM.

References

1. Vapnik, V.: The Nature of Statistical Learning Theory. Springer (1995)
2. Guyon, I.: Web page on svm applications (1999), http://www.clopinet.com/isabelle/Projects/-SVM/app-list.html
3. MacQueen, J.: Some methods for classification and analysis of multivariate observations. In: Proceedings of 5th Berkeley Symposium on Mathematical Statistics and Probability, vol. 1, pp. 281–297. University of California Press, Berkeley (1967)
4. Asuncion, A., Newman, D.: UCI repository of machine learning databases (2007)
5. LeCun, Y., Boser, B., Denker, J., Henderson, D., Howard, R., Hubbard, W., Jackel, L.: Back-propagation applied to handwritten zip code recognition. Neural Computation 1(4), 541–551 (1989)
6. LeCun, Y., Bottou, L., Bengio, Y., Haffner, P.: Gradient-based learning applied to document recognition. Proceedings of the IEEE 86(11), 2278–2324 (1998)
7. van der Maaten, L.: A new benchmark dataset for handwritten character recognition (2009), http://homepage.tudelft.nl/19j49/Publications_files/characters.zip
8. Cristianini, N., Shawe-Taylor, J.: An Introduction to Support Vector Machines: And Other Kernel-based Learning Methods. Cambridge University Press, New York (2000)
9. Platt, J.: Fast training of support vector machines using sequential minimal optimization. In: Schölkopf, B., Burges, C., Smola, A. (eds.) Advances in Kernel Methods – Support Vector Learning, pp. 185–208 (1999)
10. Vapnik, V.: Principles of risk minimization for learning theory. In: Advances in Neural Information Processing Systems 4 [NIPS Conference, Denver, Colorado, USA, December 2-5, 1991], pp. 831–838 (1991)
11. Bottou, L., Vapnik, V.: Local learning algorithms. Neural Computation 4(6), 888–900 (1992)
12. Vapnik, V., Bottou, L.: Local algorithms for pattern recognition and dependencies estimation. Neural Computation 5(6), 893–909 (1993)
13. OpenMP Architecture Review Board: OpenMP application program interface version 3.0 (2008)
14. Chang, C.C., Lin, C.J.: LIBSVM: a library for support vector machines. ACM Transactions on Intelligent Systems and Technology 2(27), 1–27 (2011)
15. Lin, C.: A practical guide to support vector classification (2003)
16. Yu, H., Yang, J., Han, J.: Classifying large data sets using SVMs with hierarchical clusters. In: Proc. of the ACM SIGKDD Intl. Conf. on KDD, pp. 306–315. ACM (2003)
17. Do, T.N., Poulet, F.: Towards high dimensional data mining with boosting of psvm and visualization tools. In: Proc. of 6th Intl. Conf. on Entreprise Information Systems, pp. 36–41 (2004)
18. Boser, B., Guyon, I., Vapnik, V.: An training algorithm for optimal margin classifiers. In: Proc. of 5th ACM Annual Workshop on Computational Learning Theory, pp. 144–152. ACM (1992)
19. Osuna, E., Freund, R., Girosi, F.: An improved training algorithm for support vector machines. In: Principe, J., Gile, L., Morgan, N., Wilson, E. (eds.) Neural Networks for Signal Processing VII, pp. 276–285 (1997)
20. Mangasarian, O., Musicant, D.: Lagrangian support vector machines. Journal of Machine Learning Research 1, 161–177 (2001)
21. Fung, G., Mangasarian, O.: Proximal support vector classifiers. In: Proc. of the ACM SIGKDD Intl. Conf. on KDD, pp. 77–86. ACM (2001)
22. Mangasarian, O.: A finite newton method for classification problems. Technical Report 01-11, Data Mining Institute, Computer Sciences Department, University of Wisconsin (2001)

23. Suykens, J., Vandewalle, J.: Least squares support vector machines classifiers. Neural Processing Letters 9(3), 293–300 (1999)
24. Shalev-Shwartz, S., Singer, Y., Srebro, N.: Pegasos: Primal estimated sub-gradient solver for svm. In: Proceedings of the Twenty-Fourth International Conference ON Machine Learning, pp. 807–814. ACM (2007)
25. Bottou, L., Bousquet, O.: The tradeoffs of large scale learning. In: Platt, J., Koller, D., Singer, Y., Roweis, S. (eds.) Advances in Neural Information Processing Systems, vol. 20, pp. 161–168 (2008), http://books.nips.cc
26. Do, T.N., Poulet, F.: Incremental svm and visualization tools for bio-medical data mining. In: Proc. of Workshop on Data Mining and Text Mining in Bioinformatics, pp. 14–19 (2003)
27. Do, T.N., Poulet, F.: Classifying one billion data with a new distributed svm algorithm. In: Proc. of 4th IEEE Intl. Conf. on Computer Science, Research, Innovation and Vision for the Future, pp. 59–66. IEEE Press (2006)
28. Fung, G., Mangasarian, O.: Incremental support vector machine classification. In: Proc. of the 2nd SIAM Int. Conf. on Data Mining (2002)
29. Poulet, F., Do, T.N.: Mining very large datasets with support vector machine algorithms. In: Camp, O., Filipe, J., Hammoudi, S., Piattini, M. (eds.) Enterprise Information Systems V, pp. 177–184 (2004)
30. Do, T.: Parallel multiclass stochastic gradient descent algorithms for classifying million images with very-high-dimensional signatures into thousands classes. Vietnam J. Computer Science 1(2), 107–115 (2014)
31. Do, T.-N., Nguyen, V.-H., Poulet, F.: Speed up SVM algorithm for massive classification tasks. In: Tang, C., Ling, C.X., Zhou, X., Cercone, N.J., Li, X. (eds.) ADMA 2008. LNCS (LNAI), vol. 5139, pp. 147–157. Springer, Heidelberg (2008)
32. Yu, H., Yang, J., Han, J.: Classifying large data sets using svms with hierarchical clusters. In: Proc. of the ACM SIGKDD Intl. Conf. on KDD, pp. 306–315. ACM (2003)
33. Do, T.N., Poulet, F.: Mining very large datasets with svm and visualization. In: Proc. of 7th Intl. Conf. on Entreprise Information Systems, pp. 127–134 (2005)
34. Boley, D., Cao, D.: Training support vector machines using adaptive clustering. In: Berry, M.W., Dayal, U., Kamath, C., Skillicorn, D.B. (eds.) Proceedings of the Fourth SIAM International Conference on Data Mining, Lake Buena Vista, Florida, USA, April 22-24, pp. 126–137. SIAM (2004)
35. Tong, S., Koller, D.: Support vector machine active learning with applications to text classification. In: Proc. of the 17th Intl. Conf. on Machine Learning, pp. 999–1006. ACM (2000)
36. Pavlov, D., Mao, J., Dom, B.: Scaling-up support vector machines using boosting algorithm. In: 15th International Conference on Pattern Recognition, vol. 2, pp. 219–222 (2000)
37. Do, T.N., Le-Thi, H.A.: Classifying large datasets with svm. In: Proc. of 4th Intl. Conf. on Computational Management Science (2007)
38. Do, T.N., Fekete, J.D.: Large scale classification with support vector machine algorithms. In: Wani, M.A., Kantardzic, M.M., Li, T., Liu, Y., Kurgan, L.A., Ye, J., Ogihara, M., Sagiroglu, S., Chen, X.W., Peterson, L.E., Hafeez, K. (eds.) The Sixth International Conference on Machine Learning and Applications, ICMLA 2007, Cincinnati, Ohio, USA, December 13-15, pp. 7–12. IEEE Computer Society (2007)
39. Freund, Y., Schapire, R.: A short introduction to boosting. Journal of Japanese Society for Artificial Intelligence 14(5), 771–780 (1999)
40. Breiman, L.: Arcing classifiers. The Annals of Statistics 26(3), 801–849 (1998)
41. Jacobs, R.A., Jordan, M.I., Nowlan, S.J., Hinton, G.E.: Adaptive mixtures of local experts. Neural Computation 3(1), 79–87 (1991)
42. Dempster, A.P., Laird, N.M., Rubin, D.B.: Maximum likelihood from incomplete data via the em algorithm. Journal of the Royal Statistical Society, Series B 39(1), 1–38 (1977)

43. Vincent, P., Bengio, Y.: K-local hyperplane and convex distance nearest neighbor algorithms. In: Advances in Neural Information Processing Systems, pp. 985–992. The MIT Press (2001)
44. Zhang, H., Berg, A., Maire, M., Malik, J.: SVM-KNN: Discriminative nearest neighbor classification for visual category recognition. In: 2006 IEEE Computer Society Conference on Computer Vision and Pattern Recognition, vol. 2, pp. 2126–2136 (2006)
45. Yang, T., Kecman, V.: Adaptive local hyperplane classification. Neurocomputing 71(13-15), 3001–3004 (2008)
46. Segata, N., Blanzieri, E.: Fast and scalable local kernel machines. Journal Machine Learning Research 11, 1883–1926 (2010)
47. Cheng, H., Tan, P.N., Jin, R.: Efficient algorithm for localized support vector machine. IEEE Transactions on Knowledge and Data Engineering 22(4), 537–549 (2010)
48. Kecman, V., Brooks, J.: Locally linear support vector machines and other local models. In: The 2010 International Joint Conference on Neural Networks (IJCNN), pp. 1–6 (2010)
49. Ladicky, L., Torr, P.H.S.: Locally linear support vector machines. In: Getoor, L., Scheffer, T. (eds.) Proceedings of the 28th International Conference on Machine Learning, ICML 2011, Bellevue, Washington, USA, June 28-July 2, pp. 985–992. Omnipress (2011)
50. Gu, Q., Han, J.: Clustered support vector machines. In: Proceedings of the Sixteenth International Conference on Artificial Intelligence and Statistics, AISTATS 2013, Scottsdale, AZ, USA, April 29-May 1. JMLR Proceedings, vol. 31, pp. 307–315 (2013)

On the Efficiency of Query-Subquery Nets with Right/Tail-Recursion Elimination in Evaluating Queries to Horn Knowledge Bases

Son Thanh Cao[1,2]

[1] Faculty of Information Technology, Vinh University
182 Le Duan street, Vinh, Nghe An, Vietnam
sonct@vinhuni.edu.vn

[2] Institute of Informatics, University of Warsaw
Banacha 2, 02-097 Warsaw, Poland

Abstract. We propose a method called QSQN-rTRE for evaluating queries to Horn knowledge bases. It is an extension of QSQN-TRE by applying elimination not only to the tail-recursive predicates but also to all the intensional predicates that appear rightmost in the bodies of the program clauses. The aim is to avoid materializing intermediate results for the mentioned cases during the processing. As a consequence, it takes the advantage of reducing the number of kept tuples and subqueries in the computer memory as well as the number of read/write operations on relations. The efficiency of our method is illustrated by empirical results.

Keywords: Horn knowledge bases, deductive databases, query processing, optimization methods, QSQ, QSQN, QSQN-TRE, QSQN-rTRE.

1 Introduction

One of the most interesting thing in database community is query optimization, which has drawn a great deal of attention from researchers for many years. There are numerous well known optimization methods and techniques that have been proposed to improve the performance of query evaluation [16,15,12,6,14]. In this work, we study optimizing query evaluation for Horn knowledge bases.

Horn knowledge bases are definite logic programs, which can be treated as an extension of Datalog deductive databases with function symbols and without range-restrictedness condition [1]. Many of the evaluation methods have been developed for Datalog deductive databases such as QSQ [18,1], QSQR [18,1,10], QoSaQ [19] and the evaluation method that combines the magic-set transformation with the improved semi-naive evaluation method [2]. In [10], Madalińska-Bugaj and Nguyen generalized the QSQR method for Horn knowledge bases. Some authors extended the magic-set technique together with the breadth-first approach [13,9] and some other authors tried to adapt computational procedures of logic programming that use tabled SLD-resolution [17,19,20] for Horn knowledge bases. In [11], Nguyen and Cao proposed a new method that is based on query-subquery nets for evaluating queries to Horn knowledge bases.

H.A. Le Thi et al. (eds.), *Advanced Computational Methods for Knowledge Engineering*,
Advances in Intelligent Systems and Computing 358, DOI: 10.1007/978-3-319-17996-4_22

An optimization technique that integrates magic-sets templates with a form of tail-recursion elimination was proposed by Ross in [14]. It improves the performance of query evaluation by not materializing the extension of intermediate views. In [6], together with Nguyen, we proposed a method that incorporates tail-recursion elimination into query-subquery nets, called QSQN-TRE. It takes the advantage of reducing the number of evaluated intermediate tuples/subqueries during the processing. The experimental results reported in [6] justify the usefulness of this method.

In this paper, we propose an optimization method called QSQN-rTRE for evaluating queries to Horn knowledge bases, which is an extension of QSQN-TRE by applying elimination not only to the tail-recursive predicates proposed in [6] but also to all the intensional predicates that appear rightmost in the bodies of the program clauses. Thus, all the rightmost intensional predicates in the bodies of the program clauses will be processed in the same way the tail-recursive predicates as stated in [6]. The aim of this method is to avoid materializing intermediate results for the mentioned cases during the processing, in order to reduce the number of kept tuples and subqueries in the computer memory as well as the number of read/write operations on relations in some cases. The efficiency of the proposed method is illustrated by empirical results.

The rest of this paper is organized as follows. Section 2 recalls the most important notation and definitions of logic programming and Horn knowledge bases. Section 3 presents our QSQN-rTRE evaluation method for Horn knowledge bases. The experimental results are provided in Section 4. Conclusions and a plan for future work are presented in Section 5.

2 Preliminaries

We assume that the reader is familiar with the notions of first-order logic such as *term, atom, predicate, expression, substitution, unification, most general unifier, computed answer, correct answer, Horn knowledge bases* and related ones. We refer the reader to [1] for further reading.

A predicate can be either *intensional* or *extensional*. A *generalized tuple* is a tuple of terms, which may contain function symbols and variables. A *generalized query atom* is a formula $q(\overline{t})$, where q is an intensional predicate and $\overline{t}$ is a generalized tuple. A *generalized relation* is a set of generalized tuples of the same arity.

A *positive program clause* is a formula of the form $\forall(A \vee \neg B_1 \vee \ldots \vee \neg B_n)$ with $n \geq 0$, written as $A \leftarrow B_1, \ldots, B_n$, where $A, B_1, \ldots, B_n$ are atoms. A is called the *head*, and $B_1, \ldots, B_n$ the *body* of the program clause.

A *positive* (or *definite*) *logic program* is a finite set of program clauses. From now on, by a "program" we will mean a positive logic program. A *goal* is a formula of the form $\forall(\neg B_1 \vee \ldots \vee \neg B_n)$, written as $\leftarrow B_1, \ldots, B_n$, where $B_1, \ldots, B_n$ are atoms and $n \geq 0$. If $n = 0$ then the goal is referred to as the *empty goal*.

Given substitutions θ and δ, the composition of θ and δ is denoted by $\theta\delta$, the domain of θ is denoted by $dom(\theta)$, the range of θ is denoted by $range(\theta)$, and the restriction of θ to a set X of variables is denoted by $\theta_{|X}$. The empty substitution is denoted by ε. The *term-depth* of an expression (resp. a substitution) is the maximal nesting depth of function symbols occurring in that expression (resp. substitution). Given a list/tuple α of terms or atoms, by $Vars(\alpha)$ we denote the set of variables occurring in α.

A *Horn knowledge base* is defined as a pair (P, I), where P is a positive logic program (which may contain function symbols and not be "range-restricted") for defining intensional predicates, and I is a *generalized extensional instance*, which is a function mapping each extensional n-ary predicate to an n-ary generalized relation.

Definition 1 (Tail-recursion). A program clause $\varphi_i = (A_i \leftarrow B_{i,1}, \ldots, B_{i,n_i})$, for $n_i > 0$, is said to be *recursive* whenever some $B_{i,j}$ $(1 \leq j \leq n_i)$ has the same predicate as A_i. If B_{i,n_i} has the same predicate as A_i then the clause is *tail-recursive*, in this case, the predicate of B_{i,n_i} is a *tail-recursive predicate.* $\triangleleft$

Definition 2. A predicate p is called *a rightmost-predicate* w.r.t. a positive logic program if it is an intensional predicate and appears rightmost in the body of some program clause of that program. $\triangleleft$

Definition 3. We say that a predicate p directly depends on a predicate q w.r.t. a positive logic program if p and q are intensional predicates and p appears in the head of a program clause and q is a *rightmost-predicate* of that program clause. A predicate p is called depends on a predicate q w.r.t. a positive logic program if p is either reflectively or transitively depends on q of that program. $\triangleleft$

3 QSQ-Nets with Right/Tail-Recursion Elimination

The method QSQN-TRE [6] is used to evaluate queries to a given Horn knowledge base (P, I). In this section, we extend QSQN-TRE to obtain a new evaluation method, which eliminates all the rightmost intensional predicates that appear in the bodies of the program clauses. Particularly, all the rightmost intensional predicates in the bodies of the program clauses will be processed in a similar way the tail-recursive predicates as proposed in [6]. Thus, from now on, a program clause is said to be *right/tail-recursive* if it is either a tail-recursive clause or a clause with a rightmost-predicate.

Let P be a positive logic program and $\varphi_1, \ldots, \varphi_m$ be all the program clauses of P, with $\varphi_i = (A_i \leftarrow B_{i,1}, \ldots, B_{i,n_i})$ and $n_i \geq 0$, for $1 \leq i \leq m$. The following definition shows how to make a QSQ-Net structure with right/tail-recursion elimination from the given positive logic program P.

Definition 4 (QSQN-rTRE structure). A *query-subquery net structure with right/tail-recursion elimination* (QSQN-rTRE structure for short) of P is a tuple (V, E, T) such that:

- T is a pair (T_{edb}, T_{idb}), called the *type* of the net structure.
- T_{idb} is a function that maps each intensional predicate to *true* or *false*. (If $T_{idb}(p)$ = *true* then the intensional relation p will be computed using right/tail-recursion elimination. T_{edb} will be explained shortly).

- V is a set of nodes that includes:
  - *input_p* and *ans_p*, for each intensional predicate p of P,
  - $\mathit{pre\_filter}_i$, $\mathit{filter}_{i,1}, \ldots, \mathit{filter}_{i,n_i}$, for each $1 \leq i \leq m$,
  - $\mathit{post\_filter}_i$ if (φ_i is not right/tail-recursive) or ($T_{idb}(p) = \mathit{false}$) or ($n_i = 0$), for each $1 \leq i \leq m$, where p is the predicate of A_i.
- E is a set of edges that includes:
  - $(\mathit{input\_p}, \mathit{pre\_filter}_i)$, for each $1 \leq i \leq m$, where p is the predicate of A_i,
  - $(\mathit{pre\_filter}_i, \mathit{filter}_{i,1})$, for each $1 \leq i \leq m$ such that $n_i \geq 1$,
  - $(\mathit{filter}_{i,1}, \mathit{filter}_{i,2}), \ldots, (\mathit{filter}_{i,n_i-1}, \mathit{filter}_{i,n_i})$, for each $1 \leq i \leq m$,
  - $(\mathit{filter}_{i,n_i}, \mathit{post\_filter}_i)$, for each $1 \leq i \leq m$ such that $n_i \geq 1$ and $\mathit{post\_filter}_i$ exists,
  - $(\mathit{pre\_filter}_i, \mathit{post\_filter}_i)$, for each $1 \leq i \leq m$ such that $n_i = 0$,
  - $(\mathit{post\_filter}_i, \mathit{ans\_p})$, for each $1 \leq i \leq m$ such that $\mathit{post\_filter}_i$ exists, where p is the predicate of A_i, and
    $(\mathit{post\_filter}_i, \mathit{ans\_p}_k)$, for each $1 \leq i \leq m$, $1 \leq k \leq m$ and $k \neq i$ such that $\mathit{post\_filter}_i$ exists and the predicate p_k of A_k depends on the predicate of A_i,
  - $(\mathit{filter}_{i,j}, \mathit{input\_p})$, for each $1 \leq i \leq m$ and $1 \leq j \leq n_i$ such that p is the predicate of $B_{i,j}$ and is an intensional predicate,
  - $(\mathit{ans\_p}, \mathit{filter}_{i,j})$, for each intensional predicate p and each $1 \leq i \leq m$, $1 \leq j \leq n_i$ such that: either ($1 \leq j < n_i$) or ($j = n_i$ and $T_{idb}(p) = \mathit{false}$) and $B_{i,j}$ is an atom of p.
- T_{edb} is a function that maps each $\mathit{filter}_{i,j} \in V$ such that the predicate of $B_{i,j}$ is extensional to *true* or *false*. If $T_{edb}(\mathit{filter}_{i,j}) = \mathit{false}$ then subqueries for $\mathit{filter}_{i,j}$ are always processed immediately without being accumulated at $\mathit{filter}_{i,j}$. ◁

From now on, $T(v)$ denotes $T_{edb}(v)$ if v is a node $\mathit{filter}_{i,j}$ such that $B_{i,j}$ is an extensional predicate, and $T(p)$ denotes $T_{idb}(p)$ for an intensional predicate p. Thus, T can be called a *memorizing type* for extensional nodes (as in QSQ-net structures), and a *right/tail-recursion-elimination type* for intensional predicates. We call the pair (V, E) the QSQN-rTRE topological structure of P with respect to T_{idb}.

Example 1. The upper part of Figure 1 illustrates a logic program and its QSQN-TRE topological structure w.r.t. T_{idb} with $T_{idb}(q) = \mathit{true}$ and $T_{idb}(p) = T_{idb}(s) = \mathit{false}$, where q, p, s are intensional predicates, t is an extensional predicate, x, y, z are variables and a is a constant symbol. The lower part depicts the QSQN-rTRE topological structure of the same program w.r.t. the T_{idb} with $T_{idb}(q) = T_{idb}(p) = \mathit{true}$ and $T_{idb}(s) = \mathit{false}$. ◁

The following definition specifies the notion QSQN-rTRE and the related ones. In addition, we also show how the data is transferred through the edges in a QSQN-rTRE.

Definition 5 (QSQN-rTRE). A *query-subquery net with right/tail-recursion elimination* (QSQN-rTRE for short) of P is a tuple $N = (V, E, T, C)$ such that

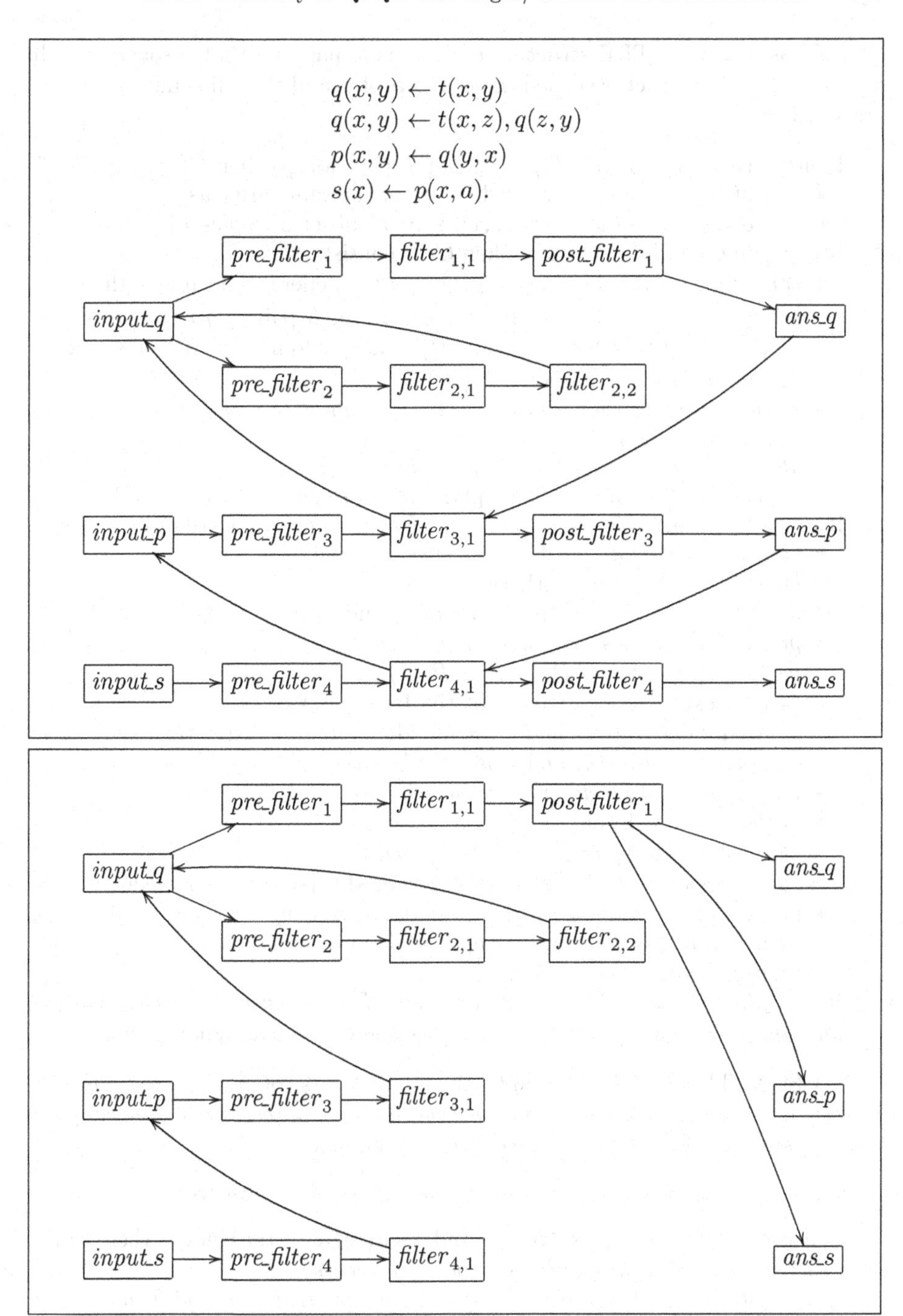

Fig. 1. QSQN-TRE and QSQN-rTRE topological structures

(V, E, T) is a QSQN-rTRE structure of P, C is a mapping that associates each node $v \in V$ with a structure called the *contents* of v, and the following conditions are satisfied:

- If either ($v = input\_p$ and $T(p) = false$) or ($v = ans\_p$) then $C(v)$ consists of:
  - $tuples(v)$: a set of generalized tuples of the same arity as p,
  - $unprocessed(v, w)$ for each $(v, w) \in E$: a subset of $tuples(v)$.
- If $v = input\_p$ and $T(p) = true$ then $C(v)$ consists of:
  - $pairs(v)$: a set of pairs $(\bar{t}, q(\bar{t}'))$, where $\bar{t}$ is a generalized tuple of the same arity as p, $q(\bar{t}')$ is a generalized query atom. A pair $(\bar{t}, q(\bar{t}'))$ means that we are trying to solve the atom $p(\bar{t})$, and any found answer substitution should generate an answer for $q(\bar{t}')$,
  - $unprocessed(v, w)$ for each $(v, w) \in E$: a subset of $pairs(v)$.
- If $v = pre\_filter_i$ then $C(v)$ consists of:
  - $atom(v) = A_i$ and $post\_vars(v) = Vars((B_{i,1}, \ldots, B_{i,n_i}))$.
- If $v = post\_filter_i$ then $C(v)$ is empty, but we assume $pre\_vars(v) = \emptyset$.
- If $v = filter_{i,j}$ and p is the predicate of $B_{i,j}$ then $C(v)$ consists of:
  - $kind(v) = extensional$ if p is extensional, and $kind(v) = intensional$ otherwise,
  - $pred(v) = p$ (called the predicate of v) and $atom(v) = B_{i,j}$,
  - $pre\_vars(v) = Vars((B_{i,j}, \ldots, B_{i,n_i}))$ and $post\_vars(v) = Vars((B_{i,j+1}, \ldots, B_{i,n_i}))$,
  - $subqueries(v)$: a set of pairs of the form $(q(\bar{t}), \delta)$, where $q(\bar{t})$ is a generalized query atom and δ is an idempotent substitution such that $dom(\delta) \subseteq pre\_vars(v)$ and $dom(\delta) \cap Vars(\bar{t}) = \emptyset$,
  - $unprocessed\_subqueries(v) \subseteq subqueries(v)$,
  - in the case p is intensional:
    $unprocessed\_subqueries_2(v) \subseteq subqueries(v)$,
    $unprocessed\_tuples(v)$: a set of generalized tuples of the same arity as p,
  - if $v = filter_{i,n_i}$, $kind(v) = intensional$, $pred(v) = p$ and $T(p) = true$ then $unprocessed\_subqueries(v)$ and $unprocessed\_tuples(v)$ are empty (hence, we can ignore them).
- If $v = filter_{i,j}$, $kind(v) = extensional$ and $T(v) = false$ then $subqueries(v)$ and $unprocessed\_subqueries(v)$ are empty (hence, we can ignore them).

A QSQN-rTRE of P is *empty* if all the sets of the form $pairs(v)$, $tuples(v)$, $subqueries(v)$, $unprocessed\_subqueries(v)$, $unprocessed\_subqueries_2(v)$, $unprocessed\_tuples(v)$ or $unprocessed(v, w)$ are empty. $\triangleleft$

If $(v, w) \in E$ then w is referred to as a *successor* of v. Observe that:

- if v is pre_filter_i or (v is $filter_{i,j}$ and $kind(v) = extensional$) then v has exactly one successor, which we denote by $succ(v)$,
- if v is $filter_{i,n_i}$ with $kind(v) = intensional$, $pred(v) = p$ and $T(p) = true$ then v has exactly one successor, which we denote by $succ_2(v) = input\_p$,
- if v is $filter_{i,j}$ with $kind(v) = intensional$, $pred(v) = p$ and either $j < n_i$ or $T(p) = false$ then v has exactly two successors: $succ(v) = filter_{i,j+1}$ if $j < n_i$; $succ(v) = post\_filter_i$ otherwise; and $succ_2(v) = input\_p$,

– if v is $post\_filter_i$ then v may have more than one successor.

By a *subquery* we mean a pair of the form $(q(\overline{t}), \delta)$, where $q(\overline{t})$ is a generalized query atom and δ is an idempotent substitution such that $dom(\delta) \cap Vars(\overline{t}) = \emptyset$. The set $unprocessed\_subqueries_2(v)$ (resp. $unprocessed\_subqueries(v)$) contains the subqueries that were not transferred through the edge $(v, succ_2(v))$ (resp. $(v, succ(v))$ – when it exists).

For an intensional predicate p with $T(p) = true$, the intuition behind a pair $(\overline{t}, q(\overline{t}')) \in pairs(input\_p)$ is that:

– $\overline{t}$ is a usual input tuple for p, but the intended goal at a higher level is $\leftarrow q(\overline{t}')$,
– any correct answer for $P \cup I \cup \{\leftarrow p(\overline{t})\}$ is also a correct answer for $P \cup I \cup \{\leftarrow q(\overline{t}')\}$,
– if a substitution θ is a computed answer of $P \cup I \cup \{\leftarrow p(\overline{t})\}$ then we will store the tuple $\overline{t}'\theta$ in ans_q instead of storing the tuple $\overline{t}\theta$ in ans_p.

We say that a pair $(\overline{t}, q(\overline{t}'))$ is *more general* than $(\overline{t}_2, q(\overline{t}'_2))$, and $(\overline{t}_2, q(\overline{t}'_2))$ is an *instance* of $(\overline{t}, q(\overline{t}'))$, if there exists a substitution θ such that $(\overline{t}, q(\overline{t}'))\theta = (\overline{t}_2, q(\overline{t}'_2))$.

For $v = filter_{i,j}$ and p being the predicate of A_i, the meaning of a subquery $(q(\overline{t}), \delta) \in subqueries(v)$ is as follows: if $T(p) = false$ (resp. $T(p) = true$) then there exists $\overline{s} \in tuples(input\_p)$ (resp. $(\overline{s}, q(\overline{s}')) \in pairs(input\_p)$) such that for processing the goal $\leftarrow p(\overline{s})$ using the program clause $\varphi_i = (A_i \leftarrow B_{i,1}, \ldots, B_{i,n_i})$, unification of $p(\overline{s})$ and A_i as well as processing of the subgoals $B_{i,1}, \ldots, B_{i,j-1}$ were done, amongst others, by using a sequence of mgu's $\gamma_0, \ldots, \gamma_{j-1}$ with the property that $q(\overline{t}) = p(\overline{s})\gamma_0 \ldots \gamma_{j-1}$ (resp. $q(\overline{t}) = q(\overline{s}')\gamma_0 \ldots \gamma_{j-1}$) and $\delta = (\gamma_0 \ldots \gamma_{j-1})_{|Vars((B_{i,j},\ldots,B_{i,n_i}))}$. Informally, a subquery $(q(\overline{t}), \delta)$ transferred through an edge to v is processed as follows:

– if $v = filter_{i,j}$, $kind(v) = extensional$ and $pred(v) = p$ then, for each $\overline{t}'' \in I(p)$, if $atom(v)\delta = B_{i,j}\delta$ is unifiable with a fresh variant of $p(\overline{t}'')$ by an mgu γ then transfer the subquery $(q(\overline{t})\gamma, (\delta\gamma)_{|post\_vars(v)})$ through $(v, succ(v))$,
– if $v = filter_{i,j}$, $kind(v) = intensional$, $pred(v) = p$ and either $(T(p) = false)$ or $(T(p) = true$ and $j < n_i)$ then
  • if $T(p) = false$ then transfer the input tuple $\overline{t}'$ such that $p(\overline{t}') = atom(v)\delta = B_{i,j}\delta$ through $(v, input\_p)$ to add its fresh variant to $tuples(input\_p)$,
  • if $T(p) = true$ and $j < n_i$ then transfer the input pair $(\overline{t}', p(\overline{t}'))$ such that $p(\overline{t}') = atom(v)\delta = B_{i,j}\delta$ through $(v, input\_p)$ to add its fresh variant to $pairs(input\_p)$,
  • for each currently existing $\overline{t}' \in tuples(ans\_p)$, if $atom(v)\delta = B_{i,j}\delta$ is unifiable with a fresh variant of $p(\overline{t}')$ by an mgu γ then transfer the subquery $(q(\overline{t})\gamma, (\delta\gamma)_{|post\_vars(v)})$ through $(v, succ(v))$,
  • store the subquery $(q(\overline{t}), \delta)$ in $subqueries(v)$, and later, for each new $\overline{t}'$ added to $tuples(ans\_p)$, if $atom(v)\delta = B_{i,j}\delta$ is unifiable with

a fresh variant of $p(\bar{t}')$ by an mgu γ then transfer the subquery $(q(\bar{t})\gamma, (\delta\gamma)_{|post\_vars(v)})$ through $(v, succ(v))$.
- if $v = filter_{i,n_i}$, $kind(v) = intensional$, $pred(v) = p$, $T(p) = true$ then transfer the input pair $(\bar{t}', q(\bar{t}))$ such that $p(\bar{t}') = atom(v)\delta = B_{i,n_i}\delta$ through $(v, input\_p)$ to add its fresh variant to $pairs(input\_p)$,
- if $v = post\_filter_i$ then transfer the subquery $(q(\bar{t}), \varepsilon)$ through $(post\_filter_i, ans\_q)$ to add $\bar{t}$ to $tuples(ans\_q)$.

Formally, the processing of a subquery, an input/answer tuple or an input pair in a QSQN-rTRE is designed so that:

- every subquery or input/answer tuple or input pair that is subsumed by another one or has a term-depth greater than a fixed bound l is ignored,
- the processing is divided into smaller steps which can be delayed at each node to maximize adjustability and allow various control strategies,
- the processing is done set-at-a-time (e.g., for all the unprocessed subqueries accumulated in a given node).

The Algorithm 1 (on page 251) repeatedly selects an active edge and fires the operation for the edge. Due to the lack of the space, the related functions and procedures used for Algorithm 1 are omitted in this paper. The details of these functions and procedures can be found in [4]. In particular, the Algorithm 1 uses the function `active-edge`(u, v), which returns *true* if the data accumulated in u can be processed to produce some data to transfer through the edge (u, v). If `active-edge`(u, v) is *true*, procedure `fire`(u, v) processes the data accumulated in u that has not been processed before and transfers appropriate data through the edge (u, v). This procedure uses the procedures `add-tuple`, `add-pair`, `add-subquery`, `compute-gamma` and `transfer`. The procedure `transfer`(D, u, v) specifies the effects of transferring data D through the edge (u, v) of a QSQN-rTRE. We omit to present the properties on soundness, completeness and data complexity of the Algorithm 1 because they are analogous to the ones given in [11] for QSQN.

Note: Regarding the QSQN-rTRE topological structure of the logic program given in Example 1, for the query $s(x)$, after producing a set of (answer) tuples, the Algorithm 1 only adds these tuples to $tuples(ans\_s)$. Thus, no tuple is added to $tuples(ans\_p)$ as well as $tuples(ans\_q)$. In this case, we can exclude the nodes ans_p, ans_q and the related edges from the net, but we keep them for answering other queries of the form $p(\ldots)$ or $q(\ldots)$.

4 Preliminary Experiments

This section presents our experimental results related to the effectiveness of the QSQN-rTRE method in comparison with the QSQN-TRE method. The comparison is made w.r.t. the number of read/write operations on relations as well as the maximum number of kept tuples/subqueries in the computer memory.

Algorithm 1. for evaluating a query $(P, q(\overline{x}))$ on an extensional instance I.

```
1  let (V, E, T) be a QSQN-rTRE structure of P;
   // T can be chosen arbitrarily or appropriately
2  set C so that N = (V, E, T, C) is an empty QSQN-rTRE of P;
3  let x̄' be a fresh variant of x̄;
4  if T(q) = false then
5  |   tuples(input_q) := {x̄'};
6  |   foreach (input_q, v) ∈ E do unprocessed(input_q, v) := {x̄'}
7  else
8  |   pairs(input_q) := {(x̄', q(x̄'))};
9  |   foreach (input_q, v) ∈ E do unprocessed(input_q, v) := {(x̄', q(x̄'))}
10 while there exists (u, v) ∈ E such that active-edge(u, v) holds do
11 |   select (u, v) ∈ E such that active-edge(u, v) holds;
   |   // any strategy is acceptable for the above selection
12 |   fire(u, v)
13 return tuples(ans_q)
```

As mentioned earlier, the QSQN-TRE and QSQN-rTRE methods allow various control strategy. We use the IDFS control strategy [5] for both methods. All the experiments have been implemented in Java codes [4] and extensional relations stored in a MySQL database. We assume that the computer memory is large enough to load all the involved relations. During the processing, for each operation of reading from a relation (resp. writing a set of tuples to a relation), we increase the counter of read (resp. write) operations on this relation by one. For counting the maximum number of kept tuples/subqueries in the memory, we increase (resp. decrease) the counter of kept tuples by two if a pair is added to (resp. removed from) *pairs*(*input_p*), otherwise we increase (resp. decrease) it by one. The returned value is the maximum value of this counter.

Test 1. This test uses the program P as in Example 1, the query is $s(x)$ and the extensional instance I for t is as follows, where a, a_i, b_i are constant symbols:

$$\begin{aligned} I(t) = \{(a, a_1)\} \cup \{(a_i, a_{i+1}) \mid 1 \leq i < 20\} \cup \{(a_{20}, a_1)\} \cup \\ \{(a, b_1)\} \cup \{(b_i, b_{i+1}) \mid 1 \leq i < 20\} \cup \{(b_{20}, b_1)\}. \end{aligned}$$

Test 2. This test uses the following logic program P, where p, q are intensional predicates, t_1, t_2 are extensional predicates, and x, y, z are variables. In this test, we consider the case p and q are mutually dependent on each other.

$$\begin{aligned} p(x, y) &\leftarrow t_1(x, y) \\ p(x, y) &\leftarrow t_1(x, z), q(z, y) \\ q(x, y) &\leftarrow t_2(x, y) \\ q(x, y) &\leftarrow t_2(x, z), p(z, y). \end{aligned}$$

Table 1. A comparison between the QSQN-TRE and QSQN-rTRE methods w.r.t. the number of read/write operations. The third column means the number of read operations on *input*/*answer*/*supplement*/*extensional* relations, respectively. Similarly, the fourth column means the number of write operations on *input*/*answer*/*supplement* relations, respectively. The last column shows the maximum number of kept tuples in the computer memory.

Tests	Methods	Reading (times)	Writing (times)	Max No. of
		inp_/ans_/sup_/edb	*inp_/ans_/sup_*	kept tuples
Test 1	QSQN-TRE	122 (46+7+47+22)	48 (23+3+22)	288
	QSQN-rTRE	114 (46+1+45+22)	46 (23+1+22)	209
Test 2	QSQN-TRE	1094 (201+396+395+102)	396 (100+197+99)	5248
	QSQN-rTRE	503 (201+2+198+102)	201 (100+2+99)	497
Test 3	QSQN-TRE	37 (14+4+13+6)	15 (7+2+6)	376
	QSQN-rTRE	33 (14+1+12+6)	14 (7+1+6)	362
Test 4	QSQN-TRE	126 (48+8+46+24)	49 (23+4+22)	566
	QSQN-rTRE	118 (48+2+44+24)	47 (23+2+22)	474

The query is $q(a_1, x)$ and the extensional instance I for t_1 and t_2 is as follows, where a_i $(1 \leq i \leq 100)$ are constant symbols:

$$I(t_1) = \{(a_2, a_3), (a_4, a_5), (a_6, a_7), \ldots, (a_{96}, a_{97}), (a_{98}, a_{99})\}$$
$$I(t_2) = \{(a_1, a_2), (a_3, a_4), (a_5, a_6), \ldots, (a_{97}, a_{98}), (a_{99}, a_{100})\}.$$

Test 3. This test concerns the case with function symbols. Consider the following positive logic program, where *path*, *query* are intensional predicates, *edge* is an extensional predicate, x, y, z, w are variables, *nil* is a constant symbol standing for the empty list, and *cons* is a function symbol standing for the list constructor:

$$path(x, y, cons(x, cons(y, nil))) \leftarrow edge(x, y)$$
$$path(x, y, cons(x, z)) \leftarrow edge(x, w), path(w, y, z)$$
$$query(x, y) \leftarrow path(x, d, y).$$

An atom $path(x, y, z)$ stands for "z is a list representing a path from x to y". This test uses the following extensional instance I, where a - m are constant symbols: $I(edge) = \{(a, b), (a, h), (b, c), (b, f), (c, d), (d, e), (e, c), (f, g), (g, f), (h, i), (i, i), (j, c), (j, k), (k, d), (k, l), (l, m), (m, j)\}$. The query is $query(x, y)$. We use the term-depth bound $l = 5$ for this test (which can be changed for [4]).

Test 4. This test is taken from [3,11]. Consider the following program P:

$$q_1(x, y) \leftarrow r_1(x, y)$$
$$q_1(x, y) \leftarrow r_1(x, z), q_1(z, y)$$
$$q_2(x, y) \leftarrow r_2(x, y)$$
$$q_2(x, y) \leftarrow r_2(x, z), q_2(z, y)$$
$$p(x, y) \leftarrow q_1(x, y)$$
$$p(x, y) \leftarrow q_2(x, y).$$

where p, q_1 and q_2 are intensional predicates, t_1 and t_2 are extensional predicates, and x, y, z are variables. Let $p(a_0, x)$ be the query, and let the extensional instance I for t_1 and t_2 be as follows, using $m = n = 10$:

$$I(t_1) = \{(a_i, a_{i+1}) \mid 0 \leq i < m\},$$
$$I(t_2) = \{(a_0, b_{1,j}) \mid 1 \leq j \leq n\} \cup$$
$$\{(b_{i,j}, b_{i+1,j}) \mid 1 \leq i < m-1;\ 1 \leq j \leq n\} \cup \{(b_{m-1,j}, a_m) \mid 1 \leq j \leq n\}.$$

Table 1 shows the comparison between the QSQN-TRE and QSQN-rTRE evaluation methods w.r.t. the number of accesses to the relations as well as the maximum number of kept tuples/subqueries in the computer memory. As can be seen in this table, by not representing intermediate results during the computation for the right/tail-recursive cases, the QSQN-rTRE method usually outperforms the QSQN-TRE method for the mentioned cases, especially for the case when the intensional predicates are mutually dependent on each other as in Test 2.

5 Conclusions

We have proposed the QSQN-rTRE method for evaluating queries to Horn knowledge bases. It extends the QSQN-TRE method by eliminating all the intensional predicates that appear rightmost in the bodies of the program clauses with respect to T_{idb}. This allows to avoid materializing intermediate results during the processing.

The experimental results reported in Table 1 show that, due to right/tail-recursion elimination, the QSQN-rTRE method is often better than the QSQN-TRE method for a certain class of queries that depend on right/tail-recursive predicates, especially for the program as in Test 2. The preliminary comparison between QSQN, QSQR and Magic-Set reported in [3], between QSQN and QSQN-TRE reported in [6] indicates the usefulness of QSQN, QSQN-TRE and hence also the usefulness of QSQN-rTRE. As a future work, we will compare the methods in detail with other experimental measures. Particularly, we will estimate these methods with respect to the number of accesses to the secondary storage when the computer memory is limited, as well as apply our method to Datalog-like rule languages of the Semantic Web [7,8].

Acknowledgments. This work was supported by Polish National Science Centre (NCN) under Grants No. 2011/02/A/HS1/00395 as well as by Warsaw Center of Mathematics and Computer Science. I would like to express my special thanks to Dr.Hab. Linh Anh Nguyen from the University of Warsaw for very helpful comments and suggestions.

References

1. Abiteboul, S., Hull, R., Vianu, V.: Foundations of Databases. Addison Wesley (1995)
2. Bancilhon, F., Maier, D., Sagiv, Y., Ullman, J.D.: Magic sets and other strange ways to implement logic programs. In: Proceedings of PODS 1986, pp. 1–15. ACM (1986)
3. Cao, S.T.: On the efficiency of Query-Subquery Nets: an experimental point of view. In: Proceedings of SoICT 2013, pp. 148–157. ACM (2013)
4. Cao, S.T.: An Implementation of the QSQN-rTRE Evaluation Methods (2014), `http://mimuw.edu.pl/~sonct/QSQNrTRE15.zip`
5. Cao, S.T., Nguyen, L.A.: An Improved Depth-First Control Strategy for Query-Subquery Nets in Evaluating Queries to Horn Knowledge Bases. In: van Do, T., Le Thi, H.A., Nguyen, N.T. (eds.) Advanced Computational Methods for Knowledge Engineering. AISC, vol. 282, pp. 281–295. Springer, Heidelberg (2014)
6. Cao, S.T., Nguyen, L.A.: An Empirical Approach to Query-Subquery Nets with Tail-Recursion Elimination. In: Bassiliades, N., Ivanovic, M., Kon-Popovska, M., Manolopoulos, Y., Palpanas, T., Trajcevski, G., Vakali, A. (eds.) New Trends in Database and Information Systems II. AISC, vol. 312, pp. 109–120. Springer, Heidelberg (2015)
7. Cao, S.T., Nguyen, L.A., Szalas, A.: The Web Ontology Rule Language OWL 2 RL+ and Its Extensions. T. Computational Collective Intelligence 13, 152–175 (2014)
8. Cao, S.T., Nguyen, L.A., Szalas, A.: WORL: a nonmonotonic rule language for the Semantic Web. Vietnam J. Computer Science 1(1), 57–69 (2014)
9. Freire, J., Swift, T., Warren, D.S.: Taking I/O seriously: Resolution reconsidered for disk. In: Naish, L. (ed.) Proc. of ICLP 1997, pp. 198–212. MIT Press (1997)
10. Madalińska-Bugaj, E., Nguyen, L.A.: A generalized QSQR evaluation method for Horn knowledge bases. ACM Trans. on Computational Logic 13(4), 32 (2012)
11. Nguyen, L.A., Cao, S.T.: Query-Subquery Nets. In: Nguyen, N.-T., Hoang, K., Jędrzejowicz, P. (eds.) ICCCI 2012, Part I. LNCS, vol. 7653, pp. 239–248. Springer, Heidelberg (2012)
12. Ramakrishnan, R., Beeri, C., Krishnamurthy, R.: Optimizing existential datalog queries. In: Proceedings of PODS 1988, pp. 89–102. ACM (1988)
13. Ramakrishnan, R., Srivastava, D., Sudarshan, S.: Efficient bottom-up evaluation of logic programs. In: Vandewalle, J. (ed.) The State of the Art in Computer Systems and Software Engineering. Kluwer Academic Publishers (1992)
14. Ross, K.A.: Tail recursion elimination in deductive databases. ACM Trans. Database Syst. 21(2), 208–237 (1996)
15. Srivastava, D., Sudarshan, S., Ramakrishnan, R., Naughton, J.F.: Space optimization in deductive databases. ACM Trans. DB. Syst. 20(4), 472–516 (1995)
16. Sudarshan, S., Srivastava, D., Ramakrishnan, R., Naughton, J.F.: Space optimization in the bottom-up evaluation of logic programs. In: Proceedings of SIGMOD 1991, pp. 68–77. ACM Press (1991)
17. Tamaki, H., Sato, T.: OLD resolution with tabulation. In: Shapiro, E. (ed.) ICLP 1986. LNCS, vol. 225, pp. 84–98. Springer, Heidelberg (1986)
18. Vieille, L.: Recursive axioms in deductive databases: The query/subquery approach. In: Proceedings of Expert Database Conf., pp. 253–267 (1986)
19. Vieille, L.: Recursive query processing: The power of logic. Theor. Comput. Sci. 69(1), 1–53 (1989)
20. Zhou, N.-F., Sato, T.: Efficient fixpoint computation in linear tabling. In: Proceedings of PPDP 2003, pp. 275–283. ACM (2003)

Parallel Multiclass Logistic Regression for Classifying Large Scale Image Datasets

Thanh-Nghi Do[1] and François Poulet[2]

[1] College of Information Technology
Can Tho University, 92100-Cantho, Vietnam
dtnghi@cit.ctu.edu.vn
[2] University of Rennes I - IRISA
Campus de Beaulieu, 35042 Rennes Cedex, France
francois.poulet@irisa.fr

Abstract. The new parallel multiclass logistic regression algorithm (PAR-MC-LR) aims at classifying a very large number of images with very-high-dimensional signatures into many classes. We extend the two-class logistic regression algorithm (LR) in several ways to develop the new multiclass LR for efficiently classifying large image datasets into hundreds of classes. We propose the balanced batch stochastic gradient descend of logistic regression (BBatch-LR-SGD) for trainning two-class classifiers used in the one-versus-all strategy of the multiclass problems and the parallel training process of classifiers with several multi-core computers. The numerical test results on ImageNet datasets show that our algorithm is efficient compared to the state-of-the-art linear classifiers.

Keywords: Logistic regression (LR), Stochastic gradient descent (SGD), Multi-class, Parallel algorithm, Large scale image classification.

1 Introduction

The image classification is to automatically assign predefined categories to images. Its applications include handwriting character recognition, zip code recognition for postal mail sorting, numeric entries in forms filled up by hand, fingerprint recognition, face recognition, auto-tagging images and so on. The image classification task involves two main steps as follows: extracting features and building codebook, training classifiers. Recently, the local image features and bag-of-words (BoW) models are used at the first step of state-of-the-art image classification. The most popular local image features are scalable invariant feature transform descriptors - SIFT [1], speeded up robust Features - SURF [2] and dense SIFT - DSIFT [3]. These feature extraction methods are locally based on the appearance of the object at particular interest points, invariant to image scale, rotation and also robust to changes in illumination, noise, occlusion. And then, k-means algorithm [4] performs the clustering task on descriptors to form visual words from the local descriptors. The representation of the image for classification is the bag-of-words is constructed from the counting of the occurrence of words in a histogram like fashion. The step of the feature extraction and the BoW representation leads to datasets with very large number of dimensions (e.g. thousands of dimensions). The support vector machine algorithms [5] are suited for dealing with very-high-dimensional datasets.

H.A. Le Thi et al. (eds.), *Advanced Computational Methods for Knowledge Engineering*,
Advances in Intelligent Systems and Computing 358, DOI: 10.1007/978-3-319-17996-4_23

However there are also many classes in the images classification (e.g. Caltech with 101 classes [6], Caltech with 256 classes [7] having hundreds of classes) and ImageNet dataset [8] with more than 14 million images in 21,841 classes. It makes the complexity of image classification become very hard. This challenge motivates us to study an efficient algorithm in both computation time and classification accuracy. In this paper, we propose the extensions of the stochastic gradient descend (SGD [9], [10]) to develop the new parallel multiclass SGD of logistic regression (PAR-MC-LR) for efficiently classifying large image datasets into many classes. Our contributions include:

1. the balanced batch stochastic gradient descend of logistic regression (BBatch-LR-SGD) for very large number of classes,
2. the parallel training process of classifiers with several multi-core computers.

The numerical test results on ImageNet datasets [8] show that our PAR-MC-LR algorithm is efficient compared to the state-of-the-art linear classifiers.

The remainder of this paper is organized as follows. Section 2 briefly presents the stochastic gradient descend algorithm of logistic regression (LR-SGD) for two-class problems. Section 3 describes how to extend the LR-SGD algorithm for dealing very large number of classes. Section 4 presents evaluation results, before the conclusions and future work.

2 Logistic Regression for Two-Class Problems

Let us consider a linear binary classification task with m datapoints x_i $(i = 1, \ldots, m)$ in the n-dimensional input space R^n, having corresponding labels $y_i = \pm 1$. Logistic regression (LR) tries to learn classification models (i.e. parameter vector $w \in R^n$) to maximize the likelihood of the data. The probability of a datapoint being drawn from the positive class is:

$$p(y_i = +1/x_i) = \frac{1}{1 + e^{-(w.x_i)}} \tag{1}$$

And then, the probability of a datapoint being drawn from the negative class is:

$$p(y_i = -1/x_i) = 1 - p(y_i = +1/x_i) = 1 - \frac{1}{1 + e^{-(w.x_i)}} = \frac{1}{1 + e^{(w.x_i)}} \tag{2}$$

The probabilities in (1) and (2) are rewritten in:

$$p(y_i/x_i) = \frac{1}{1 + e^{-y_i(w.x_i)}} \tag{3}$$

And then the log likelihood of data is as follow:

$$log(p(y_i/x_i)) = log(\frac{1}{1 + e^{-y_i(w.x_i)}}) = -log(1 + e^{-y_i(w.x_i)}) \tag{4}$$

The regularized LR method aims to use Tikhonov regularization in a trade-off with maximizing likelihood and the over-confident generalization. The regularization strategy is to impose a penalty on the magnitude of the parameter values w. The LR algorithm simultaneously tries to maximize the log likelihood of the data and minimize the L_2 norm of the parameter vector w. And then, it yields an unconstrained problem (5):

$$\min \Psi(w,[x,y]) = \frac{\lambda}{2}\|w\|^2 + \frac{1}{m}\sum_{i=1}^{m} log(1+e^{-y_i(w.x_i)}) \quad (5)$$

The LR formula (5) uses the logistic loss function $L(w,[x_i,y_i]) = log(1+e^{-y_i(w.x_i)})$. The solution of the unconstrained problem (5) can be also obtained by the stochastic gradient descent method [9], [10]. The stochastic gradient descent for the logistic regression is denoted by (LR-SGD). The LR-SGD updates w on T iterations (epochs) with a learning rate η. For each iteration t, the LR-SGD uses a single randomly datapoint (x_t,y_t) to compute the sub-gradient $\nabla_t\Psi(w,[x_t,y_t])$ and update w_{t+1} as follows:

$$w_{t+1} = w_t - \eta_t\nabla_t\Psi(w,[x_t,y_t]) = w_t - \eta_t(\lambda w_t + \nabla_t L(w,[x_t,y_t])) \quad (6)$$

$$\nabla_t L(w,[x_t,y_t]) = \nabla_t log(1+e^{-y_t(w.x_t)}) = -\frac{y_t x_t}{1+e^{y_t(w.x_t)}} \quad (7)$$

The LR-SGD using the update rule (6) is described in algorithm 1.

Algorithm 1. LR-SGD algorithm

---

input :
    training dataset D
    positive constant $\lambda > 0$
    number of epochs T
output:
    hyperplane w

1 **begin**
2   init $w_1 = 0$
3   **for** $t \leftarrow 1$ **to** T **do**
4   randomly pick a datapoint $[x_t,y_t]$ from training dataset D
5   set $\eta_t = \frac{1}{\lambda t}$
6   update $w_{t+1} = w_t - \eta_t(\lambda w_t - \frac{y_t x_t}{1+e^{y_t(w.x_t)}})$
7   **end**
8   return w_{t+1}
9 **end**

---

3 Extentions of Logistic Regression to Large Number of Classes

There are several extensions of a binary classification LR to multi-class (k classes, $k \geq 3$) classification tasks. The state-of-the-art multi-class are categorized into two types of approaches. The first one is considering the multi-class case in one optimization problem [11], [12], [13]. The second one is decomposing multi-class into a series of binary classifiers, including one-versus-all [5], one-versus-one [14] and Decision Directed Acyclic Graph [15].

In practice, one-versus-all, one-versus-one are the most popular methods due to their simplicity. Let us consider k classes ($k > 2$). The one-versus-all strategy builds k different classifiers where the ith classifier separates the ith class from the rest. The one-versus-one strategy constructs $k(k-1)/2$ classifiers, using all the binary pairwise combinations of the k classes. The class is then predicted with a majority vote.

When dealing with very large number of classes, e.g. hundreds of classes, the one-versus-one strategy is too expensive because it needs to train many thousands of binary classifiers. Therefore, the one-versus-all strategy becomes popular in this case. And then, our multiclass LR-SGD algorithms also use the one-versus-all approach to train independently k binary classifiers. Therefore, the multiclass LR-SGD algorithms using one-versus-all lead to the two problems:

1. the LR-SGD algorithms deal with the imbalanced datasets for building binary classifiers,
2. the LR-SGD algorithms also take very long time to train very large number of binary classifiers in sequential mode using a single processor.

Due to these problems, we propose two ways for creating the new multiclass LR-SGD algorithms being able to handle very large number of classes in the high speed. The first one is to build balanced classifiers with sampling strategy. The second one is to parallelize the training task of all classifiers with several multi-core machines.

3.1 Balanced Batch of Logistic Regression

In the one-versus-all approach, the learning task of LR-SGD is try to separate the ith class (positive class) from the $k-1$ others classes (negative class). For very large number of classes, e.g. 100 classes, this leads to the extreme unbalance between the positive and the negative class. The problem is well-known as the class imbalance. The problem of LR-SGD comes from **line 4** of algorithm 1. For dealing with hundreds classses, the probability for a positive datapoint sampled is very small (about 0.01) compared with the large chance for a negative datapoint sampled (e.g. 0.99). And then, the LR-SGD concentrates mostly on the errors produced by the nagative datapoints. Therefore, the LR-SGD has difficulty to separate the positive class from the negative class.

As summarized by the review papers of [16], [17], [18] and the very comprehensive papers of [19], [20], solutions to the class imbalance problems were proposed both at the data and algorithmic level. At the data level, these algorithms change the class distribution, including over-sampling the minority class [21] or under-sampling the majority class [22], [23]. At the algorithmic level, the solution is to re-balance the error rate by weighting each type of error with the corresponding cost.

Our balanced batch LR-SGD (denoted by LR-BBatch-SGD) belongs to the first approach. For separating the ith class (positive class) from the rest (negative class), the class prior probabilities in this context are highly unequal (e.g. the distribution of the positive class is 1% in the 100 classes classification problem), and then over-sampling the minority class is very expensive. We propose the LR-BBatch-SGD algorithm using under-sampling the majority class (negative class). Our modification of algorithm 1 is to use a balanced batch (instead of a datapoint at line 4 of algorithm 1) to update the w at epoch t. We also propose to modify the updating rule (**line 6** of algorithm 1) using the skewed costs as follows:

$$w_{t+1} = \begin{cases} w_t - \eta_t \lambda w_t + \eta_t \frac{1}{|D_+|} \frac{y_t x_t}{1+e^{y_t (w.x_t)}} & if\ y_t = +1 \\ w_t - \eta_t \lambda w_t + \eta_t \frac{1}{|D_-|} \frac{y_t x_t}{1+e^{y_t (w.x_t)}} & if\ y_t = -1 \end{cases} \quad (8)$$

where $|D_+|$ is the cardinality of the positive class D_+ and $|D_-|$ is the cardinality of the negative class D_-.

The LR-BBatch-SGD (as shown in algorithm 2) is to separate the ith class (positive class) from the rest (negative class), using under-sampling the negative class (balanced batch) and the skewed costs in (8).

Algorithm 2. LR-BBatch-SGD algorithm

input :
- training data of the positive class D_+
- training data of the negative class D_-
- positive constant $\lambda > 0$
- number of epochs T

output:
- hyperplane w

```
1  begin
2      init w_1 = 0
3      for t ← 1 to T do
4          creating a balanced batch Bbatch by sampling without replacement D'_- from
           dataset D_- (with |D'_-| = sqrt(|D_-|/|D_+|)) and a datapoint from dataset D_+
5          set η_t = 1/(λt)
6          for [x_i, y_i] in Bbatch do
7              update w_{t+1} using rule 8
8          end
9      end
10     return w_{t+1}
11 end
```

3.2 Parallel LR-BBatch-SGD Training

Although LR-BBatch-SGD deals with very large dataset with high speed, but it does not take the benefits of HPC. Furthermore, LR-BBatch-SGD trains independently k binary classifiers for k classes problems. This is a nice property for parallel learning. Our investigation aims at speedup training tasks of multi-class LR-BBatch-SGD with several multi-processor computers. The idea is to learn k binary classifiers in parallel.

The parallel programming is currently based on two major models, Message Passing Interface (MPI) [24] and Open Multiprocessing (OpenMP) [25]. MPI is a standardized and portable message-passing mechanism for distributed memory systems. MPI remains the dominant model (high performance, scalability, and portability) used in high-performance computing today. However, one MPI process loads the whole dataset into memory during learning tasks, thus the parallel algorithm with k MPI processes requires k main memory rooms for storing k datasets, making it wasteful. The simplest development of parallel LR-BBatch-SGD algorithm is based on the shared memory multiprocessing programming model OpenMP that does not require large amount of memory (although the parallel in MPI is more efficient than OpenMP). The parallel learning for LR-BBatch-SGD is described in algorithm 3.

Algorithm 3. Parallel LR-BBatch-SGD training

input : D the training dataset with k classes
output: LR-SGD model

1 **Learning:**
2 ***#pragma omp parallel for***
3 **for** $c_i \leftarrow 1$ **to** k **do** /* class c_i */
4 | *training LR-BBatch-SGD*($c_i - vs - all$)
5 **end**

4 Evaluation

In order to evaluate the performance of the new parallel multiclass logistic regression algorithm (PAR-MC-LR) for classifying large amounts of images into many classes, we have implemented PAR-MC-LR in C/C++ using the SGD library [10]. Our comparison is reported in terms of correctness and training time. We are interested in two recent algorithms, LIBLINEAR (a library for large linear classification [26]) and OCAS (an optimized cutting plane algorithm for SVM [27]) because they are well-known as highly efficient standard linear SVM.

LIBLINEAR and OCAS use the default parameter value $C = 1000$, our Par-MC-LR is trained the balanced batch stochastic gradient descend of logistic regression using $T = 50$ epochs and regularization term $\lambda = 0.0001$.

All experiments are run on machine Linux Fedora 20, Intel(R) Core i7-4790 CPU, 3.6 GHz, 4 cores and 32 GB main memory.

4.1 Datasets

The PAR-MC-LR algorithm is designed for the large number of images with many classes, so we have evaluated its performance on the three following datasets.

ImageNet 10
This dataset contains the 10 largest classes from ImageNet [8], including 24,807 images with size 2.4 GB). In each class, we sample 90% images for training and 10% images for testing (with random guess 10%). First, we construct BoW of every image using dense SIFT descriptor (extracting SIFT on a dense grid of locations at a fixed scale and orientation) and 5,000 codewords. Then, we use feature mapping from [28] to get the high-dimensional image representation in 15,000 dimensions. This feature mapping has been proven to give a good image classification performance with linear classifiers [28]. We end up with 2.6 GB of training data.

ImageNet 100
This dataset contains the 100 largest classes from ImageNet [8], including 183,116 images with size 23.6 GB. In each class, we sample 50% images for training and 50% images for testing (with random guess 1%). We also construct BoW of every image using dense SIFT descriptor and 5,000 codewords. For feature mapping, we use the same method as we do with ImageNet 10. The final size of training data is 8 GB.

ILSVRC 2010
This dataset contains 1,000 classes from ImageNet [8], including 1.2M images ($\sim$ 126 GB) for training, 50 K images ($\sim$ 5.3 GB) for validation and 150 K images ($\sim$ 16 GB) for testing. We use BoW feature set provided by [8] and the method reported in [29] to encode every image as a vector in 21,000 dimensions. We take roughly 900 images per class for training dataset, so the total training images is 887,816 and the training data size is about 12.5 GB. All testing samples are used to test SVM models. Note that the random guess performance of this dataset is 0.1%.

4.2 Classificaton Results

Firstly, we are interested in the performance comparison in terms of training time. We try to vary the number of OpenMP threads (1, 4, 8 threads) for all training tasks of our parallel algorithm PAR-MC-LR. Due to the PC (Intel(R) Core i7-4790 CPU, 4 cores) used in the experimental setup, the PAR-MC-LR is fastest by setting 8 OpenMP threads.

Table 1. Training time (minutes) on ImageNet 10

	# OpenMP threads		
Algorithm	1	4	8
OCAS	106.67		
LIBLINEAR	2.02		
PAR-MC-LR	2.24	0.80	0.76

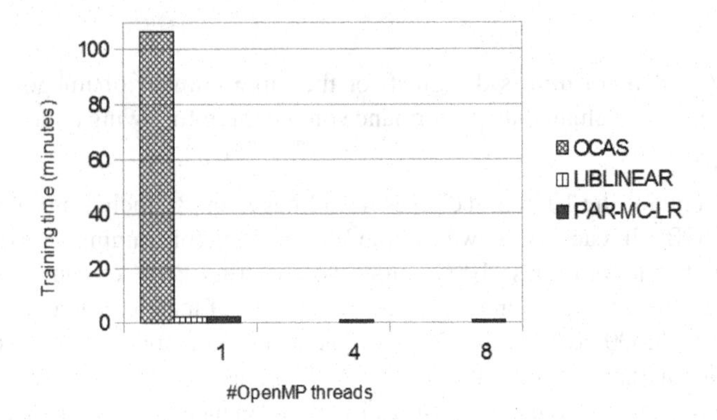

Fig. 1. Training time (minutes) on ImageNet-10

Table 2. Training time (minutes) on ImageNet 100

Algorithm	# OpenMP threads 1	4	8
OCAS	1016.35		
LIBLINEAR	30.41		
Par-MC-LR	23.40	8.09	6.55

For the small multi-class dataset as ImageNet 10, the training time of algorithms in table 1 and figure 1 show that our PAR-MC-LR with 8 OpenMP threads is 139.7 times faster than OCAS and 2.64 times faster than LIBLINEAR.

Table 2 and figure 2 present the training time on ImageNet 100 with large number of classes. Once again, the PAR-MC-LR achieves a significant speed-up in learning process using 8 OpenMP threads. It is 155.1 times faster than OCAS and 4.64 times faster than LIBLINEAR.

Table 3. Training time (minutes) on ILSVRC 2010.

Algorithm	# OpenMP threads 1	4	8
OCAS	N/A		
LIBLINEAR	3106.48		
Par-MC-LR	153.58	40.59	37.91

ILSVRC 2010 has large amount of images (more than 1 million images) and very large number of classes (1000 classes). Therefore, OCAS has not finished the learning task in several days. LIBLINEAR takes 3,106.48 minutes (about 2 days and 4 hours)

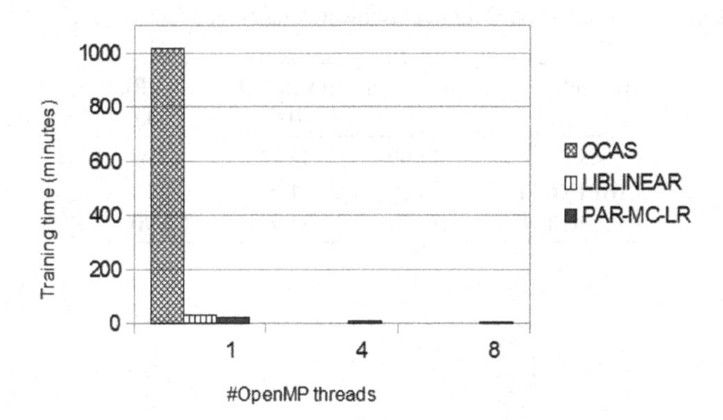

Fig. 2. Training time (minutes) on ImageNet-100

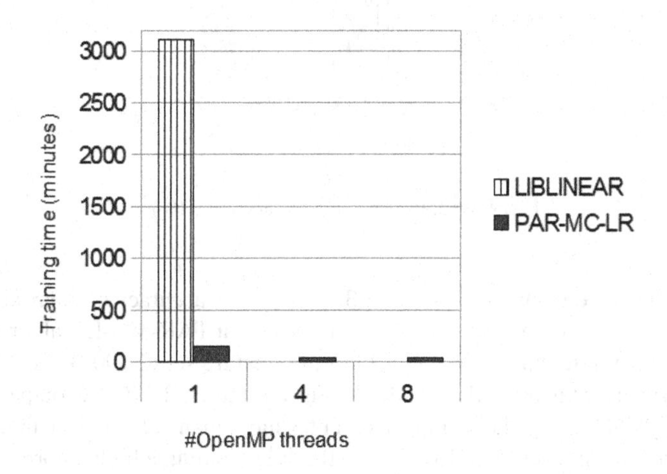

Fig. 3. Training time (minutes) on ImageNet-1000

to train the classification model for this dataset. Our PAR-MC-LR algorithm with 8 OpenMP threads performs the learning task in 37.59 minutes. This indicates that the PAR-MC-LR is 81.95 times faster than LIBLINEAR.

The classification results in terms of accuracy are presented in table 4 and figure 4. On the small datasets ImageNet 10 and medium dataset ImageNet 100, The Par-MC-LR outperforms OCAS in the classification accuracy. The PAR-MC-LR achieves very competitive performances compared to LIBLINEAR. It is more accurate than LIBLINEAR on ImageNet 10 while making more classification mistakes than LIBLINEAR on ImageNet 100.

ILSVRC 2010 is a large dataset (with more than 1 million images and 1,000 classes). Thus, it is very difficult for many state-of-the-art algorithms to obtain a high rate in classification performance. In particular, with the feature set provided by ILSVRC 2010

Table 4. Overall classification accuracy (%)

Dataset	ImageNet 10	ImageNet 100	ILSVRC 1000
OCAS	72.07	52.75	N/A
LIBLINEAR	75.09	54.07	21.11
Par-MC-LR	75.21	52.91	21.90

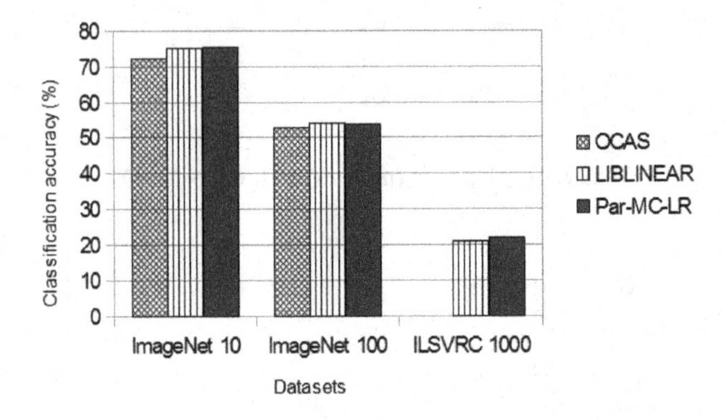

Fig. 4. Overall classification accuracy (%)

competition the state-of-the-art system [8,30] reports an accuracy of approximately 19 % (it is far above random guess, 0.1 %). And now our PAR-MC-LR algorithm gives a higher accuracy rate than [8,30] with the same feature set (21.90 % vs. 19 %). The relative improvement is more than 15 %. Moreover, the PAR-MC-LR outperforms LIBLINEAR (+0.79 %, the relative improvement is more than 3.7 %). Note that the PAR-MC-LR learns much faster than LIBLINEAR while yielding a higher correctness rate. These results show that our PAR-MC-LR has a great ability to scale-up to full ImageNet dataset.

5 Conclusion and Future Works

We have presented the new parallel multiclass logistic regression algorithm that achieves high performances for classifying large amounts of images intomany classes. The balanced batch stochastic gradient descend of logistic regression is proposed for trainning two-class classifiers used in the multiclass problems. The parallel multiclass algorithm is also developped for efficiently classifying large image datasets into very large number of classes on multi-core computers. Our algorithm is evaluated on the 10, 100 largest classes of ImageNet and ILSVRC 2010 dataset. By setting the number of OpenMP threads to 8 on PC (Intel Core i7-4790 CPU, 3.6 GHz), our algorithm achieves a significant speedup in training time without (or very few) compromise the classification accuracy. It is a roadmap towards very large scale visual classification.

In the future, we will develop a hybrid MPI/OpenMP parallel logistic regression for efficiently dealing with large scale multiclass problems and also intend to provide more empirical test on full dataset with 21,000 classes of ImageNet.

References

1. Lowe, D.G.: Distinctive image features from scale-invariant keypoints. International Journal of Computer Vision 60(2), 91–110 (2004)
2. Bay, H., Ess, A., Tuytelaars, T., Gool, L.J.V.: Speeded-up robust features (SURF). Computer Vision and Image Understanding 110(3), 346–359 (2008)
3. Bosch, A., Zisserman, A., Muñoz, X.: Image classification using random forests and ferns. In: International Conference on Computer Vision, pp. 1–8 (2007)
4. MacQueen, J.: Some methods for classification and analysis of multivariate observations. In: Proceedings of 5th Berkeley Symposium on Mathematical Statistics and Probability, vol. 1, pp. 281–297. University of California Press, Berkeley (1967)
5. Vapnik, V.: The Nature of Statistical Learning Theory. Springer, New York (1995)
6. Li, F.F., Fergus, R., Perona, P.: Learning generative visual models from few training examples: An incremental bayesian approach tested on 101 object categories. Computer Vision and Image Understanding 106(1), 59–70 (2007)
7. Griffin, G., Holub, A., Perona, P.: Caltech-256 Object Category Dataset. Technical Report CNS-TR-2007-001, California Institute of Technology (2007)
8. Deng, J., Berg, A.C., Li, K., Fei-Fei, L.: What does classifying more than 10,000 image categories tell us? In: Daniilidis, K., Maragos, P., Paragios, N. (eds.) ECCV 2010, Part V. LNCS, vol. 6315, pp. 71–84. Springer, Heidelberg (2010)
9. Shalev-Shwartz, S., Singer, Y., Srebro, N.: Pegasos: Primal estimated sub-gradient solver for svm. In: Proceedings of the Twenty-Fourth International Conference Machine Learning, pp. 807–814. ACM (2007)
10. Bottou, L., Bousquet, O.: The tradeoffs of large scale learning. In: Platt, J., Koller, D., Singer, Y., Roweis, S. (eds.) Advances in Neural Information Processing Systems, vol. 20, pp. 161–168. NIPS Foundation (2008), http://books.nips.cc
11. Ben-Akiva, M., Lerman, S.: Discrete Choice Analysis: Theory and Application to Travel Demand. The MIT Press (1985)
12. Weston, J., Watkins, C.: Support vector machines for multi-class pattern recognition. In: Proceedings of the Seventh European Symposium on Artificial Neural Networks, pp. 219–224 (1999)
13. Guermeur, Y.: Svm multiclasses, théorie et applications (2007)
14. Kreßel, U.: Pairwise classification and support vector machines. In: Advances in Kernel Methods: Support Vector Learning, pp. 255–268 (1999)
15. Platt, J., Cristianini, N., Shawe-Taylor, J.: Large margin dags for multiclass classification. Advances in Neural Information Processing Systems 12, 547–553 (2000)
16. Japkowicz, N. (ed.): AAAI'Workshop on Learning from Imbalanced Data Sets. Number WS-00-05 in AAAI Tech. Report (2000)
17. Weiss, G.M., Provost, F.: Learning when training data are costly: The effect of class distribution on tree induction. Journal of Artificial Intelligence Research 19, 315–354 (2003)
18. Visa, S., Ralescu, A.: Issues in mining imbalanced data sets - A review paper. In: Midwest Artificial Intelligence and Cognitive Science Conf., Dayton, USA, pp. 67–73 (2005)
19. Lenca, P., Lallich, S., Do, T.-N., Pham, N.-K.: A Comparison of Different Off-Centered Entropies to Deal with Class Imbalance for Decision Trees. In: Washio, T., Suzuki, E., Ting, K.M., Inokuchi, A. (eds.) PAKDD 2008. LNCS (LNAI), vol. 5012, pp. 634–643. Springer, Heidelberg (2008)

20. Pham, N.K., Do, T.N., Lenca, P., Lallich, S.: Using local node information in decision trees: coupling a local decision rule with an off-centered. In: International Conference on Data Mining, pp. 117–123. CSREA Press, Las Vegas (2008)
21. Chawla, N.V., Lazarevic, A., Hall, L.O., Bowyer, K.W.: SMOTEBoost: Improving prediction of the minority class in boosting. In: Lavrač, N., Gamberger, D., Todorovski, L., Blockeel, H. (eds.) PKDD 2003. LNCS (LNAI), vol. 2838, pp. 107–119. Springer, Heidelberg (2003)
22. Liu, X.Y., Wu, J., Zhou, Z.H.: Exploratory undersampling for class-imbalance learning. IEEE Transactions on Systems, Man, and Cybernetics, Part B 39(2), 539–550 (2009)
23. Ricamato, M.T., Marrocco, C., Tortorella, F.: Mcs-based balancing techniques for skewed classes: An empirical comparison. In: ICPR, pp. 1–4 (2008)
24. MPI-Forum: MPI: A message-passing interface standard
25. OpenMP Architecture Review Board: OpenMP application program interface version 3.0 (2008)
26. Fan, R., Chang, K., Hsieh, C., Wang, X., Lin, C.: LIBLINEAR: a library for large linear classification. Journal of Machine Learning Research 9(4), 1871–1874 (2008)
27. Franc, V., Sonnenburg, S.: Optimized cutting plane algorithm for large-scale risk minimization. Journal of Machine Learning Research 10, 2157–2192 (2009)
28. Vedaldi, A., Zisserman, A.: Efficient additive kernels via explicit feature maps. IEEE Transactions on Pattern Analysis and Machine Intelligence 34(3), 480–492 (2012)
29. Wu, J.: Power mean svm for large scale visual classification. In: IEEE Computer Society Conference on Computer Vision and Pattern Recognition, pp. 2344–2351 (2012)
30. Berg, A., Deng, J., Li, F.F.: Large scale visual recognition challenge 2010. Technical report (2010)

Statistical Features for Emboli Identification Using Clustering Technique

Najah Ghazali[1,2,*] and Dzati Athiar Ramli[1]

[1] Intelligent Biometric Research Group (IBG),
School of Electrical and Electronic Engineering,
Universiti Sains Malaysia, Engineering Campus,
14300 Nibong Tebal,
Pulau Pinang, Malaysia
najahghazali@unimap.edu.my, dzati@usm.my
[2] Institute of Engineering Mathematics (IMK),
Universiti Malaysia Perlis,
02600 Pauh Putra, Arau,
Perlis, Malaysia

Abstract. Microembolus signals (MES) detected by transcranial Doppler (TCD) ultrasound are similar to the short duration transient signals. In previous researches, an embolus was tracked by using a supervised technique to discriminate the embolus from the background. However, the classification results were found to be affected by many factors and limited under experimental setup conditions. Therefore, a detection system based on the *k*-means clustering technique (unsupervised learning) is proposed for emboli detection. In order to verify the proposed technique, the signal data sets are also be computed and compared with SVM classifier. The features selected are the measured embolus-to-blood ratio (MEBR), peak embolus-to-blood ratio (PEBR) and statistical features. Five independent data sets of different transmitted frequency, probe location and different depths are identified to evaluate the feasibility of this new proposed method. The overall result show that *k*-means is better than SVM in term of robustness aspect. This work also revealed the feasibility of the automatic detection of the features-based emboli in which it is very imperative in assisting the experts to monitor the stroke patients.

Keywords: Classification, Embolus detection, *K*-means Clustering, Transcranial Doppler Ultrasound.

1 Introduction

Research on stroke occurrence has revealed variable findings on methodology and recent technology. The area which has interested researchers is the implementation of the Transcranial Doppler Ultrasound (TCD) ultrasound in the detection of the

* Corresponding author.

H.A. Le Thi et al. (eds.), *Advanced Computational Methods for Knowledge Engineering*,
Advances in Intelligent Systems and Computing 358, DOI: 10.1007/978-3-319-17996-4_24

microembolic signal (MES). The MES or emboli are tiny particles consisting of plaque, fat, thrombus, or gaseous which travel in the blood vessel of a human body [1]. An embolization can occur in various situations such as carotid stenosis and carotid endarterectomy. The plaque or blood clot will eventually block the small arteries in the brain once it is disintegrated from the wall.

The emboli detection using the TCD is a common treatment to be carried out to the patients especially after carotid endarterectomy surgery. The analysis of the emboli from the TCD signal is done offline by human observer as a gold standard which is time consuming. Moreover, the ability to select, classify and abstract significance information by human observer is due to his sense. Hence, classification and prediction of TCD signal are among the techniques which gain the interest among researchers nowadays.

1.1 Limitation of Classification Technique

Recently, supervised classification methods such as binary decision tree [2], artificial neural network (ANN), *K*-Nearest Neighbors Rule (KNNR) [4,5] and support vector machines (SVM) [4] were proposed for the embolus classification. The results yielded in a sensitivity and specificity of greater than 80% in tracking the solid and emboli composition. On the other hand, Fan et. al [5] developed a real time identification of micro-emboli Doppler signals by using knowledge-based DSP system which used a DSP board to display, detect and store embolic signals in real time. As the results, the sensitivity of 93.55% and specificity of 99.24% were achieved which were higher than the detection results done by human experts under laboratory conditions. The characteristics of the features using peak frequency of specific signals and the time delay which were introduced by Kem [6] had proved that the offline system is able to produce a sensitivity of 97% and a specificity of 98% which give a promising result for the clinical applications. However, these results need to be further confirmed.

A drawback of current techniques is that they base on supervised learning in which the training datasets are needed. In real clinical situation, the characteristic of the signals produced by the TCD device varies due to depth in cerebral artery and presence of noises produced by the probe and TCD settings [7]. Other information (i.e solid emboli or gas emboli) also remains uncertain [2]. Hence, the experimental setup which mimicks the real human blood had been introduced for that purposed [2,3]. Although many studies performing the supervised technique are able to gain promising results, the developed model cannot fully been confirmed yet to be utilized for the clinical usage. The reasons for this discrepancy are due to many factors such as the differences on the equipment settings, signal identification criteria and ultrasonic carrier frequency [8]. For example, previous studies [7,8][10] reported that the measurements of the emboli and blood are affected by the frequency levels. Evan [11] illustrated the variation in the power backscattered by an embolus at the frequency of 2.0 MHz and 2.5 MHz. It indicates that the designed classification model would provide an optimum result on the selected frequency of the ultrasonic backscatter. However, it does not guarantee that the systems would give a consistence or robust performance if the new datasets were applied in different frequency to the other part in middle cerebral artery. Therefore, the unsupervised learning seems possible to be applied to any type of device and frequency and may potentially correspond to the robustness aspect as it does not rely on the previous datasets unlike the supervised technique.

In this paper, a new technique using the unsupervised clustering analysis is introduced as a potential technique to differentiate embolus from the background signal. Clustering is an unsupervised learning method which carries only the features to characterize an emboli from the background without providing the training sets or model in order to obtain a good result. Since the objective of this research is to develop a system without depending on the frequency, the *k*-means was employed as it offers promising technique for the future MES detection analysis for any types of parameters setup.

2 Methodology

There are three objectives highlighted in this study. The first objective of this study is to collect datasets of five different targeted groups of embolic signals from the TCD simulator. The groups are distinguished based on frequency transmitted, probe location and depth within the internal artery. The second objective is to perform feature extraction analysis. There are five features identified: MEBR, PEBR, standard deviation, maximum peak and entropy. The features are analyzed using statistical analysis to determine combination of features group.

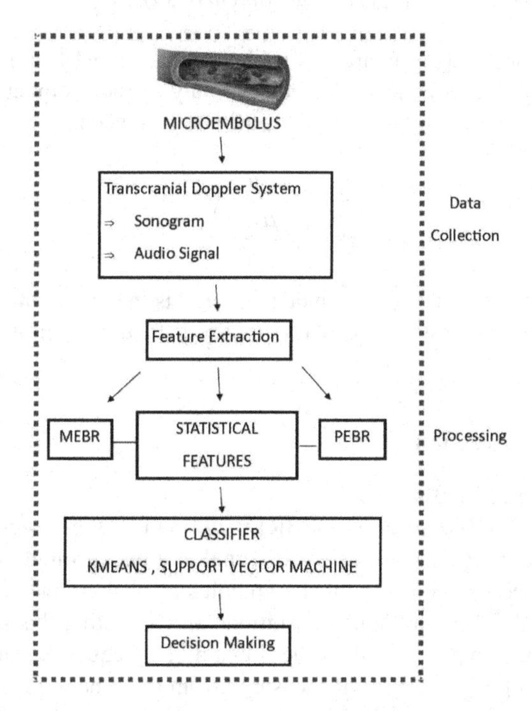

Fig. 1. Flow chart of data collection and signal processing of emboli detection

For the third objective, *k*-means algorithms (unsupervised classifier) are applied to discriminate the emboli from the background signal by using the identified features group. The total datasets consist of 1,400 samples of five targeted group. The similar data sets are then compared to SVM classifier which represented by the supervised learning. The group data sets are fed into the classifier where each session trains four groups and the remaining group is being tested. The procedures will be repeated for five times until all groups are modeled. The study is summarized in Figure 1.

2.1 *k*-Means Algorithm

The *k*-Means algorithm is a partitional clustering algorithm, which is used to distribute points in feature space among a predefined number of classes [12]. An implementation of the algorithm applied, in pseudocode, can be summarized as follows :

Initialize the centroids of the classes, i=1, … , k at random $\mu_i(t)$ *where t=0.*
Repeat

for the centroids $\mu_i(t)$ *of all classes, locate the data points whose Euclidian distance from* $\mu_i(t)$ *is minimal. Set* $\mu_i(t+1)$ *equal to the new center of mass of* x_i *,* $\mu_i(t+1)= x_i/n_i$ *,*
$t=t+1$
until $(|\mu_i(t-1)- \mu_i(t)| \leq$ *error, for all classes i).*

The above algorithm [12] is applied to the *n*-dimensional features set to be fused. The centroids of the classes are selected randomly in the *n*-dimensional features set and the algorithm evolves by moving the position of the centroids, so that the following quantity is minimized (1):

$$E = \sum_{j} \left(x_j - \mu_{c(x_i)} \right)^2 \tag{1}$$

where *j* is the number of data points index and $\mu_{c(x)}$ is the class centroid closest to data point x_j. The above algorithm could be employed to utilize information from TCD signal to achieve emboli classification.

2.2 Features Extraction

2.2.1 MEBR and PEBR

Measured embolus-to-blood ratio (MEBR) is defined as the targeted feature due to the difference in intensity of the emboli signal and background signal. The MEBR is the feature based on the ratios of the embolus to background signal within given short time interval. There are some definitions on computing this ratio, for example, the MEBR can be computed in the time domain [1], frequency domain [13] and the wavelet transform [14]. The MEBR measured from the time domain representation is due to the intensity of the signal which can be clearly observed on the detected signal.

The increase in the intensity is observed because of the passage of the embolus relative to the background signal. It can be measured by using equation (2).

$$\mathrm{MEBR} = 10\log_{10}\left(\frac{I_{\mathrm{emb+blood}}}{I_{\mathrm{blood}}}\right) \tag{2}$$

where $I_{\mathrm{emb+blood}}$ is the intensity of the embolus and the background signal and I_{blood} is the intensity of the background signal. The value of intensity is defined as the power of two local maximum of the signal peaks by looking at Figure 2.

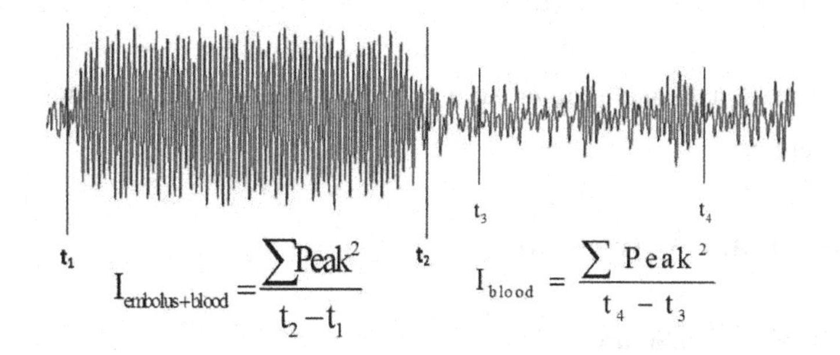

Fig. 2. Calculation of intensity obtained from the audio signal

Peak embolus-to-blood ratio (PEBR) is computed by replacing the average peak with the maximum peak of the signal in order to measure any significant difference between these two features. The values obtained by the MEBR and PEBR vary depending on the transmitted frequency, length of the attenuation of sample volume and focusing area.

2.2.2 Statistical Features

Three statistical features comprising different characteristics describing the embolus signal are investigated. The analyzed features are the entropy, standard deviation and maximum peak. A hypothesis that the white noise (background signal of TCD) is a projection of a system in thermodynamic equilibrium into a signal is normally made to investigate the relationship between the entropy and emboli signal [15]. The noise results in having high entropy which is mainly due to its periodic sound while the emboli signal produces significantly lower entropy value than the noise since it is more organized and requires an extra energy to be yielded in such organized form [15]. According to the above pre-assumption, the entropy can be employed in a signal processing to separate the useful signal from the background noise. The calculation for the entropy is stated as in equation (3):

$$H(x) = -\sum_{i=1}^{N} p(x_i)\log_{10} p(x_i), \tag{3}$$

where $x=\{x_1,x_2,...,x_N\}$ is a set of random signal and $p(x_i)$ is a probability of a random signal x_i.

The other statistical features such as the standard deviation and maximum peak are also evaluated because of the randomness and measured distances of certain value for each sampling frequency of the signal. The standard deviation as written in equation (4), measures the fluctuation of the signal from the mean of the signal

$$\sigma=\frac{1}{N-1}\left[\sum_{i=0}^{N-1}x_i^2-\frac{1}{N}\left(\sum_{i=0}^{N-1}x_i\right)^2\right] \quad (4)$$

where x_i is the signal amplitude in the time domain and N is the total number of frequency sampling.

3 Result and Discussion

3.1 Experimental Setup

In order to evaluate the detection system, TCD simulator is used to detect spontaneous MES within the middle cerebral artery as a probable risk of future stroke [16]. A transducer was placed on the right and left side of the transtemporal (region) for the patients who were having more than 80% of symptomatic internal carotid artery stenosis. The sampling frequency was 8 KHz. Five independent data sets of patients' group with different probe position and transmitted frequency (1 MHz - 3 MHz) defined in table 1. All groups had the recording time of up to 40 minutes with the Doppler instrument's sample volume depth is placed near the skull between 58 mm to 62 mm.

Table 1. The parameters for simulated ultrasound time series for ultrasound probe

Group	Frequency Setup	Probe location	SVL Depth
1	2 MHz	Right Side MCA	60 mm
2	1.5 MHz	Right Side MCA	60 mm
3	2 MHz	Left Side MCA	58 mm
4	1 MHz	Left Side MCA	62 mm
5	3 MHz	Left Side MCA	62 mm

*MCA : Middle Cerebral Artery; SVL : Sample Volume Length.

When an observer perceived the Doppler sonogram and suspected the emboli event, the permanent storage of the contents of the buffer was activated. As a result, a total of 244 MES signals had been detected. Each signal was selected according to the chirping sound and sonogram image of the velocity profile (Figure 3). The desired signal was then chosen and randomly cropped with the time error estimation between 5 μs to 7 μs within the time delay. The reason of considering the time delay in the targeted signal is to mimic the real situation in which the error in a measurement would occur and unexpected calibration would therefore has to be encountered. The signals of the extracted features were then analyzed offline.

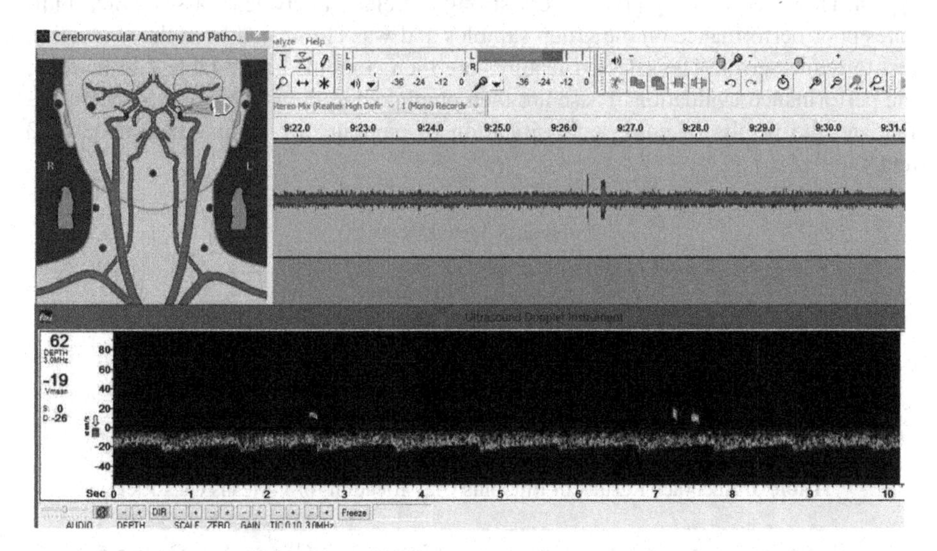

Fig. 3. Analysis of embolic signals was performed using TCD simulator equipped with the probe location and audacity audio recording. The embolic signal and background window were manually selected to estimate the features value.

3.2 Statistical Analysis

3.2.1 One Way ANOVA

The statistical significance of all features involved in all patients having stenosis was studied by using the analysis of variance (ANOVA) test. The significant threshold was set at $p<0.05$. The ANOVA result shows that there are differences between the targetted groups (embolus and noise) for all features indicated ($p<0.05$). Therefore, it ensures the probability of achieving high clustering accuracy.

3.2.2 Correlation Based-Features

Correlation based-features selection (CFS) test was evaluated by experiments on the group datasets. CFS measures correlation between independent features. The test eliminates irrelevant and redundant data and can be used as a features selector to combine features[17].

3.3 Results

Table 2 summarizes the features combination based on the CFS test. The feature of standard deviation (std) is found to give negative correlation between the two classes for each group datasets. It indicates std as a good features to differentiate between classes within the group. Hence, the feature is included in type 1, type 2, type 3 and type 4. However, entropy gives a very strong correlation between classes which indicates poor performance on the group samples and was eliminated from the evaluation test to reduce error. The selected features are then combined as in table 2 to increase the performance evaluation. These are then employed to the *k*-mean and SVM classifier and the results of correct classification of embolus are summarized in table 3 and 4.

Table 2. Selected Features

	Features
Type 1	Std, Max Peak
Type 2	Std, Max Peak, PEBR
Type 3	Std, Max Peak, MEBR
Type 4	Std, Max Peak, MEBR, PEBR

Table 3. Accuracy Results of Embolus Classification for *k*-Means Classifier

Features	**Group 1**	**Group 2**	**Group 3**	**Group 4**	**Group 5**
Type 1	88.52 %	92.62 %	85.24 %	82.38 %	82.79 %
Type 2	93.85 %	97.95 %	80.73 %	83.2 %	77.05 %
Type 3	93.03 %	96.31 %	77.05 %	83.2 %	76.64 %
Type 4	93.85 %	97.54 %	80.33 %	82.79 %	76.64 %

Table 4. Accuracy Results for Embolus Classification for SVM Classifier

Features	**Group 1**	**Group 2**	**Group 3**	**Group 4**	**Group 5**
Type 1	50.81 %	50.00 %	79.09 %	69.67 %	95.49 %
Type 2	48.36 %	83.6 %	69.67 %	74.18 %	68.44 %
Type 3	27.87 %	84.43 %	89.75 %	81.87 %	73.36 %
Type 4	46.72 %	90.98 %	90.57 %	83.6 %	86.48 %

For *k*-means (Table 3), all the features types provide satisfactory results of correct classification among groups (> 80%) except for group 5 (mean accuracy 78.28%). Contrary to SVM, the accuracy results (Table 4) show a low consistency performance rate among the group. Correct classifications of embolus are highly recorded (>90%) for group 2 (type 4), group 3 (type 4) and group 5 (type 1). However, there are also low classification rate obtained from group 1, group 2 (type 1), group 3 (type 2), group 4 (type 1 and 2) and group 5 (type 2 and 3).

3.4 Discussion

There are two important issues that are brought out for discussion. The first one is the real datasets in any TCD Ultrasound signal keep changing, follow to some parameters as discussed earlier. For that purpose, each group samples are determined as independent data sets (differ by the probe location and depth). Cross validation data are then been performed where the first four data sets are trained and the remaining group data sets had been tested. The algorithms train over 976 groups' samples and were repeated five times until all remaining groups had been tested. This will also ensure that the subject tested is not related to the model developed. Hence, we hypothesized that the model designed can be further analyzed into unknown data sets but still within the area tested near the skull by using TCD simulator. As predicted, the accuracy rate given earlier in (3) and (4) are fluctuated among the groups and the features type which later affect the overall accuracy stated in Table 5. The overall accuracy performance of SVM classifier are decreased among features selection (<80%) and groups selection (<85%) due to the inconsistency performance of each samples towards the designated model. For *k*-Means classifier, overall accuracy among features are performing reasonably within 85%-90% of overall accuracy rate. It is not the value of accuracy rate that we want to emphasize here but the fluctuation trend which affects the robustness of the model designed. This leads to the second issue of robustness. Figure 4 and Figure 5 show the error bounds between the two classifiers. The error bounds, which correspond to the standard deviations, reported for *k*-Means classifier as ±4.29, ±8.92, ±9.06, ±9.01 which are smaller than ±19.34, ±12.91, ±25.08, ±18.67 for SVM classifier among features types (Table 5). For *k*-Means classifier, the processing datasets are independent from the other group datasets. It indicates that *k*-Means classifier can overcome the limitation of embolus discrimination for clinical usage.

Table 5. Mean and Standard Deviations (SD) of Some Parameters for *k*-Means and SVM

		k-Means		SVM	
		Mean (%)	SD	Mean (%)	SD
Features	Type 1	86.31	4.29	69.02	19.34
	Type 2	86.56	8.92	68.85	12.91
	Type 3	85.25	9.06	71.48	25.08
	Type 4	86.23	9.01	79.67	18.67
Groups	1	92.32	2.56	43.44	10.51
	2	96.11	2.42	77.25	18.47
	3	80.84	3.37	82.27	9.89
	4	82.89	0.39	77.36	6.57
	5	78.28	3.01	80.94	12.33

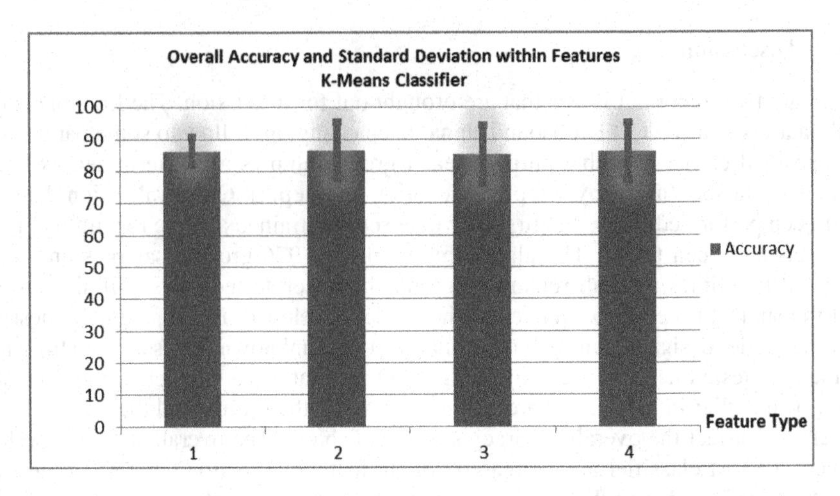

Fig. 4. Overall features performance with error estimation for *k*-means classifier

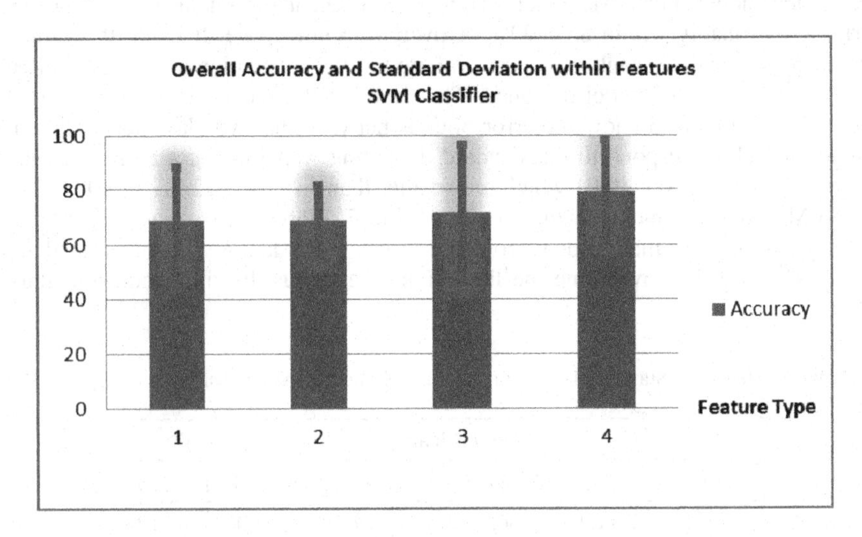

Fig. 5. Overall features performance with error estimation for SVM classifier

4 Conclusion

Although clustering can offer a standardized technique to discriminate the MES, there is also a limitation of implementing this method. If the initial cluster centroid position is wrongly chosen, the *k*-means converges to the unexpected local optimum leading to the misclassification. However, there is another way to overcome this obstacle which is by performing replication in which a new set of initial cluster centroid is repeatedly selected.

In conclusion, to further improve the embolus detection study, we suggest to go beyond the unsupervised technique in the acquisition of the TCD signal and to introduce advanced signal-processing methods.

Acknowledgment. The authors would like to thank the financial support provided by Fundamental Research Grant of Ministry of Education of Malaysia, 203/PELECT/6071266 (USM) and Professor Dr. Jafri Malim Abdullah from Hospital Universiti Sains Malaysia for all forms of assistance during this research work.

References

1. Evans, D.H.: Limitations of Dual Frequency Measurements for Embolus Classification. In: IEEE Ultrasonics Symposium, vol. (c), pp. 457–460 (2004)
2. Darbellay, G.A., Duff, R., Vesin, J.-M., Despland, P.A., Droste, D.W., Molina, C., Serena, J., Sztajzel, R., Ruchat, P., Karapanayiotides, T., Kalangos, A., Bogousslavsky, J., Ringelstein, E.B., Devuyst, G.: Solid or gaseous circulating brain emboli: are they separable by transcranial ultrasound? J. Cereb. Blood Flow Metab. 24(8), 860–868 (2004)
3. Ferroudji, K., Benoudjit, N., Bahaz, M., Bouakaz, A.: Feature Selection Based on RF Signals and KNN Rule: Application to Microemboli Classification. In: 7th Int. Work. Syst. Signal Process. their Appl., pp. 251–254 (2011)
4. Ferroudji, K., Bahaz, M., Benoudjit, N., Bouakaz, A.: Microemboli Classification using Non-linear Kernel Support Vector Machines and RF signals. Rev. des Sci. la Technol. 3, 79–87 (2012)
5. Fan, L., Evans, D.H., Naylor, A.R., Tortoli, P.: Real-time identification and archiving of micro-embolic Doppler signals using a knowledge-based DSP system. Med. Biol. Eng. Comput. 42(2), 193–200 (2004)
6. Kem, V.: Microemboli Detection by Transcranial Doppler Ultrasound. Semmelweis University (2005)
7. Angell, E.L., Evans, D.H.: Limits of uncertainty in measured values of embolus-to-blood ratio due to Doppler sample volume shape and location. Ultrasound Med. Biol. 29(7), 1037–1044 (2003)
8. Dittrich, R., Ritter, M.A., Droste, D.W.: Microembolus detection by transcranial doppler sonography. Eur. J. Ultrasound 16(1-2), 21–30 (2002)
9. Moehring, M.A., Spencer, M.P., Demljth, R.P., Davis, D.L.: Exploration of the Embolus to Blood Power Ratio Model (EBR) for Characterizing Microemboli Detected in Middle Cerebral Artery. In: IEEE Ultrason. Symp., pp. 1531–1535 (1995)
10. Bollinger, B.R.: Dual-Freaquency Ultrasound Detection and Sizing of Microbubbles for Studying Decompression Sickness. Dartmouth College (2008)
11. Evans, D.H., Gittins, J.: Limits of uncertainty in measured values of embolus-to-blood ratios in dual-frequency TCD recordings due to nonidentical sample volume shapes. Ultrasound Med. Biol. 31(2), 233–242 (2005)
12. Stergiopoulos, S.: Advanced Signal Processing Handbook: Theory and Implementation for Radar, Sonar, and Medical Imaging Real Time Systems - CRC Press Book, p. 752 (2000)
13. Martin, M.J., Chung, E.M.L., Ramnarine, K.V., Goodall, A.H., Naylor, A.R., Evans, D.H.: Thrombus size and Doppler embolic signal intensity. Cerebrovasc. Dis. 28(4), 397–405 (2009)
14. Serbes, G., Sakar, B.E., Aydin, N., Gulcur, H.O.: An Emboli Detection System Based on Dual Tree Complex Wavelet Transform, vol. 41, pp. 819–822. Springer International Publishing, Cham (2014)
15. Pavelka T., "Entropy And Entropy-based Features In Signal Processing," (2004)
16. Zuromskis, T., Wetterholm, R., Lindqvist, J.F., Svedlund, S., Sixt, C., Jatuzis, D., Obelieniene, D., Caidahl, K., Volkmann, R.: Prevalence of micro-emboli in symptomatic high grade carotid artery disease: a transcranial Doppler study. Eur. J. Vasc. Endovasc. Surg. 5, 534–540 (2008)
17. Hall, M.A.: Correlation-based Feature Selection for Machine Learning (1999)

Twitter Sentiment Analysis Using Machine Learning Techniques

Bac Le[1] and Huy Nguyen[2]

[1] Department of Computer Science, VNUHCM-University of Science, Viet Nam
lhbac@fit.hcmus.edu.vn
[2] Department of Information Science, Sai Gon University, Viet Nam
huynguyen@sgu.edu.vn

Abstract. Twitter is a microblogging site in which users can post updates (tweets) to friends (followers). It has become an immense dataset of the so-called sentiments. In this paper, we introduce an approach to selection of a new feature set based on Information Gain, Bigram, Object-oriented extraction methods in sentiment analysis on social networking side. In addition, we also proposes a sentiment analysis model based on Naive Bayes and Support Vector Machine. Its purpose is to analyze sentiment more effectively. This model proved to be highly effective and accurate on the analysis of feelings.

Keywords: Twitter, sentiment analysis, sentiment classification.

1 Introduction

Twitter is a popular microblogging service in which users post status messages, called "tweets", with no more than 140 characters. The million of statuses appear on socical networking everyday. In most cases, its users enter their messages with much fewer characters than the limit established. Twitter represents one of the largest and most dynamic datasets of user generated content approximately 200 million users post 400 million tweets per day. Tweets can express opinions on different topics, which can help to direct marketing campaigns so as to share consumers' opinions concerning brands and products, outbreaks of bullying, events that generate insecurity, polarity prediction in political and sports discussions, and acceptance or rejection of politicians, all in an electronic word-of-mouth way. In such application domains, one deals with large text corpora and most often "formal language". At least two specific issues should be addressed in any type of computer-based tweet analysis: firstly, the frequency of misspellings and slang in tweets is much higher than that in other domains. Secondly, Twitter users post messages on a variety of topics, unlike blogs, news, and other sites, which are tailored to specific topics. Big challenges can be faced in tweet sentiment analysis: *a*) neutral tweets are way more common than positive and negative ones. This is different from other sentiment analysis domains (e.g. product reviews), which tend to be predominantly positive or negative; *b*) there are linguistic representational challenges, like those that arise from feature engineering issues; and *c*) tweets are very short and often show limited sentiment cues.

H.A. Le Thi et al. (eds.), *Advanced Computational Methods for Knowledge Engineering*,
Advances in Intelligent Systems and Computing 358, DOI: 10.1007/978-3-319-17996-4_25

In this paper, we used a data set about 200000 tweets for training classifiers. Table 1 is an example about the types of Twitter. We built a model which classified tweets collected from Twitter APIs into the positive class or the negative class. The model runs on three steps: a classifier categorizes tweets into objective tweets or subjective tweets, another classifier organizes subjective tweets into positive or negative and finally, the system summarizes tweets into a vitual graph. For training, we applied on three kinds of features: Unigram, Bigram, Object-oriented. The training set contains tweets without emoticons. We evaluated that emoticons were noisy when classifiers analyzed tweets. Our experiments proved to be highly accurate. Related work on tweet sentiment analysis is rather limited, but the initial results are promising

Our main contributions can be summarized as follows: *a*) We extracted a new feature set and appraised effectively based on Information Gain, Bigram, Object-oriented extraction methods. *b*) We built a sentiment analysis model based on supervised learning sush as Naive Bayes and Support Vector Machine for enhancing effective classification.

2 Related Works

There were many studies in sentiment analysis but almost those focused on a part of texts or critiques. A tweet is only limited to 140 characters, so it is as different as a critique. Bing Liu [3] (2010), Tang and colleagues [11] (2009) expressed an overview in sentiment analysis in which analyzed the strong points and the weak points of sentiment analysis and they gave many research ways of sentiment analysis. Pang and Lee [1] [2] (2004, 2008) compared many classifiers on movie reviews and gave a vision of insight and comprehension in sentiment analysis and opinion mining. Authors also used star rating as a feature for classification. Go et al [8] (2009) studied on Bigram and POS. They removed emoticons out from their training data for classification and compared with Naive Bayes, MaxEnt, Support Vector Machine (SVM). They evaluated that SVM outperforms others. Barbosa and Feng [9] (2010) pointed that N-gram is slow, so they researched on Microblogging features. Agarwal et al [7] (2011) approached Microblogging, POS and Lexicon features, also they built tree kernel to classify tweets and applied on POS and N-Gram. Akshi Kumar and Teeja Mary Sebastian [5] (2012) approached a dictionary method for analyzing the sentiment polarity of tweets. On the other hand, Stanford University (2013)[1] performs a twitter sentiment classifier based on Maximum Entropy and they built a Recursive Deep Model with a Sentiment Tree Bank but they applied on Movie Reviews, not Twitter.

3 Our Approach

Our approach used classifiers to categorize sentiment into positive or negative and we applied a good feature extractor for enhancing accuracy. The classifiers which are built into a model to effectively classify are Naive Bayes (NB) and Support Vector Machine (SVM).

[1] `http://nlp.stanford.edu/sentiment/index.html`

3.1 Pre-processing of Data

The language model of Twitter has its properties. We pre-processed data for enhancing accuracy and removing noisy features. we pre-processed data on the following steps: *a*) we converted tweets to lower case. In every tweet, we splitted sentences. We removed stopwords out from tweets. In addition, we removed spaces and words which do not start with the alphabet letters. Characters which repeat in a word sush as "huuuuuuuuuuungry", we converted into "huungry". *b*) Usernames. They are attached in tweets to transfer a status to other usernames. An username which attached in a tweet starts with "@". We replaced usernames by "AT_USER". *c*) URLs. Users attach urls in tweets. Urls look like "http://tinyurl.com/cvvg9a". We replaced those into URLs. *d*) #hashtag. Hashtags might provide useful information for us, so we only removed "#" at hashtags. For example, from "#nice" to "nice". *e*) Emoticons. We removed emoticons from tweets because we believed that those are noisy when we analyzed sentiment. For example, we have a following tweet: "@soundwav2010 Reading my kindle2... Love it... Lee childs is good read. #kindle". We converted the tweet into: "AT_USER reading my kindle2 love it lee childs is good read kindle".

3.2 Feature Extraction

We identified the object words of tweets which have sentiment polarity and we extracted those into a feature set. Its name is the object-oriented feature. On the other hand, we detected that a number of words are meaningless or insignificant. So, we have to remove them. We used the Information Gain method to extract these words. Another consideration is that positive words mean negative if they stay behind a negative preposition and in a contrary way. This is an important problem that we used Bigram to resolve. We chose following features for increasing the accuracy of classifiers:

a) We believe that there are an expression of moral of an entity or a keyword which could be extracted from tweets. Objects which belong to a group sush as "company", "person", "city" that they describe an object as "Steve Jobs", "Vodafone", "London" are significant. They have correlation to positive or negative. This is the sentiment polarity of objects. We used this object-oriented feature for increasing the accuracy of classifiers. We used an open data set of Alchemy API to extract these objects. The data set was evaluated by Rizzo and Troncy [10] (2011) on five data sets: AlchemyAPI[2], DBPedia Spotlight[3],Extractiv[4], OpenCalais[5] and Zemanta[6]. Their research showed that AlchemyAPI outperforms others. In additon, Hassan Saif, Yulan He and Harith Alani

[2] `http://www.alchemyapi.com/`

[3] `http://dbpedia.org/spotlight/`

[4] `http://wiki.extractiv.com/w/page/29179775/Entity-Extraction`

[5] `http://www.opencalais.com/`

[6] `http://www.zemanta.com/`

[12] also used AlchemyAPI to extract objects from Twitter and they evaluated that AlchemyAPI got the accuracy of 73.97% of extracting objects. we believe that the object-oriented feature is significant for enhancing the accuracy of sentiment analysis, especially tweets do not contain verbs, adjectives and we used the sentiment polarity of object-oriented features for classifying tweets. We chose the AlchemyAPI data set for extracting this feature.

Table 1. The accuracy of object extraction tools on 500 tweet from Stanford Corpus

Data sets	extracted object-oriented features	Accuracy (%)
AlchemyAPI	108	73.9
Zemanta	70	71
OpenCalais	65	68

b) However, verbs and adjectives are also important that we must extract. Information Gain (IG) is a good method for extracting features (term-goodness) in Machine Learning. IG measures the number of information of bits obtained to predict classifications by the presence or the absence of features in documents. Another problem that the number of words are meaningless or insignificant but classifiers have to assign them into any class. So, we have to remove them from feature sets. Information Gain resolves this problem. We chose Unigram words which are highly significant for classifying. The formula of Information Gain is as follows:

$$IG(t) = -\sum_{i=1}^{m} Pr(c_i)logPr(c_i) + Pr(t)\sum_{i=1}^{m} Pr(c_i|t)logPr(c_i|t) + Pr(\bar{t})\sum_{i=1}^{m} Pr(c_i|\bar{t})logPr(c_i|\bar{t}) \quad (1)$$

In which: *i)* $Pr(c_i)$: the probability of class c_i happens. *ii)* P(t): the probability of feature t happens. *iii)* $P(\bar{t})$: the probability of feature t do not happen. *iv)* $Pr(c_i|t)$ the probability of feature t appears in class c_i. *v)* $Pr(c_i|\bar{t})$: probability of feature t do not appear in class c_i.

c) However, when we used Unigram, classifiers are only learned single words. For example, if classifiers process a sentence: "I do not like Iphone", this sentence is categorized into the negative class because word "not" might more weight than word "like", but a sentence: "It is not bad", classifiers categorize this sentence into the negative class because word "not" and word "bad" are negative. We have to train classifiers multiple words. We measured the correlation of two words by statistics χ^2 method and we also chose Bigram multiple words which have high signification.

We extracted 10000 unigram, 100 bigram and 200 object-oriented features to built the list of feature sets. Because when we extracted more features, the accuracy did not change. The list of feature sets on the following steps: *a)* For Naive

Bayes classifier, we used Information Gian and statistics χ^2 to extract Unigam words and Bigram multiple words from Ravikiran Janardhana data set [13] and used AlchemyAPI data set to extract object words from Ravikiran Janardhana data set [13] which have sentiment polarity. *b*) For Support Vector Machine classifier, the process is same, we also used Information Gian and statistics χ^2 to extract Unigam words and Bigram multiple words from Sentiment Stanford data set [13] and used AlchemyAPI data set to extract object words from Sentiment Stanford data set [13] which have positive or negative signification . *c*) For the incorporation of object-oriented feature into Language model, we incorporated the object-oriented feature with the following augmentation method:

$$|V'| = |V| + |S| \tag{2}$$

In which: *i*) V': The new vocabulary set. *ii*) V: The original vocabulary set. *iii*) S: The additional vocabulary set.

d) After extracting features for classifiers, we standardized the feature sets following: *i*) For Unigram, Bigram, Object-oriented features extracted, We saved into two databases. The first database is words, multiple words which are labeled positive. The second database is words, multible words which are labeled negative. *ii*) We mixed the above-mentioned two databases into a database with (key, value) in which key is the vocabulary set of a sentence and value is positive or negative label of this vocabulary set. *iii*) Finally, we converted the vocabulary set into a dictionary set with binary occurences for training classifiers.

3.3 Classification Model

We used Twitter APIs as a library tool to collect tweets from internet for sentiment analysis and we built a system based on Naive Bayes (NB) and Support Vector Machine (SVM). We trained classifiers and we classified tweets collected from internet on the following steps: *a*) We built a Naive Bayes classifier to categorize subjective tweets and objective tweets. The subjective training set is sentences labeled subjective or objective and we applied Unigram, Bigram, Object-oriented features for training. *b*) For the SVM classifer, it classifies the subjective tweets into the positive class or the negative class. The sentiment training set is sentences labeled positive or negative and we applied Unigram, Bigram, object-oriented features for training. *c*) The system also draws a graph after the analysis of feelings. Picture 1 is sentiment analysis steps of the model based on Unigram, Bigram and Object-oriented features.

Naive Bayes (NB). Naive Bayes is a classification method based on statistics using Machine Learning. The idea of this approach is conditional probability among words, phrases and classes to predict statistics of them belonging to a class. A special point of this method is the appearance of words which is independent. Naive Bayes do not evaluate the dependance of words with any class. We used Multinomial Naive Bayes. Naive Bayes assign tweet d with class c. the formula is as follows:

$$C* = argmac_c P_{NB}(c|d) \tag{3}$$

$$P_{NB}(c|d) = \frac{(P(c) \sum_{i=1}^{m} P(f|c)^{n_i(d)})}{P(d)} \tag{4}$$

In which: *i)* f: describing a feature. *ii)* $n_i(d)$: the number of features f_i are found in tweets. There are m features.

We used a Scikit-Learn tool[7] with a Multinomial Naive Bayes model to classify subjective tweets and objective tweets.

Support Vector Machine (SVM). Another technique which is popular in Machine Learning is Support Vector Machine. We also used a SVM Scikit-Learn tool with Linear kernel. A training set which is for Support Vector Machine are two vector sets with m size. m is all of features. An element of the vector describes the presence or the absence of that feature . For instance, A feature is a word found in tweet by the Information Gain method. If the feature is present, its value is 1. In contrary, its value is 0. Bigram and Object-oriented features are same. Fig.1 describes the system including Naive Bayes classifier and Support Vector Machine classifier for classifying tweets from Twitter.

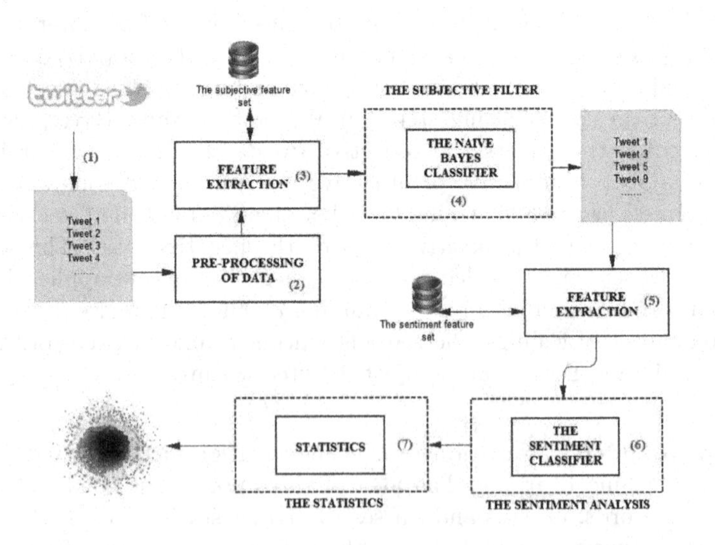

Fig. 1. The overview struture of the system

[7] `http://scikit-learn.org/stable/`

The operations of classifiers perform through the following steps:

a) Tweets are collected based on a topic from Twitter through Twitter APIs and saved into a database. For example, the topic is "Iphone", the system will downloads tweets from Twitter relating "Iphone". *b*) Pre-processing tweets collected. *c*) After the pre-processing step, the system extracts subjective features based on Information Gain, Bigram and use AlchemyAPI data set to extract the object-oriented feature. *d*) Thanks to the subjective feature set, the Naive Bayes classifier categorizes tweets into the subjective class or the objective class. Objective tweets are then removed. *e*) For the subjective tweets, the system extracts sentiment features based on Information Gain, Bigram and use AlchemyAPI data set to extract the object-oriented feature. *f*) The system puts tweets extracted features into the Support Vector Machine classifier and the classifier classify tweets into the negative class or the positive class. *g*) The system draws a graph based on tweets which is classified.

4 Experiments and Evalutions

4.1 Experiments

We trained Support Vector Machine classifier based on a Sentiment Stanford training set containing 200000 tweets in which 100000 tweets labeled positive and 100000 tweets labeled negative. Thanks to the training set 20000 tweet collected through Twitter APIs by Ravikiran Janardhana[13]. We labeled subjective or objective for training the Naive Bayes classifier. We extracted three types of features containing Unigram, Bigram, Object-oriented features for training and experiments. We used the testing data of Stanford University containing 500 tweets of neutral, negative and positive labeled manual. We built three types of the testing data set: *a*) A data set is extracted features through the Information Gain method with only unigram feature. *b*) A data set is extracted features through Bigram extraction with only Bigram feature. *c*) A data set is extracted three features. They are object-oriented extraction, Information Gain and Bigram. Table 4 describes the accuracy of the system on above-mentioned data sets.

In addition, we also evaluated the accuracy of classifiers on 200000 sentiment tweets of Standford University and 20000 subjective tweets of Ravikiran Janardhana[13] based on 10-fold method. The experiment shows that the Support Vector Machine classifier outferforms the Naive bayes classifier. However, we chose the Naive Bayes classifier to classify subjective tweets because the Naive Bayes classifier measures statistics on independent sentences, so the classifer easily removes the objective sentences of a tweet before the system puts them into the Support Vector Machine classifier for analyzing positive or negative.

4.2 Evalutions

We evaluated the accuracy of classifiers on 200000 sentiment tweets of Standford University and 20000 subjective tweets of Ravikiran Janardhana[13] to compare between classifiers. We used 10-fold method to test classifiers. However, we chose the training data of Ravikiran Janardhana[13] for classifying subjective. Because the training data of Ravikiran Janardhana[13] have subjective and objective tweets which are significant. Table 2 describes the accuracy of the classifiers on 20000 tweets of Ravikiran Janardhana[13].

Table 2. The accuracy of classifiers using Unigram, Bigram and Object-oriented features on 20000 tweets

Classifiers	Features	Accuracy (%)
Naive Bayes	Object-oriented, Unigram, Bigram features	80
Support Vector Machine	Object-oriented, Unigram, Bigram features	80

Table 3 describes the accuracy of the classifiers on 200000 tweets of Standford University.

Table 3. The accuracy of classifiers using Unigram, Bigram and Object-oriented features on 200000 tweets

Classifiers	Features	Accuracy (%)
Naive Bayes	Object-oriented, Unigram, Bigram features	79.54
Support Vector Machine	Object-oriented, Unigram, Bigram features	79.58

As can be seen, Support Vector Machine classifier outferforms the Naive bayes classifier on 200000 tweets of Standford University. Because Naive Bayes classifier measures statistics on independent sentences, but Support Vector Machine evaluates negative or positive on whole tweet which may have many sentences in. However, the classifiers just got 80 percent of accuracy. Because the grammar of tweet gathering from Twitter is not good. Many words are wrong and not necessary. Those have effects on feature selection. Table 4 describes the experiment result on three above-mentioned data sets using the system. The system take average of Naive Bayes and Support Vector Machine. The data set which is used Object-oriented, Unigram, Bigram features are better than others.

Table 4. The accuracy of the system on three data sets

Data sets	Features	Accuracy (%)
1	Object-oriented, Unigram, Bigram features	79.5
2	Unigram	77
3	Bigram	77

Table 5 and table 6 express the detail of Precision, Recall and Accuracy of classifiers on 200000 tweets of Standford University and 20000 tweets of Ravikiran Janardhana[13] using 10-fold method:

Table 5. Precision, Recall and F-Score of the Naive Bayes subjective classifier on 20000 tweets of Ravikiran Janardhana[13] using 10-fold method

Average			
Precision	Recall	F-score	Accuracy (%)
0.80	0.79	0.79	80

Table 6. Precision, Recall and F-Score of the SVM sentiment classifier on 200000 tweets of Standford University using 10-fold method

Average			
Precision	Recall	F-score	Accuracy (%)
0.79	0.79	0.79	79.5

The paper also compared the experiment result with other papers which also used data set of Standford University. Fig.2 describes the graph of accuracy of the papers.

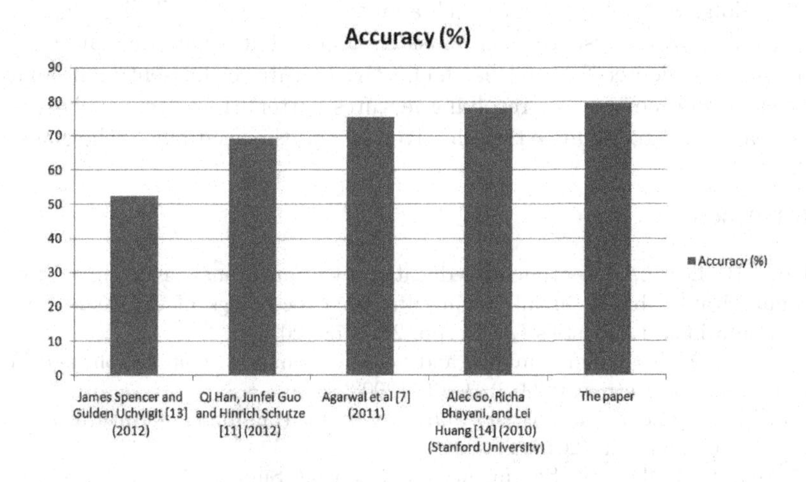

Fig. 2. Comparision with other papers

In our view, the paper outferforms others because we extracted fully significant features for sentiment analysis. We also resolved many comparison tweets. This helps to increase the accuracy of the classifiers. In addition, Fig. 3 describes a graph which the system create for expressing analysis tweets collecting from Twitter based on keyword in the most recent seven days.

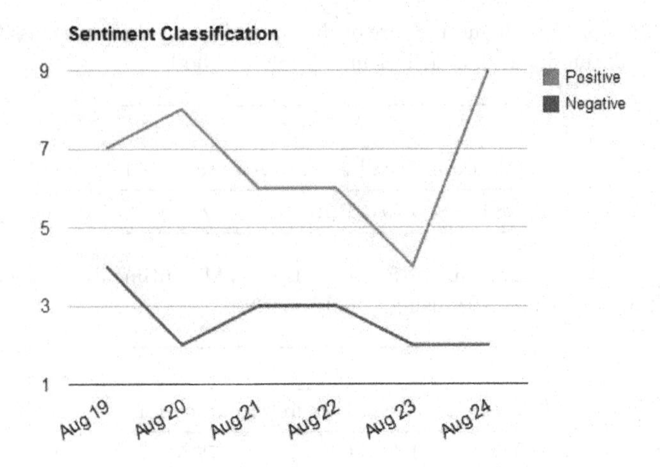

Fig. 3. A graph of analyzing tweets collected from Twitter on keyword "Iphone" in the most recent seven days

5 Conclusions

In conclusion, we built a model which analyzes sentiment on Twitter using Machine Leaning Techiques. Another consideration is that we applied Bigram, Unigram , Object-oriented features as an effective feature set for sentiment analysis. We used a good memory for resolving features better. However, we chose an effective feature set to enhance the effectiveness and the accuracy of the classifiers.

References

1. Pang, B., Lee, L.: A sentimental education: Sentiment analysis using subjectivity summarization based on minimum cuts. In: Proceedings of the Association for Computational Linguistics (ACL), pp. 271–278 (2004)
2. Pang, B., Lee, L.: Opinion mining and sentiment analysis. Foundations and Trends in Information Retrieval 2(1-2), 1–135 (2008)
3. Liu, B.: Sentiment Analysis and Subjectivity. In: Handbook of Natural Language Processing, 2nd edn. (2010)
4. Mullen, T., Collier, N.: Sentiment Analysis using Support Vector Machines with Diverse Information Sources. In: Proc. of Conference on Empirical Methods in Natural Language Processing (EMNLP 2004) (2004)
5. Kumar, A., Sebastian, T.M.: Sentiment Analysis on Twitter. IJCSI International Journal of Computer Science Issues 9(4(3)) (2012)
6. Saif, H., He, Y., Alani, H.: Alleviating: Data Sparsity for Twitter Sentiment Analysis. In: Proceedings of the 2nd Workshop on Making Sense of Microposts (#MSM2012): Big Things Come in Small Packages: in Conjunction with WWW 2012 (2012)
7. Agarwal, A., Xie, B., Vovsha, I., Rambow, O., Passonneau: Sentiment analysis of twitter data. In: Proc. ACL 2011 Workshop on Languages in Social Media, pp. 30–38 (2011)

8. Go, A., Bhayani, R., Huang: Twitter sentiment classification using distant supervision. CS224N Project Report, Stanford (2009)
9. Barbosa, L., Feng, J.: Robust Sentiment detection on twitter from biased and noisy data. In: Proceedings of COLING, pp. 36–44 (2010)
10. Rizzo, G., Troncy, R.: Nerd: Evaluating named entity recognition tools in the web of data. In: Workshop on Web Scale Knowledge Extraction (WEKEX 2011), vol. 21 (2011)
11. Tang, H., Tan, S., Cheng, X.: A survey on sentiment detection of reviews. Expert Systems with Applications 36(7), 10760–10773 (2009)
12. Saif, H., He, Y., Alani, H.: Semantic Sentiment Analysis of Twitter. In: Cudré-Mauroux, P., et al. (eds.) ISWC 2012, Part I. LNCS, vol. 7649, pp. 508–524. Springer, Heidelberg (2012)
13. Janardhana, R.: Twitter Sentiment Analysis and Opinion Mining (2010)

Video Recommendation Using Neuro-Fuzzy on Social TV Environment

Duc Anh Nguyen and Trong Hai Duong*

School of Computer Science and Engineering,
International University, Vietnam National University HCMC, Vietnam
haiduongtrong@gmail.com

Abstract. Prior collaborative filtering (CF) methods based on neighbors' ratings to predict a target user's rating. In this work, we consider recommendation on the context of Social TV (STV). The watchers/users may either share, comment, rate, or tag videos they are interested in. Each video must be watched and rated by many users. For these assumptions, we proposed a novel model-based collaborative filtering using a fuzzy neural network to learn user's social web behaviors for video recommendation on STV. We use netflix data-set to evaluate the proposed method. The result shown that the proposed approach is a significant effective method.

Keywords: Ontology, Smart TV, Video, Recommendation system, and Neural network.

1 Introduction

The trend for using online social networks to talk about TV programmes and to share their opinions with others, is increasing. This reflected with the dissemination of platforms designed for Social TV [1]. The NoTube [1] brings the social web and TV closer to consumers. The social TV is able to provide users' social context that personalize users' TV program and video with both of content- and collaborative-based filtering manners. Content-based filtering (CBF)[4] relies on the description of previously preferred items of a target user and generates recommended items with content are similar to those the target user has preferred in the past, without directly relying on the preferences of other users. Collaborative filtering (CF) [5] relies on the basis of previously preferred items of a large group of users' rating information and make recommended items to a target user based on the items that similar users have preferred in the past, without relying on any information about the items themselves other than their ratings. According to algorithms of CF, CF can be grouped into two types: (a) Memory-based collaborative filtering methods recommend items are those that were previously preferred by users who share similar preferences as the target user [6]. These algorithms require all ratings, items, and users to be stored in memory. (b) Model-based collaborative filtering methods recommend items based on models that are trained by using the collection of ratings to identify patterns in the input data [7]. The memory-based

* Corresponding author.

H.A. Le Thi et al. (eds.), *Advanced Computational Methods for Knowledge Engineering*,
Advances in Intelligent Systems and Computing 358, DOI: 10.1007/978-3-319-17996-4_26

collaborative filtering store the training data in memory that is delayed until a recommendation is made to the system, as opposed to model-based collaborative filtering, where the system tries to generalize a model using the training data before recommendation making. The advantage of memory-based methods is deal with less parameters to be tuned, while the disadvantage is that the approach cannot deal with data sparsity in a principled manner [9].

In Social TV, recommendation systems have been developed to help users access the TV programs that are appropriate to their preferences by learning from viewing history data, mapping social users' preferences and TV program attributes [15,16,9]. Authors [9] proposed hybrid approach combining content-based methods with those based on collaborative filtering for TV program recommendation. To eliminate the overloading computation of collaborative filtering, singular value decomposition technique [17] is applied in order to reduce the dimension of the user-item representation, and afterwards, how this low-rank representation can be employed in order to generate item-based prediction, which has shown a good behavior in the TV domain. Authors [10] proposed a framework for adaptive news recommendation in social media by utilizing user comments. User comments are collected to build a topic profile using a weighted graph. To generate weighted importance of topics, the standard TF×IDF model [11] and variant of the PageRank algorithms [12] are applied. With the topic profile constructed, it can be used to select relevant news from a collection of news articles in the database by constructing a retrieval module using combination of the strengths of two state-of-the-art news retrievers time factor [13] and language model [14].

In this work, we consider recommendation on the context of Social TV (STV). The watchers/users can either share, comment, rate, or tag videos they are interested in. Each video can be watched and rated by many users. For the assumption, we proposed a novel model approach using fuzzy neural network for video recommendation on STV.

2 User Behaviors-Based CF Using Neuro-Fuzzy Network

2.1 Profile Modeling

There are more than 40 genres in the whole netflix dataset, but there are several terms which are the same meaning, such as *Sport* and *Sports*, etc. So in total, the Netflix dataset has only 37 different genres, in following: *Action, Adult, Adventure,Animation,Anime, Biography, Children, Classics, Comedy, Crime, Documentary, Drama, Faith, Family, Fantasy, FilmNoir, Fitness, Foreign, Game Show, Gay, History, Horror, Independent, Lesbian, Music, Mystery, Romance, Sci-Fi, Short, Special Interest, Spirituality, Sport, Talk Show, Television, Thriller, War, Western*. For each user, we create a corresponding profile represented by a feature vector including 37 genres as its components. The user profile is represented by using a weighted vector defined as follows:

Definition 1 (Profile Feature). *Let p_i be a profile of an user u_i. The profile feature p_i is defined as follows: $C^{p_i} = \{(c_1^i, w_1^i), (c_2^i, w_2^i), \ldots, (c_{p_i}^i, w_{p_i}^i)\}$ is a set of pairs of concept and its weight.*

The $tf–idf$ weight (term frequency inverse document frequency) is a weight often used in information retrieval and text mining. This weight is a statistical measure used to evaluate how important a word is to a document in a collection or corpus. Here we use traditional vector space model(tf-idf) to define the feature of the documents [18]. Where, a document is considered as an abstract of a movie. A Collection is whole movie database. Each user maybe watch a lots of movies which are used to generated the user profile.

2.2 Content-Based Filtering Using Neuro-Fuzzy Network

We assume that user u_i interests in a set $M^i = \{m_1, m_2, \ldots, m_n\}$ of movies (he/she watched and made a rating to them). The user's profile can be considered as a feature vector $C^i = \{(c^i_1, w^i_1), (c^i_2, w^i_2), \ldots, (c^i_k, w^i_k)\}$, where c^i_j is a genre j^{th} from movies in M^i and w^i_1 is a corresponding weight generated by using vector space tf/idf. We assume that each movie $m_j, j = 1..n$ also can be interested by $h - 1$ other users $U^j = \{u_1, u_2 \ldots, u_h\}$. The rating-score of the movie $m_j \in M^i$ with respect to the user u_i can be denoted as y^j_i, so the rating-score set of a movie m_j with respect to U^j is denoted by $Y^j = \{y^j_1, y^j_2, \ldots, y^j_n\}$. For each movie m_j, we consider a black-box-typed model expressing a mathematical relationship between a input of feature vectors of users in $U^j = \{u_1, u_2 \ldots, u_n\}$, denoted by $\{C^1, C^2 \ldots, C^k\}$ and a output of the rating-score space,$Y^j = \{y^j_1, y^j_2, \ldots, y^j_n\}$, based on a given data set of corresponding users U^j as follows: $(\overrightarrow{C^{p_j}}, y^j_i), i = 1..k$ where $\overrightarrow{C^{p_i}}$ is the feature vector of ith user of the data set U^{m_j} and y^j_i is rating-score from user u_i to movie m_j, as an output. This work can be seen as system-identifying process, in which the model works as a mathematical function f expressed by a mapping as follows:

$$\begin{gathered} f : \Re^n \rightarrow \Re^1 \\ \overrightarrow{C^i} \mapsto y^j_i | y^j_i = f(\overrightarrow{C^i}) \end{gathered} \tag{1}$$

In this paper, the above mathematical model is expressed by a fuzzy-neuron structure(FNS), which is a combination of a fuzzy inference system (FIS) and a neural network structure (NNS).

Relating to the FIS, it can be summarized as follows. The FIS is built based on the algorithm establishing an adaptive neuro-fuzzy system, ENFS [3]. By using data-driven method, the same features or characteristics of the object are expressed by hyperbox-typed data clusters, which can be considered as a structure upon which fuzzy sets and membership functions are established to build the FIS. In the FIS, the fuzzy deducing rules are built based on constituting clauses depicting the fuzzy relationships typing MISO as following:

$$IF\ \chi_1 = B^{(i)}_1\ and\ \chi_2 = B^{(i)}_2\ and\ \ldots\ and\ \chi_M = B^{(i)}_M\ THEN\ \gamma = C^{(i)} \tag{2}$$

where χ_j is language variables expressing the result of clustering process; $B^{(i)}_k, k = 1...M$ is maximum membership value of the ith sample in k^{th}-labeled data clusters, which is used to establish the corresponding hyperbox value $y^{(i)}_1$ of this sample; γ

is constituting rule; and C^i is the constituting value, which is used to calculate the predicting value of the ith sample.

We consider a set of the patterns T_t covered by the $t^t h$ min-max hyperbox HB_t. The HB_t is determined using two vertexes, the max vertex $\bar{\omega_t} = [\omega_{t1}, \omega_{t2} \ldots, \omega_{tn}]$ and the min vertex$(\bar{v_t}) = [v_{t1}, v_{t2} \ldots, v_{tn}]$, where $\omega_{tj} = max(x_{ij}|(\bar{x_i}) \in T_t)$ and $v_t j = min(x_{ij}|(\bar{x_i}) \in T_t)$. If T_t consists of the patterns associated with the cluster labeling m only, then the HB_t will be considered as a pure hyperbox labeling m, and denoted $pHB_t^{(m)}$. An HB can be considered as a crisp frame on which different types of membership functions (MFs) can be adapted. Here, the original Simpson's MF is adopted, in which the slope outside the HB is established by the value of the fuzziness parameter γ.

$$\mu_{pHB_t^m}(\bar{x_i}) = \frac{1}{n}\sum_{j=1}^{n}[1 - f(x_{ij} - \omega_{tj}, \gamma) - f(v_{tj} - x_{ij}, \gamma)]$$

$$f(x,y) = \begin{cases} 1 & \text{if xy} \leq 1 \\ xy & \text{if } 0 < \text{xy} < 1 \\ 0 & \text{if xy} \geq 0 \end{cases} \quad (3)$$

where $t = R_m$; R_m is the number of pure hyperboxes labeling m. Several pHB can be associated with the same cluster labeling m, thus the overall input MF, $\mu_{(B_i^{(m)})}(\bar{x_i})$, is calculated as follows:

$$\mu_{B_i^{(m)}}(\bar{x_i}) = max\{\mu_{pHB_1^{(m)}}(\bar{x_i}), ..., \mu_{pHB_{R_m}^{(m)}}(\bar{x_i})\} \quad (4)$$

The process of the ANFIS can be summarized as follows:

Choose the number of neurons of the hidden layer N_0.

Step 1. Separate the data set $\{(\overrightarrow{C^{p_j}}, y_i^j), i = 1..k\}$ (1) to build data clusters $\chi_i, i = 1..m$

Using the algorithm for parting data space, PDS [2], the given data set (1) is separated into hyperbox-typed data clusters in the input space and hyperplanes, $\chi_i, i = 1..M$, in the output data space. Where, M is optimal number of data clusters established by the clustering process.

Step 2. Build a new data set, named NN-set, for training the NN

The NN-set has k samples with input-output samples depicted by (1).

Step 3. Train the NN

The NN-set is used for train the NN based on the algorithm Le-venberg-Marquardt.

- Calculate values of MFs based on equations (3) and (4);
- The output of the neuro-fuzzy network is calculated as following equations:

$$\hat{y_i} = \frac{\sum_{i-1}^{k} \mu_{B_i^{(m)}}(\bar{x_i}) * y_{mi}\bar{x_i}}{\sum_{i-1}^{k} \mu_{B_i^{(m)}}(\bar{x_i})} \quad (5)$$

$$y_{mi} = \sum_{j=1}^{n} a_j^{(k)} x_{mj} + a_0^{(k)}, \; k = 1...M \quad (6)$$

Step 4. Check for stopping condition

Calculate error between output of the NN-set and corresponding depicting output of the NN

$$E_N = \frac{1}{p}\sqrt{\sum_{i=1}^{P}(\hat{y}_i - y_i)^2}$$

- If $E_N \leq [E]$: the structure FNS based on the NN is chosen;
- If $E_N > [E]$: N=N+1 then return to Step 3.

3 Experiments

3.1 Data Set

To evaluate our approach, we use netflix data set [19,20] that contains 14,707,483 ratings which performed by 460,936 anonymous NetFlix's customers over 17770 movies, from 1999-11-11 to 2005-12-31. The rating scale has 5 values including 5 be excellent, 4 be very good, 3 be good, 2 be fair, and 1 be poor. We randomly take 5 movies corresponding to movies' id 2464, 2548, 2848, 30. For each movie, we collect 1100 users who made a rating to it. In which, 1000 users are used for training the ANFIS, other are used for testing.

There are more than 40 genres in the whole dataset, but some has same meaning, such as "Sport" and "Sports", etc. So in total, the Netflix dataset has only 37 different genres, in following: *Action, Adult, Adventure,Animation,Anime, Biography, Children, Classics, Comedy, Crime, Documentary, Drama, Faith,Family, Fantasy, FilmNoir, Fitness, Foreign, Game Show, Gay, History,Horror, Independent, Lesbian, Music, Mystery, Romance, Sci-Fi, Short, Special Interest, Spirituality, Sport, Talk Show, Television, Thriller, War, Western.* For each user, we create a corresponding profile represented by a feature vector including 37 genres as its components.

3.2 Evaluation Methods

Accuracy Metrics. Here, we briefly review notable metrics metrics , Mean Absolute Error (MAE) and Root Mean Squared Error (RMSE), that are used to measure the quality of our proposed recommendation. Denote r_{ij} is the true rating on object j by the target user i, $\widetilde{r}_{ij}$ is the predicted rating, MAE and RMS are defined as follows:

$$MAE = \frac{1}{|E^{u_i}|} \sum_{m_{ij} \in E^{u_i}} |r_{ij} - \widetilde{r}_{ij}| \quad (7)$$

$$MAE = \left(\frac{1}{|E^{u_i}|} \sum_{m_{ij} \in E^{u_i}} (r_{ij} - \widetilde{r}_{ij})^2 \right)^{1/2} \quad (8)$$

Lower MAE and RMSE correspond to higher prediction accuracy. Since RMSE squares the error before summing it, it tends to penalize large errors more heavily. As

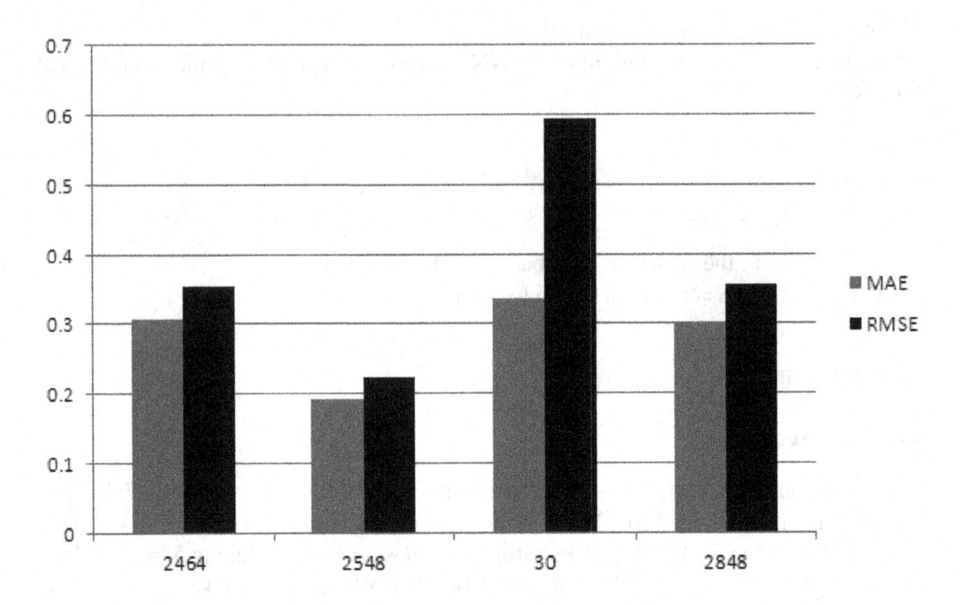

Fig. 1. MAE and RMSE of movies 30, 2464, 2548 and 2848

these metrics treat all ratings equally no matter what their positions are in the recommendation list, they are not optimal for some common tasks such as finding a small number of objects that are likely to be appreciated by a given user (Finding Good Objects). Yet, due to their simplicity, RMSE and MAE are widely used in the evaluation of recommender systems.

3.3 Evaluation Results

We used 100 testing samples corresponding to movies 30, 2464, 2548, 2848. The comparison of MAE and RMSE between the movies are shown in $Fig.$ 1.

With movie's id 30, the samples with rating-scores 1 and 2 are very small in comparison with samples with rating-scores 3, 4 and 5, thus the model have wrong prediction for tests with rating-scores 1 and 2. Therefore, the difference between MAE and RMSE of movie 30 are significant since distance between real rating and predicted rating of testing samples with rating-scores 1 and 2 are very big. With movies 2464 and 2848, the predicted value is more closed with the real value in comparison with movie 30's. The reason here is that rating-scores 1,2,3,4 and 5 of samples in the training set used for 3 movies 2464, 2848 are significant balance to train for the ANFIS. Therefore, the test result of these movies is better. The best result is the case study of the movie 2548, where the rating-scores of these samples are more balance than aforementioned samples' for movies 30, 2464 and 2848.

From these experimental results, we can conclude that the samples' rating-scores are more balance, the result are more accurate.

4 Conclusion

The idea of this proposed method is to adjust users' social web behavior to their owning ratings dual with a target video. In particular, a user profile is learned by the user's social web behavior. This user profile is presented by a vector. For each target video, we collect all users' profiles who rated on the target video. Each user's profile are considered as an input vector and his/her corresponding rating-score is as output value of the fuzzy neural network. The trained neural network is used to predict the rating of a user to the target video. We use netflix data-set to evaluate the proposed method. The result shown that the proposed method is a significant effective approach. However, the result will go down hill if the rating-scores of training set are not balance.

Acknowledgment. This research is funded by Vietnam National Foundation for Science and Technology Development (NAFOSTED) under Grant No. 102.01-2013.12

References

1. Aroyo, L., Nixon, L., Dietze, S.: Television and the future internet: the No-Tube project. In: Future Internet Symposium (FIS) 2009, Berlin, Germany, September 1-3 (2009)
2. Nguyen, S.D., Ngo, K.N.: An Adaptive Input Data Space Parting Solution to the Synthesis of Neuro-Fuzzy Models. International Journal of Control, Automation, and Systems, IJ-CAS 6(6), 928–938 (2008)
3. Nguyen, S.D., Choi, S.B.: A new Neuro-Fuzzy Training Algorithm for Identifying Dynamic Characteristics of Smart Dampers. Smart Materials and Structures 21 (2012)
4. Pazzani, M.J., Billsus, D.: Content-based recommendation systems. In: Brusilovsky, P., Kobsa, A., Nejdl, W. (eds.) Adaptive Web 2007. LNCS, vol. 4321, pp. 325–341. Springer, Heidelberg (2007)
5. Resnick, P., Iacovou, N., Suchack, M., Bergstrom, P., Riedl, J.T.: GroupLens: an open architecture for collaborative filtering of netnews. In: Proceedings of the ACM Conference on Computer Supported Cooperative Work, pp. 175–186 (1994)
6. Breese, J.S., Heckerman, D., Kadie, C.: Empirical analysis of predictive algorithms for collaborative filtering. In: Proceedings of the 14th Conference on Uncertainty in Artificial Intelligence, pp. 43–52 (1999)
7. Goldberg, K., Roeder, T., Gupta, D., Perkins, C.: Eigentaste: a constant time collaborative filtering algorithm. Information Retrieval 4(2), 133–151 (2001)
8. Smeaton, A.F., Over, P., Kraaij, W.: Evaluation campaigns and TRECVid. In: Proc. of the ACM MIR, pp. 321–330 (2006)
9. Barragáns-Martínez, A.B., Costa-Montenegro, E., Burguillo-Rial, J.C., Rey-López, M., Mikic-Fonte, F.A., Peleteiro-Ramallo, A.: A hybrid content-based and item-based collaborative filtering approach to recommend TV programs enhanced with singular value decomposition. Presented at Inf. Sci., 4290–4311 (2010)
10. Li, Q., Wang, J., Chen, Y.P., Lin, Z.: User comments for news recommendation in forum-based social media. Presented at Inf. Sci., 4929–4939 (2010)
11. Baeza-Yates, R., Ribeiro-Neto, B.: Modern information retrieval. Addison Wesley Longman Publisher (1999)
12. Brin, S., Motwani, R., Page, L., Winograd, T.: What can you do with a Web in your pocket. Bulletin of the IEEE Computer Society Technical Committee on Data Engineering, 37–47 (1998)

13. Del Corso, G.M., Gullí, A., Romani, F.: Ranking a stream of news. In: Proceedings of the 14th International Conference on World Wide Web (WWW), pp. 97–106 (2005)
14. Lavrenko, V., Schmill, M., Lawrie, D., Ogilvie, P., Jensen, D., Allan, J.: Language models for financial news recommendation. In: Proceedings of the Ninth International Conference on Information and Knowledge Management (CIKM), pp. 389–396 (2000)
15. Ardissono, L., Gena, C., Torasso, P., Bellifemine, F., Difino, A., Negro, B.: User modeling and recommendation techniques for personalized electronic program guides. In: Personalized Digital Television-Targeting Programs to Individual Viewers. Human-Computer Interaction Series, vol. 6, ch. 1, pp. 3–26. Kluwer Academic Publishers (2004)
16. Baudisch, P., Brueckner, L.: TV scout: Lowering the entry barrier to personalized TV program recommendation. In: De Bra, P., Brusilovsky, P., Conejo, R. (eds.) AH 2002. LNCS, vol. 2347, pp. 58–68. Springer, Heidelberg (2002)
17. Golub, G., Loan, C.V.: Matrix Computations, 3rd edn. Johns Hopkins Studies in Mathematical Sciences, Baltimore (1996)
18. Duong, T.H., Uddin, M.N., Li, D., Jo, G.S.: A Collaborative Ontology-Based User Profiles System. In: Nguyen, N.T., Kowalczyk, R., Chen, S.-M. (eds.) ICCCI 2009. LNCS, vol. 5796, pp. 540–552. Springer, Heidelberg (2009)
19. Netflix Prize: Forum/Dataset README file (Editor 2006), http://www.netflixprize.com/
20. Netflix, netflix dataset N. Prize, Editor 2011, lifecrunch.biz

Part IV

Knowledge Information System

A Two-Stage Consensus-Based Approach for Determining Collective Knowledge

Van Du Nguyen and Ngoc Thanh Nguyen

Department of Information Systems, Faculty of Computer Science and Management, Wroclaw University of Technology,
Wyb. St. Wyspianskiego 27, 50-370 Wroclaw, Poland
{van.du.nguyen,ngoc-thanh.nguyen}@pwr.edu.pl

Abstract. Generally, the knowledge of a collective is understood as a representative for a set of knowledge states of a collective. In this paper, we present a two-stage consensus-based approach for determining the knowledge of a large collective. For this aim, k-means algorithm is used to cluster the collective into smaller ones. The representatives of new smaller collectives are determined in the first stage of consensus choice. Next, each representative is assigned a weight value depending on the number of members in the corresponding collective, which is formed a new collective (clustered-collective). Then the second stage of consensus choice serves for determining the representative for the clustered-collective. The experimental results reveal that the weighted approach is helpful in reducing the difference between the two-stage and the single-stage consensus choice in determining the knowledge of a large collective in reference to non-weighted approach.

Keywords: consensus method, collective knowledge, two-stage consensus-based.

1 Introduction

The problem of determining a representative of knowledge states of a collective (i.e. collective knowledge) consisting of autonomous sources is a very important and difficult task because the knowledge states often contain inconsistency, incompleteness and uncertainty [21]. Collective knowledge is determined on the basic of knowledge states given by collective members and understood as the common knowledge of whole collective [20].

Although determining the knowledge of a collective is an important task, the aspect its quality is also very essential, especially, when the knowledge is determined through two or more-stage consensus choice. The quality reflects how good the collective knowledge is. It can be improved by adding or removing some knowledge states. In this paper, we assume that the quality is based on the distances from the collective knowledge to the knowledge states of members in a collective [2, 6, 18, 19, 22]. This paper aims at combining k-means algorithm and consensus choice in determining a representative of a large collective through two-stage. K-means algorithm is

H.A. Le Thi et al. (eds.), *Advanced Computational Methods for Knowledge Engineering*,
Advances in Intelligent Systems and Computing 358, DOI: 10.1007/978-3-319-17996-4_27

one of the most popular clustering methods in data mining which aims at partitioning n objects into k groups (clusters) [9]. In this approach, K-means algorithm is used to cluster a large collective into smaller ones. The representatives of new smaller collectives are determined in the first stage of consensus choice. Next, each representative is assigned a weight value depending on the number of members in the corresponding cluster (weight approach) or considered as the same importance role (no weight approach). Then the second stage of consensus choice is used to determine the representative for these representatives. For this aim, based on Euclidean space, we experiment with different numbers of collective members. We will compare the results between weighted and non-weighted approaches. This serves for proving the hypothesis whether weight values are helpful in reducing the difference between two-stage and single-stage consensus choice. This approach is missing in the literature.

The rest of the paper is set out as follows. Section 2 presents some basic concepts about consensus choice, collective of knowledge states, knowledge of a collective and the quality of collective knowledge. The proposed method is presented in Section 3. Some experimental results and their evaluation are mentioned in Section 4. Finally, Section 5 draws our conclusions.

2 Preliminaries

2.1 Collective of Knowledge States

By U we denote a set of objects representing the potential elements of knowledge referring to a concrete real world. The elements of U can represent, for example, logic expressions, tuples, etc. Symbol 2^U denotes the powerset of U that is the set of all subsets of U. By $\Pi_k(U)$ we denote the set of all k-element subsets (with repetitions) of set U for $k \in N$ (N is the set of natural numbers), and let

$$\Pi(U) = \bigcup_{k=1}^{\infty} \Pi_k(U)$$

Thus $\Pi(U)$ is the set of all non-empty finite subsets with repetitions of set U. A set $X \in \Pi(U)$ can represent the knowledge of a collective where each element $x \in X$ represents knowledge of a collective member.

In this work, based on Euclidean space we perform experiments with different numbers of collective members. A collective of knowledge states (or collective for short) is described as follows:

$$X = \{x_i = (x_{i1}, x_{i2}, \dots, x_{im}): i = 1,2,\dots,n\}$$

where $x_{ik} \in [0,1]$, $k = 1,2,\dots,m$.

Each element is a multi-dimensional vector which randomly generated.

2.2 Consensus Choice

A consensus choice has usually been understood as a general agreement in situations where parties have not agreed on some matter [5]. It is not required a full degree of agreement i.e. unanimity [10] because the unanimity is difficult to achieve in practice. The oldest consensus model was worked out by Condorcet, Arrow and Kemeny [1]. With consensus choice, there exist two well-known criteria for determining the representative of knowledge from different sources such as: the minimal sum of distances (O_1)/squared distances (O_2) from the collective knowledge to the collective members respectively. These criteria are well-known in Consensus Theory [13, 18, 19] and play an important role in determining the representative of a collective. Because of satisfying these criteria, it implies satisfying the majority of other criteria [16, 18].

2.3 Knowledge of Collective

Let the consensus of members' knowledge states be the knowledge of a collective (*or collective knowledge*). Thus, the collective knowledge is a knowledge state that determined from the basic of knowledge states of a collective. It is considered as a representative of a collective. There exist a lot of algorithms proposed for determining the representative of a collective for difference knowledge structure such as: ordered partitions [4], relational structure [17, 18], interval numbers [14, 15], hierarchical structure [11], ontology [18], and etc. In this work, we assume the knowledge of a collective is a knowledge state that satisfies:

- Criterion O_1 if :

$$d(x^*, X) = min_{y \in U} d(y, X)$$

- Criterion O_2 if:

$$d^2(x^*, X) = min_{y \in U} d^2(y, X)$$

where x is the knowledge of collective X and $d(x^*, X)$ is the sum of distances from x to members of collective X.

2.4 Quality of Collective Knowledge

The quality of collective knowledge reflects how good a collective knowledge is. There are three common methods for measuring the quality such as:

- Based on distances between the knowledge states of a collective [3, 7, 8],
- Based on distances from the collective knowledge to the knowledge states of a collective [2, 6,18, 19, 22],
- Based on distance from the collective knowledge to the real knowledge state [18, 19].

In this work, the quality of collective knowledge is based on distances from the collective knowledge to the knowledge states of collective members. In other words, it is measured by taking into account the sum of the distances from the collective knowledge to the knowledge states of a collective. We assume that collective knowledge of collective X is x^*. Then the quality is defined as follows:

$$\partial(x^*, X) = 1 - \frac{d(x^*, X)}{|X|}$$

where $|X|$ is the size of collective X.

3 The Proposed Method

In order to determine how the two-stage consensus choice influences the quality of collective knowledge. For each collective, k-means algorithm is used to cluster it into smaller collectives. Next, the first stage of consensus choice is used to determined representatives for these smaller collectives. In this stage, each member has the same importance value (weight). These representatives are formed a new collective (clustered-collective). Then, in second stage of consensus choice, representative for clustered-collective is determined. In this stage, both weighted and non-weighted approaches are used. In other words, each representative is assigned a weight value depending on the number of members in the corresponding cluster (weighted approach) or considered as the same importance role (non-weighted approach). The procedure for combining clustering methods and consensus choice is described as follows:

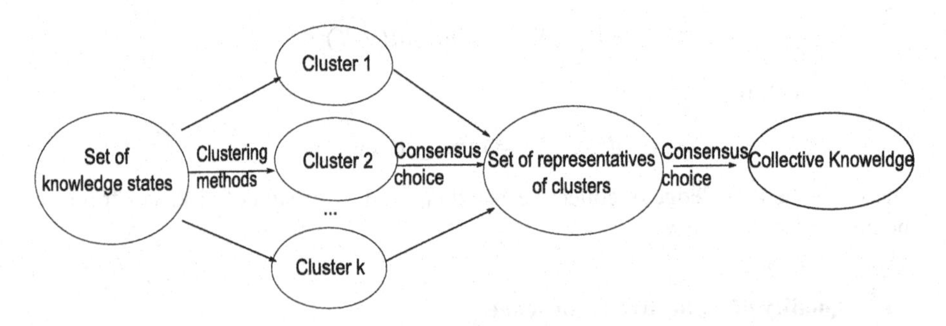

Fig. 1. The proposed method

For realizing that aim, we use the following parameters related to the number of clusters and collective members in the following table:

Table 1. Parameters of the proposed method

Number of clusters	Collective members
2, 3, .., 15	50, 75, 100, 200

According to the proposed method, the algorithm for determining the representatives of clustered-collectives is described as follows:

Algorithm 1: Determining the knowledge of a collective

Input: X – collective with n members
m – number of clusters
Output: The set of representatives of k clusters

Initial phase:
$R=\phi$ – distances to the knowledge of X
Determine the knowledge of X
$Y=\phi$ – set of representatives
BEGIN

For $k=2$ **do** m
Cluster X into smaller ones ($X'=\{X_1, X_2, \ldots, X_k\}$)
$Y_i=\phi$ - set of representatives of k clusters
For each X_i from X' **do**
Determine x_i as the knowledge of X_i
Assign weight value to x_i
$Y_i=Y_i \cup x_i$
Determine y_i as the knowledge of Y_i
$Y = Y \cup y_i$
END

Remark 1. In a collective, the knowledge satisfies criterion O_1 or O_2 has the maximal quality [18]. Thus, in two-stage approach, the better the quality of collective knowledge, the closer the distance from the collective knowledge to the collective knowledge determined from the single-stage approach.

4 Experiments and Evaluation

4.1 Experimental Results

For objective purpose, we perform experiments with different collectives and different number of collective members. In this section, however, we present the experimental results of algorithm 1 with collective of 100 members using criterion O_1 for determining the knowledge of the corresponding collective. In addition, with each cluster we perform weighted and non-weighted approaches. In both approaches, the distances from single-stage to two-stage consensus choice (weighted and non-weighted approaches) are calculated *(NonWeight and Weight columns respectively).*

Table 2. Experimental results

Clusters	NonWeight	Weight
2	0.3808	0.1772
3	0.0985	0.0860
4	0.1	0.01
5	0.06	0.1044
6	0.0412	0.0316
7	0.0922	0.0223
8	0.0539	0.0223
9	0.0860	0
10	0.02	0
11	0.0721	0.01
12	0.1104	0.01
13	0.0141	0
14	0.1063	0.01
15	0.1360	0.0141

From table 2, we can see that almost cases the result of two-stage consensus choice with weighted approach is closer to the single-stage than non-weighted approach. This is because of the difference between the number of members of each cluster. In other word, each representative of a smaller collective (cluster) is represent for a number of members of the corresponding cluster. However, the numbers of members of clusters are not identical. Meaning the importance value (weight) of each representative is different and this leads the result of the two-stage consensus choice far from the representative of the single-stage approach. For concrete, we consider the following figures:

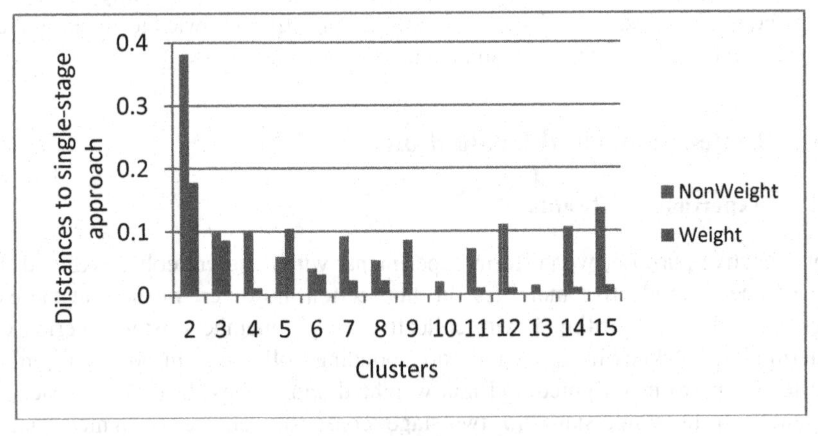

Fig. 2. Collective of 100 members

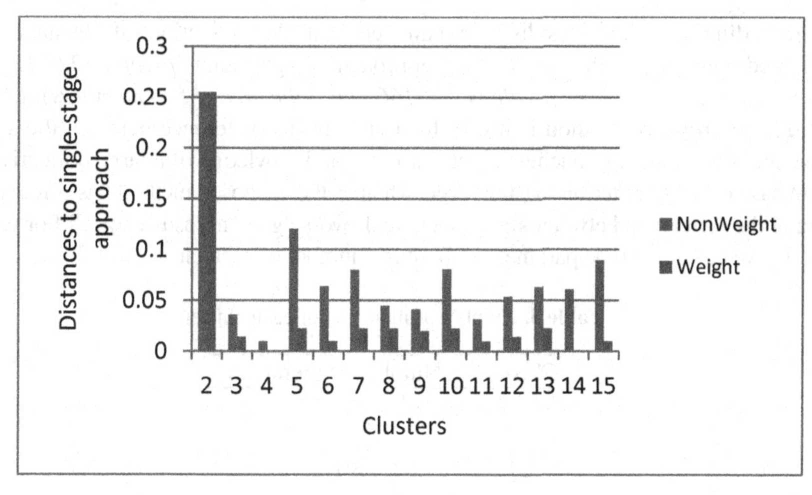

Fig. 3. Collective of 500 members

Clearly, from above figures, the weighted approach is helpful in reducing the difference between the knowledge of a collective determined from the single-stage approach to the two-stage approach. This reduction will be verified in the next section by a statistical test.

4.2 Experimental Evaluation

In this section, we present a statistical analysis to evaluate whether weighted approach have effect on reducing the difference between collective knowledge which determined through two-stage consensus choice in reference to that determined by single-stage consensus choice. The data from our experiments do not come from a normal distribution (according to the Shapiro-Wilk tests). Therefore, we choose the Wilcoxon test as the method for evaluating the influence of two-stage approach on the quality of collective knowledge.

```
> wilcox.test(NonWeight, Weight, paired=TRUE)

        Wilcoxon signed rank test with continuity correction
data:  NonWeight and Weight
V = 99, p-value = 0.001709
alternative hypothesis: true location shift is not equal to 0

> wilcox.test(NonWeight, Weight, paired=TRUE)

        Wilcoxon signed rank test with continuity correction
data:  NonWeight and Weight
V = 91, p-value = 0.001662
alternative hypothesis: true location shift is not equal to 0
```

According to above results, we can see that the reduction of distance from weighted approach to the single-stage approach is significant (*p-value=0.001709 for collective of 100 members, p-value=0.001662 for collective of 500 members*). In other words, the proposed method is useful for reducing the difference between the single-stage and two-stage approaches in determining the knowledge of a large collective.

In case of using criterion O_2, however, whether the proposed method has effect on reducing the difference between single-stage and two-stage consensus choice? For easy to follow, we consider some parameters about the number of each cluster as follows:

Table 3. Number of members of each cluster

Cluster	Number of members
c_1	l_1
c_2	l_2
...	...
c_k	l_k

In single-stage, the collective knowledge (satisfies criterion O_2) of a collective is:

$$r_1 = \frac{1}{n}\left(\sum_{i=1}^{n} x_{i1}, \sum_{i=1}^{n} x_{i2}, \dots, \sum_{i=1}^{n} x_{im}\right)$$

In two-stage, the collective knowledge with non-weighted approach is:

$$r_2 = \frac{1}{k}\left(\frac{1}{l_1}\sum_{i=1}^{l_1} x_{i1} + \frac{1}{l_2}\sum_{i=1}^{l_2} x_{i1} + \dots + \frac{1}{l_k}\sum_{i=1}^{l_k} x_{i1}, \frac{1}{l_1}\sum_{i=1}^{l_1} x_{i2} + \frac{1}{l_2}\sum_{i=1}^{l_2} x_{i2} + \dots + \frac{1}{l_k}\sum_{i=1}^{l_k} x_{i2}, \dots, \frac{1}{l_1}\sum_{i=1}^{l_1} x_{im} + \frac{1}{l_2}\sum_{i=1}^{l_2} x_{im} + \dots + \frac{1}{l_k}\sum_{i=1}^{l_k} x_{im}\right)$$

Similarly, the collective knowledge of weighted approach is:

$$r_3 = \left(\frac{1}{l_1}\alpha_1\sum_{i=1}^{l_1} x_{i1} + \frac{1}{l_2}\alpha_2\sum_{i=1}^{l_2} x_{i1} + \dots + \frac{1}{l_k}\alpha_k\sum_{i=1}^{l_k} x_{i1}, \frac{1}{l_1}\alpha_1\sum_{i=1}^{l_1} x_{i2} + \frac{1}{l_2}\alpha_2\sum_{i=1}^{l_2} x_{i2} + \dots + \frac{1}{l_k}\alpha_k\sum_{i=1}^{l_k} x_{i2}, \frac{1}{l_1}\alpha_1\sum_{i=1}^{l_1} x_{im} + \frac{1}{l_2}\alpha_2\sum_{i=1}^{l_2} x_{im} + \dots + \frac{1}{l_k}\alpha_k\sum_{i=1}^{l_k} x_{im}\right)$$

where $\alpha_1, \alpha_2, \dots, \alpha_k$ are weight values for each representative of the corresponding cluster and n is the total number of members of the large collective.

Then, we have the following theorems related to using criterion O_2 for determining the knowledge of a collective.

Theorem 1. If $l_1 = l_2 = \cdots = l_k$, then $r_1 = r_2$.

Proof for theorem 1 is not included because of page limitation.

According to theorem 1, when clusters have identical number of members, then there is no difference between the knowledge determined from single-stage and two-stage approaches. In case the numbers of members of clusters are different, we consider the following theorem:

Theorem 2. If $\alpha_1 = \frac{l_1}{n}, \alpha_2 = \frac{l_2}{n}, \dots, \alpha_k = \frac{l_k}{n}$, then $r_1 = r_3$

Proof for theorem 1 is not included because of page limitation.

According to theorem 2, if the ratio between the number of members of each cluster and the number of collective members is considered as a weight value for the representative of the corresponding cluster, there is also no difference between the collective knowledge which determined from single-stage and two-stage approaches.

5 Conclusions

In this paper we have proposed an approach to reduce the difference between the two-stage and the single-stage consensus choice. In the proposed method, the ratio between the number of members of each cluster and the total members of a collective is considered as the weight value for the representative of the corresponding cluster. Additionally, two theorems about the relationship between the number of members of each cluster and the difference between two approaches also have been investigated.

The future works should investigate the problem with other structures such as: binary vector, relational structure, etc. In addition, we also aim to propose a mathematical model to prove the influence of the number of collective members on the quality of its knowledge. To the best of our knowledge, paraconsistent logics could be useful [12].

Acknowledgement. This research is funded by Vietnam National University Ho Chi Minh City (VNU-HCM) under grant number C2014-26-05.

References

1. Arrow, K.J.: Social choice and individual values. Wiley, New York (1963)
2. Ben-Arieh, D., Zhifeng, C.: Linguistic-labels aggregation and consensus measure for autocratic decision making using group recommendations. IEEE Transactions on Systems, Man and Cybernetics, Part A: Systems and Humans 36(3), 558–568 (2006)
3. Bordogna, G., Fedrizzi, M., Pasi, G.: A linguistic modeling of consensus in group decision making based on OWA operators. IEEE Transactions on Systems, Man and Cybernetics, Part A: Systems and Humans 27(1), 126–133 (1997)

4. Danilowicz, C., Nguyen, N.T.: Consensus-based partitions in the space of ordered partitions. Pattern Recognition 21, 269–273 (1988)
5. Day, W.H.E.: The Consensus Methods as Tools for Data Analysis. In: Block, H.H. (ed.) IFC 1987: Classifications and Related Methods of Data Analysis, pp. 317–324. Springer, Heidelberg (1988)
6. Duong, T.H., Nguyen, N.T., Jo, G.-S.: A Method for Integration of WordNet-Based Ontologies Using Distance Measures. In: Lovrek, I., Howlett, R.J., Jain, L.C. (eds.) KES 2008, Part I. LNCS (LNAI), vol. 5177, pp. 210–219. Springer, Heidelberg (2008)
7. Herrera, F., Herrera-Viedma, E., Verdegay, J.L.: Linguistic measures based on fuzzy coincidence for reaching consensus in group decision making. International Journal of Approximate Reasoning 16(3-4), 309–334 (1997)
8. Kacprzyk, J., Fedrizzi, M.: A 'soft' measure of consensus in the setting of partial (fuzzy) preferences. European Journal of Operational Research 34(3), 316–325 (1988)
9. Kanungo, T., Mount, D.M., Netanyahu, N.S., Piatko, C.D., Silverman, R., Wu, A.Y.: An Efficient k-Means Clustering Algorithm: Analysis and Implementation. IEEE Trans. Pattern Anal. Mach. Intell. 24(7), 881–892 (2002)
10. Kline, J.A.: Orientation and group consensus. Central States Speech Journal 23(1), 44–47 (1972)
11. Maleszka, M., Mianowska, B., Nguyen, N.T.: A method for collaborative recommendation using knowledge integration tools and hierarchical structure of user profiles. Knowledge-Based Systems 47, 1–13 (2013)
12. Nakamatsu, K., Abe, J.: The paraconsistent process order control method. Vietnam J. Comput. Sci. 1(1), 29–37 (2014)
13. Nguyen, N.T.: Using consensus for solving conflict situations in fault-tolerant distributed systems. In: Proceedings of the First IEEE/ACM International Symposium on Cluster Computing and the Grid, pp. 379–384 (2001)
14. Nguyen, N.T.: Consensus-based Timestamps in Distributed Temporal Databases. The Computer Journal 44(5), 398–409 (2001)
15. Nguyen, N.T.: Representation choice methods as the tool for solving uncertainty in distributed temporal database systems with indeterminate valid time. In: Monostori, L., Váncza, J., Ali, M. (eds.) IEA/AIE 2001. LNCS (LNAI), vol. 2070, pp. 445–454. Springer, Heidelberg (2001)
16. Nguyen, N.T.: Using consensus for solving conflict situations in fault-tolerant distributed systems. In: Proceedings of the First IEEE/ACM International Symposium on Cluster Computing and the Grid, pp. 379–384 (2001)
17. Nguyen, N.T.: Consensus system for solving conflicts in distributed systems. Journal of Informatioin Sciences 147(1), 91–122 (2002)
18. Nguyen, N.T.: Advanced methods for inconsistent knowledge management. Springer, London (2008)
19. Nguyen, N.T.: Inconsistency of knowledge and collective intelligence. Cybernetics and Systems 39(6), 542–562 (2008)
20. Nguyen, N.T.: Processing inconsistency of knowledge in determining knowledge of collective. Cybernetics and Systems 40(8), 670–688 (2009)
21. Nguyen, V.D., Nguyen, N.T.: A method for temporal knowledge integration using indeterminate valid time. Journal of Intelligent and Fuzzy Systems 27(2), 667–677 (2014)
22. Spillman, B., Bezdek, J., Spillman, R.: Development of an instrument for the dynamic measurement of consensus. Communication Monographs 46(1), 1–12 (1979)

Context in Ontology for Knowledge Representation

Asmaa Chebba[1], Thouraya Bouabana-Tebibel[1], and Stuart H. Rubin[2]

[1] École Nationale Supérieure d'Informatique, LCSI Laboratory, Algiers, Algeria
{a_chebba,t_tebibel}@esi.dz
[2] SSC-PAC 71730, San Diego, CA 92152-5001
stuart.rubin@navy.mil

Abstract. The ontology represents a rich model for knowledge representation that is gaining popularity over years. Its expressivity and computability promote its use in a large scale of applications. In this paper, we first present a meta-model that describes the ontology basic elements and the relationships among them. A new element is introduced to the ontology model called "Context". This new conceptual element improves the representation quality and allows for the modeling of more complex knowledge efficiently. The context is next integrated to the proposed meta-model.

Keywords: ontology, meta-model, context, knowledge representation.

1 Introduction

The ontology model has known an increasing interest in many application areas such as intelligent information integration, semantic web and knowledge-based systems, where it has proven to be useful [1]. Nowadays, ontology is becoming the focus of the knowledge representation and management research domains. It is expressive and computable in the same time. It organizes information on both schema level and instance level. Optionally combined with other pieces of knowledge, ontology is used to take into account the reasoning process. According to the guide [2], ontology finds utility in the following purposes: (1) to share common understanding of the structure of information among people or software agents; (2) to enable reuse of the domain knowledge; (3) to make the domain assumptions explicit; (4) to separate the domain knowledge from the operational knowledge; (5) to analyze the domain knowledge.

Many definitions were given to the ontology and are most commonly related to its use domain. However, ontology is a term in philosophy and its meaning is "theory of existence". Formally, in artificial intelligence, it is defined as a formal and explicit specification of a shared conceptualization [3]. The "formal" term refers to the "machine-readable" property characterizing the ontology. As for the "explicit" term, it refers to its well defined constructs. Furthermore, "shared" is due to the semantic consensus provided by the ontology about knowledge of a particular domain of interest. The sharing property allows for the processes of reuse, integration and combination of ontology knowledge. An ontology is also defined as the common terms and concepts (meaning) used to describe and represent an area of knowledge. We mention

H.A. Le Thi et al. (eds.), *Advanced Computational Methods for Knowledge Engineering*,
Advances in Intelligent Systems and Computing 358, DOI: 10.1007/978-3-319-17996-4_28

the following definition which was adopted from the RFP (Request for Proposal) [13]: “An ontology can range in expressivity from a Taxonomy (knowledge with minimal hierarchy or a parent/child structure), to a Thesaurus (words and synonyms), to a Conceptual Model (with more complex knowledge), to a Logical Theory (with very rich, complex, consistent, and meaningful knowledge).”

Several description languages are used for expressing ontologies. OWL [6] is the current standard that came with three flavors: owl lite, owl DL and owl full. This language has been standardized by the W3C consortium and is an extension of a restricted use of RDF [7] and RDFS.

Despite the well recognized expressivity power of ontologies, we found that they do not include the context construct which is a very significant notion. The context defined here refers to a representative element that allows for the description of the external environment of a concept. The internal description of a concept derived from its attributes with the external knowledge form an integrated and complete image of the concept. In this work, we, first, propose a meta-model for the ontology. We, next, define the ontological notion of context and integrate it into the ontology meta-model, thus enhancing the latter expressivity.

The following sections explain the proposed context element by highlighting first, some of works about the ontology meta-modeling and previous attempts to model the context in section 2. Next, section 3 explains the adopted meta-model for the ontological model and section 4 presents the context element, explains its integration in the previous meta-model and underlines its different cases of use.

2 Related Works

Research on ontology construction has gained much attention. The ontology has stepped many usage evolutions starting from controlling vocabulary and organizing taxonomies. It is a raising research area in many fields such as data mining and pattern classification. To make use of this model efficiently by automated processing, the implementation of ontology is proposed through diverse ontology description languages. In this perspective, a specification [4] of OMG describes the Ontology Definition Metamodel (ODM). ODM is a family of MOF (Meta Object Facility [12]) meta-models that reflect the abstract syntax of several standard knowledge representations and conceptual modeling languages that have either been recently adopted by other international standard bodies (e.g., RDF and OWL by the W3C), are in the process of being adopted (e.g., Common Logic and Topic Maps by the ISO), or are considered industry de facto standards (non-normative ER and DL appendices) [4].

The meta-models are exposed as packages, each of which describing a part of the considered language, either for the classes part or for the properties part. In our work, another treated aspect is the context. This notion has been significantly studied in literature [5] and was associated with diverse definitions. There is a major difference between the context considered in our case and that described in these definitions. Indeed, the latter refer to the context through an external environment of a specific framework or domain in a general manner. As far as we are concerned, we define the context as an external environment that is related to each concept of the domain. An example is the interoperability of exchanges treated in [8] where the authors use ontology in order to share information about the services provided by communicating organizations. In addition, the authors in [9] present an extensible CONtext ONtology (CONON) for context modeling in pervasive computing environments. The authors state that location, user, activity and computational entity are the most fundamental context for capturing the information about the executing situation. They propose a context model divided into upper ontology and specific ontology. The upper ontology is a high-level ontology which captures general features of basic contextual entities. The specific ontology is a collection of ontology sets which define the details of general concepts and their features in each sub-domain. Authors in [10] argue that no attempt was made to design and develop generic context models. In addition, approaches not tied to specific application domains usually represent a limited amount of context information types. Therefore, they intend to propose uniform context models, representation and query languages, as well as reasoning algorithms that facilitate context sharing and interoperability of applications. As a final example, we cite the work presented in [11] where Context OWL (C-OWL) is proposed. C-OWL is a language whose syntax and semantics have been obtained by extending the OWL syntax and semantics to allow for the representation of contextual ontologies. The work is motivated by the fact that OWL cannot model certain situations like the directionality of information flow, interpretation of ontologies in local domains and context mappings. In our work, we consider the context notion at its finer level of granularity, on concepts, unlike the listed works.

3 Ontology Metamodel

Ontology is still subject to various studies in respect to what is to be considered in ontology and how and where it should be used. In our study, we consider the ontology for knowledge representation. It is a formal description consisting of a set of concepts, relations, and axioms, which correspond to a domain of interest. Concepts are also called (and implemented as) classes, which have properties describing diverse features and characteristics of the concept. Constraints or restrictions are another aspect supported by the ontology. They allow for the expression of multiple situations like cardinalities on relations and limitation of the range values that can take a property. With a set of individual instances of classes, the ontology general knowledge constitutes the knowledge base that describes the domain of interest.

To highlight the fundamental modeling concepts and rules for the construction of the ontology, we introduce here a meta-model for the ontology model. The aim of a meta-model is to describe the basic model elements and the relationships between these elements as well as their semantics. In addition, it specifies the rules to be applied on the elements of the corresponding language. Figure-1 shows the relationships existing between the meta-models adopted by the OMG standard (series of Ontology

Definition Metamodel (ODM)). These meta-models are developed following the MOF that provides "a meta-model management framework and a set of meta-model services to enable the development and interoperability of model and metadata driven systems" [4]. As shown in figure-1, we state that the meta-model of OWL inherits from that of RDF like it is the case for their respective languages.

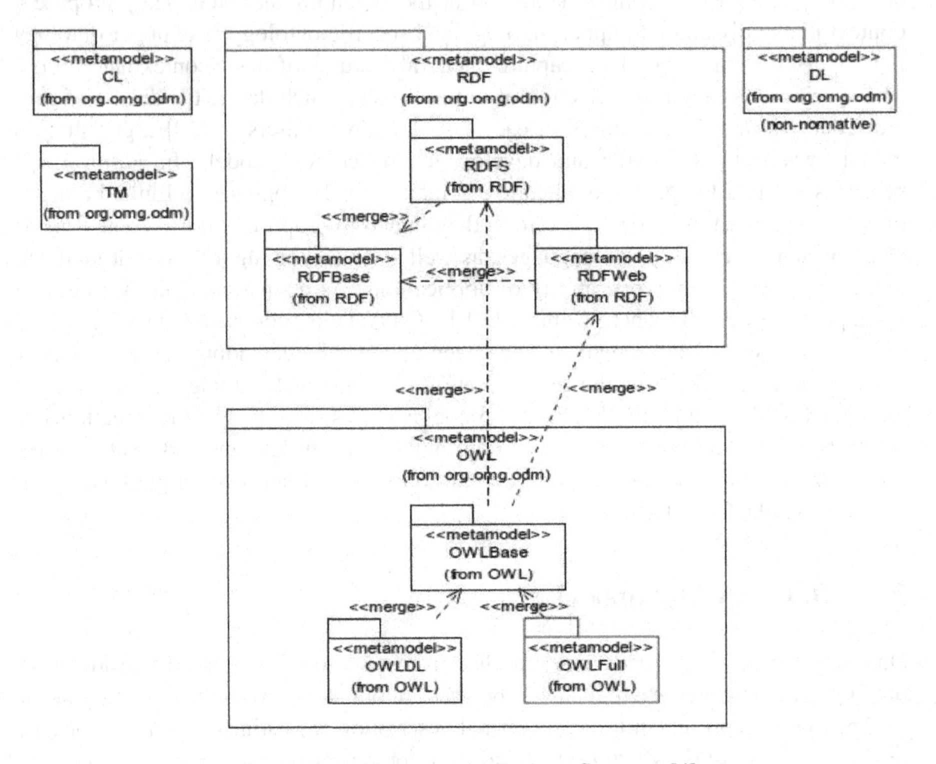

Fig. 1. ODM Meta-models: Package Structure [4]

Based on the adopted ODM and following the ontology model definition, we derived the ontology meta-model described in figure-2. The principal goal from this proposition is illustrated in next. It concerns the integration of a new concept, namely the context, in the ontology. Integration in the meta-model allows us to limit the scope of this new element, fixe its role in the model and clarify its adequate use.

The meta-model proposed in this section describes the main elements of the ontology and the relationships among them. We summarize the components the follows.

- A universe for the ontology that contains all ontology elements and represents then the knowledge about the considered domain of interest.
- Classes that implement the concepts of the domain. They are described through their properties and can be created by applying several restrictions or composition processes (union, complement, etc.).

- Data-types that are the list of possible types that can take an attribute as value. In addition, with the "collection of data" element, a data-type can be expressed as a range of data. The type of the range of data is specified by one of Data-type elements.
- Literals, Lists, …, are elements of the ontology allowing the expression of complex knowledge and are applied combined together with the other elements.
- Instances represent the low level of information since they are specific knowledge of the domain. The set of instances in ontology is equivalent to the ABox in logic.
- Constraints can be applied on different elements of the ontology. We can express constraint on relations, on values of attributes, on classes and on combination of a set of diverse components which enable the expression of complex knowledge.

We note that the created meta-model is designed with EMF eCore (Eclipse Metamodel Framework) which takes into account the specification MOF of OMG standard.

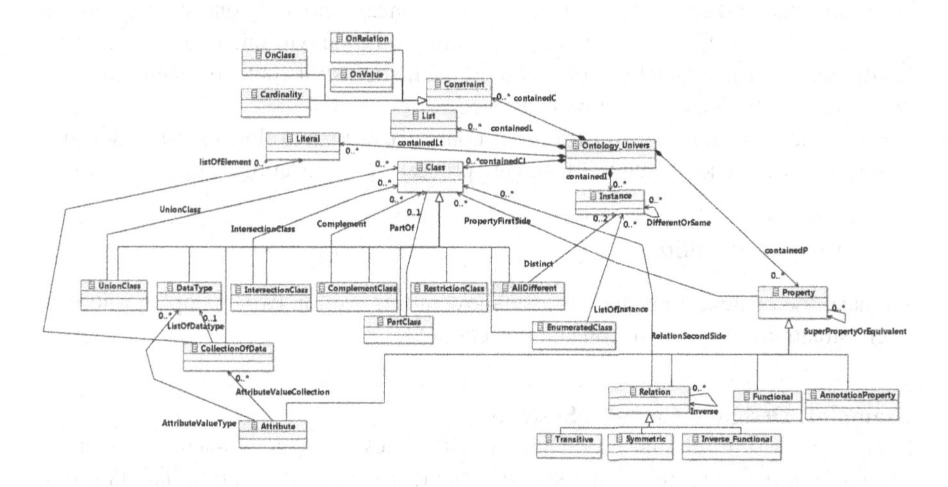

Fig. 2. Ontology Meta-model

4 Context

The "context", as referred to through its name, is a representation of the situations where a concept can be found. It describes the external environment of the concept. A concept can be used to express different "things" and has then different characteristics based on its current situation.

In fact, a concept is defined by its attributes; but, this representation remains incomplete since it describes only the internal aspect of the concept independently of its environment. Hence, to get a complete representation of a concept, we must define its internal knowledge as well as external knowledge. The latter is composed of the different contexts of the concept. A context defines the interaction of the concept with the other concepts as well as its role or its semantics in the corresponding situation. Accordingly, representation of both internal and external contexts provides a global and precise view of the concept.

At the conceptual level of the ontology definition, there doesn't exist an element which makes reference to the "context" as it is defined above. In fact, the ontology allows representing, formally, the concepts and the relationships between them and is limited consequently to the static knowledge only, missing the dynamicity of all existing situations.

For the listed reasons, in addition to concepts and relations already existing in the ontology, we propose a new element called "context" to be integrated in this rich model. The context is attached to each concept in order to describe the possible situations in which it could be present. The context is in form of list of situations, each of which modeled by a combination of concepts that are in direct interaction with the corresponding represented concept. Significance of the context notion appears clearly in systems where there is no general use of the concept (not only one static definition) which may have different definitions depending on its environment and case of use. In addition, we can attach to a concept a probabilistic list of contexts where the occurrence probability of each situation is defined.

In the next section, we illustrate the context use in two domains: the design of complex systems and the natural language processing (translation, semantics, etc).

4.1 Usage and Utility

To make the context usage clearer, we show in this section two examples where different situations of the same concept present.

Example1: Design of Complex Systems

Figure-3 shows an Internal Bloc Diagram (IBD) describing a two-stage refrigeration system. In order to represent the knowledge included in this diagram into an ontology, we transform each component into a concept of the created ontology. One of these concepts is “compressor” which is used in this example two times referring to two different situations. We note that for each situation, connections to this component are relatively different and the role of the latter changes depending on these connections. To represent the knowledge contained in IBD using ontology, we need to manage the dynamic knowledge in form of context. Indeed, the context lists all situations (connections, role, etc. for each situation) where a concept is used. Hence, the definition of a concept is henceforth specific to each situation. For example, with the two contexts of “compressor”, this latter has two definitions that are represented in the ontology.

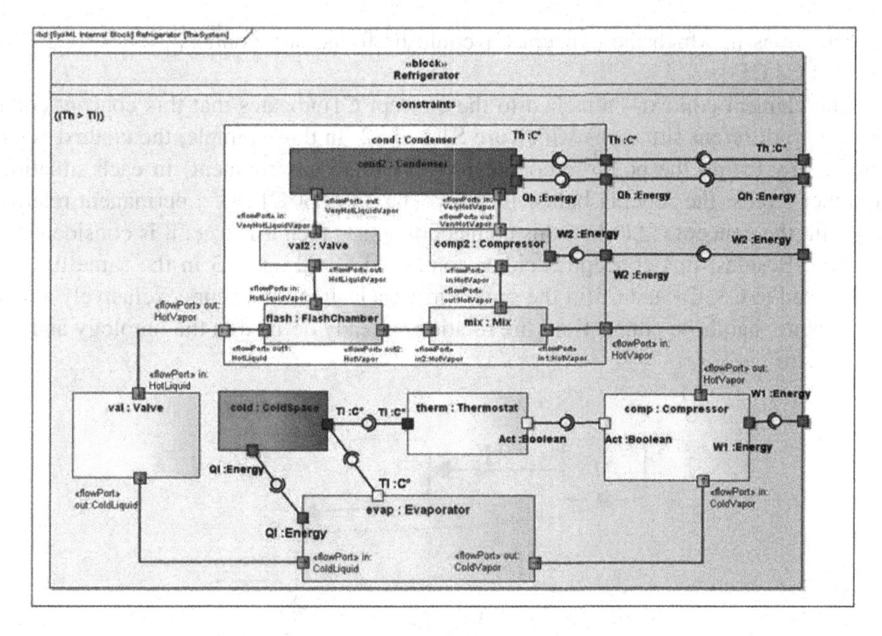

Fig. 3. IBD of a two-stage refrigeration system

Example2: Treatment of Natural Languages

We explain in this example the major role for context in natural languages treatment. Indeed, adding context to a concept makes the semantics of this latter more precise and the reasoning about it more efficient. For instance, the synonyms of a word (considered in an ontology as a concept) are defined independently of the context of use of this word. However, using context element, we can specify a range of synonyms according to many criteria which make the precision on the relative semantics better. Applying the context notion on natural languages allows improving the translation process since the semantics is more precise and is specific to a context.

The context element allows hence improving the representation expressivity by given the possibility of representing different knowledge about a concept in different situation. This permits getting a complete image of this concept. In addition, with context, the semantics gains in precision which improves the concepts reuse.

4.2 Integration to the Ontology Model

In order to integrate the element "context" in the ontological model, we must, first, integrate it in the meta-model of the ontology to guarantee consistency of the model with the introduced changes and also to limit the scope of the added element by specifying its usage mode and its domains of application.

Figure-4 represents a schematization of the following example. Let C1, C2, C3, C4 and C5 be classes. Let also the set of relations (C1 R1 C2), (C4 R2 C5), (C4 R3 C3) be the existing relationships among these concepts.

Situations in which the concept C1 could be found are: $context_{C1}$ = {S1:(C2 C3), S2:(C3 C4 C5)}

The element $context_{C1}$ attached to the concept C1indicates that this concept could be in two different situations which are S1 and S2. In this example, the context is expressed by listing the possible connections (external environment) in each situation. In other words, the schema below means: "The concept C1 has a permanent relation R1 with the concept C2 (when this connection exists then it is true; it is considered as axiom). Besides, this concept is either connected to C2 and C3 in the same time or connected to C3, C4 and C5 in the same time; each situation occurs exclusively at one time. Note that these connections are relations already defined in the ontology and not listed here".

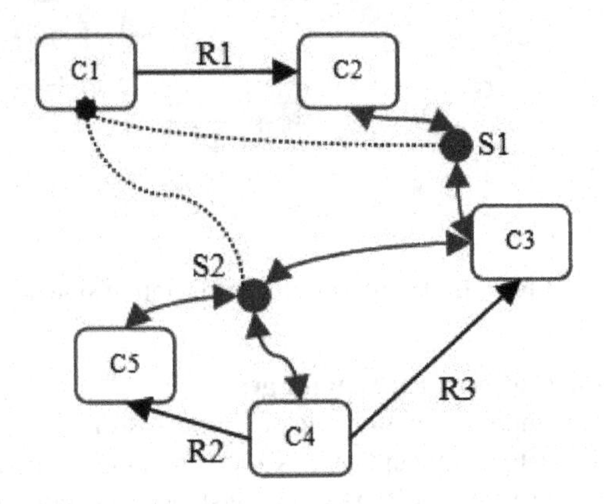

Fig. 4. The interactions of context with the other ontology basic elements

Integration of the context element in the ontology meta-model is shown in figure-5. This element is modeled as a property (relation) that has its second side connection related to a list (the element ListOfCombinations). This list can contain a number of situations that are modeled as different combinations of classes. As shown in the meta-model, the combinations that contain multiple classes describe the correspondent situation with associated information. Note that the semantics of list requires that only one of its elements can be true at one time (exclusive or).

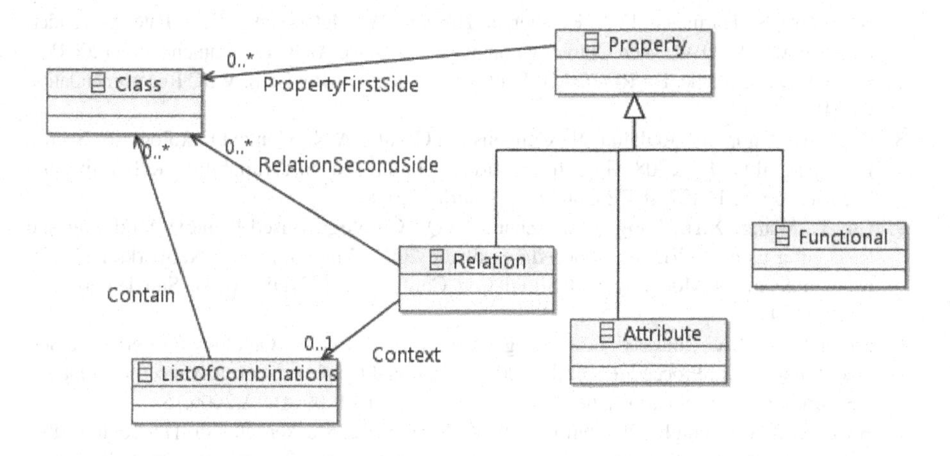

Fig. 5. Integration of context in the ontology meta-model

5 Conclusion

We reviewed in this study the definition of the ontology model and listed its diverse elements allowing for the representation of knowledge with expressiveness and computability. We, therefore, proposed a meta-model for ontology in order to highlight the bounds of this model expressiveness and note its drawbacks. This led us to propose enhancing the meta-model with a new conceptual element, namely, the context. Attached to concepts, the context improves the representation efficiency to include dynamic knowledge and makes the expressed semantics more accurate and complete, which allows for a better reuse of concepts depending on their contexts. We integrated this new element to the proposed meta-model taking care to preserve consistency and integrity of the ontological model. For future work, we propose to extend OWL language to include the context concept.

References

1. Staab, S., Maedche, A.: Axioms are Objects, too – Ontology Engineering beyond the Modeling of Concepts and Relations. In: 14th European Conference on Artificial Intelligence; Workshop on Applications of Ontologies and Problem-Solving Method, Berlin, Germany (2000)
2. Noy, N.F., McGuinness, D.L.: Ontology Development 101: A Guide to Creating Your First Ontology. Stanford University, Stanford (2001)
3. Gruber, T.R.: A Translation Approach to Portable Ontology Specifications. Knowledge Acquisition 5, 199–220 (1993)
4. OMG, "Ontology definition metamodel", OMG adopted specification, Version 1.0, OMG Document Number: formal/2009-05-01 (2009)
5. Zainol, Z., Nakat, K.: Generic context ontology modelling: A review and framework. In: IEEE 2nd International Conference on Computer Technology and Development (ICCTD 2010), Piscataway, NJ, pp. 126–130 (2010)

6. Bechhofer, S., Harmelen, F.V., Hendler, J., Horrocks, I., McGuinness, D.L., Patel-Schneider, P.F., Stein, L.A.: OWL Web Ontology Language Reference. W3C Recommendation (2004)
7. Beckett, D., McBride, B.: RDF/XML Syntax Specification (Revised). W3C Recommendation (2004)
8. Coma, C., Cuppens-Boulahia, N., Cuppens, F., Cavalli, A.N.: Context Ontology for Secure Interoperability. In: 2008 Third International Conference on Availability, Reliability and Security, pp. 821–827. IEEE Computer Security (2008)
9. Gu, T., Wang, X.H., Pung, H.K., Zhang, D.Q.: Ontology Based Context Modeling and Reasoning using OWL. In: Proceedings of the 2004 Communication Networks and Distributed Systems Modeling and Simulation Conference (CNDS 2004), San Diego, CA, USA (2004)
10. Strimpakou, M.A., Roussaki, L.G., Anagnostou, M.E.: A Context Ontology for Pervasive Service Provision. In: Proceedings of the 20th International Conference on Advanced Information Networking and Applications, pp. 775–779 (2006), doi:10.1109/AINA.2006.15
11. Bouquet, P., Giunchiglia, F., Harmelen, F.V., Serafini, L., Stuckenschmidt, H.: Contextualizing ontologies. Web Semantics: Science, Services and Agents on the World Wide Web 1(4), 325–343 (2004), doi:10.1016/j.websem.2004.07.001
12. OMG, "Meta Object Facility (MOF) Core Specification Version 2.0", `http://www.omg.org/spec/MOF/`
13. OMG, "Ontology Definition Metamodel Request for Proposal", OMG Document ad/2003-03-40 in [4] (2003)

Designing a Tableau Reasoner for Description Logics

Linh Anh Nguyen

Division of Knowledge and System Engineering for ICT,
Ton Duc Thang University, No. 19, Nguyen Huu Tho Street,
Tan Phong Ward, District 7, Ho Chi Minh City, Vietnam
nguyenanhlinh@tdt.edu.vn
Institute of Informatics, University of Warsaw
Banacha 2, 02-097 Warsaw, Poland
nguyen@mimuw.edu.pl

Abstract. We present the design of a tableau reasoner, called TGC 2, that uses global caching for reasoning in description logics. We briefly describe the ExpTime tableau methods used by TGC 2. We then provide the design principles and some important optimization techniques for increasing the efficiency of the reasoner.

1 Introduction

An ontology is a formal description of a common vocabulary for a domain of interest and the relationships between terms built from the vocabulary. Ontologies have been applied in a wide range of practical domains such as medical informatics, bio-informatics and the Semantic Web. They play a key role in data modeling, information integration, and the creation of semantic Web services, intelligent Web sites and intelligent software agents [2,1,33]. Ontology languages based on Description Logics (DLs) like OWL 2 are becoming increasingly popular due to the availability of DL reasoners, which provide automated support for visualization, debugging, and querying of ontologies.

Ontology classification and consistency checking of ontologies are basic reasoning services of ontology reasoners [3,2,1]. The goal of classification is to compute a hierarchical relation between concepts, which is used, e.g., for browsing ontologies, for user interface navigation, and for suggesting more (respectively, less) restricted queries. Consistency checking is a very important task for ontology engineering. It is essentially the problem of checking satisfiability of the knowledge base representing the ontology in DL. Most of other reasoning problems for ontologies, including instance checking and ontology classification, are reducible to this one.

The most efficient decision procedures for the mentioned problems are usually based on tableaux [20,3], as many optimization techniques have been developed for them. However, the satisfiability problem in DLs has high complexity (ExpTime-complete in $\mathcal{ALC}$ and N2ExpTime-complete in $\mathcal{SROIQ}$) and existing ontology reasoners (including tableau-based ones) are not yet satisfactory, especially when dealing with large ontologies.

H.A. Le Thi et al. (eds.), *Advanced Computational Methods for Knowledge Engineering*,
Advances in Intelligent Systems and Computing 358, DOI: 10.1007/978-3-319-17996-4_29

OWL is a standardized Web ontology language, with the first and second versions recommended by W3C in 2004 and 2009, respectively. These versions, OWL 1 and OWL 2, have some sublanguages (species/profiles):

- OWL 1 DL and OWL 2 DL support those users who want maximum expressiveness without losing computational completeness.
- OWL 1 Lite, OWL 2 EL, OWL 2 QL and OWL 2 RL are restricted versions with PTime data complexity of the "DL" sublanguages.
- OWL 1 Full and OWL 2 Full support those users who want maximum expressiveness and the syntactic freedom of RDF with no computational guarantees.

OWL 1 DL is based on the DL $\mathcal{SHOIN}$ [18], while OWL 2 DL is based on a more expressive DL $\mathcal{SROIQ}$ [15]. Other well-known DLs are $\mathcal{ALC}$, $\mathcal{SH}$, $\mathcal{SHI}$, $\mathcal{SHIQ}$, $\mathcal{SHIO}$, $\mathcal{SHOQ}$ and $\mathcal{SHOIQ}$. DLs are fragments of classical first-order logic and are variants of modal logics, used to describe the domain of interest by means of individuals, concepts and roles. A concept is interpreted as a set of individuals, while a role is interpreted as a binary relation between individuals. In the basic DL $\mathcal{ALC}$, concepts are constructed from atomic concepts using complement ($\neg C$), union ($C \sqcup D$), intersection ($C \sqcap D$), value restriction ($\forall R.C$, where R is a role and C is a concept), and existential restriction ($\exists R.C$). More advanced DLs are extensions of $\mathcal{ALC}$. In particular,

- $\mathcal{SH}$ extends $\mathcal{ALC}$ with transitive roles and role hierarchies
- $\mathcal{SHI}$, $\mathcal{SHIQ}$, $\mathcal{SHIO}$, $\mathcal{SHOQ}$ and $\mathcal{SHOIQ}$ extend $\mathcal{SH}$ with the features corresponding to the letters occurring in the name: $\mathcal{I}$ indicates inverse roles, $\mathcal{Q}$ indicates qualified number restrictions ($\geq n\,R.C$ and $\leq n\,R.C$), and $\mathcal{O}$ indicates nominals (if o is a nominal then $\{o\}$ is a concept)
- $\mathcal{SHOIN}$ (of OWL 1 DL) extends $\mathcal{SHIO}$ with unqualified number restrictions ($\geq n\,R$ and $\leq n\,R$)
- $\mathcal{SROIQ}$ (of OWL 2 DL) extends $\mathcal{SHOIQ}$ with role axioms of the form $R_1 \circ \ldots \circ R_k \sqsubseteq S$, irreflexive roles, disjointness of roles, the universal role, negative individual assertions, the concept constructor $\exists R.\mathit{Self}$, together with some restrictions to guarantee decidability.

The satisfiability checking problem is ExpTime-complete in $\mathcal{ALC}$, $\mathcal{SH}$, $\mathcal{SHI}$, $\mathcal{SHIQ}$, $\mathcal{SHIO}$ and $\mathcal{SHOQ}$; NExpTime-complete in $\mathcal{SHOIN}$ and $\mathcal{SHOIQ}$; and N2ExpTime-complete in $\mathcal{SROIQ}$.

According to the recent survey [20], the third generation ontology reasoners that support $\mathcal{SHIQ}$ or $\mathcal{SROIQ}$ are FaCT, FaCT++, RACER, Pellet, KAON2 and HermiT. The reasoners FaCT, FaCT++, RACER and Pellet are based on tableaux, KAON2 is based on a translation into disjunctive Datalog, and HermiT is based on hypertableaux. That is, all of the listed reasoners except KAON2 are tableau-based. According to [21], KAON2 provides good performance for ontologies with rather simple TBoxes, but large ABoxes[1]; however, for ontologies

[1] A TBox is a set of terminology axioms, including role axioms. In the literature of DLs, sometimes role axioms are grouped into an RBox. An ABox is a set of individual assertions.

with large and complex TBoxes, existing tableau-based reasoners still provide superior performance. The current version of KAON2 does not support nominals and cannot handle large numbers in cardinality statements.

The reasoners FaCT, FaCT++, RACER and Pellet use traditional tableau decision procedures like the ones for $\mathcal{SHI}$ [16], $\mathcal{SHIQ}$ [19], $\mathcal{SHIO}$ [14], $\mathcal{SHOQ}$ [17], $\mathcal{SHOIQ}$ [18], $\mathcal{SROIQ}$ [15]. These procedures use backtracking to deal with disjunction (e.g., the union constructor). Their search space is an "or"-tree of "and"-trees. Despite advanced blocking techniques (e.g., anywhere blocking), their complexities are non-optimal:

- NExpTime instead of ExpTime for $\mathcal{SHI}$, $\mathcal{SHIQ}$, $\mathcal{SHIO}$ and $\mathcal{SHOQ}$
- N2ExpTime instead of NExpTime for $\mathcal{SHOIQ}$
- N3ExpTime instead of N2ExpTime for $\mathcal{SROIQ}$.

The decision procedure of HermiT also has a non-optimal complexity [7].

The tableau decision procedure given in [4] for $\mathcal{ALC}$ uses a combination of blocking and caching techniques and is complexity-optimal (ExpTime). A more general technique for developing complexity-optimal tableau decision procedures is based on global caching. Global caching means that, instead of a search space of the form "or"-tree of "and"-trees, we use an "and-or" graph whose nodes are labeled by unique sets of formulas. The idea of global caching comes from Pratt's paper on PDL (propositional dynamic logic) [32].

In this work we present the design of a tableau reasoner, called TGC 2, that uses global caching for reasoning in description logics (DLs). The core reasoning problem is to check whether a knowledge base is satisfiable. We briefly describe the tableau methods used by TGC 2. We then provide the design principles and some important optimization techniques for increasing the efficiency of the reasoner. The system is designed for automated reasoning for a large class of modal and description logics, including $\mathcal{ALC}$, $\mathcal{SH}$, $\mathcal{SHI}$, $\mathcal{SHIQ}$, $\mathcal{SHIO}$, $\mathcal{SHOQ}$, REG (regular grammar logics), REG^c (REG with converse), PDL, CPDL (PDL with converse), and CPDL_{reg} (the combination of CPDL and REG^c). It is designed to work also for combinations of the listed logics in the case when there are no interactions between $\mathcal{O}$ (nominals), $\mathcal{I}$ (inverse roles) and $\mathcal{Q}$ (qualified number restrictions). That is, it does not fully support $\mathcal{SHOIQ}$ nor $\mathcal{SROIQ}$, but it allows PDL-like role constructors and advanced role inclusion axioms. TGC 2 is based on ExpTime tableau methods using global caching and its complexity is ExpTime when numbers (in qualified number restrictions) are coded in unary.

Implementation of TGC 2 is very time-consuming and still in progress. At the moment of writing this paper, it contains more than 15000 lines of code in C++, Flex and Bison. We estimate that for the first version of TGC 2 that offers basic functionalities we have to write about 3000 more lines of code. The reasoner will be available soon. For thorough testing, evaluation, extended functionalities and further optimizations, we will need more time.

2 Tableaux with Global Caching

A tableau with global caching for the basic DL $\mathcal{ALC}$ is an "and-or" graph, where each of its nodes is labeled by a unique set of formulas, which are ABox assertions in the case the node is a "complex" node and concepts in the case when the node is a "simple" node. The formulas in the label of a node are treated as requirements to be realized (satisfied) for the node. They are realized by expanding the graph using tableau rules. Expanding a node using a static rule causes the node to become an "or"-node, expanding a node using the transitional rule causes the node to become an "and"-node. Usually, the transitional rule has a lower priority than the static rules. An "and"-node is also called a state, while an "or"-node is called a non-state. States in tableaux with global state caching for advanced DLs like $\mathcal{SHIQ}$ [24,26] or $\mathcal{SHOQ}$ [29,28] are more sophisticated than "and"-nodes. For an introduction to the tableau method using global caching and its relationship to the other tableau methods, we refer the reader to [10]. To see the differences between the tableau method using global caching and the traditional tableau method for DLs, the reader may also consult [29, Section 3].

In [8,9] together with Goré we formalized global caching and applied it to REG (regular grammar logics) and the DL $\mathcal{SHI}$ to obtain the first ExpTime tableau decision procedures for these logics. Later, together with Szałas we extended the method to give the first ExpTime tableau decision procedures for CPDL [30] and REG^c [31]. In [5] together with Dunin-Kęplicz and Szałas we used global caching to develop a tableau calculus leading to the first ExpTime tableau decision procedure for CPDL$_{reg}$. In [9,30,31,5] analytic cuts are used to deal with inverse roles and converse modal operators. As cuts are not efficient in practice, Goré and Widmann developed the first cut-free ExpTime tableau decision procedures, based on global state caching, for the DL $\mathcal{ALCI}$ (which extends $\mathcal{ALC}$ with inverse roles) [12] and CPDL [13]. Global state caching is a variant of global caching in which only states are globally cached for the constructed graph during the process of checking satisfiability of a knowledge base.

The reasoner TGC 2 uses a tableau method that unifies our tableau decision procedures developed recently for CPDL$_{reg}$ [23,25] and the DLs $\mathcal{SHIQ}$ [24,26], $\mathcal{SHOQ}$ [28,29] and $\mathcal{SHIO}$ [27]. We briefly describe below these tableau decision procedures.

- The tableau decision procedure for CPDL$_{reg}$ given in [23,25] improves the one of [5] by eliminating cuts to give the first cut-free ExpTime (optimal) tableau decision procedure for CPDL$_{reg}$. This procedure uses global state caching, which modifies global caching for dealing with converse without using cuts. It also uses local caching for non-states of tableaux. As cut rules are "or"-rules for guessing the "future" and they are usually used in a systematic way and generate a lot of branches, eliminating cuts is very important for efficient automated reasoning in CPDL$_{reg}$.
- The tableau decision procedure given in [24,26] is for checking satisfiability of a knowledge base in the DL $\mathcal{SHIQ}$. It has complexity ExpTime when numbers are coded in unary. It is based on global state caching and integer

linear feasibility checking. Both of them are essential for the procedure in order to have the optimal complexity. Global state caching can be replaced by global caching plus cuts, which still guarantee the optimal complexity. However, we chose global state caching to avoid inefficient cuts although it makes our procedure more complicated. In contrast to Farsiniamarj's method of exploiting integer programming for tableaux [6], to avoid nondeterminism we only check feasibility of the considered set of constraints, without finding and using their solutions. As far as we know, we are the first one who applied integer linear feasibility checking to tableaux.

- The tableau decision procedure for the DL $\mathcal{SHOQ}$ given in the joint works with Golińska-Pilarek [29,28] has complexity ExpTime when numbers are coded in unary. The DL $\mathcal{SHOQ}$ differs from $\mathcal{SHIQ}$ in that nominals are allowed instead of inverse roles. The procedure exploits the method of integer linear feasibility checking [24,26] for dealing with number restrictions. Without inverse roles, it uses global caching instead of global state caching to allow more cache hits. It also uses special techniques for dealing with nominals and their interaction with number restrictions.
- The work [27] presents the first tableau decision procedure with an ExpTime (optimal) complexity for checking satisfiability of a knowledge base in the DL $\mathcal{SHIO}$. It exploits global state caching and does not use blind (analytic) cuts. It uses a new technique to deal with the interaction between inverse roles and nominals.

Recall that $\mathcal{SHIQ}$, $\mathcal{SHOQ}$, $\mathcal{SHIO}$ are the three most well-known expressive DLs with ExpTime complexity. Due to the interaction between the features $\mathcal{I}$, $\mathcal{Q}$ and $\mathcal{O}$, the complexity of the DL $\mathcal{SHOIQ}$ is NExpTime-complete. The reasoner TGC 2 allows knowledge bases that use all the features $\mathcal{I}$, $\mathcal{O}$, $\mathcal{Q}$ together, but it does not fully support the DL $\mathcal{SHOIQ}$. In the case of interaction between $\mathcal{I}$, $\mathcal{O}$ and $\mathcal{Q}$, the reasoner tries to avoid the difficulty by exploring other branches in the constructed graph in hope for overcoming the problem.

3 Design Principles

3.1 Avoiding Costly Recomputations by Caching

Recall that a tableau with global caching is like an "and-or" graph, where each node is labeled by a unique set of formulas. Furthermore, a node may be re-expanded at most once (when dealing with inverse roles or nominals). By constructing such a tableau in a deterministic way instead of searching the space of the form "or"-tree whose nodes are "and"-trees as in the traditional tableau method, the complexity is guaranteed to be ExpTime. The essence of this is to divide the main task to appropriate subtasks (like expansion of a node) so that each subtask is executed only once and the result is cached to avoid recomputations. (In the traditional tableau method, nodes in "and"-trees are not cached across branches of the main "or"-tree, which means that some kind of recomputation occurs and that causes the non-optimal complexity.)

We apply the idea of avoiding costly recomputations by caching also to other entities/tasks as described below.

- Formulas are normalized and globally cached. Furthermore, certain tasks involved with formulas like computing the negation, the size or the weight of a formula are also cached in order to avoid recomputations.
- As the reasoner TGC 2 allows PDL-like role constructors and roles are similar to formulas, they are also normalized and cached. Tasks like computing the inverse of a role, the set of subroles of a role, the size or the weight of a role are cached to avoid recomputations.
- Due to hierarchies of roles and merging nodes in the constructed graph, an edge outgoing from a state may be labeled by a set of roles instead of a single role. For this reason, sets of roles are also cached. Consequently, tasks like computing the union of two sets of roles or the inverse of an edge label are also cached to avoid recomputations.
- In the case the underlying logic allows qualified number restrictions, a state in the constructed graph may have a set of integer linear constraints (ILCS). Checking feasibility of such a set may be very costly. Furthermore, different states may have the same ILCS. So, we also cache ILCS' and the results of their feasibility checking to avoid costly recomputations.

3.2 Memory Management

As TGC 2 may use an exponential number of nodes, formulas and other entities, efficient memory management is very important. It is a critical matter of the approach. The earlier version TGC [22] for reasoning in the basic DL $\mathcal{ALC}$ manages memory usage very well. It shows that, for hard instances of reasoning in the basic DL $\mathcal{ALC}$, execution time but not memory lack is the cause of not being able to solve the instance. For TGC 2, in hope for satisfying this claim, we also pay a special effort in memory management.

Caching is good for saving execution time, but it may increase memory usage. On one hand, when identical objects are cached by using only one representative in the computer memory, we save memory usage. On the other hand, if we cache too much, we may use too much memory. To reduce memory usage and costly recomputations by increasing the number of cache hits, i.e. by increasing the quality of caching, objects must be normalized before being kept in the memory or searching for an existing representative. Furthermore, the normalization should be good, i.e. as strong as possible without using too much time.

Technically, for each kind of objects to be cached, TGC 2 uses a catalog of objects of that kind, which is either a *map* or a *multimap* that uses hash values as keys. Only normalized objects are kept in the catalog. When an object is created not only for temporary usage, for example, to be used as an attribute of another object, we normalize it and search for its representative in the catalog, store it in the catalog if not found, and then return the pointer to the representative for further usages. An object in the catalog is a "hard-resource", other references to the object are "soft-resources". We store each object in a catalog together with its usage count. The count is initialized to 1 (for the occurrence of the object in the catalog). When a reference to an object is used as an attribute of another object, the usage count of the former is increased by 1. We release that "soft-resource" by decreasing the usage count by 1. When the usage count of an object is decreased to 1, the object is not used anywhere out of the catalog, and if it is not for permanent caching, the "hard-resource" corresponding to the object will be released (i.e., deleted from the catalog) in the next reduction round of the catalog. Important objects are permanently kept in the catalog, including formulas with the status "satisfiable" or "unsatisfiable" as well as basic instances of the integer linear feasibility problem. Less important objects may also be permanently cached, depending on the used options. catalog reduction is done periodically, depending on parameters and the effect of the previous reduction. The discussed technique is similar to the techniques of smart pointers of C++ and garbage collection of Java. The differences are that we want to cache objects by using catalogs and some objects are permanently cached.

Another principle of reducing memory usage of TGC 2 relies on that we apply dynamic and lazy allocation as well as greedy deallocation for memory management. The types of additional attributes of an object, which may be complex objects themselves, may be known only "on-the-fly". Furthermore, such attributes may receive values or remain empty. To save memory usage, we allocate memory for an object (or an attribute of an object) only when we really have to, and in an minimal amount, depending on its dynamic type. Apart from that, in most of cases, we release memory used by an object as soon as possible. An exception is reduction of catalogs, which is done periodically. The discussed approach of saving memory increases execution time a bit and makes the code more complicated.

3.3 Search Strategies

There are technical problems for which we do not know which from the available solutions is the best. Sometimes, it may be just about selecting a good value for a parameter. Furthermore, different classes of instances of the reasoning problem may prefer different solutions. For this reason, we implement different potentially good solutions for encountered technical problems and parameterize the choices. When everything has neatly been implemented, evaluation will help us in tuning parameters. One of the most important problems is: what search strategy should be used?

Recall that the core reasoning problem is to check whether a given knowledge base is satisfiable. For that, TGC 2 constructs a rooted graph which is like an "and-or" graph. Formulas from the labels of the nodes are requirements to be realized and the task is to compute the status of the root, which is either "unsatisfiable" or "satisfiable" or "unsolvable" (caused by interaction between the features $\mathcal{I}$, $\mathcal{O}$, $\mathcal{Q}$ or by exceeding the time limit). In short, the task can be realized as follows: starting from the root, expand the graph, compute and propagate the statuses of the nodes. Propagating known statuses of nodes is done appropriately and has the highest priority. The status of a node may be computed by detecting a direct inconsistency involved with the node (inconsistency means "unsatisfiability" in a certain sense). There are two kinds of inconsistencies: local inconsistency is directly caused by the node and its neighbors, while global inconsistency is caused by impossibility of realizing a requirement with a constructor like existential star modal operators of PDL. When direct detection does not provide a known status for a node, the node may be chosen for expansion, and its status is computed from the statuses of its successors. Mutual dependences between the statuses of nodes may occur. At the end, when all essential computations that may affect the status of the root have been done, if the status of the root is neither "unsatisfiable" nor "unsolvable", then the status becomes "satisfiable". Detection of the status "satisfiablity" for nodes in a fragment of the graph may be done earlier when it is possible to "close" the fragment.

As discussed above, when no more propagations can be done, we have to choose an unexpanded node and expand it, or choose a node that should be re-expanded and re-expand it (each node may be re-expanded at most once, for dealing with inverse roles or nominals). Re-expansion of a node is done by a specific rule, in a deterministic manner. When an unexpanded node is chosen for expansion, there are choices for how to expand it. The preferences are: unary static rules have a higher priority than non-unary static rules, which in turn have a higher priority than transitional rules. A unary (resp. non-unary) static rule causes the node to become a non-state (an "or"-node) with only one (resp. more than one) successor. A transitional rule causes the node to become a state. There is in fact only one transitional rule, but we break it into two phases and treat them as two transitional rules, the partial one and the full one [24,29] (this is discussed further in the next section).

Which among the applicable unary static rules should be chosen is rather not important. One can imagine that they are applied in a sequence, and the order in the sequence does not really matter. A non-unary static rule realizes a specific requirement in the label of a node (which is called the principal formula of the rule). Which requirement from the label of a node should be chosen for realization first? A general answer is to choose a requirement such that realizing it allows more applications of unary static rules to the successors and the descendants of the node, i.e., so that more simplifications can be done for the the successors and the descendants. Among the best candidates (requirements) in respect to this criterion, we choose the "heaviest" one in order to cause the node to have much lighter successors. The weight of a node is determined by the weight of its label. The weight of a formula is more sophisticated than the size of the formula. It is a subject for optimization and investigation.

For TGC 2, we adopt the strategy that, apart from propagations (of various kinds), unary static rules have the highest priority globally. This means that we treat nodes to which some unary static rules are applicable as unstable, and a non-unary static rule or a transitional rule is applied to a node only when all nodes of the graph are stable w.r.t. unary static rules (i.e., no unary static rule is applicable to any node of the graph).

Now assume that no more propagations can be done and all nodes of the current graph are stable w.r.t. unary static rules. TGC 2 chooses a node for expansion or re-expansion. The choice is made by using an expansion queue, which is a priority queue of nodes waiting for expansion or re-expansion. For the priority queue, the reasoner uses a comparison function that depends on attributes of the compared nodes and the used options. Important attributes of nodes and options for determining the priority of a node are the following:

- SOS (set-of-supports) – whether the label of the node contains a formula originated from the query: for a query like checking whether an individual is an instance of a concept w.r.t. a knowledge base KB or whether a concept is subsumed by another w.r.t. KB, the reasoner checks unsatisfiability of the knowledge base that extends KB with the negation of the query; this technique comes from Otter and other SAT solvers for classical propositional logic;
- the depth of the node in a skeleton tree of the constructed graph: this is used to emphasize depth-first-search (DFS);
- the current limit for the depths of nodes: this is used to emphasize a kind of iterative deepening search; the limit is multiplied by a constant greater than 1 after each deepening round;
- the weight of the label of the node;
- a random number generated for the node.

Each combination of options on how to use the above mentioned attributes provides a search strategy. For example, if the depths of nodes are compared first and the bigger the better, then we have a kind of strong DFS.

In our opinion, strategies based on strong DFS are potentially good as they allow the reasoner to "close" fragments of the graph "on-the-fly" (also note that the graph is finite). However, we do not exclude other strategies. In general, at the current stage of investigation, we use a *mixed expansion queue*, which mixes different strategies together to obtain a binary tree of priority queues. Each leaf of the tree is a normal priority queue. Each inner node of the tree is labeled by an option and an execution-time-balance parameter. The left edge outgoing from a node turns on the option, while the right edge turns off the option. The execution-time-balance parameter specifies the percentage of choosing the left edge out from the two edges to follow. All mixed strategies work on the current graph. For the first version of TGC 2, we do not exploit concurrent computing, but the approach is very suitable for incorporating concurrency.

4 Some Other Optimization Techniques

The earlier version TGC [22] for reasoning in the basic DL $\mathcal{ALC}$ uses a set of optimizations that co-operates very well with global caching and various search strategies on search spaces of the form "and-or" graph. Apart from normalization, caching objects and efficient memory management, it also incorporates the techniques: literals elimination, propagation of unsatisfiability for parent nodes and sibling nodes using "unsat-cores", and cutoffs. These optimization techniques are very useful for TGC and hence adopted for TGC 2, with the exception that computing "unsat-cores" is complicated and for the very first version of TGC 2 will be not yet available.

Some other optimization techniques have already been incorporated into the tableau decision procedures [23,24,29,27] that are used for TGC 2, including:

- using finite automata to deal with PDL-like roles;
- using global state caching instead of "pairwise" global state caching for $\mathcal{SHIQ}$ and $\mathcal{SHIO}$; caching nodes in the "local subgraph" of a state in the case of using global state caching;
- using global caching instead of global state caching for $\mathcal{SHOQ}$;
- integer linear feasibility checking for dealing with the feature $\mathcal{Q}$;
- techniques for dealing with interaction between the features $\mathcal{I}$ and $\mathcal{Q}$, between the features $\mathcal{O}$ and $\mathcal{Q}$, and between the features $\mathcal{I}$ and $\mathcal{O}$.

We discuss below three other optimization techniques of TGC 2.

4.1 Delaying Time-Consuming Subtasks

A good practice for students taking part in a competition or an exam is to try to solve easy tasks first, as quickly as possible, and then concentrate on harder tasks. This is due to the time limit. In general, easy tasks are preferred over harder tasks, and the latter are usually delayed in comparison to the former. The strategy adopted for TGC 2 is similar. Trying to solve "easy" subtasks first takes less time and this may establish statuses for some nodes of the constructed graph, which may be then propagated backward to determine the final status of the root of the graph. TGC 2 has options for specifying whether potentially time-consuming subtasks should be delayed. There are two kinds of such subtasks:

- checking global inconsistencies,
- checking feasibility of a set of integer linear constraints.

As the constructed graph may be exponentially large, checking global inconsistencies may be very time-consuming. One can adopt the technique of "on-the-fly" checking for detecting global inconsistencies as proposed by Goré and Widmann in [11] for PDL. This probably increases efficiency of the reasoner, as global inconsistencies are detected as soon as possible. However, in our opinion, too much more memory is needed, as using this technique each node may contain information about exponentially many nodes in the graph, and this may affect scalability of the reasoner. Our approach for TGC 2 is to check global inconsistencies periodically. As strong DFS tries to "close" fragments of the constructed graph and delete them as soon as possible, in combination with strong DFS (which is one among the mixed search strategies) checking global inconsistencies may involve only relatively small fragments of the graph.

Checking feasibility of a set of integer linear constraints is NP-hard in the size of the feasibility problem. It makes sense to allow an option for delaying such operations in an appropriate manner. The manner relies on doing partial checks first and delaying full checks. TGC 2 uses two kinds of partial checks. A number restriction $\geq n\, r.C$ with $n \geq 1$ requires at least satisfaction of $\exists r.C$. So, as the first kind of partial checks, if a number restriction $\geq n\, r.C$ belongs to the label of a node, we add the requirement $\exists r.C$ to the label of the node, and we expand a state first only by the transitional partial-expansion rule, which ignores number restrictions (of both kinds $\geq n\, r.C$ and $\leq m\, s.D$), delaying application of the transitional full-expansion rule to the node, which deals also with number restrictions. As the second kind of partial checks, we relax an integer linear feasibility problem by assuming that the numbers may be real (continuous). It is known that a linear programming problem is solvable in polynomial time.

4.2 Converting TBoxes

The way of treating a TBox axiom $C \sqsubseteq D$ as the global assumption $\neg C \sqcup D$ and a TBox axiom $C \equiv D$ as the global assumption $(\neg C \sqcup D) \sqcap (\neg D \sqcup C)$ causes that the global assumption is added to the label of each node in the constructed graph as a requirement to be realized. This creates a lot of "or"-branchings for dealing with $\sqcup$. TGC 2 preprocesses a given TBox $\mathcal{T}$ to obtain two parts $\mathcal{T}_1$ and $\mathcal{T}_2$ such that $\mathcal{T}_1$ is an acyclic TBox and $\mathcal{T}_2$ is a set of global assumptions. The aim is to make $\mathcal{T}_2$ as lightest as possible. Then, for dealing with $\mathcal{T}_1$, TGC 2 applies the well-known absorption technique that lazily replaces a defined concept by its definition. As example, TGC 2 is able to convert the TBox consisting of the following axioms to an acyclic TBox:

$$
\begin{array}{llll}
A_1 \sqsubseteq \neg A_2 & A_2 \sqsubseteq \neg A_3 & C \sqsubseteq \forall R.A_1 & C \sqcap \exists R.A_1 \sqsubseteq D \\
A_1 \sqsubseteq \neg A_3 & A_3 \sqsubseteq \neg A_1 & C \sqsubseteq \forall R.A_2 & C \sqcap \exists R.A_2 \sqsubseteq D \\
A_2 \sqsubseteq \neg A_1 & A_3 \sqsubseteq \neg A_2 & C \sqsubseteq \forall R.A_3 & C \sqcap \exists R.A_3 \sqsubseteq D
\end{array}
$$

4.3 Ontology Classification

As mentioned earlier, the goal of ontology classification is to compute a hierarchical relation between concepts. Given an ontology specified by a knowledge base KB (which may be just a TBox), the task is compute the set of inclusion axioms $A \sqsubseteq B$ such that A and B are atomic concepts (i.e., concept names) and $KB \models A \sqsubseteq B$. For a specific axiom $A \sqsubseteq B$, checking whether $KB \models A \sqsubseteq B$ can be done by checking unsatisfiability of the extended knowledge base $KB' = KB \cup \{\tau : A \sqcap \neg B\}$, where τ is a fresh (new) individual name. To increase efficiency, one can apply a strategy that reduces the number of such inclusion axioms to be checked. For the first version of TGC 2, we just apply a simple but effective optimization technique: the part consisting of simple nodes is common for all of the checks. That is, all of the created simple nodes remain through all of the checks (for different inclusion axioms $A \sqsubseteq B$). Thus, TGC 2 avoids costly recomputations. Complex nodes of the constructed graph may be deleted after each check in order to save memory.

5 Conclusions

We have presented the design of TGC 2, a tableau reasoner that uses global caching for reasoning in DLs. Despite that implementation of TGC 2 is still in progress and its efficiency must be verified, our design principles and optimization techniques presented in this paper are ideas that have been carefully analyzed and chosen in order to guarantee high efficiency. Recall that global caching guarantees an optimal complexity for ExpTime modal and description logics. This distinguishes TGC 2 from the existing tableau reasoners for DLs.

Acknowledgments. This work was supported by the Polish National Science Center (NCN) under Grant No. 2011/01/B/ST6/02759.

References

1. Antoniou, G., van Harmelen, F.: A Semantic Web Primer. MIT Press, Cambridge (2004)
2. Baader, F., Calvanese, D., McGuinness, D.L., Nardi, D., Patel-Schneider, P.F.: The description logic handbook: theory, implementation, and applications. Cambridge University Press, New York (2003)
3. Baader, F., Sattler, U.: An overview of tableau algorithms for description logics. Studia Logica 69, 5–40 (2001)
4. Donini, F., Massacci, F.: ExpTime tableaux for $\mathcal{ALC}$. Artificial Intelligence 124, 87–138 (2000)
5. Dunin-Kęplicz, B., Nguyen, L.A., Szałas, A.: Converse-PDL with regular inclusion axioms: A framework for MAS logics. Journal of Applied Non-Classical Logics 21(1), 61–91 (2011)
6. Farsiniamarj, N.: Combining integer programming and tableau-based reasoning: a hybrid calculus for the description logic SHQ. Master's thesis, Concordia University (2008)
7. Glimm, B., Horrocks, I., Motik, B.: Optimized description logic reasoning via core blocking. In: Giesl, J., Hähnle, R. (eds.) IJCAR 2010. LNCS, vol. 6173, pp. 457–471. Springer, Heidelberg (2010)
8. Goré, R., Nguyen, L.A.: A tableau calculus with automaton-labelled formulae for regular grammar logics. In: Beckert, B. (ed.) TABLEAUX 2005. LNCS (LNAI), vol. 3702, pp. 138–152. Springer, Heidelberg (2005)
9. Goré, R., Nguyen, L.A.: EXPTIME tableaux with global caching for description logics with transitive roles, inverse roles and role hierarchies. In: Olivetti, N. (ed.) TABLEAUX 2007. LNCS (LNAI), vol. 4548, pp. 133–148. Springer, Heidelberg (2007)
10. Goré, R., Nguyen, L.A.: ExpTime tableaux for ALC using sound global caching. J. Autom. Reasoning 50(4), 355–381 (2013)
11. Goré, R., Widmann, F.: An optimal on-the-fly tableau-based decision procedure for PDL-satisfiability. In: Schmidt, R.A. (ed.) CADE 2009. LNCS (LNAI), vol. 5663, pp. 437–452. Springer, Heidelberg (2009)
12. Goré, R., Widmann, F.: Sound global state caching for ALC with inverse roles. In: Giese, M., Waaler, A. (eds.) TABLEAUX 2009. LNCS, vol. 5607, pp. 205–219. Springer, Heidelberg (2009)

13. Goré, R., Widmann, F.: Optimal and cut-free tableaux for propositional dynamic logic with converse. In: Giesl, J., Hähnle, R. (eds.) IJCAR 2010. LNCS, vol. 6173, pp. 225–239. Springer, Heidelberg (2010)
14. Hladik, J., Model, J.: Tableau systems for SHIO and SHIQ. In: Proc. of Description Logics 2004. CEUR Workshop Proceedings, vol. 104 (2004)
15. Horrocks, I., Kutz, O., Sattler, U.: The even more irresistible $\mathcal{SROIQ}$. In: Doherty, P., Mylopoulos, J., Welty, C.A. (eds.) Proc. of KR 2006, pp. 57–67. AAAI Press (2006)
16. Horrocks, I., Sattler, U.: A description logic with transitive and inverse roles and role hierarchies. J. Log. Comput. 9(3), 385–410 (1999)
17. Horrocks, I., Sattler, U.: Ontology reasoning in the SHOQ(D) description logic. In: Nebel, B. (ed.) Proc. of IJCAI 2001, pp. 199–204. Morgan Kaufmann (2001)
18. Horrocks, I., Sattler, U.: A tableau decision procedure for $\mathcal{SHOIQ}$. J. Autom. Reasoning 39(3), 249–276 (2007)
19. Horrocks, I., Sattler, U., Tobies, S.: Reasoning with individuals for the description logic SHIQ. In: McAllester, D. (ed.) CADE 2000. LNCS, vol. 1831, pp. 482–496. Springer, Heidelberg (2000)
20. Mishra, R.B., Kumar, S.: Semantic web reasoners and languages. Artif. Intell. Rev. 35(4), 339–368 (2011)
21. Motik, B., Sattler, U.: A comparison of reasoning techniques for querying large description logic ABoxes. In: Hermann, M., Voronkov, A. (eds.) LPAR 2006. LNCS (LNAI), vol. 4246, pp. 227–241. Springer, Heidelberg (2006)
22. Nguyen, L.A.: An efficient tableau prover using global caching for the description logic $\mathcal{ALC}$. Fundamenta Informaticae 93(1-3), 273–288 (2009)
23. Nguyen, L.A.: A cut-free ExpTime tableau decision procedure for the logic extending Converse-PDL with regular inclusion axioms. arXiv:1104.0405 (2011)
24. Nguyen, L.A.: ExpTime tableaux for the description logic $\mathcal{SHIQ}$ based on global state caching and integer linear feasibility checking. arXiv:1205.5838 (2012)
25. Nguyen, L.A.: Cut-free ExpTime tableaux for Converse-PDL extended with regular inclusion axioms. In: Proc. of KES-AMSTA 2013. Frontiers in Artificial Intelligence and Applications, vol. 252, pp. 235–244. IOS Press (2013)
26. Nguyen, L.A.: A tableau method with optimal complexity for deciding the description logic SHIQ. In: Nguyen, N.T., van Do, T., Thi, H.A. (eds.) ICCSAMA 2013. SCI, vol. 479, pp. 331–342. Springer, Heidelberg (2013)
27. Nguyen, L.A.: ExpTime tableaux with global state caching for the description logic SHIO. Neurocomputing 146, 249–263 (2014)
28. Nguyen, L.A., Golińska-Pilarek, J.: An ExpTime tableau method for dealing with nominals and qualified number restrictions in deciding the description logic SHOQ. Fundam. Inform. 135(4), 433–449 (2014)
29. Nguyen, L.A., Golińska-Pilarek, J.: ExpTime tableaux with global caching for the description logic SHOQ. CoRR, abs/1405.7221 (2014)
30. Nguyen, L.A., Szałas, A.: An optimal tableau decision procedure for Converse-PDL. In: Nguyen, N.-T., Bui, T.-D., Szczerbicki, E., Nguyen, N.-B. (eds.) Proc. of KSE 2009, pp. 207–214. IEEE Computer Society (2009)
31. Nguyen, L.A., Szałas, A.: A tableau calculus for regular grammar logics with converse. In: Schmidt, R.A. (ed.) CADE 2009. LNCS (LNAI), vol. 5663, pp. 421–436. Springer, Heidelberg (2009)
32. Pratt, V.R.: A near-optimal method for reasoning about action. J. Comp. Syst. Sci. 20(2), 231–254 (1980)
33. Semantic web case studies and use cases, `http://www.w3.org/2001/sw/sweo/public/UseCases/`

Granular Floor Plan Representation for Evacuation Modeling*

Wojciech Świeboda, Maja Nguyen, and Hung Son Nguyen

Institute of Mathematics, The University of Warsaw,
Banacha 2, 02-097, Warsaw Poland

Abstract. In this paper we describe the architecture of a simple evacuation model which is based on a graph representation of the scene. Such graphs are typically constructed using Medial Axis Transform (MAT) or Straight Medial Axis Transform (S-MAT) transformations, the former being a part of the Voronoi diagram (Dirichlet tessellation) of the floor plan. In our work we construct such graphs for floor plans using Voronoi diagram along with the dual Delaunay triangulation of a set of points approximating the scene. Information supplied by Delaunay triangulation complements the graph in two ways: it determines capacities of some paths associated with edges, and provides a bijection between graph vertices and a set of regions forming a partition of the floor plan. We call the representation granular for this reason.

In our earlier work we discussed a simplified model coupling a partial behavioral model (accounting for typical pre-movement times) and a traffic model. In this expanded paper, we provide a clearer exposition of the representation of fire scene and briefly discuss the applicability of network flow models (e.g. the Ford-Fulkerson method or the push-relabel method) in our setting.

1 Introduction

ICRA project (`http://icra-project.org`) aims to build modern engineering tools to support fire commanders during fire operations. In this paper we focus on the problem of modeling evacuation process. While there are various questions one may want to ask regarding evacuation, e.g. related to finding optimal (static or dynamic) evacuation plans, we consider the problem of evacuation time estimation and bottleneck analysis. In fact, four ICRA modules relevant to evacuation modelling are being developed.

- **fire and smoke localization module**, which determines the placement and spread of fire and smoke,

* This work was partially supported by the Polish National Centre for Research and Development (NCBiR) – grant O ROB/0010/03/001 under Defence and Security Programmes and Projects: "Modern engineering tools for decision support for commanders of the State Fire Service of Poland during Fire&Rescue operations in buildings", and by the Polish Na tional Science Centre grants DEC-2011/01/B/ST6/03867 and DEC-2012/05/B/ST6/03215.

H.A. Le Thi et al. (eds.), *Advanced Computational Methods for Knowledge Engineering*,
Advances in Intelligent Systems and Computing 358, DOI: 10.1007/978-3-319-17996-4_30

- **occupant localization module**, which aims to estimate occupant density in various parts of a buildings or determine occupant locations,
- **building topology construction**, which builds a geometric network of the fire scene,
- **evacuation module**, which estimates the egress time, determines bottlenecks and is used in setting priorities during find and rescue operations.

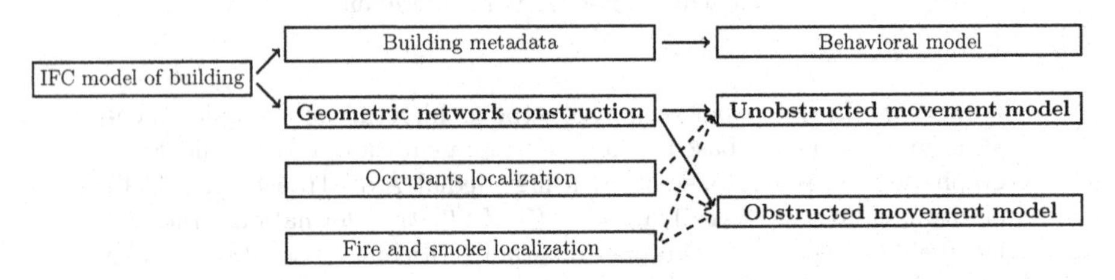

Fig. 1. Various models of evacuation module

Figure 1 summarizes the inputs to various models in the evacuation module. For simplicity we treat various localization modules as direct inputs. The focus of this paper is on geometric network construction and movement models.

The paper is organized as follows: in Section 2 we present the main notions and prerequisites of the evacuation process. Section 3 describes the representation of the fire rescue scene which is the main goal of the paper. Section 4 presents the algorithm that constructs the augmented graph which is the Geometric Network Representation of the evacuation plan for the given floor plan. In Section 5 we present the summaries and conclusions.

2 The Evacuation Process

Figure 2 shows the typical evacuation process model from fire protection engineering literature. Alarming times and pre-movement times may be different for different occupants. For a single occupant, the pre-movement time Δt_{pre} could be modeled by a behavioral model, whereas travel time Δt_{trav} could be independently modeled by a movement model. The differences in Δt_a and Δt_{pre} among occupants, as well as interactions among occupants may suggest that the problem leads to occupant-level modeling.

In this paper we look at a simplified model of the process and treat Δt_a and Δt_{pre} globally, rather than on an individual (occupant) basis, by considering two scenarios. As [1] points out, if the building is sparsely populated, the (close to) worst-case scenario may be conveniently modeled by considering the 99-th percentile of pre-movement times, and adjusting it by a conservative estimate of unimpeded walking time, thus $t_{evac}^{sparse} = \Delta t_{pre}^{99th,sparse} + \Delta t_{walk}$. If the building is densely populated, the evacuation process differs in that a queue may form

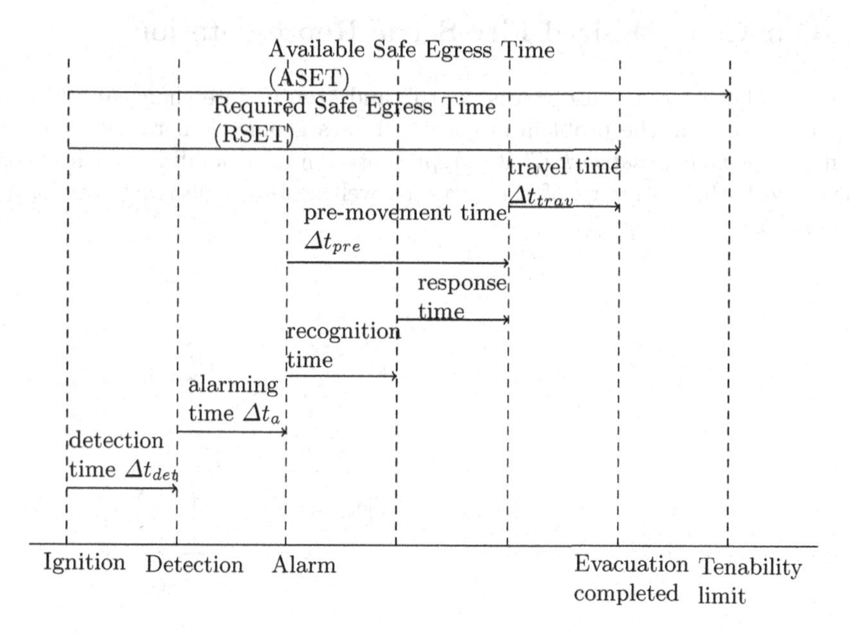

Fig. 2. The typical evacuation process model from fire protection engineering literature [6]

relatively early as occupants start leaving the building, thus the evacuation time may be modeled as $t_{evac}^{dense} = \Delta t_{pre}^{1st,dense} + \Delta t_{walk} + \Delta t_{flow}$.

Thus, the results of this specific model may be reported in the form of Table 1, which summarizes the possible output of the system. We stress that pre-movement times are defined differently in the sparse and the dense scenario; these definitions need to be taken into account when analysing the output of the model.

In the remaining parts of this paper we will discuss the calculation if Δt_{walk} and Δt_{flow}, along with the algorithm that determines building topology based on floor plans.

Table 1. The evacuation time is separated into (global) alarm time, pre-movement time and movement time

scenario	Δt_a	Δt_{pre}	$t_{\mathrm{trav(walk)}}$	$t_{\mathrm{trav(flow)}}$
sparse	$2-5$	>20	10	0
dense	$2-5$	>10	10	15

3 Our Goal: Desired Fire Scene Representation

In what follows we will use words "path" and "route" interchangeably. In this paper we focus on the problem of constructing a geometric network of the fire scene. A *geometric network* [7] is a representation of a building which encompasses both the geometry of the scene as well as the underlying topology of available routes.

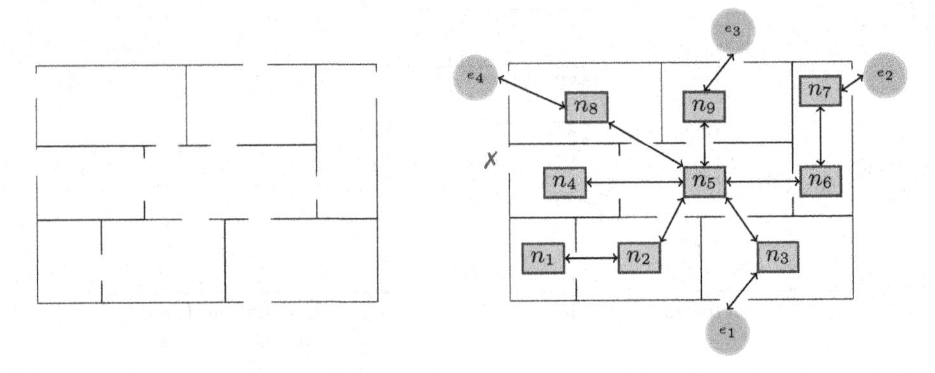

Fig. 3. An example of a floor plan and a network of available evacuation routes

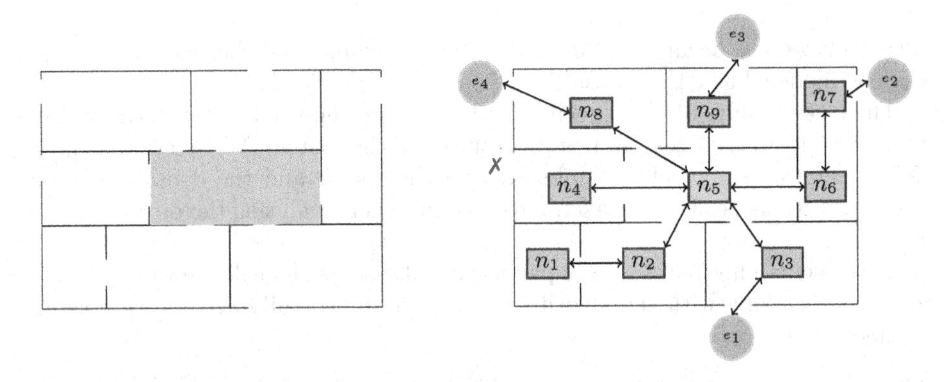

Fig. 4. The duality of geometric and topological representations induces a bijection between a set of regions on the floor plan and nodes in the graph

Figure 3 shows such a graph for an example floor plan. The figure on the left shows a planar floor plan reconstructed from an IFC file. Additional information that we need to reconstruct from IFC files are parameters of stairs: Tread and Riser. The figure on the right shows an idealized graph simplifying the underlying topology of the building. Vertices in this graph include *transition points.* The graph may be refined by incorporating real-time dynamics (unavailability of certain routes), like the X mark suggests.

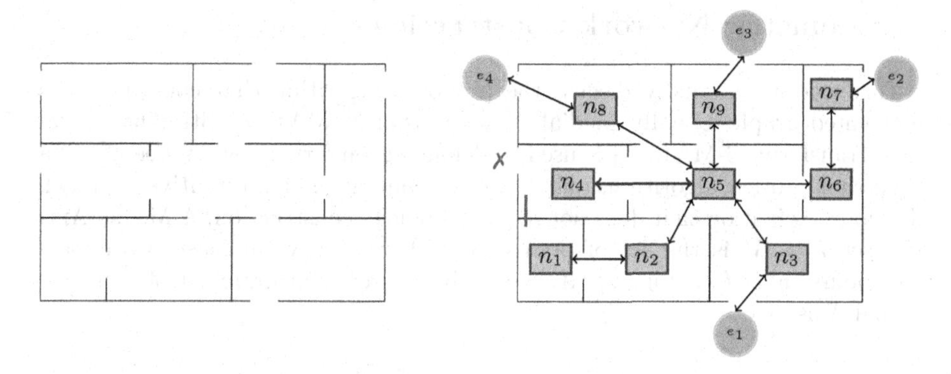

Fig. 5. Edges in the graph are augmented with path lengths and door/corridor widths. These widths determine capacities in flow modeling.

Geometric networks necessarily by their nature store route lengths of available routes, thus we augment edges in the graph by the underlying route lengths (see Figure 5). Furthermore, we wish to (i) provide a means to map vertices in the network bijectively to regions that form a partition of the floor plan (as shown on Figure 4), and (ii) determine widths of corridors or doors so as to determine edge capacities in the underlying graph (Figure 5).

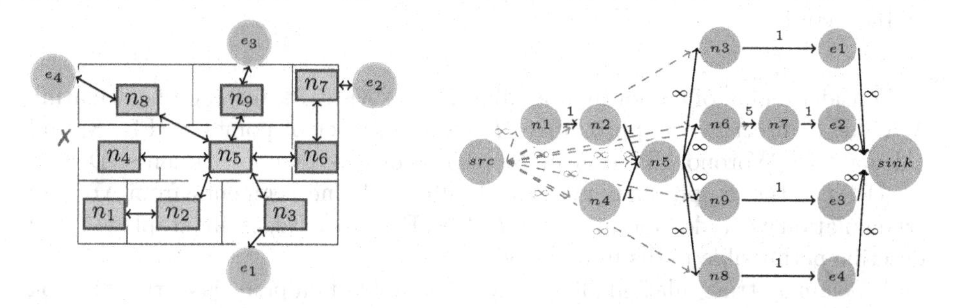

Fig. 6. Searching for the evacuation routes from a node to the available escapes

Using the graph augmented by route lengths alone lets us determine worst-case scenarios in uncongested scenarios by using Dijkstra algorithm. Similarly, using corridor or door widths we can determine capacities associated with edges (see [6] for discussion of maximum flows for horizontal and vertical movement).

Modeling occupant traffic using a flow network. Initially each non-empty room is linked from the source with an edge of infinite capacity. As soon as all occupants move out of the room, the corresponding edge is removed. The flow can be calculated using e.g. the Ford-Fulkerson method or push-relabel method [5],[2].

4 Geometric Network Construction

In this section we finally discuss the proposed algorithm that constructs the augmented graph. Usually Medial Axis Transform (MAT) or Straight-Medial Axis Transform (S-MAT) [7] is used to define a geometric network. See [9] for a discussion of other algorithms and [8] for an example of an alternative approach of indoor navigation that does not require topology construction. A Medial Axis of a set $F \subseteq V$ is the set of points $M \subseteq V$ that have at least two closest neighbours in F (see Fig. 7). If F is finite, Voronoi diagram [4] of F is the Medial Axis of F.

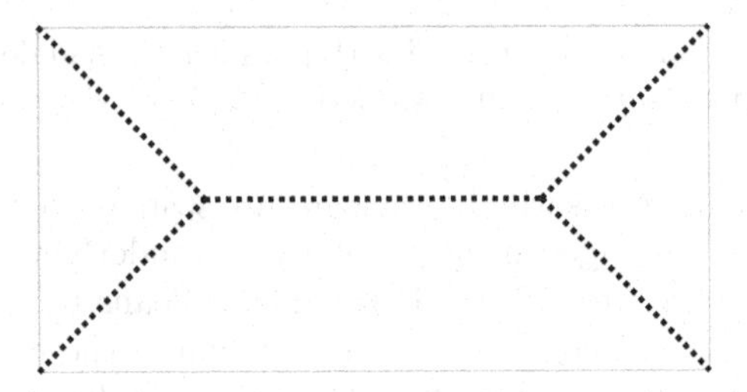

Fig. 7. Medial Axis Transform of the gray rectangle consists of five line segments shown on the picture

Consider a plan of a floor in a building $F \subset \mathbb{R}^2$ (Fig. 8). Instead of calculating MAT of F directly, we approximate F by a finite set of points S (Fig. 8) and calculate the Voronoi diagram of S, which consists of line segments. Denote by (V', E') the graph that consists of subset of line segments from Voronoi triangulation that do not intersect F (see Fig. 10). Edges in graph (V', E') describe permissible paths in our model.

A Delaunay triangulation [3] for a set S of points in a plane is a triangulation $DT(S)$ such that no point in S is inside the circumcircle of any triangle in $DT(S)$. Delaunay triangulations maximize the minimum angle of all the angles of the triangles in the triangulation; they tend to avoid skinny triangles.

Delaunay triangulation of S is the dual graph of the Voronoi diagram of S (Fig. 9). If a Delaunay triangle contains several points associated with vertices $v \in V'$, we further contract these vertices into one vertex.

We utilize Delaunay triangulation in two ways:

1. Firstly, we use it to approximate the minimum width of certain routes (associated with $e \in E'$) by the shortest line segment in Delaunay triangulation that intersects e, and thus set capacities of $e \in E'$.
2. Secondly, triangles in Delaunay triangulation are assigned to vertices $v \in V'$ and provide a partitioning of the geometric view of the scene. Consider a set of regions initially consisting of Delaunay triangles.

If a region contains a point $v \in V$, we assign it to corresponding $v' \in V'$, otherwise we merge this region with any of its neighbouring regions that is not separated by a line segment defining minimum width of an edge $e \in E'$. By repeating this process we provide a bijection between region and vertices $v' \in V'$.

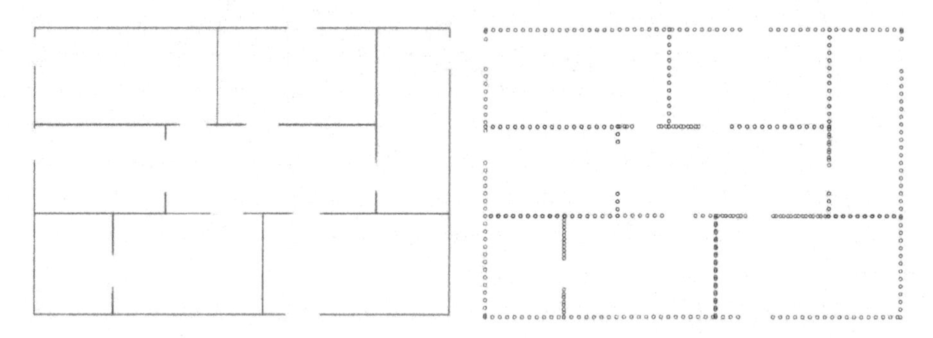

Fig. 8. A 2-dimensional slice of the building that represents a single floor (with doors removed) and the approximation of line segments by a set of points

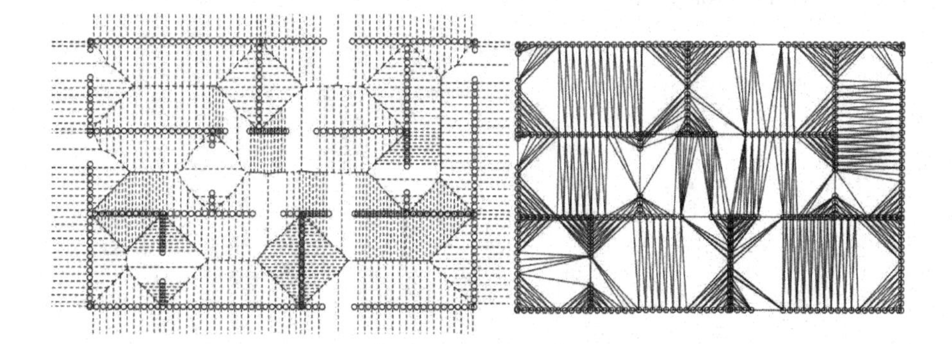

Fig. 9. Voronoi tessellation and Delaunay triangulation of a floor plan approximated by a set of points

It is important to notice the fact that both algorithms of construction of Vonoroi tesellation as well as Delaunay triangulation for a given set of points in the plane can be implemented in $O(n \log n)$ steps. Thus the proposed solution is efficient, even for the large and complicated floor plans.

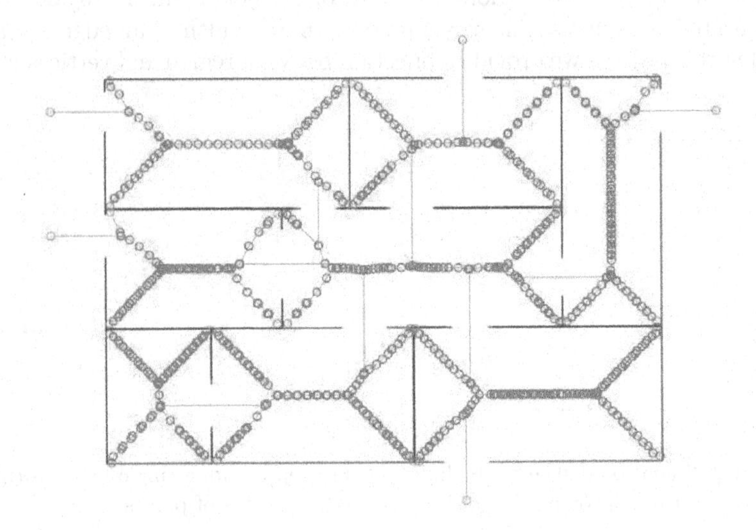

Fig. 10. A subset of line segments from Voronoi tessellation determines permissible paths. This is the initial graph (V', E').

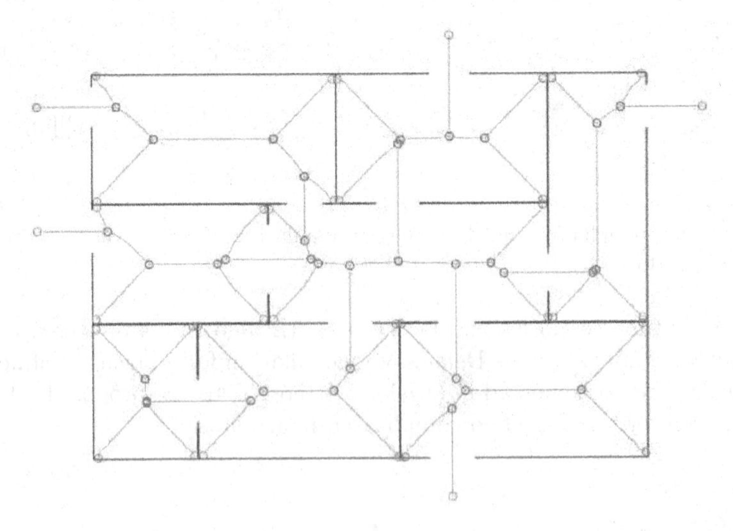

Fig. 11. A graph (V, E) resulting from contraction of edges of degree 2 in graph (V', E')

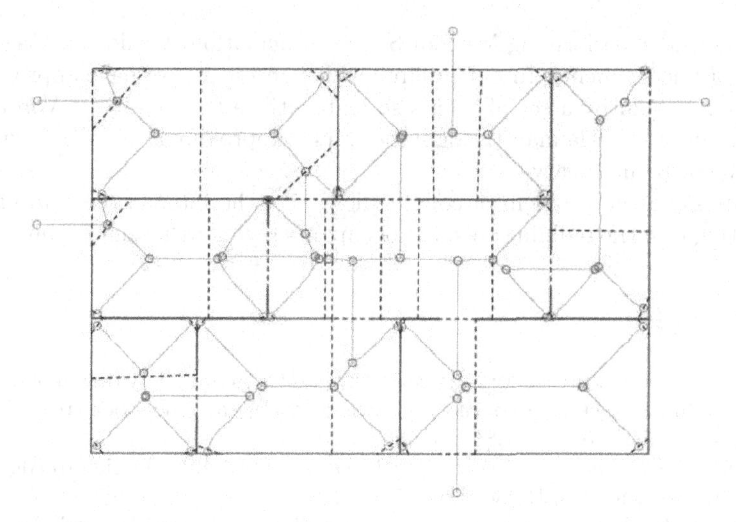

Fig. 12. Selected edges (dashed, blue) from Delaunay triangulation determine widths of paths described by edges in the graph. Delaunay triangulation also provides a mapping of points on the floor plan to corresponding vertices (though some triangles may contain a few vertices $v \in V$). Vertices in corners have very small regions assigned to them and the ratio of width to area of such vertices is relatively high, thus they are not bottlenecks.

5 Conclusions and Future Work

In this paper we have discussed a framework for evacuation modeling based on building topology extraction from the building. We supplemented the typical graph representation of a floor plan derived from MAT by information obtained from Delaunay triangulation of the point set that approximates a single floor: path widths and areas assigned to vertices.

We have mentioned various areas of future research throughout our paper:

- Design of evacuation models based on different assumptions of occupant localization. Our current research focuses on localization of people within buildings.
- More refined models, e.g. taking into account effects of fire or smoke.
- Staircases are often bottlenecks during evacuations, which suggests a more detailed modeling of the effect of different types of staircases and their parameters on movement speeds of crowds.

Other possible areas of future research are:

- Specification of evacuation scenarios in a dialogue with the user (during the ride to fire scene). The dialogue necessarily needs to be very limited, but it could aid the commander in specifying initial plans of action better than passive information delivery. On rare occasions some hints may be also provided by the operator of Fire Command Center, e.g. unavailability of certain exits.

- We consider performing MAT or S-MAT calculations for floor plans consisting of line segments. In our preliminary experiments we have approximated the floor plan by a set of points so as to utilize the duality of Voronoi tessellation and Delaunay triangulation. The approximation by a discrete set may not be necessary.
- Addressing the problem of counter-flows, i.e. the interaction of fire-fighters getting into the building with the occupants trying to get out of the building.

References

1. The application of fire safety engineering principles to fire safety design of buildings: part 6: Human factors: Life safety strategies: occupant evacuation, behaviour and condition (sub-system 6). BSI, London (2004)
2. Cormen, T.H., Stein, C., Rivest, R.L., Leiserson, C.E.: Introduction to Algorithms, 2nd edn. McGraw-Hill Higher Education (2001)
3. Delaunay, B.: Sur la sphère vide. A la mémoire de Georges Voronoï. Bulletin de l'Académie des Sciences de l'URSS (6), 793–800 (1934)
4. Dirichlet, G.L.: Über die Reduktion der positiven quadratischen Formen mit drei unbestimmten ganzen Zahlen. Journal für die Reine und Angewandte Mathematik 40, 209–227 (1850)
5. Ford, L.R., Fulkerson, D.R.: Maximal Flow through a Network. Canadian Journal of Mathematics, 399–404
6. Gwynne, S.M.V., Rosenbaum, E.: Employing the Hydraulic Model in Assessing Emergency Movement. In: DiNenno, P.J., Drysdale, D., Beyler, C.L. (eds.) SFPE Handbook of Fire Protection Engineering, 4th edn., sec. 3, pp. 373–395. National Fire Protection Association, Quincy (2008)
7. Lee, J.: A spatial access-oriented implementation of a 3-D GIS topological data model for urban entities. GeoInformatica 8(3), 237–264 (2004)
8. Liu, L., Zlatanova, S.: A 'door-to-door' path-finding approach for indoor navigation. In: Altan, Backhause, Boccardo, Zlatanova (eds.) International Archives ISPRS XXXVIII, 7th Gi4DM (June 2011)
9. Meijers, N.P.M., Zlatanova, S.: 3D geo-information indoors: Structuring for evacuation. In: Proceedings of the Joint International ISPRS, EuroSDR, and DGPF Workshop on Next Generation 3D City Models, pp. 21–22 (2005)

Integrated Assessment Model on Global-Scale Emissions of Air Pollutants

Thanh Binh Nguyen

International Institute for Applied Systems Analysis (IIASA),
Schlossplatz 1, A-2361 Laxenburg, Austria
nguyenb@iiasa.ac.at

Abstract. In order to analyse recent trends and future world emission scenarios, a global analytical tool, namely GAINS-IAM (integrated assess model), has been proposed and developed. In this context, a global-scale multidimensional data cubes has been specified and devloped based on the Semantic Collective Cubing platform, which has also been introduced in our studying literature. In order to prove of our concepts, typical examples have been illustrated to present how the GAINS-IAM system supports the development of global science-driven policies.

Keywords: GAINS (Greenhouse Gas - Air Pollution Interactions and Synergies), IAM (integrated assess model), FDWA (Federated Data Warehousing Application) framework, Semantic Collective Cubing platform.

1 Introduction

In recent years, new scientific insights revealed linkages and mechanisms that raise interest in the global evolution of air pollutant emissions [1]. To develop a global perspective of emission analysis, according to [20], the "representative concentration pathway" (RCP) studies of global integrated assessment models have recently added future emission scenarios of air pollutants to their projections of long-lived GHGs. Integrated assessment models have been proposed to identify portfolios of measures that improve air quality and reduce greenhouse gas emissions at least cost at global scales [1]. Such models bring together scientific knowledge and quality-controlled data on future socio-economic driving forces of emissions, on the technical and economic features of the available emission control options, on the chemical transformation and dispersion of pollutants in the atmosphere, and the resulting impacts on human health and the environment [1].

In our previous research [15,16,17], we have been focusing on regional federated data warehousing application (FDWA) framework in the context of the GAINS (Greenhouse gas – Air Pollution Interactions and Synergies) model, in which data marts are designed for regional analytical requirements and a data warehouse contains underline data organized in relational multidimensional data models. In [15] we have also presented a semantic metadata approach to generate as well as operate and

H.A. Le Thi et al. (eds.), *Advanced Computational Methods for Knowledge Engineering*,
Advances in Intelligent Systems and Computing 358, DOI: 10.1007/978-3-319-17996-4_31

manage linked data cubes, which are specified by using the FDWA framework [14] and the Collective Cubing platform [6,7] with various levels of regional data granularities.

In this paper, in order to analyse recent trends and future world emission scenarios, a global analytical tool, namely GAINS-IAM (integrated assess model), have been proposed and developed. By inhering the data structure and functions from the FDWA framework [14], the global-scale multidimensional schema can be defined with three global version-specific dimensions, i.e. *iam_region*, *iam_fuelactivity*, and *iam_sector* which are specified from GAINS *region*, *fuel activity* and *sector* dimensions [18]. Afterwards, the current existing regional data sets in the GAINS data warehouse and linked data marts are preprocessed and integrated into GAINS-IAM data mart, i.e. *IAM Activity Pathway* data cube, according to the mapping between GAINS-IAM dimensional structures and other GAINS dimensions, e.g. *GAINS region* dimension is aggregated to *iam_region*. These improvements would make the GAINS-IAM model more suitable for calculating global-scale emission in terms of *IAM Emission* data cube by using the Semantic Collective Cubing Platform [14]. To illustrate the concepts, some typical examples have been presented. The utilization of the global integrated assess model approach provides a feasible and effective method to improve the ability of building, managing as well as analysis of global emission analytical tool.

The rest of this paper is organized as follows: section 2 introduces some approaches and projects related to our work; after an introduction of GAINS IAM concepts in section 3. Section 4 will present our implementation results. At last, section 5 gives a summary of what have been achieved and future works.

2 Related Work

The characters of the proposed approach can be rooted in several research areas of federated data warehousing technologies [4,9,13] and Business Intelligence (BI) [3,19,22], including the trends and concepts of BI solutions, the combined use of mathematical models and federated data warehousing technologies for supporting BI, as well as their utilization in GAINS-IAM model.

In [21], the notion of BI [5,7,8,16,22] that was "invented" in the context of (early) data mining as a circumscription of the fact that business can improve or enhance their "intelligence" regarding customers and revenues by analyzing and "massaging" their data to discover the unknown will now enter the next level. In this context, data cubes are used to support data analysis, in which the data are thought of as a multidimensional array with various measures of interest [10,11,12]. However, business requirements and constraints frequently change over time and ever-larger volumes of data and data modification, which may create a new level of complexity [6], in which Collective Cubing platform is proposed to explore sample data and enables the data users to quickly identify the most appropriate set of cube definitions in the warehouse so that they optimize two costs: the query processing cost and cube maintenance cost [11,12].

Our research area of integrated assessment related to climate change is an example of these issues, where the affects of future climate on social, economic and ecosystems are investigated [1,2]. As input for our work on scenarios, we have looked especially into existing global scenarios which include a description of long-term (i.e. at least several decades) trends in air pollutant emissions. In this area, especially the GAINS scenarios are well known and are used in various ongoing research projects [2].

This paper focuses on integrating FDWA framework [14,15] and Collective Cubing platform [6] to consider the potential approach of the developments of data cubes in global emission analytical context. The work to be developed in this research is aimed to achieve an efficient approach for mapping data and its multidimensional schema from GAINS regional data marts to the global integrated assess model, i.e. GAINS-IAM.

3 GAINS-IAM Concepts

3.1 GAINS-IAM Global Emission Concepts

In GAINS-IAM global scenarios for future emissions are made of essentially three building blocks: (a) integrated activity projections, such as an integration of energy use in different sectors, economic output of different regional data marts etc. (b) global control strategies, i.e. application rates of control/mitigation technologies (these control strategies represent the various environmental policies), and (c) sets of emission factors and cost factors. As illustrated in Fig. 1, a global scenario, namely *ECLIPSE_V5_CLE_base*, is a combination of a dataset for each of these three components, and the scenario can be configured not only for different countries, but also certain classes of macro sectors individually.

SCEN	ACT_TYPE	PATH_ABB	CON_STRAT	REGION
ECLIPSE_V5_CLE_base	AGR	FAO12_2050_E_V5	KORN_WHOL_cle_Ev5	KORN_WHOL
		FAO12_2050_E_V5	KORS_NORT_cle_Ev5	KORS_NORT
		FAO12_2050_E_V5	KORS_NORT_cle_Ev5	KORS_PUSA
		FAO12_2050_E_V5	KORS_NORT_cle_Ev5	KORS_SEOI
		FAO12_2050_E_V5	KORS_NORT_cle_Ev5	KORS_SOUT
		FAO12_2050_E_V5	RUSS_ASIA_cle_Ev5	RUSS_ASIA
		FAO12_2050_E_V5	RUSS_EURO_cle_Ev5	RUSS_EURO
	ENE	ETP_2012_6C	KORN_WHOL_cle_Ev5	KORN_WHOL
		ETP_2012_6C	KORS_NORT_cle_Ev5	KORS_NORT
		ETP_2012_6C	KORS_NORT_cle_Ev5	KORS_PUSA
		ETP_2012_6C	KORS_NORT_cle_Ev5	KORS_SEOI
		ETP_2012_6C	KORS_NORT_cle_Ev5	KORS_SOUT
		ETP_2012_6C	RUSS_ASIA_cle_Ev5	RUSS_ASIA
		ETP_2012_6C	RUSS_EURO_cle_Ev5	RUSS_EURO
	MOB	ETP_2012_6C	KORN_WHOL_cle_Ev5	KORN_WHOL
		ETP_2012_6C	KORS_NORT_cle_Ev5	KORS_NORT
		ETP_2012_6C	KORS_NORT_cle_Ev5	KORS_PUSA
		ETP_2012_6C	KORS_NORT_cle_Ev5	KORS_SEOI
		ETP_2012_6C	KORS_NORT_cle_Ev5	KORS_SOUT
		ETP_2012_6C	RUSS_ASIA_cle_Ev5	RUSS_ASIA
		ETP_2012_6C	RUSS_EURO_cle_Ev5	RUSS_EURO

Fig. 1. An example of GAINS-IAM global scenarios for future emissions

As illustrated in Fig. 2, first the FDWA framework [14,15,17] is applied to specify a data warehouse, which data are integrated from multiple data sources and a class of virtual and manifest data marts. In this context, the platform is also used to generate data of the GAINS-IAM data mart. Afterwards, the Semantic Collective Cubing platform [14] is used to generate GAINS-IAM linked data cubes.

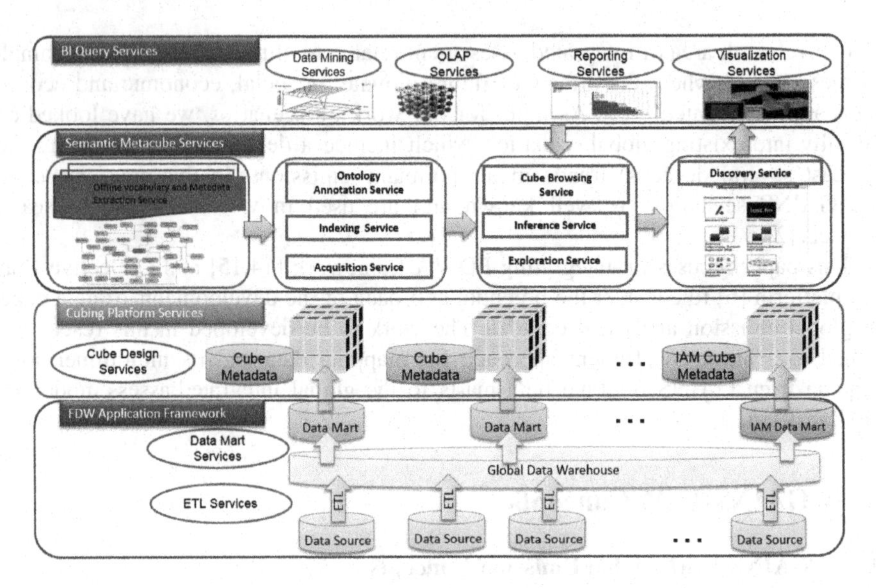

Fig. 2. FDWA Framework used to generate GAINS- IAM Data Mart Data

3.2 GAINS IAM Data Model

The GAINS IAM Data Model is defined based on the FDWA framework, therefore it could also be specified as follows:

$$GAINS - IAM =< \{Dim_n\}, \{Fact_m\}, \{Cube_q\}\} >$$, where:

- $\{Dim_n\}$={*iam_region,pollutant,iam_sector,iam_fuelactivity,technology,year, scenario*} is a set of dimensions.
- $\{Fact_m\}$ is a set of decision variables or facts.
- $\{Cube_q\}$= is a set of global emission data cubes, i.e. *IAM_Pathway* and *IAM_Emission* data cubes.

GAINS-IAM Dimensions

The GAINS-IAM data model has *iam_region, iam_sector, iam_fuel-activity, pollutant, year* dimensions.

- *iam_region dimension.* The regional division for the regional averages is the following:

All-IAM-Regions->IAM-RegionGroup->IAM-Region

- *IAM Sector Dimension (iam_sector) and IAM Fuel Activity Dimension (iam_fuel activity).* GAINS–IAM also covers a number of sectors, and each sector may be associated with a number of different activities. The *iam_sector* and *iam_fuel activity* dimensions covered by GAINS-IAM are organized as illustrated in Fig. 4 a+ b.

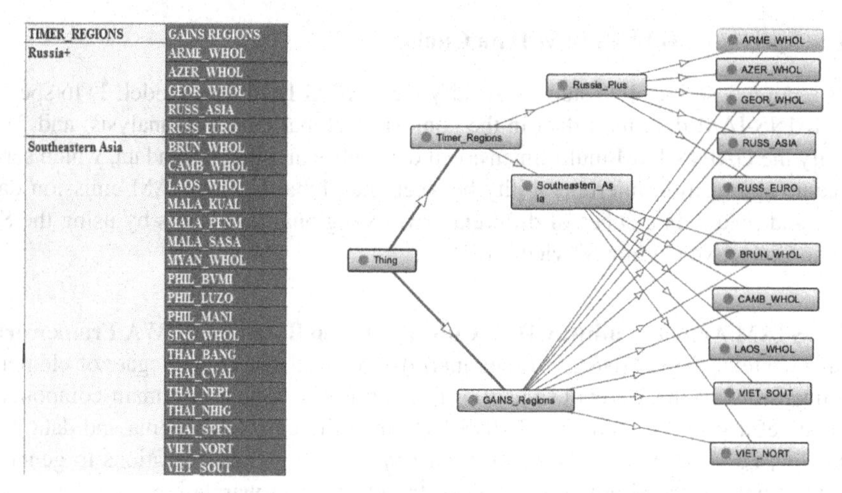

TIMER_REGIONS	GAINS REGIONS
Russia+	ARME_WHOL
	AZER_WHOL
	GEOR_WHOL
	RUSS_ASIA
	RUSS_EURO
Southeastern Asia	BRUN_WHOL
	CAMB_WHOL
	LAOS_WHOL
	MALA_KUAL
	MALA_PENM
	MALA_SASA
	MYAN_WHOL
	PHIL_BVMI
	PHIL_LUZO
	PHIL_MANI
	SING_WHOL
	THAI_BANG
	THAI_CVAL
	THAI_NEPL
	THAI_NHIG
	THAI_SPEN
	VIET_NORT
	VIET_SOUT

Fig. 3. a + b. Mapping between *iam_region* and GAINS *region* dimensions

All_IAM-Sectors->IAM-Sector
All_IAM_Fuel Activities->IAM-Activity

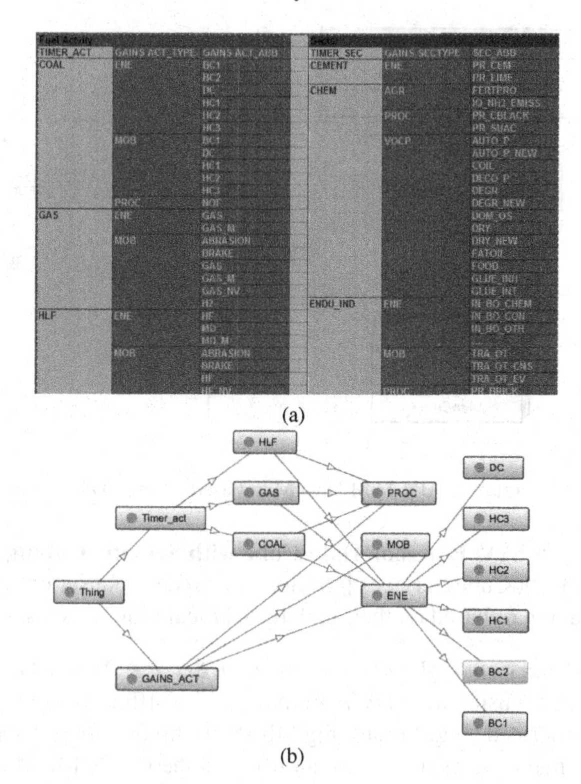

Fuel Activity

TIMER_ACT	GAINS ACT_TYPE	GAINS ACT_ABB
COAL	ENE	BC1
		BC2
		DC
		HC1
		HC2
		HC3
	MOB	BC1
		DC
		HC1
		HC2
		HC3
	PROC	NOF
GAS	ENE	GAS
		GAS_M
	MOB	ABRASION
		BRAKE
		GAS
		GAS_M
		GAS_NV
		H2
HLF	ENE	HF
		MD
		MD_M
	MOB	ABRASION
		BRAKE
		HF
		HF_NV

Sector

TIMER_SEC	GAINS SECTYPE	SEC_ABB
CEMENT	ENE	PR_CEM
		PR_LIME
CHEM	AGR	FERTPRO
		IO_NH3_EMISS
	PROC	PR_CBLACK
		PR_SUAC
	VOCP	AUTO_P
		AUTO_P_NEW
		COIL
		DECO_P
		DEGR
		DEGR_NEW
		DOM_OS
		DRY
		DRY_NEW
		FATOIL
		FOOD
		GLUE_IND
		GLUE_INT
ENDU_IND	ENE	IN_BO_CHEM
		IN_BO_CON
		IN_BO_OTH
	MOB	TRA_OT
		TRA_OT_CNS
		TRA_OT_LV
	PROC	PR_BRICK

(a)

(b)

Fig. 4. a+b..Mapping between *iam_fuel activity* and GAINS *fuel activity* dimensions

3.3 Specifying GAINS IAM Data Cubes

This section describes two steps to specify the GAINS-IAM data model: 1) to specify the GAINS-IAM data mart data in the context of global emission analysis; and 2) to specify the GAINS-IAM multidimensional data cubes and their metadata, which serve dynamic aggregation interoperability between the global GAINS-IAM emission data cubes and linked data cubes of different GAINS regional data marts by using the Semantic Collective Cubing Services [15].

GAINS IAM Activity Pathway Data Cube Definition Based on FDWA Framework
In this context, the GAINS IAM data mart data is a subset, an aggregate of elements in dimensions of the warehouse as illustrated in Fig. 5. There two main components will be defined in this step: the GAINS IAM multidimensional schema and data mart data [14], i.e. *IAM Activity Pathway Data Cube*. Many of the operations to generate this mart will be the simple projection of the global data warehouse onto the chosen dimensional levels, but in the GAINS IAM case, these operations involve aggregation, or linking existing multiple regional mart data, potentially with additional data/sources of information to fulfill the global requirements.

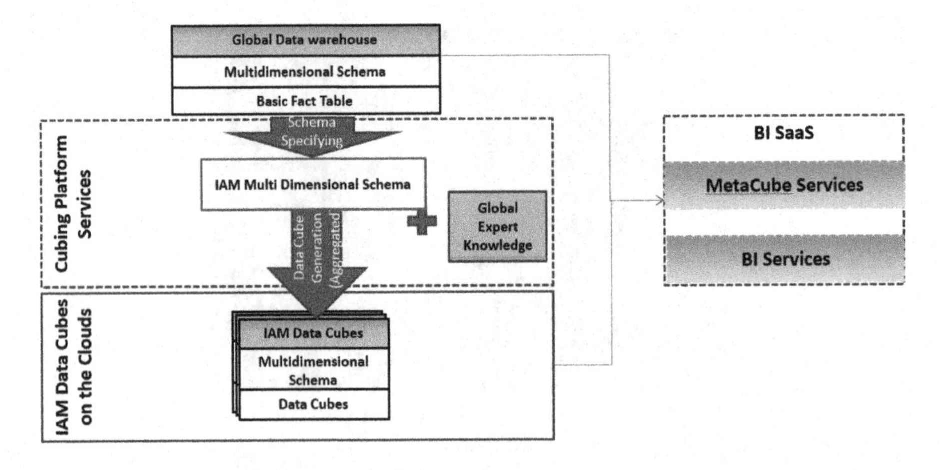

Fig. 5. Generating GAIN-IAM Data Mart based on the FDWA framework

Specifying GAINS IAM Emission Data Cube with Sematic Cubing Platform
In this step, the Data Cube Services [15] are used to define a set of data cubes of the GAINS-IAM data mart based on their global analytical requirements.

- *Offline vocabulary and Metadata Extraction Service.* When GAINS-IAM data mart is defined by using the FDWA framework, its offline vocabulary and metadata are also generated. Fig. 3b and Fig. 4b show the ontology mapping, including dimension domain values and metadata between the GAINS-IAM and the GAINS models.

- *Data Cube Services.* The platform supports an interface that receive request and presents the result by picking cube terms by hierarchical browsing or through keyword/ontology search.
- *Arbitrary cube to cube browsing.* Cubing platform supports in exploring more detail for a particular element in a cube reporting and data analysis session or moving to a cube in another location in the data space altogether. The inference engine turns the request into a particular cube selection/definition and returns that cube to display results.

4 Implementing Results of the GAINS-IAM Data Model

The GAINS model [17,18] has been employed to estimate—for exogenous projections of energy consumption and other emission generating activities—the possible range of future air pollutant emissions, using a bottom-up approach that explicitly accounts for detailed emission control legislation in 162 world regions. With this model, it is possible to explore, for different energy and climate pathways, the scope for future air pollutant emissions that could be achieved by different levels of policy interventions. For each of these 162 regions, more than 1,500 specific emission control measures have been considered [1,2].

4.1 Global Activity Pathway Data

In the GAINS-IAM, the global *Activity Pathway* data is an integration of activity pathway data from regional linked data marts by mapping of 162 world regions to *iam_region* dimension (Fig. 3a +b), of regional GAINS *fuel_activity-sector* combinations to *iam_fuel* activity and *iam_sector* dimensions as shown in the following figure:

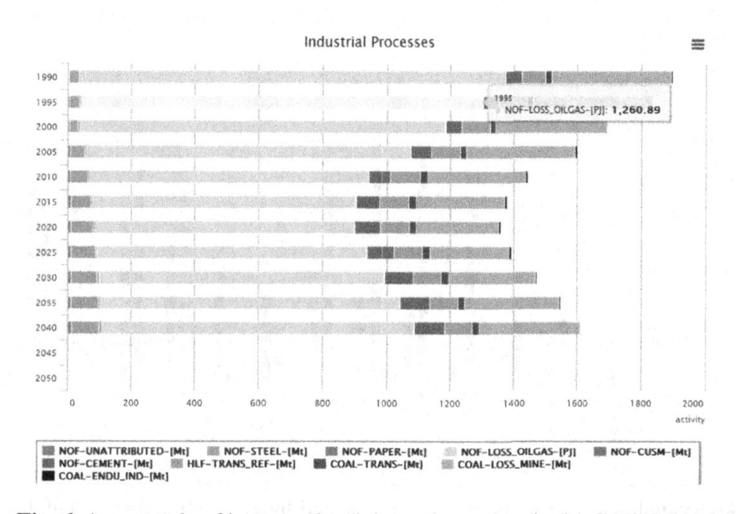

Fig. 6. An example of integrated activity pathway data in the GAINS-IAM

4.2 Global Emission Data Cube

More explicit spatial emissions for future years are derived through downscaling approaches that apportion emissions calculated for the aggregated world regions, i.e. *iam_region*(s) to individual countries or grid cells on the basis of historic emission patterns and generic rules that reflect urbanization dynamics and economic concentration processes. Thereby, future emissions are first determined for a limited number of world regions with an assumed development of regional mean emission factors, and then, the result is downscaled to specific locations based on certain rules. Using the data of the *IAM Activity Pathway* data cube, global emissions scenarios in term of *IAM Emission* data cube could be developed with the GAINS-IAM model. These scenarios consider the implementation schedules of currently agreed national emission control legislation and explore, for different energy and climate policy scenarios, the scope for emission reductions that could be achieved with additional air pollution control measures that are beyond current policies.

The divergence of regional emission trends for different sectors is highlighted in Fig. 7 as a typical example of the GAINS-IAM application results. Between 1990 and 2050, for China, current legislation is expected to alleviate the situation to some extent, although high emission densities are expected to prevail over large areas, especially in the northeastern section. By contrast, given the lack of emission control legislation and the expected growth in power generation from coal, the Indian subcontinent will emerge as a new global hot spot of SO_2 emissions.

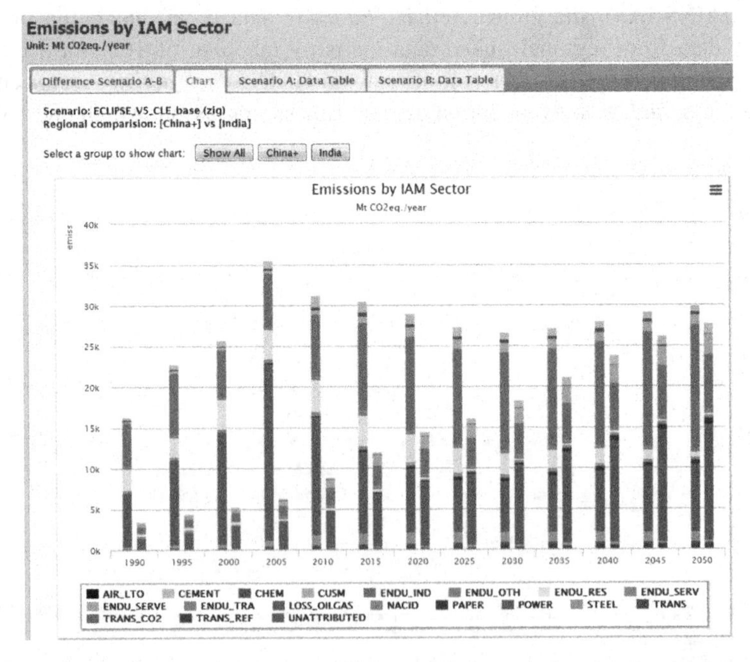

Fig. 7. An example of comparison between China and India emission data in a global emission scenario

5 Conclusion

This paper introduced, described and proposed a tool to analyse recent trends and future global emission scenarios in term of a global analytical model, namely GAINS-IAM (integrated assess model). The model has been developed and presented. This global analytical tool and its global concepts and requirements have been introduced with some additional development needs, i.e. mapping GAINS regional multidimensional schemas to global aggregated ones, aggregating data from existing GAINS data marts to global levels of data granularities in term of linked data cubes [6,14,15]. In this context, three global version-specific dimensions, i.e. *iam_region*, *iam_fuelactivity*, and *iam_sector* which are generalized from GAINS *region*, *fuel activity* and *sector* dimensions are described together with the mapping information. Afterwards, the current existing regional data sets in the GAINS data warehouse as well as in linked data marts are preprocessed and integrated into GAINS-IAM model, according to the dimension mapping information. As a result, GAINS-IAM model could be suitable for calculating global-scale emission in term of *IAM Emission* data cube, which is defined by using the Semantic Collective Cubing Platform [15]. Some typical examples are presented to illustrate the proposed concepts.

Future work of our approach could then be able to support users in building data cubes in cost efficient and elastic manner that spans all aspects of cube building lifecycle, i.e. cube definition, cube computation, cube evolution as well as cube sharing. Furthermore, we will focus on the implementation of the global cost analysis among linked emission data cubes to make use of our concepts.

References

1. Amann, M., Klimont, Z., Wagner, F.: Regional and global emissions of air pollutants: Recent trends and future scenarios. Annual Review of Environment and Resources 38, 31–55 (2013)
2. Amann, M., Bertok, I., Cofala, J., Heyes, C., Klimont, Z., Nguyen, T.B., R., Posch, M., Rafaj, P., Sandler, R., Schöpp, W., Wagner, F., Winiwarter, W.: Cost-effective control of air quality and greenhouse gases in Europe: modeling and policy applications. Environmental Modelling & Software (2011)
3. full360: LESSON- Managed Software-as-a-Service Business Intelligence: A Model that works. TDWI-The Data Warehousing Institute (2010)
4. Furtado, P.: A survey of parallel and distributed data warehouses. International Journal of Data Warehousing and Mining 5(2), 57–77
5. Gangadharan, G.R., Swami, S.N.: Business Intelligence Systems: Design and Implementation Strategies. In: Proc. of the 26th International Conference Information Technology Interfaces, ITI 2004, Croatia, pp. 139–144 (2004)
6. Hoang, D.T.A., Ngo, N.S., Nguyen, B.T.: Collective Cubing Platform towards Definition and Analysis of Warehouse Cubes. In: Nguyen, N.-T., Hoang, K., Jędrzejowicz, P. (eds.) ICCCI 2012, Part II. LNCS, vol. 7654, pp. 11–20. Springer, Heidelberg (2012)
7. Hoang, A.D.T., Nguyen, T.B.: An Integrated Use of CWM and Ontological Modeling Approaches towards ETL Processes. In: ICEBE 2008, pp. 715–720 (2008)

8. Hoang, A.D.T., Nguyen, T.B.: State of the Art and Emerging Rule-driven Perspectives towards Service-based Business Process Interoperability. In: RIVF 2009, pp. 1–4 (2009)
9. Jindal, R., Acharya, A.: Federated data warehouse architecture, in White Paper
10. Le, P.D., Nguyen, T.B.: OWL-based data cube for conceptual multidimensional data model. In: Proceedings of the First International Conference on Theories and Applications of Computer Science ICTACS 2006, Ho Chi Minh City, Vietnam, pp. 247–260 (2006) ISBN: 978-981-270-063-6
11. Messaoud, R.B.E.N., Boussaid, O., Rabaseda, S.L.: A Multiple Correspondence Analysis to Organize Data Cubes. Information Systems Frontiers 1, 133–146 (2007)
12. Meissner, A.: A simple distributed reasoning system for the connection calculus. Vietnam Journal of Computer Science 1(4), 231–239 (2014)
13. Nguyen, N.T.: Processing inconsistency of knowledge in determining knowledge of a collective. Cybernetics and Systems 40(8), 670–688 (2009)
14. Nguyen, T.B.: Cloud-based data warehousing application framework for modeling global and regional data management systems. In: Nguyen, N.T., van Do, T., Thi, H.A. (eds.) ICCSAMA 2013. SCI, vol. 479, pp. 319–327. Springer, Heidelberg (2013)
15. Nguyen, T.B., Ngo, N.S.: Semantic Cubing Platform enabling Interoperability Analysis among Cloud-based Linked Data Cubes. In: Proceedings of the 8th International Conference on Research and Pratical Issues of Enterprise Information Systems, CONFENIS 2014. ACM International Conference Proceedings Series (2014)
16. Nguyen, T.B., Tjoa, A.M., Wagner, R.: Conceptual multidimensional data model based on metaCube. In: Yakhno, T. (ed.) ADVIS 2000. LNCS, vol. 1909, pp. 24–33. Springer, Heidelberg (2000)
17. Nguyen, T.B., Wagner, F.: Collective intelligent toolbox based on linked model framework. Journal of Intelligent and Fuzzy Systems 27(2), 601–609 (2014)
18. Nguyen, T.B., Wagner, F., Schoepp, W.: GAINS-BI: Business Intelligent Approach for Greenhouse Gas and Air Pollution Interactions and Synergies Information System. In: Proc. of the International Organization for Information Integration and Web-based Application and Services, IIWAS 2008, Linz (2008)
19. Tvrdikova, M.: Support of Decision Making by Business Intelligence Tools. In: Proc. of the 6th International Conference on Computer Information Systems and Industrial Management Applications, pp. 364–368 (2007)
20. Van Vuuren, D.P., Edmonds, J.A., Kainuma, M., Riahi, K., Thomson, A.M., et al.: The representative concentration pathways: an overview. Clim. Change 109, 5–31
21. Vossen, G.: Big data as the new enabler in business and other intelligence. Vietnam Journal of Computer Science, 3–14, http://dx.doi.org/10.1007/s40595-013-0001-6
22. Watson, H.J., Wixom, B.H.: The current state of business intelligence. Computer 40, 96–99

Query-Subquery Nets with Stratified Negation

Son Thanh Cao[1,2]

[1] Faculty of Information Technology, Vinh University
182 Le Duan street, Vinh, Nghe An, Vietnam
sonct@vinhuni.edu.vn
[2] Institute of Informatics, University of Warsaw
Banacha 2, 02-097 Warsaw, Poland

Abstract. In this paper, we incorporate the concept of stratified negation into query-subquery nets to obtain an evaluation method called QSQN-STR for dealing with stratified knowledge bases. Particularly, we extend query-subquery nets by allowing negated atoms in the bodies of some program clauses. The empirical results illustrate the usefulness of the method.

Keywords: Horn knowledge bases, deductive databases, query processing, stratified negation, QSQN, QSQN-STR.

1 Introduction

In the recent years, rule-based query languages, including languages related to Datalog, have attracted increasing attention from researchers although they have been studied for more than three decades, especially as rule languages are now applied in areas such as the Semantic Web [4,6,11,12,14,19,23]. Since deductive databases are widely used in practical applications, it is worth doing further research on this topic.

Positive logic programs can express only monotonic queries. As many queries of practical interest are non-monotonic, it is desirable to consider normal logic programs, which allow negation to occur in the bodies of program clauses. A number of interesting semantics for normal logic programs have been defined, for instance, stratified semantics [2] (for stratified logic programs), stable-model semantics [16] and well-founded semantics [15]. The survey [3] also provides a good source for references on these semantics. This work studies query processing for stratified knowledge bases, which is based on query-subquery nets.

A normal logic program is stratifiable if it can be divided into strata such that if a negative literal of a predicate p occurs in the body of a program clause in a stratum, then the clauses defining p must belong to an earlier stratum. A stratified knowledge base consists of a stratified logic program (for defining intensional predicates) and an instance of extensional predicates. Programs in this class have a very intuitive semantics and have been considered in [2,5,17,18,20].

There is an approach for evaluating stratified programs (which evaluates stratum-by-stratum): (*i*) evaluating intensional predicates in lower strata first;

H.A. Le Thi et al. (eds.), *Advanced Computational Methods for Knowledge Engineering*,
Advances in Intelligent Systems and Computing 358, DOI: 10.1007/978-3-319-17996-4_32

(*ii*) after that, treating them as extensional instances for the intensional predicates in higher strata.

Magic-set transformation together with the improved semi-naive evaluation method [5,13], which we will call the Magic-Sets method, takes the advantage of reducing irrelevant facts and restricting the search space. It combines the pros of top-down and bottom-up strategies. However, the improved semi-naive evaluation method uses breadth-first search, and as discussed in [7,21] it is not always efficient.

Horn knowledge bases are a generalization of Datalog deductive databases without the range-restrictedness and function-free conditions [1]. In [21], together with Nguyen we proposed a new method called QSQN that is based on query-subquery nets for evaluating queries to Horn knowledge bases. The preliminary experimental results reported in [7,9,10] justify the usefulness of this method and its extensions. In this paper, we incorporate the concept of stratified negation into query-subquery nets to obtain an evaluation method called QSQN-STR for dealing with stratified knowledge bases. In particular, we extend query-subquery nets by allowing negated atoms in the bodies of some program clause in order to deal with the class of stratified logic programs. We use the Improved Depth-First Control Strategy (IDFS) [9] together with QSQN-STR for evaluating queries to stratified knowledge bases. It is a top-down approach that processes stratum-by-stratum w.r.t. the intensional predicates. The essence of the IDFS is to enter deeper cycles in the considered QSQN-STR first and keep looping along the current "local" cycle as long as possible. This allows to accumulate as many as possible tuples or subqueries at a node, and tries to compute all possible answers of an intensional relation before processing the next ones. Additionally, by applying the advantages of a top-down approach and processing the predicates stratum-by-stratum, at the time of processing a negative subquery, this subquery contains no variables. Thus, this takes the advantage of reducing the number of generated tuples in intensional relations corresponding to negated atoms. The empirical results illustrate the usefulness of our method.

The rest of this paper is structured as follows. Section 2 recalls some of the most important concepts and definitions that are related to our work. Section 3 presents our QSQN-STR evaluation method for dealing with stratified knowledge bases. The experimental results are provided in Section 4. Conclusions and a plan for future work are presented in Section 5.

2 Preliminaries

In this section, we only recall the most important definitions of first-order logic that are needed for our work and refer the reader to [1,17,21] for further reading.

A *predicate* can be classified either as *intensional* or as *extensional*. A *generalized tuple* is a tuple of terms, which may contain function symbols and variables.

A *program clause* is a formula of the form $A \leftarrow B_1, \ldots, B_n$, where A is an atom, $B_1, \ldots, B_n$ are literals (i.e., either atoms or negated atoms). A is called the *head*, and $B_1, \ldots, B_n$ the *body* of the program clause. A *goal* is a formula of the form $\leftarrow B_1, \ldots, B_n$, where $B_1, \ldots, B_n$ are literals.

In order to ensure the results of programs are finite, program clauses are required to satisfy the following *safety* conditions: (i) *range-restrictedness*, which means every variable occurring in the head of a program clause must also occur in the body of that program clause; (ii) *safety*, which means every variable appearing in a negative literal in the body of a program clause also appears in some positive literal in the body of that program clause. Note here that, the order of literals in the body of a program clause is not important under the semantics of the program, thus, from now on we will put negative literals after positive ones in a program clause. Therefore, when a negative subquery is processed, it contains no variables.

A *normal logic program* is a finite set of program clauses. A program which allows negation to appear only in the bodies and only before atoms of extensional predicates is called *semi-positive* program. A program which disallows recursion through negation is called *stratified program.* In this paper, we only consider the semi-positive and stratified programs. Figure 1 shows examples of them.

(a) $q(x, y) \leftarrow t(x, y)$
$q(x, y) \leftarrow t(x, z), not(s(x)), q(z, y)$
$p(x, y) \leftarrow q(y, x).$

(b) $p(x) \leftarrow q(x), not(r(x))$
$p(x) \leftarrow q(x), not(t(x))$
$r(x) \leftarrow s(x), not(t(x)).$

Fig. 1. (a) A semi-positive program; (b) A stratified program

Let θ and δ denote substitutions, $\theta\delta$ denotes the composition of θ and δ, $dom(\theta)$ denotes the domain of θ, $range(\theta)$ denotes the range of θ, and $\theta_{|X}$ denotes the restriction of θ to a set X of variables. Given a list/tuple α of terms or atoms, the set of variables occurring in α is denoted by $Vars(\alpha)$. The *empty substitution* is denoted by ε.

3 Query-Subquery Nets with Stratified Negation

The method QSQN [21] is used for evaluating queries to Horn knowledge bases. The empirical results given in [7,9] indicate the usefulness of this method. In this section, we extend QSQN to obtain a new evaluation method called QSQN-STR for dealing with stratified knowledge bases.

We now present the notions of QSQN-STR structure and QSQN-STR together with detailed explanations. Let P be a logic program and $\varphi_1, \ldots, \varphi_m$ be all the program clauses of P, with $\varphi_i = (A_i \leftarrow B_{i,1}, \ldots, B_{i,n_i})$, for $1 \leq i \leq m$ and $n_i \geq 0$. From a logic program P, the QSQ-Net structure with stratified negation is defined as follows:

Definition 1 (QSQN-STR structure). A *query-subquery net structure with stratified negation* of P, denoted by QSQN-STR structure, is a tuple (V, E, T), where V is a set of nodes, E is a set of edges, T is a function called the *memorizing type* of the net structure, in particular:

- V is composed of:
  - *input_p* and *ans_p*, for each intensional predicate p of P,
  - pre_filter_i, $filter_{i,1}$, ..., $filter_{i,n_i}$, $post\_filter_i$, for each $1 \leq i \leq m$.
- E is composed of:
  - $(filter_{i,1}, filter_{i,2})$, ..., $(filter_{i,n_i-1}, filter_{i,n_i})$, for each $1 \leq i \leq m$,
  - $(input\_p, pre\_filter_i)$ and $(post\_filter_i, ans\_p)$ for each $1 \leq i \leq m$, where p is the predicate of A_i,
  - $(pre\_filter_i, filter_{i,1})$ and $(filter_{i,n_i}, post\_filter_i)$ for each $1 \leq i \leq m$ such that $n_i \geq 1$,
  - $(pre\_filter_i, post\_filter_i)$, for each $1 \leq i \leq m$ such that $n_i = 0$,
  - $(filter_{i,j}, input\_p)$ and $(ans\_p, filter_{i,j})$, for each intensional predicate p and each $1 \leq i \leq m$ and $1 \leq j \leq n_i$ such that $B_{i,j}$ is an atom of p.
- T maps each $filter_{i,j} \in V$ such that the predicate of $B_{i,j}$ is extensional to *true* or *false*. If $T(filter_{i,j}) = false$ then subqueries for $filter_{i,j}$ are always processed immediately without being accumulated at $filter_{i,j}$.

If $(v, w) \in E$ then we call w a *successor* of v. Here, V and E are uniquely specified by P. The pair (V, E) is called the QSQN-STR topological structure of P. ◁

Example 1. This example is a stratified program, which defines whether a node is connected another, but not vice versa of a directed graph. It is taken from [22]. Figure 2 illustrates the program and its QSQN-STR topological structure, where *path* and *acyclic* are intensional predicates, *edge* is an extensional predicate, x, y and z are variables. ◁

We now present the notion of QSQN-STR. It is specified by the following definition.

Definition 2 (QSQN-STR). A *query-subquery net with stratified negation* of P, denoted by QSQN-STR, is a tuple $N = (V, E, T, C)$ such that (V, E, T) is a QSQN-STR structure of P, and C is a mapping that associates each node $v \in V$ with a structure called the *contents* of v with the following properties:

- If $v \in \{input\_p, ans\_p\}$ then $C(v)$ is composed of:
  - $tuples(v)$: a set of generalized tuples of the same arity as p,
  - $unprocessed(v, w)$ for $(v, w) \in E$: a subset of $tuples(v)$.
- If $v = pre\_filter_i$ then $C(v)$ is composed of:
  - $atom(v) = A_i$ and $post\_vars(v) = Vars((B_{i,1}, \ldots, B_{i,n_i}))$.
- If $v = post\_filter_i$ then $C(v)$ is empty, but we assume $pre\_vars(v) = \emptyset$,
- If $v = filter_{i,j}$ and p is the predicate of $B_{i,j}$ then $C(v)$ is composed of:
  - $neg(v) = true$ if $B_{i,j}$ is a negated atom, and $neg(v) = false$ otherwise,
  - $kind(v) = extensional$ if p is extensional, and
    $kind(v) = intensional$ otherwise,
  - $pred(v) = p$ (called the predicate of v) and $atom(v) = B_{i,j}$,
  - $pre\_vars(v) = Vars((B_{i,j}, \ldots, B_{i,n_i}))$ and
    $post\_vars(v) = Vars((B_{i,j+1}, \ldots, B_{i,n_i}))$,

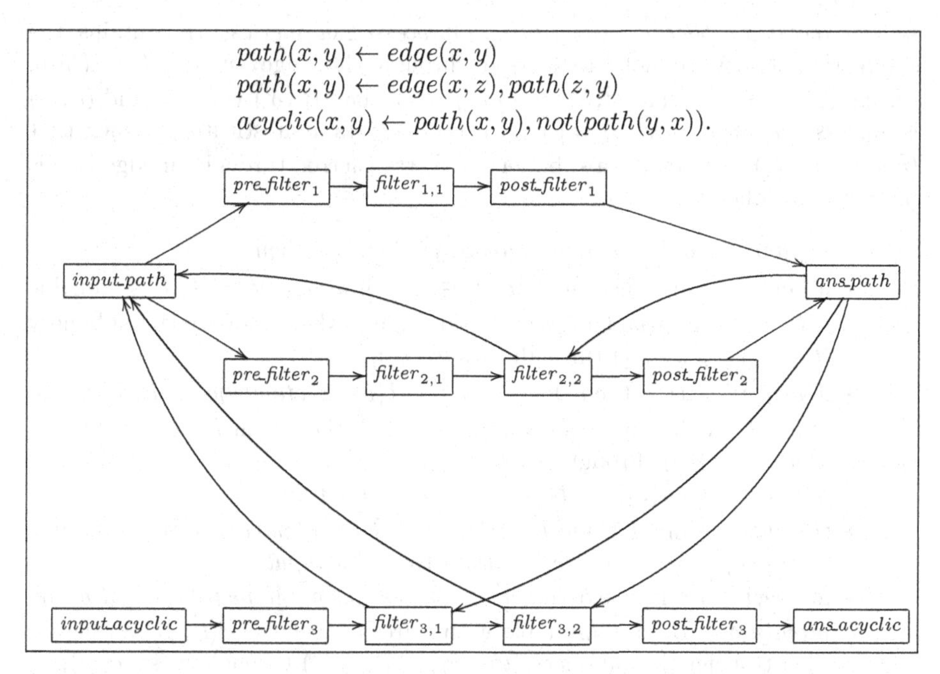

Fig. 2. The QSQN-STR topological structure of the program given in Example 1

- $subqueries(v)$: a set of pairs of the form $(\bar{t}, \delta)$, where $\bar{t}$ is a generalized tuple of the same arity as the predicate of A_i and δ is an idempotent substitution such that $dom(\delta) \subseteq pre\_vars(v)$ and $dom(\delta) \cap Vars(\bar{t}) = \emptyset$,
- $unprocessed\_subqueries(v) \subseteq subqueries(v)$,
- in the case p is intensional:
  $unprocessed\_subqueries_2(v) \subseteq subqueries(v)$,
- in the case p is intensional and $neg(v) = false$:
  $unprocessed\_tuples(v)$: a set of generalized tuples of the same arity as p.

– If $v = filter_{i,j}$, $kind(v) = extensional$ and $T(v) = false$ then $subqueries(v)$ and $unprocessed\_subqueries(v)$ are empty (thus, we can ignore them).

A QSQN-STR of P is *empty* if all the sets of the form $tuples(v)$, $subqueries(v)$, $unprocessed\_subqueries(v)$, $unprocessed\_subqueries_2(v)$, $unprocessed\_tuples(v)$ or $unprocessed(v, w)$ are empty. ◁

We can observe that, for each $(v, w) \in E$:

– if v is either pre_filter_i or $post\_filter_i$ or ($filter_{i,j}$ and $kind(v) = extensional$) then v has exactly one successor, denoted by $succ(v)$;
– if v is $filter_{i,j}$ with $kind(v) = intensional$, $pred(v) = p$ then v has exactly two successors: $succ(v) = filter_{i,j+1}$ if $n_i > j$; $succ(v) = post\_filter_i$ otherwise; and $succ_2(v) = input\_p$.

A *subquery* is a pair of the form $(\bar{t}, \delta)$, where $\bar{t}$ is a generalized tuple and δ is an idempotent substitution such that $dom(\delta) \cap Vars(\bar{t}) = \emptyset$. The set *unprocessed_subqueries*$_2(v)$ (resp. *unprocessed_subqueries*(v)) contains the subqueries that were not transferred through the edge $(v, succ_2(v))$ (resp. $(v, succ(v))$). We say that $(\bar{t}, \delta)$ is *more general* than $(\bar{t}', \delta')$ w.r.t. v, and $(\bar{t}', \delta')$ is an *less general* than $(\bar{t}, \delta)$ w.r.t. v, if there exists a substitution θ such that $(\bar{t}, \delta)\theta = (\bar{t}', \delta')$. Informally, a subquery $(\bar{t}, \delta)$ transferred through an edge to v is processed as follows:

- if $v = filter_{i,j}$, $kind(v) = extensional$, $pred(v) = p$ then
  - if $neg(v) = false$ then, for each $\bar{t}' \in I(p)$, if $atom(v)\delta = B_{i,j}\delta$ is unifiable with a fresh variant of $p(\bar{t}')$ by an mgu γ then transfer the subquery $(\bar{t}\gamma, (\delta\gamma)_{|post\_vars(v)})$ through $(v, succ(v))$,
  - if $neg(v) = true$ then, for every $\bar{t}' \in I(p)$, if $atom(v)\delta = B_{i,j}\delta$ is not unifiable with any fresh variant of $p(\bar{t}')$ then transfer the subquery $(\bar{t}, \delta_{|post\_vars(v)})$ through $(v, succ(v))$.
- if $v = filter_{i,j}$, $kind(v) = intensional$, $pred(v) = p$ then
  - transfer the input tuple $\bar{t}'$ such that $p(\bar{t}') = atom(v)\delta = B_{i,j}\delta$ through $(v, input\_p)$ to add its fresh variant to $tuples(input\_p)$,
  - for each currently existing $\bar{t}' \in tuples(ans\_p)$, if $neg(v) = false$ and $atom(v)\delta = B_{i,j}\delta$ is unifiable with a fresh variant of $p(\bar{t}')$ by an mgu γ then transfer the subquery $(\bar{t}\gamma, (\delta\gamma)_{|post\_vars(v)})$ through $(v, succ(v))$,
  - for every currently existing $\bar{t}' \in unprocessed(ans\_p, v)$, if $neg(v) = true$ and $atom(v)\delta = B_{i,j}\delta$ is not unifiable with any fresh variant of $p(\bar{t}')$ then transfer the subquery $(\bar{t}, \delta_{|post\_vars(v)})$ through $(v, succ(v))$,
  - store the subquery $(\bar{t}, \delta)$ in $subqueries(v)$, and later, for each new $\bar{t}'$ added to $tuples(ans\_p)$, if $neg(v) = false$ and $atom(v)\delta = B_{i,j}\delta$ is unifiable with a fresh variant of $p(\bar{t}')$ by an mgu γ then transfer the subquery $(\bar{t}\gamma, (\delta\gamma)_{|post\_vars(v)})$ through $(v, succ(v))$.
- if $v = post\_filter_i$ and p is the predicate of A_i then transfer the answer tuple $\bar{t}$ through $(post\_filter_i, ans\_p)$ to add it to $tuples(ans\_p)$.

The Algorithm 1 (on page 361) repeatedly selects an active edge and fires the operation for the edge. Due to the lack of the space, the related functions and procedures used for Algorithm 1 are omitted in this paper. The details of these functions and procedures can be found in [8]. In particular, the Algorithm 1 uses the function `active-edge`(u, v), which returns *true* if the data accumulated in u can be processed to produce some data to transfer through the edge (u, v). If `active-edge`(u, v) is *true*, procedure `fire`(u, v) processes the data accumulated in u that has not been processed before and transfer appropriate data through the edge (u, v). This procedure uses the procedure `transfer`(D, u, v), which specifies the effects of transferring data D through the edge (u, v) of a QSQN-STR. We omit to present the properties on soundness, completeness and data complexity of the Algorithm 1 as they are similar to the ones given in [21] for QSQN.

Algorithm 1. for evaluating a query $(P, q(\overline{x}))$ on an extensional instance I.

1 let (V, E, T) be a QSQN-STR structure of P;
  `// T can be chosen arbitrarily or appropriately`
2 set C so that $N = (V, E, T, C)$ is an empty QSQN-STR of P;
3 let $\overline{x}'$ be a fresh variant of $\overline{x}$;
4 $tuples(input\_q) := \{\overline{x}'\}$;
5 **foreach** $(input\_q, v) \in E$ **do** $unprocessed(input\_q, v) := \{\overline{x}'\}$
6 **while** *there exists* $(u, v) \in E$ *such that* `active-edge`(u, v) *holds* **do**
7   select $(u, v) \in E$ such that `active-edge`(u, v) holds according to the IDFS control strategy proposed in [9];
8   `fire`(u, v);
9 **return** $tuples(ans\_q)$

4 Preliminary Experiments

In this section, we present the experimental results of the QSQN-STR method by applying the IDFS control strategy [9] and a discussion on its performance. A comparison between the QSQN-STR method and Datalog Educational System (DES) [24] is also estimated w.r.t. the number of generated tuples in intensional relations. All the experiments for QSQN-STR have been performed using our Java codes [8] and extensional relations stored in a MySQL database. Besides, the logic programs for testing on DES are also included in [8]. For counting the number of generated tuples for intensional relations on each below test in DES, we use the following commands:

- */consult* < *file_name* >, for consulting a program.
- */trace_datalog* < *a_query* >, for tracing a query.

In order to make a comparison between the QSQN-STR method and DES, we use the following tests:

Test 1. This test is taken from Example 1, the query is $acyclic(a, x)$ and the extensional instance I for *edge* is as follows, where a, a_i, b_i, c_i, d_i are constant symbols and $n = 30$:

$$\begin{aligned} I(edge) = {} & \{(a, a_1)\} \cup \{(a_i, a_{i+1}) \mid 1 \le i < n\} \cup \{(a_n, a_1)\} \cup \\ & \{(a, b_1)\} \cup \{(b_i, b_{i+1}) \mid 1 \le i < n\} \cup \{(b_n, b_1)\} \cup \\ & \{(c_i, c_{i+1}), (d_i, d_{i+1}) \mid 1 \le i < n\} \cup \{(c_n, c_1), (d_n, d_1)\}. \end{aligned}$$

Test 2. This test is a semi-positive program and taken from [17]. It computes pairs of nodes which are not directly linked, but are still reachable from one another. In this program, *reachable* and *indirect* are intensional predicates, *link* is an extensional predicate, x, y and z are variables.

$$\begin{aligned} &reachable(x,y) \leftarrow link(x,y)\\ &reachable(x,y) \leftarrow link(x,z), reachable(z,y)\\ &indirect(x,y) \leftarrow reachable(x,y), not(link(x,y)). \end{aligned}$$

Let the query be $indirect(a,x)$ and the extensional instance I for $link$ be as follows, where a, a_i, b_i are constant symbols and $n = 30$:

$$\begin{aligned} I(link) = \{(a,a_1)\} \cup \{(a_i,a_{i+1}) \mid 1 \leq i < n\} \cup \{(a_n,a_1)\} \cup\\ \{(b_i,b_{i+1}) \mid 1 \leq i < n\} \cup \{(b_n,b_1)\}. \end{aligned}$$

Test 3. Consider the red/green bus line program as follows, where *green_path*, *red_monopoly* are intensional predicates, *green*, *red* are extensional predicates, x, y and z are variables:

$$\begin{aligned} &green\_path(x,y) \leftarrow green(x,y)\\ &green\_path(x,y) \leftarrow green\_path(x,z), green\_path(z,y)\\ &red\_monopoly(x,y) \leftarrow red(x,y), not(green\_path(x,y)). \end{aligned}$$

In this program, $red(x,y)$ (resp. $green(x,y)$) means the red (resp. green) bus line runs from x to y, $green\_path(x,y)$ is the path from x to y using only green buses, $red\_monopoly(x,y)$ is the path from x to y using red buses, but we cannot get there on green, even changing buses. The query is $red\_monopoly(x,y)$ and the extensional instance I for *green* and *red* is as follows, where a_i, b_i, c_i are constant symbols:

$$\begin{aligned} I(green) &= \{(a_i,a_{i+1}) \mid 1 \leq i < 30\} \cup \{(b_i,b_{i+1}) \mid 1 \leq i < 20\}\\ I(red) &= \{(a_i,a_{i+1}) \mid 1 \leq i < 5\} \cup \{(a_5,c_1)\} \cup \{(c_i,c_{i+1}) \mid 1 \leq i < 10\}. \end{aligned}$$

Test 4. This test computes all pairs of disconnected nodes in a graph. It is taken from [17], where *reachable*, *node*, *un_reachable* are intensional predicates, *link* is an extensional predicate, x, y and z are variables. The query is $un\_reachable(a,x)$ and the extensional instance I for $link$ is the same as in Test 2 using $n = 30$. The program is specified as follows:

$$\begin{aligned} &reachable(x,y) \leftarrow link(x,y)\\ &reachable(x,y) \leftarrow link(x,z), reachable(z,y)\\ &node(x) \leftarrow link(x,y)\\ &node(y) \leftarrow link(x,y)\\ &un\_reachable(x,y) \leftarrow node(x), node(y), not(reachable(x,y)). \end{aligned}$$

Test 5. This test is a modified version from [7,21]. Figure 3 illustrates the program and the extensional instance I for this test. In this program, p, q_1, q_2 and s are intensional predicates, t, t_1 and t_2 are extensional predicates, a_i and b_i are constant symbols. In this figure, t_1 and t_2 are also represented by a graph. We use $m = n = 30$ for this test (i.e., t_1 has 30 records, t_2 has 900 records).

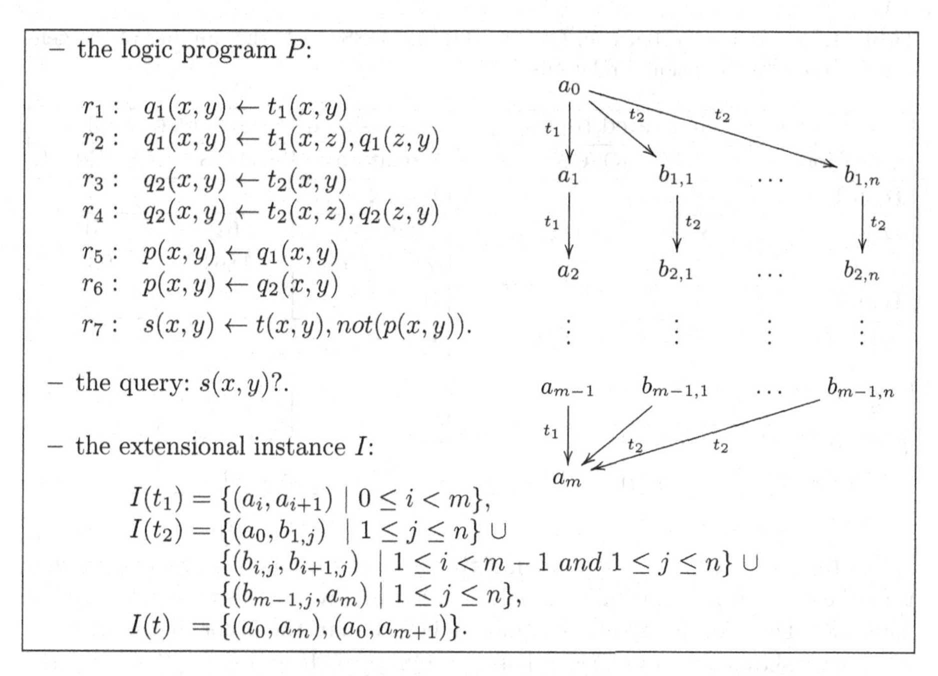

– the logic program P:

$$
\begin{aligned}
r_1 &: q_1(x, y) \leftarrow t_1(x, y) \\
r_2 &: q_1(x, y) \leftarrow t_1(x, z), q_1(z, y) \\
r_3 &: q_2(x, y) \leftarrow t_2(x, y) \\
r_4 &: q_2(x, y) \leftarrow t_2(x, z), q_2(z, y) \\
r_5 &: p(x, y) \leftarrow q_1(x, y) \\
r_6 &: p(x, y) \leftarrow q_2(x, y) \\
r_7 &: s(x, y) \leftarrow t(x, y), not(p(x, y)).
\end{aligned}
$$

– the query: $s(x, y)?$.

– the extensional instance I:

$$
\begin{aligned}
I(t_1) &= \{(a_i, a_{i+1}) \mid 0 \leq i < m\}, \\
I(t_2) &= \{(a_0, b_{1,j}) \mid 1 \leq j \leq n\} \cup \\
&\quad \{(b_{i,j}, b_{i+1,j}) \mid 1 \leq i < m-1 \text{ and } 1 \leq j \leq n\} \cup \\
&\quad \{(b_{m-1,j}, a_m) \mid 1 \leq j \leq n\}, \\
I(t) &= \{(a_0, a_m), (a_0, a_{m+1})\}.
\end{aligned}
$$

Fig. 3. The logic program P and the extensional instance I used for Test 5

Similar to a discussion in [7] for Test 5, the improved semi-naive evaluation method uses breadth-first strategy, the Magic-Sets method is inefficient for the mentioned extensional instances. The reason is that, after performing the magic-sets transformation, the improved semi-naive evaluation method constructs a list $[R_1], \ldots, [R_n]$ of equivalence classes of intensional predicates w.r.t. their dependency [1]. Next, it computes the instances (relations) of the predicates in $[R_i]$ for each $1 \leq i \leq n$ in the increasing order, treating all the predicates in $[R_j]$ with $j < i$ as extensional predicates. For computing the answers for p, since p follows both q_1 and q_2 in the mentioned list of equivalence classes, which means p is only processed after finishing the computation for q_1 and q_2, in this case, Magic-Sets does a lot of unnecessary and redundant computations.

In contrast, since the answer can only be either $true$ or $false$, for computing the answers for p in QSQN-STR, after transferring $\{(a_0, a_{30}), (a_0, a_{31})\}$ through the edge $(\mathit{filter}_{7,2}, \mathit{input\_p})$ and adding their variants to $\mathit{tuples}(\mathit{input\_p})$, $\mathit{unprocessed}(\mathit{input\_p}, \mathit{pre\_filter}_5)$, $\mathit{unprocessed}(\mathit{input\_p}, \mathit{pre\_filter}_6)$ (for r_5, r_6), the algorithm processes $\mathit{unprocessed}(\mathit{input\_p}, \mathit{pre\_filter}_5)$ (i.e., which will process r_1, r_2), which is then returned an answer $true$ for the tuple (a_0, a_{30}) (i.e., $(a_0, a_{30}) \in \mathit{tuples}(\mathit{ans\_p})$) and an answer $false$ for the tuple (a_0, a_{31}). At this point, we remove the tuple (a_0, a_{30}) (a $true$ answer) from $\mathit{unprocessed}(\mathit{input\_p}, \mathit{pre\_filter}_6)$ for not processing it again later. Next, the algorithm processes the tuple $(a_0, a_{31}) \in \mathit{unprocessed}(\mathit{input\_p}, \mathit{pre\_filter}_6)$ (i.e., which will process r_3, r_4), which is then returned an answer *false* (i.e., no tuple is

Table 1. A comparison between QSQN-STR and DES w.r.t. the number of the generated tuples for intensional relations

Intensional relations	Generated tuples DES	Generated tuples QSQN-STR
Test 1		
path	3720	1860
acyclic	60	60
Test 2		
reachable	930	930
indirect	29	29
Test 3		
green_path	635	435
red_monopoly	10	10

Intensional relations	Generated tuples DES	Generated tuples QSQN-STR
Test 4		
node	61	61
reachable	1861	930
un_reachable	31	31
Test 5		
q_1	465	30
q_2	13921	0
p	14386	1
s	1	1

added to $tuples(ans\_q_2)$). In general, if the user only wants one or some answers then the evaluation can terminate earlier. That is why, for this case, QSQN-STR using the IDFS control strategy is more efficient than the other methods.

Table 1 shows the comparison between QSQN-STR and DES [24] w.r.t. the number of generated tuples for intensional relations. As can be seen in this table, QSQN-STR and DES have the same results in intensional relations for the semi-positive program as in Test 2 as well as the relations corresponding to non-negated predicates. However, the number of generated tuples in QSQN-STR is often smaller than DES in the intensional relations corresponding to negated predicates, especially for the case with extensional instance as in Test 5. The explanation is similar to that of Test 5.

5 Conclusions

We have proposed the QSQN-STR method for dealing with stratified knowledge bases. By incorporating stratified negation into query-subquery nets together with the IDFS control strategy, this method takes the advantage of reducing the number of generated tuples in intensional relations corresponding to negated atoms and outperforms the methods that apply the breadth-first search for the case as in Test 5. Particularly, by applying the advantages of a top-down approach and processing the predicates stratum-by-stratum, at the time of processing a negative subquery, this subquery contains no variables. Similarly to QSQN, the processing in QSQN-STR is divided into smaller steps which can be delayed at each node to maximize adjustability, and QSQN-STR can be viewed as a flow control network for determining which subqueries in which nodes should be processed next, in an efficient way. The preliminary experimental results indicate the usefulness of this method. As a future work, we will compare the methods in detail and extend the QSQN-STR method for normal logic programs using the well-founded semantics.

Acknowledgments. This work was supported by Polish National Science Centre (NCN) under Grants No. 2011/02/A/HS1/00395 as well as by Warsaw Center of Mathematics and Computer Science. I would like to express my special thanks to Dr.Hab. Linh Anh Nguyen from the University of Warsaw for his valuable advice.

References

1. Abiteboul, S., Hull, R., Vianu, V.: Foundations of Databases. Addison Wesley (1995)
2. Apt, K.R., Blair, H.A., Walker, A.: Towards a Theory of Declarative Knowledge. In: Found. of Deductive Databases and Logic Programming, pp. 89–148 (1988)
3. Apt, K.R., Bol, R.: Logic programming and negation: A survey. Journal of Logic Programming 19, 9–71 (1994)
4. Bailey, J., Bry, F., Furche, T., Schaffert, S.: Semantic web query languages. In: Encyclopedia of Database Systems, pp. 2583–2586. Springer US (2009)
5. Balbin, I., Port, G.S., Ramamohanarao, K., Meenakshi, K.: Efficient bottom-up computation of queries on stratified databases. J. Log. Program. 11(3-4), 295–344 (1991)
6. Bry, F., Furche, T., Ley, C., Marnette, B., Linse, B., Schaffert, S.: Datalog relaunched: Simulation unification and value invention. In: de Moor, O., Gottlob, G., Furche, T., Sellers, A. (eds.) Datalog 2010. LNCS, vol. 6702, pp. 321–350. Springer, Heidelberg (2011)
7. Cao, S.T.: On the efficiency of Query-Subquery Nets: an experimental point of view. In: Proceedings of SoICT 2013, pp. 148–157. ACM (2013)
8. Cao, S.T.: An Implementation of the QSQN-STR Evaluation Methods (2014), `http://mimuw.edu.pl/~sonct/QSQNSTR15.zip`
9. Cao, S.T., Nguyen, L.A.: An Improved Depth-First Control Strategy for Query-Subquery Nets in Evaluating Queries to Horn Knowledge Bases. In: van Do, T., Le Thi, H.A., Nguyen, N.T. (eds.) Advanced Computational Methods for Knowledge Engineering. AISC, vol. 282, pp. 281–295. Springer, Heidelberg (2014)
10. Cao, S.T., Nguyen, L.A.: An Empirical Approach to Query-Subquery Nets with Tail-Recursion Elimination. In: Bassiliades, N., Ivanovic, M., Kon-Popovska, M., Manolopoulos, Y., Palpanas, T., Trajcevski, G., Vakali, A. (eds.) New Trends in Database and Information Systems II. AISC, vol. 312, pp. 109–120. Springer, Heidelberg (2015)
11. Cao, S.T., Nguyen, L.A., Szalas, A.: The Web Ontology Rule Language OWL 2 RL+ and Its Extensions. T. Comp. Collective Intelligence 13, 152–175 (2014)
12. Cao, S.T., Nguyen, L.A., Szalas, A.: WORL: a nonmonotonic rule language for the Semantic Web. Vietnam J. Computer Science 1(1), 57–69 (2014)
13. Chen, Y.: Magic sets and stratified databases. Int. Journal of Intelligent Systems 12(3), 203–231 (1997)
14. Eiter, T., Ianni, G., Lukasiewicz, T., Schindlauer, R.: Well-founded semantics for description logic programs in the Semantic Web. ACM Trans. Comput. Log. 12(2), 11 (2011)
15. Gelder, A.V., Ross, K.A., Schlipf, J.S.: The well-founded semantics for general logic programs. J. ACM 38(3), 619–649 (1991)
16. Gelfond, M., Lifschitz, V.: The stable model semantics for logic programming. In: Proceedings of Logic Programming Symposium, pp. 1070–1080. MIT Press (1988)

17. Green, T.J., Huang, S.S., Loo, B.T., Zhou, W.: Datalog and Recursive Query Processing. Foundations and Trends in Databases 5(2), 105–195 (2013)
18. Kerisit, J.M., Pugin, J.M.: Efficient query answering on stratified databases. In: FGCS 1988, pp. 719–726 (1988)
19. Madalińska-Bugaj, E., Nguyen, L.A.: A generalized QSQR evaluation method for Horn knowledge bases. ACM Trans. on Computational Logic 13(4), 32 (2012)
20. Naqvi, S.: A logic for negation in database system. In: Foundations of Deductive Data-bases and Logic Programming (1987)
21. Nguyen, L.A., Cao, S.T.: Query-Subquery Nets. In: Nguyen, N.-T., Hoang, K., Jędrzejowicz, P. (eds.) ICCCI 2012, Part I. LNCS, vol. 7653, pp. 239–248. Springer, Heidelberg (2012)
22. Ramamohanarao, K., Harland, J.: An Introduction to Deductive Database Languages and Systems. The VLDB Journal 3(2), 107–122 (1994)
23. Ruckhaus, E., Ruiz, E., Vidal, M.-E.: Query evaluation and optimization in the semantic web. Theory Pract. Log. Program. 8(3), 393–409 (2008)
24. Sáenz-Pérez, F., et al.: Datalog Educational System: A Deductive Database System (2014), http://des.sourceforge.net

Part V

Software Engineering

Distributed Hierarchy of Clusters in the Presence of Topological Changes

François Avril, Alain Bui, and Devan Sohier

Laboratoire PRiSM (UMR CNRS 8144), Université de Versailles, France
francois.avril@uvsq.fr, {alain.bui,devan.sohier}@prism.uvsq.fr

Abstract. We propose an algorithm that builds a hierarchical clustering in a network, in the presence of topological changes. Clusters are built and maintained by random walks, that collect and dispatch information to ensure the consistency of clusters.

We implement distributed communication primitives allowing clusters to emulate nodes of an overlay distributed system. Each cluster behaves like a virtual node, and executes the upper level algorithm. Those primitives ensure that messages sent by a cluster are received and treated atomically only once by their recipient, even in the presence of topological changes. Decisions concerning the behavior of the cluster (virtual node for the higher level algorithm) are taken by the node that owns the random walk at this time.

Based on this abstraction layer and the overlay network it defines, we present a distributed hierarchical clustering algorithm, aimed at clustering large-scale dynamic networks.

1 Introduction

We deal in this paper with the problem of clustering large dynamic networks. This study is motivated by the need to cope with topological changes in a local and efficient manner: as far as possible, we intend to take in charge a topological change inside the cluster where it happened, transparently for the rest of the network.

To achieve this, the clustering algorithm itself must be resilient to topological changes. Thus, after a topological change, the clustering is recomputed only locally: the only clusters affected by the algorithm are the cluster in which the topological change occurred and possibly adjacent clusters.

Based on the algorithm presented in [4], that computes clusters with a size bounded by a parameter K of the algorithm, we propose a distributed algorithm above this clustering, by making each cluster behave as a virtual node. We then apply this method to build a distributed hierarchical clustering.

The cluster is embodied in a special message, the *Token*: any action taken by the cluster is actually initiated by the node that holds the token, making the treatment of higher level messages atomic.

H.A. Le Thi et al. (eds.), *Advanced Computational Methods for Knowledge Engineering*,
Advances in Intelligent Systems and Computing 358, DOI: 10.1007/978-3-319-17996-4_33

The inter-cluster communication primitive $SSend$ is implemented so as to ensure that it works even in the presence of topological changes: if a message is sent by a cluster to an adjacent cluster using this primitive, it is eventually received if the clusters remain adjacent. Then the associated treatment is executed exactly once. Thus, the mobility of nodes affects only a limited portion of the system.

A distributed algorithm can then be applied on the overlay network defined by the clustering. If this algorithm is resistant to topological changes, the resulting method will also be resistant to topological changes. In particular, we apply this method to the presented clustering, leading to a distributed bottom-up hierarchical clustering algorithm.

1.1 Related Works

Designing distributed algorithms that build a nested hierarchical clustering has recently been the subject of many research works (see for example [7]). Those methods build a clustering with a standard method iterated on clusters, represented by a distinguished node called the clusterhead. First, clusterheads are elected in the network. Then, each clusterhead recruits its neighbors. This election can be done on several criteria, such as their id ([1]), their degree ([8]), mobility of nodes ([3]), a combination of those criteria ([5]) or any kind of weight ([2],[9]). The algorithm is then iterated on the graph of the clusters, involving only clusterheads. This results in creating clusters of clusters, and thus building a hierarchical nested clustering by a bottom-up approach.

These methods are very sensitive to topological changes: after a topological change, all clusters can be destroyed before a new clustering is computed, with possibly new clusterheads.

In this paper, we present an algorithm based on the clustering in [4]: a fully decentralized one-level clustering method based on random walks, without clusterhead election, and with the guarantee that reclusterization after a topological change is local, i.e. after a cluster deletion, only clusters adjacent to this cluster can be affected. We design a hierarchical clustering by allowing every cluster in every level to emulate a virtual node. This algorithm preserves properties of [4].

Previous works on virtual nodes, like [6] or [10], aim at simplifying the design of distributed algorithms. Virtual nodes are in a predetermined area of the network, and their mobility is known in advance. Several replica of the virtual machine are necessary to resist topological changes and crashes. Maintaining coherence between replica requires synchronized clocks and a total order on messages, which also allows to verify FIFO hypothesis on channels.

In this work, virtual nodes are built dynamically, with weaker hypotheses on communications: non-FIFO, asynchronous.

1.2 Model and Problem Statement

We suppose that all nodes have a unique id, and know the ids of their neighbors. In the following, we do not distinguish between a node and its id.

The distributed system is modeled by an undirected graph $G = (V, E)$. V is the set of nodes with $|V| = n$. E is the set of bidirectional communication links. The neighborhood of a node i (denoted by N_i) is the set of all nodes that can send messages to and receive messages from i.

We make no assumption on the delivery order of messages. Indeed, the protocol we define for communications between virtual nodes does not verify any FIFO property.

We suppose that all messages are received within finite yet unbounded time, except possibly when the link along which the message is sent has disappeared because of a topological change.

The algorithm is homogeneous and decentralized. In particular, all nodes use the same variables. If *var* is such a variable, we note var_i the instance of *var* owned by node i.

A random walk is a sequence of nodes of G visited by a token that starts at a node and visits other vertices according to the following transition rule: if a token is at i at time t then, at time $t + 1$, it will be at one of the neighbors of i chosen uniformly at random among all of them.

All nodes have a variable *cluster* that designates the cluster to which they belong. A boolean *core* indicates if the node is in the cluster core. All nodes know a parameter K of the algorithm, that represents an upper bound on the size of a cluster.

Definition 1. *The cluster C_x is the set of nodes $\{i \in V/cluster_i = x\}$. Its core is the set $K_x = \{i \in C_x/core_i\}$. A cluster is said complete if $|K_x| = K$.*

The clustering we are computing is aimed at giving clusters with cores of size K as far as possible. To ensure intra-cluster communications, a spanning tree of each cluster is maintained, through a *tree* variable, with $tree[i]$ the father of i in the tree.

Definition 2 (Specification). *A cluster is called consistent if: it is connected; its core is a connected dominating set of the cluster; its core has a size between 2 and K with K a parameter of the algorithm; a single token circulates in the cluster and carries a spanning tree of the cluster.*

A clustering is called correct if: all clusters are consistent; each node is in a cluster; a cluster neighboring an incomplete cluster is complete.

This, together with the property that an incomplete cluster has no ordinary node, guarantees that clusters have maximal cores with respect to their neighborhood.

The algorithm presented next eventually leads to a correct clustering, after a convergence phase during which no topological change occurs. Each topological change may entail a new convergence phase.

Topological changes result in configurations in which the values of variables are not consistent with the actual topology of the system. Then, mechanisms triggered by nodes adjacent to the topological change allow to recompute correct clusters, without affecting clusters other than the one in which the change occurred, and possibly adjacent clusters.

Additionally, we require that primitives are defined for inter-cluster communications, providing the functionalities of *send* and *upon reception of* to higher levels. The *SSend* primitive we define and implement ensures that a node holding the token can send higher level messages to adjacent clusters. Then, we guarantee that the token in charge of the recipient cluster eventually meets this message, and executes the appropriate treatment on the variables it stores. This allows to implement a distributed algorithm on the clustering, with clusters acting as nodes w.r.t. this algorithm.

We define a hierarchical clustering, in which a correct clustering of level i is a correct clustering on the overlay graph defined by the clusters of level $i-1$, with edges joining adjacent clusters.

2 Algorithm

2.1 Clustering

Each unclustered node may create a cluster, by sending a token message to a neighbor chosen at random. The token message then follows a random walk, and recruits unclustered and ordinary nodes it visits to the cluster core. It visits all nodes of the cluster infinitely often, which ensures the updating of variables on nodes and in the token message. When a token message visits a node in another cluster core, it may initiate the destruction of its own cluster with a *Delete* message if both clusters are incomplete and if the id of the token cluster is less than that of the visited cluster. Thus two adjacent clusters cannot be both incomplete once the algorithm has converged.

A node that receives a *Delete* message from its father leaves its cluster and informs all its neighbors. This initiates the destruction of the cluster by propagation of *Delete* messages along the spanning tree.

All nodes periodically inform their neighbors of their cluster by sending a *Cluster*[*cluster*] message, which ensures the coherence of *gate* variables.

Each node knows the cluster to which it belongs with the variable *cluster*; it knows if it is a core node thanks to the *core* variable. Some other technical variables are present on each node : *complete* indicates if the cluster to which the node belongs is complete, *father* is the father of the node in the spanning tree.

The algorithm uses five types of messages: *Token*[*id, topology, N, gateway, Status*], *Recruit*[*id*], *Cluster*[*id*], *Delete*[*id*] and *Transmit*[*message, path, em, dst*].

Token messages carry the following variables: *id*, the id of their cluster *id*; *N*, a list of adjacent clusters; *tree*, a spanning tree of the cluster; *size*, the size of the cluster core; *corenodes*, an array of booleans indicating which nodes are in the cluster core, *gateway*, a vector of links to adjacent clusters; and *status*, a structure containing the same variables as a node, to be used by upper levels.

Token messages follow a random walk and recruit core nodes. *Recruit*[*id*] messages recruit nodes adjacent to a core node as ordinary nodes in cluster *id*,

Delete[*id*] messages delete the cluster *id* and *Cluster*[*id*] messages inform that the sender is in cluster *id*.

A variable *var* on a message *Token* is noted *token.var*.

Initialization: When a node connects to the network, we assume that all variables are set to 0 or $\perp$, except for T, that is initialized with a random value.
***JoinCore* function:** The *JoinCore* function is used to update variables in node i when the node enters a cluster as a core node.

Algorithm 1. *JoinCore*()

$cluster \leftarrow token.id$
$core \leftarrow true$
$token.size \leftarrow token.size + 1$
$token.corenodes[id] \leftarrow true$
$complete \leftarrow (token.size \geq K)$

On Timeout. When the timeout on node i is over, it creates a new cluster, with id $x = (i, nexto)$, and increments *nexto*. Node i joins cluster x and then sends a token message to a neighbor chosen uniformly at random. This node becomes the father of i in the spanning tree. Node i also recruits all its unclustered neighbors as ordinary nodes by sending them *Recruit* messages.

Algorithm 2. *timeout*()

---

if $N \neq \emptyset$ **then**
  token =new token message
  $token.size \leftarrow 0$
  $token.id \leftarrow (id, nexto)$
  $token.tree \leftarrow$ EmptyVector
  $token.tree[id] \leftarrow id$
  UpdateStatus()
  JoinCore()
  $nexto \leftarrow nexto + 1$
  Send *Cluster*[*cluster*] to all nodes in N
  Send *Recruit*[*cluster*] to all nodes in N
  $father \leftarrow$ random value in N
  Send *token* to *father*
else
  $T \leftarrow$ random value

On Reception of a *Recruit*[*id*] Message. On reception of a *Recruit*[*id*] message, if node i is unclustered, it joins cluster *id* as an ordinary node by calling *JoinOrdinary*. The core node that sent the *Recruit* message becomes the father of i in the spanning tree.

Algorithm 3. *JoinOrdinary*(*Sender*)

---

if $(cluster = (\perp, 0)) \vee (cluster = recruit.id)$ **then**
  $cluster \leftarrow recruit.id$
  $core \leftarrow false$
  $father \leftarrow Sender$
  Send *Cluster*[*cluster*] to all nodes in N

On Reception of a *Delete*[*id*] Message from *e*. When node i receives a *Delete*[*id*] message from e, if e is its father in the spanning tree, i leaves the cluster and sends a *Delete*[*id*] message to all its neighbors.

Algorithm 4. On reception of a *Delete*[*id*] message from *e*

```
if e = father ∧ delete.id = cluster then
   LeaveCluster()
Send Delete[cluster] to all neighbors.
```

LeaveCluster() reinitializes all variables and sends a *Cluster*[$(\perp, 0)$] message to all neighbors. *Delete* messages are then propagated along the cluster spanning tree.

On Reception of a *Cluster*[*id*] Message from *e*. When node i receives a *Cluster*[*id*] message from node e, it stores the received cluster id in *gate*[*e*].

On j Leaving N (Disconnection between Current Node and a Neighbor j). The disappearance of an edge (i, j) can lead to an unconnected (and thus faulty) cluster only if the link is in the spanning tree. If a node loses connection with its father in the spanning tree, it leaves the cluster by calling *LeaveCluser*().

When a communication link (i, j) disappears, messages in transit on this link may be lost. A *Recruit*[*id*] or a *Cluster*[*id*] message is no longer necessary, since nodes are no longer adjacent. If a *Delete*[*id*] message is lost, the system will react as if it receives it when j detects the loss of connection. If a *Token* message is lost, the associated cluster is deleted: indeed, if i had sent a *Token* message to j, it considers j as its father. When i detects the loss of connection, it initiates the destruction of the cluster. Last, if a *Transmit* message is lost, then this is a topological change at the upper level, and the same arguments apply.

In any case, the configuration resulting from the loss of a message on a disappeared channel may be the result of an execution without any message loss, and the algorithm continues its operation transparently.

On Reception of a *Token*[*id, topology, N, gateway, status*] Message. When node i receives a *Token*[*id, topology, N, gateway, status*] message from node e, the following cases may occur:

Update Information on a Core Node: If i is a core node of cluster *token.id*, variables *token.topology* and *complete* are updated. Then, the treatment of upper level messages is triggered. Then, i sends *Recruit*[*token.id*] messages and *Cluster*[*token.id*] messages to all its neighbors, and sends the token to a neighbor chosen at random.

Algorithm 5. *token*(*Sender*) on a node i with $(cluster_i = token.id) \wedge (core_i)$

```
{update information - core node}
UpdateStatus()
if (token.tree[i] ≠ i) then
   {the token has not just been bounced back by another cluster}
   token.tree[e] ← id
   token.tree[id] ← id
complete ← (token.size ≥ K)
TriggerUpperLevel()
Send Recruit[token.id] to all nodes in N
Send Cluster[token.id] to all nodes in N
Choose father uniformly at random in N
Send Token[id, topo, N, gateway, status] to father
```

Update Information on an Ordinary Node: If i is an ordinary node of cluster $token.id$ and the cluster is complete ($token.size = K$), i cannot be recruited. Information on token is updated, upper level messages are processed and the token is sent back to the core node that sent it.

Algorithm 6. $token(Sender)$ on a node i with $(cluster_i = token.id) \wedge (\neg core_i) \wedge (token.size \geq K)$

{update information - ordinary node}	TriggerUpperLevel()
$UpdateStatus()$	$father \leftarrow e$
$token.tree[e] \leftarrow id$	Send $Token[id, topo, N, gateway, status]$
$token.tree[id] \leftarrow id$	to $father$

Recruit a Node to the Core: If i is an ordinary or unclustered node, it becomes a core node of cluster $token.id$. Information on the token is updated, upper level messages are processed, $Recruit[cluster]$ and $Cluster[cluster]$ messages are sent to all neighbors, and the token is sent to a neighbor chosen uniformly at random.

Algorithm 7. $token(Sender)$ on a node i with $(token.size < K) \wedge (\neg core_i)$

{recruit a node to the core}	$complete \leftarrow (token.size \geq K)$
if $cluster \neq token.id$ **then**	$TriggerUpperLevel()$
LeaveCluster()	Send $Cluster[cluster]$ to all nodes in N
$JoinCore()$	Send $Recruit[cluster]$ to all nodes in N
$UpdateStatus()$	Choose $father$ uniformly at random in N
$token.tree[e] \leftarrow id$	Send $Token[id, topo, N, gateway, status]$
$token.tree[id] \leftarrow id$	to $father$

Initiate the Destruction of the Cluster: A cluster x can destroy a cluster y when y has a size of 1, or when both clusters are non-complete and the id of x is greater than that of y. If the token cluster can be destroyed by the node cluster, the token message is destroyed and a $Delete[cluster]$ and a $Recruit[cluster]$ messages are sent to the sender, that considers i as its father in the spanning tree. This leads to the destruction of cluster $token.id$.

Algorithm 8. $token(Sender)$ on a node i with $[(token.size < K) \wedge (\neg complete_i) \wedge (cluster_i > token.id)] \vee (token.size = 1)$

{initiate the destruction of the cluster	Send $Delete[token.id]$ to e
$token.id$}	Send $Recruit[cluster]$ to e

Send the Token Back: In all other cases, the token message is sent back to the sender.

Algorithm 9. $token(Sender)$ on all other cases

```
{send the token back}
if token.tree[id] ≠ id then
   Send Token[id, topo, N, gateway, status] to e
```

Algorithm 10. $UpdateStatus()$

```
for all k with token.tree[k] = i {i the id of the node runing this function} do
   if k ∉ N then
      RemoveFromTree[k, token.topology]
for all id with token.gateway[id] ≠ ⊥ do
   (l, m) ← token.gateway[id]
   if (l ∉ token.tree) ∨ [(l = i) ∧ ((m ∉ N) ∨ (gate_i[m] ≠ id)] then
      token.gateway[id] ← ⊥
      remove id from token.N
      {This may entail a topological change w.r.t. the higher level, in which case, it triggers the higher level Leave function}
for all k ∈ N with gate[k] ≠ (⊥, 0) and cluster > gate[k] do
   token.gateway[gate[k]] ← (i, k)
   Add gate[k] to token.N
```

2.2 Virtual Nodes

We aim at making every cluster behave as a virtual node, in order to be able to execute a distributed algorithm on the overlay network defined by the clustering. In particular, this will allow to build a hierarchical clustering through a bottom-up approach. In this section, we present mechanisms that allow a cluster to mimic the behavior of a node:

- virtual nodes know their neighbors (adjacent clusters), and are able to communicate with them;
- virtual nodes execute atomically the upper level algorithm.

To ensure that clusters have a unique atomic behavior, the node that holds the *token* message is in charge of the decision process about the cluster seen as a virtual node. Variables of the virtual node are stored in $token_{id}.Status$.

Knowledge on the neighborhood is maintained by *Cluster* messages and *UpdatesStatus* function, as seen in section 2.1.

Neighborhood Observation. Since $Cluster[id]$ messages are sent infinitely often, all nodes maintain the vector *gate* indicating the cluster to which each of their neighbors belong. When a message $token_x$ visits a node in cluster x that has a neighbor in another cluster y, the link between the two nodes can be selected as a privileged inter-cluster communication link between clusters x and y, and added to $token_x.gateway$ and $token_y.gateway$. To ensure that topological changes are detected symmetrically at upper level, adjacent clusters need to agree on the gateway they use to communicate, i.e. $token_{C_x}.gateway[C_y] = token_{C_y}.gateway[C_x]$. Thus, when this link breaks, both clusters are aware of this loss of connexion between the two virtual nodes, which allows the overlay graph to remain undirected. Only the cluster with the smallest *id* can select an inter-cluster communication link between two adjacent clusters. The other cluster adds it at the reception of the first upper level message.

The *gateway* array in *Token* message, along with the spanning tree stored in the *Token* message, allows to compute a path from the node holding the *Token* to a node in any adjacent cluster. This enables virtual nodes to send upper level messages to their neighbors: see Algorithm 11.

Thus, if $token.gateway[id] = (i, j)$, then j is a node in cluster id. To send a message to an adjacent cluster id, $SSend$ computes a path to j in the spanning tree stored in *token.topology* and adds link (i, j). Then, the upper level message is encapsulated in a $Transmit$ message, and routed along this path.

Communication. $SSend$ is used when handling a topological change, or when upper level messages are received. In both cases, this treatment can be initiated only in presence of the token. Thus, *token.topology* information is available. On a higher level, the processing of a $Token$ message is triggered by the processing of the lower level token. So, when processing a token message, all lower level tokens are available.

Algorithm 11. $SSend(msg, dst, tok)$ function

$(l, m) \leftarrow tok.gateway[dst]$
$path \leftarrow ComputePath(tok.topo, l)$
$path \leftarrow (path, m)$
$lowlevmsg \leftarrow Transmit[msg, Tail(path), tok.id, dst]$
$SSend(lowlevmsg, Head(path), token)$

The recursive calls to $SSend$ end with the bottom level using $Send$ instead of $SSend$.

Upper level messages are encapsulated in $Transmit$ messages, along with the path to follow, sender and recipient virtual nodes id. Such an upper level message is forwarded along the path to a node of the recipient cluster. This node stores three pieces of information in $listmsg$: the encapsulated message; the node that sent the $Transmit$ message; the cluster that sent the encapsulated message.

This information ensures that the cluster can reply to the sender cluster; it will be used to add a link to this adjacent cluster and guarantee the symmetry of comunications. When the token visits a node in this cluster, it triggers the processing of upper level messages stored in $listmsg$ by the virtual node, and removes them from $listmsg$.

Algorithm 12. Reception of a $Transmit[msg, path, em, dst]$ message from e

if $(path = \emptyset) \wedge (cluster = dst)$ **then**
  Add (msg, e, em) to $listmsg$
else if $((cluster = em) \wedge (Head(path) \in N))$ **then**
  Send $Transmit[msg, Tail(path), em, dst]$ to $Head(path)$

Processing the Upper Level. Processing of upper level messages stored in a node is triggered when the $Token$ message visits the node. Function

TriggerUpperLevel then pops messages out of *listmsg* and emulates the reception of this message by the virtual node. *TriggerUpperLevel* also manages the triggering of function *timeout* for upper level, simulating a timer by counting *Token* hops.

If the sender virtual node is not in *token.N* and *token.gateway*, *listmsg* contains enough information to add it. Once all upper level messages are treated, it decrements the count-down on *token.T* in structure *token.status*.

Algorithm 13. *TriggerUpperLevel*()

```
while (listmsg ≠ ∅) do
  (msg, s, C) ← Pop(listmsg)
  if token.gateway[C] = ⊥ then
    token.gateway[C] ← (i, s)
    Add C to token.N
  else if token.gateway[C] ≠ (i, s) then
    run "loss of connection" procedure on the virtual node
    token.gateway[C] ← (i, s)
    Add C to token.N
  Emulate reception of msg by the virtual node using SSend.
if (token.T > 0) then
  token.T ← token.T − 1
  if (token.T = 0) then
    Stimeout()
```

Functions for processing upper levels are identical to the first level functions presented in subsection 2.1, except that they use *SSend* instead of *send*.

2.3 Hierarchical Clustering

If the upper level algorithm executed by the virtual nodes is this same clustering algorithm, the virtual nodes compute a distributed clustering resistant to topological changes. Iterating this process, we obtain a hierarchical clustering.

This provides a framework to contain the effect of topological changes inside clusters, while allowing communication between nodes in any two clusters. In particular, a modular spanning tree of the system consisting in the spanning trees of clusters of all levels is built.

3 Example

We present here an example involving three levels of clustering. Consider a clustering with C_A (a cluster of level ≥ 3) containing clusters C_x and C_y, C_B a cluster of the same level as C_A, containing C_z. C_x, C_y and C_z are composed of nodes or of clusters of a lower level. Consider that cluster C_A sends a $Cluster[(\perp, 0)]$ message to cluster C_B.

Sending the Message Out of x. C_A sends a $Cluster[(\perp, 0)]$ message to C_B. This is triggered by the processing of the $token_{C_A}$ message on a lower level. Thus, cluster C_x necessarily holds the $token_{C_A}$ message. It calls the *SSend* mechanism and uses information on $token_{C_A}$ to compute a path to C_B, here to cluster z (see figure 1). Then, since the path is xyz, cluster C_x sends a $Transmit[Cluster[(\perp, 0)], z, A, B]$ to the next cluster on the path, C_y (with

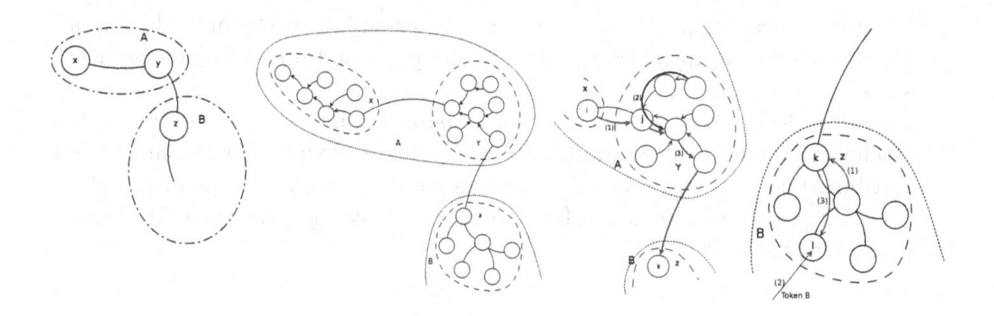

Fig. 1. Seen from $token_{C_A}$

Fig. 2. Detailed view

Fig. 3. Routing in C_y

Fig. 4. Reception of the message

the *SSend* mechanism). A node of C_y finally receives a *Transmit* message: $Transmit[Transmit[Cluster[(\bot, 0)], z, A, B], \emptyset, x, y]$.

Routing the Message in C_y.

1. Node j in C_y receives a $Transmit[Transmit[Cluster[\,(\bot, 0)], z, A, B], \emptyset, x, y]$ from node i. Since it is in C_y, and the path is empty, it stores the transmitted message $(Transmit[Cluster[\,(\bot, 0)], z, A, B], i, y)$ in its *listmsg*.
2. $token_{C_y}$ eventually visits j and treats upper level messages. Since $path = z \neq \emptyset$, node j uses *SSend* to send $Transmit[Cluster[(\bot, 0)], \emptyset, A, B]$ to cluster C_z (line 5 algorithm 12). *SSend* uses $token_{C_y}$*.topology* to compute a path to $token_{C_y}$*.gateway*$[z]$ and sends a *Transmit* message to *head*(*path*): $Transmit[\,Transmit[\,Cluster[(\bot, 0)], z, A, B], tail(path), y, z]$.
3. Every node in the path transmits the *Transmit* message until node k receives the message $Transmit[\,Transmit[\,Cluster[(\bot, 0)], \emptyset, A, B], \emptyset, y, z]$.

Reception of the Message by C_B. Node k receives a *Transmit* message from node e ($Transmit[\,Transmit[\,Cluster[(\bot, 0)], \emptyset, A, B], \emptyset, y, z]$) and stores $(Transmit[Cluster[(\bot, 0)], \emptyset, A, B], e, y)$ in *listmsg* (algorithm 12, line 2). Then:

1. $token_{C_z}$ eventually visits k during its random walk, and treats messages in $listmsg_k$. $(Cluster[(\bot, 0)], y, A)$ is stored in $token_{C_z}$*.listmsg*.
2. $token_{C_B}$ eventually visits cluster C_z during its random walk, i.e. it is eventually stored in the $listmsg_l$ variable of a node l in C_z.
3. $token_{C_z}$ eventually visits node l. Then, it processes $token_{C_B}$. *TriggerUpperLevel* is called and messages in $token_{C_Z}$*.listmsg* are treated. In particular, $(Cluster[\,(\bot, 0)], y, A)$ is processed, ie $token_{C_B}.gate[A] \leftarrow (\bot, 0)$.

4 Conclusion and Perspectives

The algorithm presented in this paper computes a size-oriented hierarchical clustering of a dynamic network. It reacts to topological changes in a local manner:

after a topological change that makes the clustering faulty, the only clusters affected are the cluster in which this event took place and possibly some adjacent clusters.

Thus, it is designed to provide a local mechanism to handle topological changes. Inter-cluster communication mechanisms are proposed that allow to implement a distributed algorithm above the clustering, and a formal proof is provided.

Proofs of this algorithm may be found at: `http://www.prism.uvsq.fr/~sode/hierclus/proof.pdf`.

References

1. Baker, D., Ephremides, A.: The architectural organization of a mobile radio network via a distributed algorithm. IEEE Transactions on Communications 29(11), 1694–1701 (1981)
2. Basagni, S.: Distributed and mobility-adaptive clustering for multimedia support in multi-hop wireless networks. In: IEEE VTS 50th Vehicular Technology Conference, VTC 1999 - Fall, vol. 2, pp. 889–893 (1999)
3. Basagni, S.: Distributed clustering for ad hoc networks. In: Proceedings of the Fourth International Symposium on Parallel Architectures, Algorithms, and Networks (I-SPAN 1999), pp. 310–315 (1999)
4. Bui, A., Kudireti, A., Sohier, D.: An adaptive random walk-based distributed clustering algorithm. International Journal of Foundations on Computer Science 23(4), 802–830 (2012)
5. Chatterjee, M., Das, S.K., Turgut, D.: Wca: A weighted clustering algorithm for mobile ad hoc networks. Journal of Cluster Computing (Special Issue on Mobile Ad hoc Networks) 5, 193–204 (2001)
6. Dolev, S., Gilbert, S., Lynch, N.A., Schiller, E., Shvartsman, A.A., Welch, J.L.: Virtual mobile nodes for mobile *ad hoc* networks. In: Guerraoui, R. (ed.) DISC 2004. LNCS, vol. 3274, pp. 230–244. Springer, Heidelberg (2004)
7. Dolev, S., Tzachar, N.: Empire of colonies: Self-stabilizing and self-organizing distributed algorithm. Theoretical Computer Science 410, 514–532 (2009), `http://www.sciencedirect.com/science/article/pii/S0304397508007548`, principles of Distributed Systems
8. Gerla, M., Chieh Tsai, J.T.: Multicluster, mobile, multimedia radio network. Journal of Wireless Networks 1, 255–265 (1995)
9. Myoupo, J.F., Cheikhna, A.O., Sow, I.: A randomized clustering of anonymous wireless ad hoc networks with an application to the initialization problem. J. Supercomput. 52(2), 135–148 (2010), `http://dx.doi.org/10.1007/s11227-009-0274-9`
10. Nolte, T., Lynch, N.: Self-stabilization and virtual node layer emulations. In: Masuzawa, T., Tixeuil, S. (eds.) SSS 2007. LNCS, vol. 4838, pp. 394–408. Springer, Heidelberg (2007), `http://dl.acm.org/citation.cfm?id=1785110.1785140`

Native Runtime Environment for Internet of Things

Valentina Manea, Mihai Carabas, Lucian Mogosanu, and Laura Gheorghe

University POLITEHNICA of Bucharest, Bucharest, Romania

Abstract. Over the past few years, it has become obvious that Internet of Things, in which all embedded devices are interconnected, will soon turn into reality. These smart devices have limited processing and energy resources and there is a need to provide security guarantees for applications running on them. In order to meet these demands, in this paper we present a POSIX-compliant native runtime environment which runs over a L4-based microkernel, a thin operating system, on top of which applications for embedded devices can run and thus be a part of the Internet of Things.

Keywords: runtime environment, POSIX, L4 microkernel, Internet of Things.

1 Introduction

Internet of Things (IoT) is a new paradigm in which objects used by people in their everyday lives will be able to communicate both with each other but also with all the devices connected to the Internet [12]. Internet of Things can be seen as a convergence of smart devices, such as laptops or tablets, with devices that use RFID (Radio-Frequency Identification) and Wireless Sensor Networks. Connecting such a diverse range of devices opens new possibilities for improving the quality of life and offer new services. Some examples of fields which Internet of Things promises to enhance are home automation [7] and medical healthcare [8].

It is estimated that by 2020, around 50 billion devices will be connected to the Internet [6]. Applications will have to be developed to support the variety of devices that will be part of IoT, with different available resources and constraints. In order to ease the development of applications the low-level details of devices should be abstracted and exposed to the developers in an unified way, on top of which they can start building features.

The fact that small devices that collect information and act accordingly will be connected to the Internet poses a security risk. For example, a home thermostat collects the temperature in a home and regulates itself to fit the home owner's preferences. A malicious hacker can break into the thermostat via Internet and change its behaviour. As such, the security guarantees provided by an operating system running on an embedded device are much more important in the context of IoT.

H.A. Le Thi et al. (eds.), *Advanced Computational Methods for Knowledge Engineering*, Advances in Intelligent Systems and Computing 358, DOI: 10.1007/978-3-319-17996-4_34

In this paper, we present a native runtime environment called VMXPOSIX, designed to run on embedded devices. Since it is partially POSIX-compliant, it offers a common platform for developers to write applications on. VMXPOSIX runs on top of the VMXL4 microkernel and is able to provide the security guarantees applications for devices in IoT need.

The paper is structured as follows. Section 2 reviews related work. We describe the native runtime environment architecture in Section 3. Technical details and implementation are presented in Section 4. Section 5 presents the experimental evaluation. Finally, Section 6 presents the conclusions of this paper.

2 Related Work

The idea of offering a common runtime environment for a variety of embedded devices has been previously explored. Examples of such environments are Contiki and RIOT, which are operating systems targeted at devices from the Internet of Things.

Contiki [10] is an operating system designed for Wireless Sensor Networks. It is implemented in the C language and operates on a series of microcontroller architectures, such as Atmel AVR.

As sensor networks are composed of thousands of nodes, Contiki aims to make downloading code at run-time, as well as loading and unloading applications, an efficient process, from a processing power and energy point of view. Another design principle is to offer a common software framework for applications to use in order to ensure their portability on a variety of platforms. However, due to the fact that sensor networks are application-specific, Contiki offers very few abstractions.

Furthermore, it is an event-driven system, in which threads have no private stack, but are implemented as event handlers that run uninterrupted until they are done. While this kind of system is appropriate for Wireless Sensor Networks, it is more difficult to use for more general purpose devices because more complex programs are not easily translated to an event-driven flow.

Contiki targets a restricted class of devices: Wireless Sensor Networks. Using Contiki for a wider range of devices, such as home automation systems, is not feasible due to its event-driven model and lack of a rich API on top of which complex applications can be built, both characteristics impairing the growth of an extensive application ecosystem.

RIOT aims to power a large scale ecosystem of heterogeneous devices that are connected with each other and with the cloud [5]. Its design objectives include power efficiency, small memory footprint, modularity and an API independent of the underlying hardware [9]. Due to these design objectives, RIOT is based on a microkernel architecture. It offers developer-friendly APIs for multi-threading, C++ libraries support and a TCP/IP stack. Its microkernel architecture makes it possible to confine failures in one module from the rest of the system thus making RIOT a robust operating system. This also makes it easy to implement distributed systems using the message API.

Another important element for certain classes of devices in the Internet of Things are real-time requirements. As such, RIOT enforces fixed durations for kernel tasks and uses a scheduler which runs in constant time. Since embedded devices have limited energy resources and must minimize consumption, RIOT features a state called "deep-sleep" for the times when the system is idle. Kernel functions have low complexity and RIOT strives to minimize the duration and occurrences of context switches. It features standard development tools such as `gcc` or `valgrind` and is partially POSIX-compliant.

Despite its strong set of features, which make RIOT appropriate for a wide range of embedded devices, it lacks one crucial characteristic: security guarantees. Many IoT scenarios have special security requirements. For example, in the case of a home automation system which adjusts thermostat temperature, it is important to prevent unauthorized access to the system. This aspect is even more important for critical systems, such as the ones deployed in automotive, where an unauthorized access can lead to serious damage and even loss of lives.

3 Design

In the following subsections we will describe the proposed native runtime environment architecture, VMXPOSIX, as well as the L4-based microkernel it runs on top of, VMXL4.

3.1 VMXL4 Microkernel

VMXL4 is a high performance microkernel which provides a minimal layer of hardware abstraction. It is built with the principle of minimality in mind: a feature is added to the microkernel only if moving it outside would impair system functionality or if including it in the microkernel would provide significant benefits [11].

The microkernel provides application isolation by confining each one into a Secure Domain. Each Secure Domain is given its own physical resources. VMXL4 allows great control over communication between domains and facilitates the implementation of desired security policies in order to meet the application's requirements. Besides improving security, Secure Domains isolate faulty applications and, thus, assure system availability. Applications residing in different Secure Domains cannot communicate directly, but only with the intervention of the microkernel.

Isolating an application into a Secure Domain gives the possibility to implement the necessary security policies. However, a great deal of effort would be required to change an application to use VMXL4 API, since the microkernel only offers mechanisms, not services.

3.2 VMXPOSIX Native Runtime Environment

VMXPOSIX makes use of mechanisms provided by VMXL4 and presents them in a more convenient way to applications. It aims at providing a POSIX-compliant

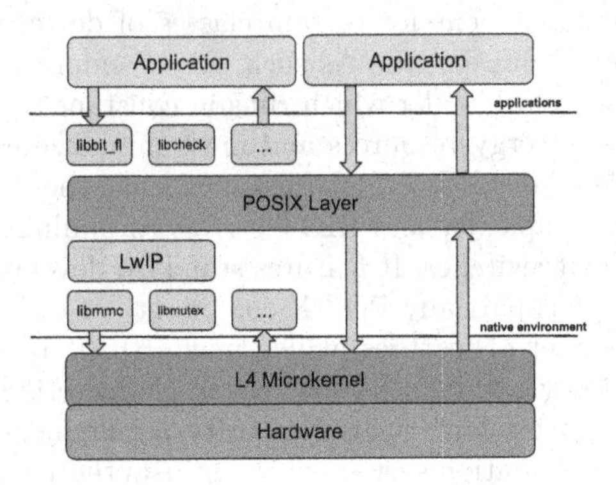

Fig. 1. VMXL4 Environment

programming interface by abstracting microkernel primitives. It is similar to GNU libc or Bionic libc, offering a C native runtime environment as a library.

As can be seen from Figure 1, the runtime environment is made up of several libraries, with the POSIX layer as the core library. Some libraries offer functionality to VMXPOSIX layer, while others use its API. Libraries using VMXPOSIX API are, among others, `libbit_fl`, a library for managing bit arrays, and `libcheck`, a framework for unit testing. However, an important part of VMXPOSIX is implemented using VMXL4 primitives; functions for memory management and process management use the microkernel API directly.

An example of library on which VMXPOSIX relies is `libmmc`. In essence, `libmmc` is a driver used for managing multimedia cards (MMC). This library provides primitives for working with files residing on multimedia cards, such as for open and read system calls.

The execution flow through the presented environment is as follows. Applications are isolated in containers, called, in VMXL4 terminology, Secure Domains or SecDoms. In order to be able to access system functionality, such as allocate memory or spawn threads, applications must go through the runtime environment. In turn, the runtime environment translates applications' request in terms of microkernel primitives. The microkernel honors the request and the results are propagated back to the applications through the runtime environment. Using a capabilities-based model and the security extensions VMXL4 provides, various security policies can be implemented and data integrity and confidentiality can be ensured.

4 Implementation

From an implementation perspective, VMXPOSIX is a library which provides a programming interface for applications to use when running on top of VMXL4 microkernel. Like POSIX, C language bindings are provided. The specifications are implemented using C as well. The library is split into two parts: header files and source files.

Header files aim to comply to the definitions included in the Base Definitions volume of POSIX.1-2008 standard [4]. In order to make use of existing code and not write the header files from scratch, VMXPOSIX uses code from existing files. As such, many headers have been imported from a minimal C library called Musl [2] due to its permissive license. VMXL4 environment includes a minimal C library, which provides some functions not defined in POSIX standard, such as memory allocation ones. In order to provide a unified native runtime environment, the C library was merged into VMXPOSIX.

System call interface definitions are provided by the Musl header files. VMXL4 API is used for implementing system calls. Some additional libraries that use the microkernel programming interface already exist, such as a MultiMediaCard (MMC) driver.

Development, as well as testing, was performed using a Pandaboard [3] development board. This platform is based on an OMAP system on a chip developed by Texas Instruments.

4.1 Functions for File Operations

An important class of operations are represented by file operations, such as creating, reading or writing a file. Linux-based systems offer a unified interface for interacting with files residing on different types of file systems and it is the kernel's task to translate these operations into the specifics of each file system type.

Because VMXL4 is a microkernel and, thus, offers minimal functionality, it is up to other layers to implement functionality for handling files. In this case, VMXPOSIX will take care of these aspects for each supported file system type. Currently, only one file system driver exists in VMXL4 environment, for FAT filesystems located on multimedia cards (MMC) but support for other file systems can be easily added. The system calls for file operations use the API provided by the mentioned driver.

The `open()` function signature is the standard POSIX one, with support for the third parameter, mode, which may or may not be provided. When opening a file, we must first initialize the MMC driver. This gives it information about the type of partition we want to work with, as well as the partition number. Because a MMC device can be emulated or exist physically in the system and we want to use the card attached to our Pandaboard, we specified that the real hardware device should be used.

Next, depending on the flags provided, different actions take place. If `O_CREAT` flag was given the file should be created if it doesn't exist already. However, if the `O_EXCL` flag was also provided and the file exists, an error is returned. If the file doesn't exist and `O_CREAT` flag was not specified, a specific error code is returned. File truncating is possible with `O_TRUNC` flag.

In the POSIX standard, `open()` system call returns a file descriptor, an integer which will be used in subsequent file operations. An opened file is described by a structure called file. The file descriptor is an index in a global struct file array pointing to the entry assigned to the file in question. The code tries to find an unoccupied entry in the files array, insert file information there, such as the callbacks for `write()` and `read()` calls, and then return the index. If no array slot is empty, the maximum number of files have been opened and an error is returned.

It should be noted that operations such as reading a file's size or truncating it are translated in actions understood by the MMC driver, with `read_file_size()`, `create_file()` and `truncate_file()` functions.

The other system calls for file operations are implemented in a similar manner.

4.2 fork() System Call

One of the basic system calls of an operating system is `fork()`, which deals with process creation. An existing process, called a parent, executes this system call in order to create a new process, called a child. The child process will inherit many of its parent's characteristics, such as its code, stack, data sections or its open file descriptors.

In VMXL4, memory segments are statically added at compile time. By default, each Secure Domain has two main memory segments: a code segment and a data segment. The initial thread in each Secure Domain, the root thread, uses these segments for the mentioned sections. Since the `fork()` system call creates a new process that will share the code section with its parent, using the code segment for both processes will yield the expected behaviour. However, for the other sections, different memory regions must be used. As such, a new segment was added for a SecDom, called a fork segment. In this segment, a child thread can place its stack and data section. For now, the fork segment has a size of 4 MB and has read, write and execute permissions. The structure of a Secure Domain, with a root thread and a child thread is shown in Figure 2.

Each Secure Domain also has a pager thread, which handles page faults. Since the stack and data sections of the parent and child have, initially, the same content, they will be mapped to the same virtual addresses. However, when the child process tries to modify any of these sections, a new page must be allocated and mapped in its address space and the content be copied to the new location. The pager thread was modified so that whenever it intercepts a write page fault from the child process to allocate a page from the fork segment, copy the contents to the new location and map the new address in the child processes' address space. A buddy allocator has been used for allocating pages in the fork segment. This mechanism is known as copy-on-write.

The child process creation is straightforward. First, a new, unused, process ID is chosen from a pool. Next, a new thread with the chosen process ID is created. The parent's registers are copied to the child registers because both processes should have the same context. Finally, the child thread is started. The child thread starts the execution from a given function. However, the `fork()` POSIX definition requires both threads to start executing code from the same point. This is achieved by first memorising the parent thread context using `setjmp()` and restoring the context in the child start function with `longjmp()`.

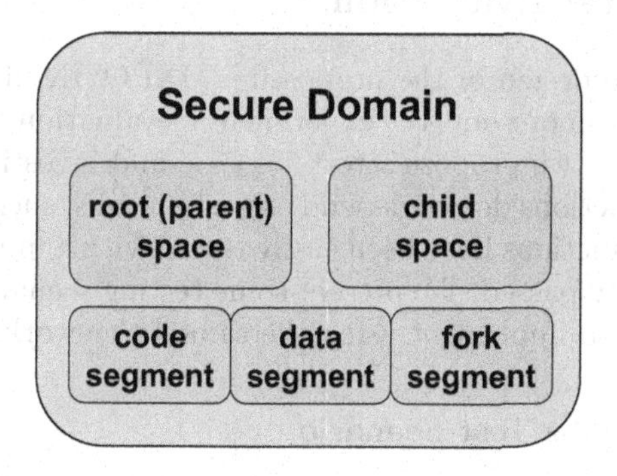

Fig. 2. Secure Domain Layout

Finally, the parent thread must not return before the child thread starts executing. Therefore, the parent thread will block until it received a confirmation from the child thread via messages, a form of inter-process communication (IPC). From that point on, both threads will return to the function from which `fork()` was called.

4.3 Standard Input Implementation

Although embedded devices do not usually provide means of communicating with the user using a keyboard, some larger ones do. The platform used for developing VMXPOSIX offers a serial port which can be used to provide input.

The typical POSIX way of reading anything, from files to keyboard input, is by using the `read()` system call and a specific file descriptor. Regular files require to first be opened and are assigned a file descriptor. On the other hand, in a POSIX environment, all processes have opened, by default, a set of special descriptors. One of them is the standard input descriptor, also known as `stdin`.

A driver for the console is already available in the microkernel for debugging purposes. In true microkernel design spirit, a console input/output server is also provided. A process will receive input from the server and transfer output to the server for printing at the console. The client and the console server will function in separate Secure Domains. The client must have a capability for sending IPC messages to the console server. Similarly, the console server must have the proper capability to send messages to the client.

First, the client must register itself to the console server. Thus, multiple clients can access the console. When waiting for input, the client will block until it receives an IPC message from the server with a specific notification bit activated. Then, the client will be able to receive input. This loop it repeated until the client receives as many bytes as it requires.

5 Experimental Evaluation

After discussing the design of the proposed VMXPOSIX library and its implementation, this section presents the experimental evaluation of our work. In order to evaluate our work, we propose a test scenario and investigate its outcome.

Many POSIX functions do not depend on system calls, such as `malloc`, `memset` or `strcpy`. Those functions have been easily tested with a unit testing framework and have successfully passed. We present some testing scenarios that would have been more difficult to implement using the same framework.

5.1 File Operations Test Scenario

For testing various features of our proposed native runtime environment, we have used several utilities from Busybox. Busybox is a software which provides a variety of utilities a user may find on most Linux systems. Because Busybox was designed to use as little resources as possible, the utilities only have their core features and are all provided inside one executable [1].

The test assumes the existence of a file named `test1.txt` on the multimedia card. Its contents are shown in Listing 1.1.

Listing 1.1. Test File Contents

```
1 Lorem ipsum dolor sit amet.
2 Nemo enim ipsam voluptatem.
```

The test scenario is as follows. First, the `echo` utility is used, which allows us to test the standard output functionality and a message is printed. Then, the `cat` utility is used and the contents of the previous mentioned file is printed, thus `open()` and `read()` system calls are tested. The `head` utility prints only the first line of the file in question. Next, the root directory's contents are listed. We rename the test file with the `mv` command. To validate the change, the directory's contents is printed again. Finally, the test finishes and an appropriate message is printed by the `echo` command.

Listing 1.2 shows the results of this run. It can be noticed that running the test program yields the expected results.

Listing 1.2. File Operations Run Output

```
1 [1-0] T0: Lorem ipsum dolor sit amet.
2 [1-0] T0: Nemo enim ipsam voluptatem.
3 [1-0] T0: Lorem ipsum dolor sit amet.
4 [1-0] T0:
5 [1-0] T0:      213718    uimage
6 [1-0] T0:        9216    test1.txt
7 [1-0] T0:
8 [1-0] T0:      213718    uimage
9 [1-0] T0:        9216    test2.txt
```

5.2 fork() System Call Test Scenario

Testing the `fork()` system call is straightforward. A parent thread calls `fork()` and both the parent thread and the child thread print a message to identify themselves, which shows that different values are returned. Afterwards, both print a common message, to signal the fact that they continue execution from a common point. The used code is shown in Listing 1.3.

Listing 1.3. `fork()` Test Scenario

```
1 if (fork() != 0)
2         printf("in parent thread\n");
3 else
4         printf("in child thread\n");
5 printf("both threads should print this\n");
```

The results of this test are shown in Listing 1.4.

Listing 1.4. `fork()` Run Output

```
1 [1-1] T0: in parent thread
2 [1-1] T0: both threads should print this
3 [1-0] forked-3: in child thread
4 [1-0] forked-3: both threads should print this
```

5.3 Standard Input Test Scenario

As with the `fork()` system call, testing standard input functionality is simple. What we actually test is the ability of the `read()` system call to read data from the `stdin` file descriptor, as shown in Listing 1.5.

Listing 1.5. Standard Input Test Scenario

```
1 read(STDIN_FILENO, buffer, 10);
2 printf("%s\n", buffer);
```

As expected, at the end of the test, `buffer` will contain what we have typed from the keyboard.

6 Conclusion and Further Work

In order to satisfy the requirements of applications for embedded devices in the Internet of Things, we proposed a POSIX-compliant programming interface called VMXPOSIX. It is designed to be an intermediate layer between applications and the VMXL4 API.

We managed to implement an important subset of the POSIX standard that can be easily extended. In order to test our work, we ran several Busybox utilities and observed the correctness of the results. We have also developed unit tests and created test scenarios.

Because of the magnitude of the POSIX standard, having a full implementation takes a considerable amount of effort. Since this project covered a subset of the standard, additional features will be implemented, such as support for environment variables, support for UNIX-style device trees or support for locales.

Acknowledgement. The work has been funded by the Sectoral Operational Programme Human Resources Development 2007-2013 of the Ministry of European Funds through the Financial Agreement POSDRU/159/1.5/S/134398.

References

1. BusyBox: The Swiss Army Knife of Embedded Linux, http://www.busybox.net/about.html (Last Access: January 2015)
2. musl libc, http://www.musl-libc.org/ (Last Access: January 2015)
3. Pandaboard Resources, http://pandaboard.org/content/resources/references (Last Access: January 2015)
4. POSIX.1-2008, http://pubs.opengroup.org/onlinepubs/9699919799/ (Last Access: January 2015)
5. RIOT Vision, https://github.com/RIOT-OS/RIOT/wiki/RIOT-Vision (Last Access: January 2015)
6. The Internet of Things [INFOGRAPHIC], http://blogs.cisco.com/news/the-internet-of-things-infographic (Last Access: January 2015)
7. The state (and future) of Internet of Things and home automation, http://www.eachandother.com/2014/11/the-state-and-future-of-internet-of-things-and-home-automation/ (Last Access: January 2015)
8. Transforming patient care with IoT, http://www.microsoft.com/en-us/server-cloud/solutions/internet-of-things-health.aspx (Last Access: January 2015)
9. Baccelli, E., Hahm, O., Günes, M., Wählisch, M., Schmidt, T., et al.: Riot os: Towards an os for the internet of things. In: The 32nd IEEE International Conference on Computer Communications, INFOCOM 2013 (2013)
10. Dunkels, A., Gronvall, B., Voigt, T.: Contiki - a lightweight and flexible operating system for tiny networked sensors. In: 29th Annual IEEE International Conference on Local Computer Networks, pp. 455–462. IEEE (2004)
11. Liedtke, J.: On Micro-Kernel Construction (1995), http://www2.arnes.si/~mmarko7/javno/printaj/6/KVP/LIEDTKE-1.PDF (Last Access: January 2015)
12. Zanella, A., Bui, N., Castellani, A.P., Vangelista, L., Zorzi, M.: Internet of things for smart cities. IEEE Internet of Things Journal (2014)

Searching for Strongly Subsuming Higher Order Mutants by Applying Multi-objective Optimization Algorithm

Quang Vu Nguyen and Lech Madeyski

Faculty of Computer Science and Management, Wroclaw University of Technology, Wybrzeze Wyspianskiego 27, 50-370 Wroclaw, Poland
{quang.vu.nguyen,Lech.Madeyski}@pwr.edu.pl

Abstract. Higher order mutation testing is considered a promising solution for overcoming the main limitations of first order mutation testing. Strongly subsuming higher order mutants (SSHOMs) are the most valuable among all kinds of higher order mutants (HOMs) generated by combining first order mutants (FOMs). They can be used to replace all of its constituent FOMs without scarifying test effectiveness. Some researchers indicated that searching for SSHOMs is a promising approach. In this paper, we not only introduce a new classification of HOMs but also new objectives and fitness function which we apply in multi-objective optimization algorithm for finding valuable SSHOMs.

Keywords: Mutation Testing, Higher Order Mutation, Higher Order Mutants, Strongly Subsuming, Multi-objective optimization algorithm.

1 Introduction

First order mutation testing (traditional mutation testing (MT)) is a technique that has been developed using two basic ideas: Competent Programmer Hypothesis ("programmers write programs that are reasonably close to the desired program") and Coupling Effect Hypothesis ("detecting simple faults will lead to the detection of more complex faults"). It was originally proposed in 1970s by DeMillo et al.[1] and Hamlet[2]. MT is based on generating different versions of the program (called mutants) by insertion (via a mutation operator) a semantic fault or change into each mutant. The process of MT can be explained by the following steps:

1. Suppose we have a program P and a set of test cases T
2. Produce mutant P1 from P by inserting only one semantic fault into P
3. Execute T on P and P1 and save results as R and R1
4. Compare R1 with R:
    4.1 If R1 ≠ R: T can detect the fault inserted and has killed the mutant.
    4.2 If R1=R: There could be 2 reasons:
    + T can't detect the fault, so we have to improve T.

H.A. Le Thi et al. (eds.), *Advanced Computational Methods for Knowledge Engineering*,
Advances in Intelligent Systems and Computing 358, DOI: 10.1007/978-3-319-17996-4_35

+ The mutant has the same semantic meaning as the original program. It's an equivalent mutant.

The ratio of killed mutants to all generated mutants is called mutation score indicator (MSI) [8][9][10][11]. MSI is a quantitative measure of the quality of test cases. It is different from mutation score (MS)that was defined [1][2] as the ratio of killed mutants to difference of all generated mutants and equivalent mutants.

Although MT is a high automation and effective technique for evaluating the quality of the test data, there are three main problems[4][11][18]: a large number of mutants (this also leads to a very high execution cost); realism and equivalent mutant problem[11][19]. A number of approaches were proposed for overcoming that problems[18] including second order mutation testing [11][12][13][14][15] in particular and higher order mutation testing[3][4][5][6] in general. These approaches use more complex changes to generate mutants by inserting two or more faults into original program.

In 2009, Jia and Harman[3] introduced six types of HOMs on a basis of two definitions:

Definition 1: The higher order mutant (HOM) is called "Subsuming HOM" if it is harder to kill than their constituent first order mutants (FOMs). If set of test cases (TCs) which kill HOM are only inside the intersection of the sets of TCs which kill FOMs, it is a "Strongly Subsuming HOM", otherwise it is a "Weakly Subsuming HOM".

Definition 2: The HOM is called "Coupled HOM" if set of TCs that kill FOMs also contains cases that kill HOM. Otherwise, it is called "De-coupled HOM".

Six types of HOMs are (1)Strongly Subsuming and Coupled; (2)Weakly Subsuming and Coupled; (3)Weakly Subsuming and Decoupled; (4)Non-Subsuming and Decoupled; (5)Non-Subsuming and Coupled; (6)Equivalent. **Strongly Subsuming and Coupled HOM** is harder to kill than any constituent FOM and it is only killed by subset of the intersection of set of test cases that kill each constituent FOM. As a result, **Strongly Subsuming and Coupled HOM** can be used to replace all of its constituent FOMs without loss of test effectiveness. Finding **Strongly Subsuming and Coupled HOMs** can help us to overcome the above-mentioned three limitations of first order mutation testing.

Jia and Harman[3] introduced also some approaches to find the Subsuming HOMs by using the following algorithms: Greedy, Genetic and Hill-Climbing. In their experiments, approximately 15% of all found Subsuming HOMs were Strongly Subsuming HOMs. In addition, Langdon et al.[7] suggested applying multi-objective optimization algorithm to find higher order mutants that represent more realistic complex faults. As a result, our research focuses on searching valuable SSHOMs by applying multi-objective optimization algorithm.

In the next section, we present our approach to classify HOMs. Section 3 describes the proposed objectives and fitness function used for searching and identifying HOMs. Section 4 presents our experiment on searching for SSHOMs by applying multi-objective optimization algorithm. And the last section is conclusions and future work.

2 HOMs Classification

Our HOMs classification approach based on the combination of a set of test cases (TCs) which kill HOM and sets of TCs which kill FOMs. Let H be a HOM, constructed from FOMs F1 and F2. We also explain some notation below (See Fig. 1).

T: The given set of test cases
$T_{F1} \subset T$: Set of test cases that kill FOM1
$T_{F2} \subset T$: Set of test cases that kill FOM2
$T_H \subset T$: Set of test cases that kill HOM generated from FOM1 and FOM2

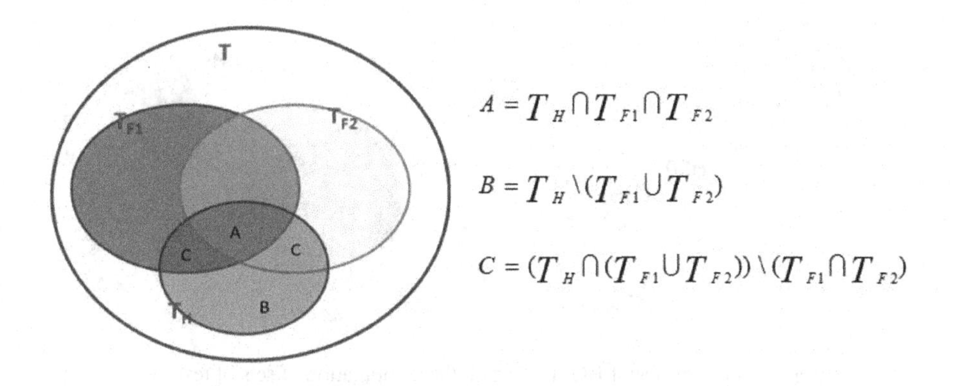

Fig. 1. The combination of sets of TCs

The categories of HOMs were named on a basis of two definitions above (see Section 1) and three new definitions below, and are illustrated in Figure 2 and Table 1.

Definition 3: The HOM that is killed by set of TCs that can kill FOM1 or FOM2 (this set of TCs $\subset$ C) is called "HOM With Old Test Cases".

Definition 4: The HOM that is killed by set of TCs that cannot kill any their constituent FOM (this set of TCs $\subset$ B) is called "HOM With New Test Cases".

Definition 5: The HOM that is killed by set of TCs which has some TCs $\subset$ B and some others TCs $\subset$ A or $\subset$ C is called "HOM With Mixed Test Cases".

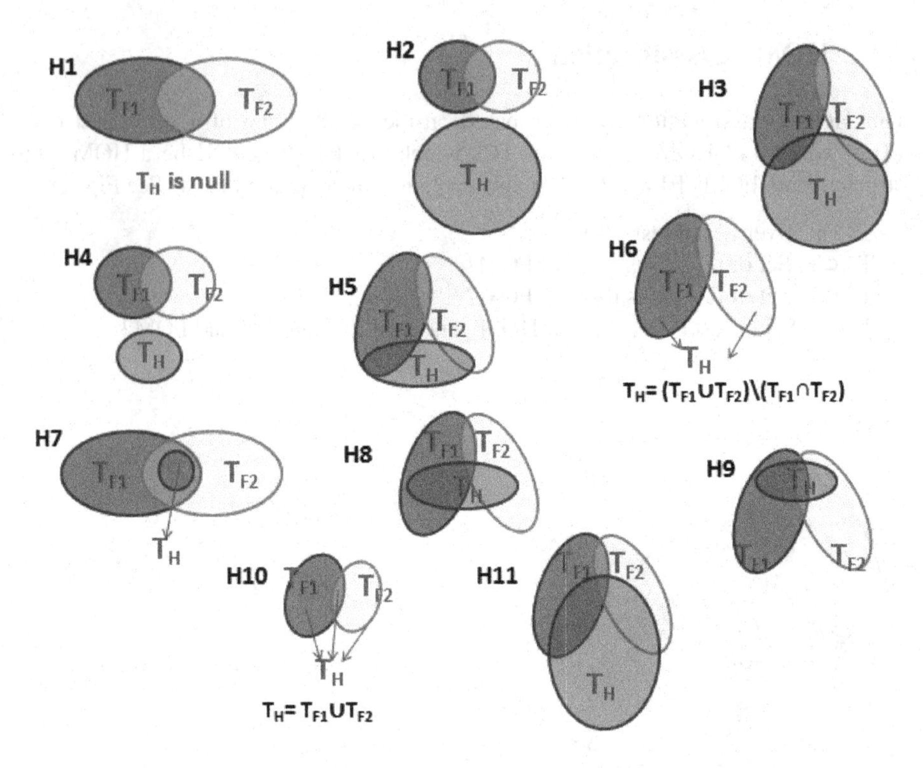

Fig. 2. 11 categories of HOM based on the combination of sets of test cases

Table 1. Elevencategoriesof HOMs

Case	HOM is
H1	Equivalent Mutant (T_H is null)
H2	Non-Subsuming, De-coupled and With New TCs.
H3	Non-Subsuming, De-coupled and With Mixed TCs.
H4	Weakly Subsuming, De-coupled and With New TCs.
H5	Weakly Subsuming, De-coupled and With Mixed TCs.
H6	Weakly Subsuming, De-coupled and With Old TCs.
H7	**Strongly Subsuming and Coupled**
H8	Weakly Subsuming, Coupled and With Mixed TCs.
H9	Weakly Subsuming, Coupled and With Old TCs.
H10	Non-Subsuming, Coupled and With Old TCs.
H11	Non-Subsuming, Coupled and With Mixed TCs.

3 Identify HOMs Based on Objective and Fitness Functions

We propose objectives and fitness function in order to identify HOMs, especially the most valuable SSHOMs. Each SSHOM represents an optimal solution that we want to find.

The set of TCs that kill HOM (TH) can be divided into 3 subsets:

- The first subset that can kill HOM and all its constituent FOMs (subset A in Fig. 1)
- The second subset that kill HOM and can kill FOM1 or FOM2 (subset C in Fig. 1)
- The third subset that only kill HOM (subset B in Fig. 1)

When the subsets B, C are empty and subset A is not empty, HOM is Strongly Subsuming HOM. In this case, the number of TCs that kill SSHOM is as small as possible. If all 3 subsets are empty, HOM is equivalent HOM. If A and C are empty, B is not empty, we call HOM is "HOM With New Test Cases". By contrast, if B is empty but A and C are not empty, we call HOM is "HOM With Old Test Cases". And if B is not empty, at least one of A or C is not empty, we call HOM is "HOM With Mixed Test Cases".

Hence, we proposed three objectives and one fitness function (see Equation 4) to apply to multi objectives optimization algorithm for searching for Strongly Subsuming HOMs.

Objective 1: Minimum number of TCs that kill HOM and also kill all its constituent FOMs (Minimum value of OB1 in Equation 1).

Objective 2: Minimum number of TCs which kill HOM but cannot kill any their constituent FOM (Minimum value of OB2 in Equation 2).

Objective 3: Minimum number of TCs which kill HOM and can kill FOM1 or FOM2 (Minimum value of OB3 in Equation 3).

$$OB1 = \frac{T_H \cap T_{F1} \cap T_{F2}}{T_H} \tag{1}$$

$$OB2 = \frac{T_H \setminus (T_{F1} \cup T_{F2})}{T_H} \tag{2}$$

$$OB3 = \frac{(T_H \cap (T_{F1} \cup T_{F2})) \setminus (T_{F1} \cap T_{F2})}{T_H} \tag{3}$$

$$fitness(H) = \frac{\#T_H}{\#(T_{F1} \cup T_{F2})} \tag{4}$$

The values of OB1, OB2, OB3 and fitness functions lie between 0 and 1. Generated HOMs are identified on a basis of that values as shown in Table 2.

Table 2. Identilfy HOMs based on objective and fitness functions

Case	Value of OB1 (V1)	Value of OB2 (V2)	Value of OB3 (V3)	Value of fitness function (F)	HOM is
H1	0	0	0	0	Equivalent Mutant (T_H is null)
H2	0	1	0	>=1	Non-Subsuming, De-coupled and With New TCs.
H3	0	0<V2<1	0<V3<1	>=1	Non-Subsuming, De-coupled and With Mixed TCs.
H4	0	1	0	<1	Weakly Subsuming, De-coupled and With New TCs.
H5	0	0<V2<1	0<V3<1	<1	Weakly Subsuming, De-coupled and With Mixed TCs.
H6	0	0	1	<=1	Weakly Subsuming, De-coupled and With Old TCs.
H7	**0<V1<=1**	**0**	**0**	**<=1**	**Strongly Subsuming and Coupled**
H8	0<V1<=1	0<V2<1	0<=V2<1	<1	Weakly Subsuming, Coupled and With Mixed TCs.
H9	0<V1<=1	0	0<V2<1	<1	Weakly Subsuming, Coupled and With Old TCs.
H10	0<V1<=1	0	0<V2<1	1	Non-Subsuming, Coupled and With Old TCs.
H11	0<V1<=1	0<V2<1	0<=V2<1	>=1	Non-Subsuming, Coupled and With Mixed TCs.

4 Experimental Evaluation

4.1 Research Questions

We posed the following research questions (RQ) that the study will address(Section 4.4 provides answers):

RQ1: What is mutation score indicator (MSI) for each project under test?

RQ2: How many HOMs are Non-Subsuming and De-coupled?

Non-Subsuming and De-coupled HOMs are the HOMs which are easier to kill than their constituent FOMs and the set of test cases which kill HOMs cannot kills simultaneously their constituent FOMs.

RQ3: What is the ratio of subsuming HOMs to total number of HOMs as well as ratio of strongly subsuming HOMs to Subsuming HOMs?

4.2 Software under Test and Supporting Tool

For our experiment, we used three open source projects (see Table 3), which are real-world software. The following projects have been selected for the experiment:

- ***BeanBin*** is a tool to make persisting EJB (Enterprise JavaBeans) 3.0 entity beans easier.
- ***Barbecue*** is a library that provides the means to create barcodes for Java applications.
- ***JWBF***(Java Wiki Bot Framework) is a library to maintain Wikis like Wikipedia based on MediaWiki and provides methods to connect, modify and read collections of articles, to help created wiki bot.

To implement our experiment, we have used Judy[11] [16], a mutation testing tool for Java language programming. It has the largest set of mutation operators, support for HOM generation, HOM execution and mutation analysis. The list of all mutation operators available in Judy [11][16] is presented in Table 4.

Table 3. Software under test

Project	NOC	LOC	# Test cases
BeanBin (http://beanbin.sourceforge.net)	72	5925	68
Barbecue (http://barbecue.sourceforge.net)	57	23996	190
JWBF (http://jwbf.sourceforge.net)	51	13572	305

Table 4. Java mutation operators available in Judy

AIR	AIR_ Add	Replaces basic binary arithmetic instructions with ADD
	AIR_ Div	Replaces basic binary arithmetic instructions with DIV
	AIR_ LeftOperand	Replaces basic binary arithmetic instructions with their left operands
	AIR_ Mul	Replaces basic binary arithmetic instructions with MUL
	AIR_ Rem	Replaces basic binary arithmetic instructions with REM
	AIR_ RightOperand	Replaces basic binary arithmetic instructions with their right operands
	AIR_ Sub	Replaces basic binary arithmetic instructions with SUB
JIR	JIR_ Ifeq	Replaces jump instructions with IFEQ (IF_ ICMPEQ, IF_ ACMPEQ)
	JIR_ Ifge	Replaces jump instructions with IFGE (IF_ ICMPGE)
	JIR_ Ifgt	Replaces jump instructions with IFGT (IF_ ICMPGT)
	JIR_ Ifle	Replaces jump instructions with IFLE (IF_ ICMPLE)
	JIR_ Iflt	Replaces jump instructions with IFLT (IF_ ICMPLT)
	JIR_ Ifne	Replaces jump instructions with IFNE (IF_ ICMPNE, IF_ ACMPNE)
	JIR_ Ifnull	Replaces jump instruction IFNULL with IFNONNULL and vice-versa
LIR	LIR_ And	Replaces binary logical instructions with AND
	LIR_ LeftOperand	Replaces binary logical instructions with their left operands
	LIR_ Or Replaces	binary logical instructions with OR
	LIR_ RightOperand	Replaces binary logical instructions with their right operands
	LIR_ Xor	Replaces binary logical instructions with XOR

Table 4. *(continued)*

SIR	SIR_ LeftOperand	Replaces shift instructions with their left operands
	SIR_ Shl	Replaces shift instructions with SHL
	SIR_ Shr	Replaces shift instructions with SHR
	SIR_ Ushr	Replaces shift instructions with USHR
Inheritance	IOD	Deletes overriding method
	IOP	Relocates calls to overridden method
	IOR	Renames overridden method
	IPC	Deletes super constructor call
	ISD	Deletes super keyword before fields and methods calls
	ISI	Inserts super keyword before fields and methods calls
Polymorphism	OAC	Changes order or number of arguments in method invocations
	OMD	Deletes overloading method declarations, one at a time
	OMR	Changes overloading method
	PLD	Changes local variable type to super class of original type
	PNC	Calls new with child class type
	PPD	Changes parameter type to super class of original type
	PRV	Changes operands of reference assignment
Java-Specific Features	EAM	Changes an access or method name to other compatible access or method names
	EMM	Changes a modifier method name to other compatible modifier method names
	EOA	Replaces reference assignment with content assignment (clone) and vice-versa
	EOC	Replaces reference comparison with content comparison (equals) and vice-versa
	JDC	Deletes the implemented default constructor
	JID	Deletes field initialization
	JTD	Deletes this keyword when field has the same name as parameter
	JTI	Inserts this keyword when field has the same name as parameter
Jumble-Based	Arithmetics	Mutates arithmetic instructions
	Jumps	Mutates conditional instructions
	Returns	Mutates return values
	Increments	Mutates increments

4.3 Multi-objective Optimization Algorithm

With the objective and fitness functions above, we use NSGA-II algorithm to generate HOMs and searching for Strongly Subsuming HOMs. NSGA-II is the second version of the Non-dominated Sorting Genetic Algorithm that was proposed by Deb et al. [17] for solving non-convex and non-smooth single and multi objective optimization problems. Its main features are: it uses an elitist principle; it emphasizes non-dominated solutions; and it uses an explicit diversity preserving mechanism.

The general scheme of NSGA-II to generate and evaluate HOMs in pseudo-code as follows:

```
Input: List of FOMs
Output: List of HOMs
Create HOMs from list of FOMs;
fitness(HOMs);
rank(HOMs)
```

```
while not (Stop Condition()) do
  Select(HOMs);
  Crossover(HOMs);
  Mutate(HOMs);
  fitness(HOMs);
  combine (HOMs parent and child);
//Recombination;//
  rank (HOMs after combined)
  select individuals (HOMs)
end while
```

4.4 Results and Analysis

The following table shows the number of generated HOMs as well as HOMs in each category (defined in Section 3) in different projects under test (using NSGA-II algorithm with the proposed objectives and fitness function).

Table 5. Number of 11 categories of generated HOMs

Project	**BeanBin**		**Barbecue**		**JWBF**	
	Number	%	Number	%	Number	%
Generated HOMs	**402**	**100**	**891**	**100**	**238**	**100**
H1	259	64.43	251	28.17	13	5.46
H2	0	0	1	0.11	0	0
H3	7	1.74	90	10.10	6	2.52
H4	38	9.45	18	2.02	0	0
H5	0	0	9	1.01	0	0
H6	55	13.68	458	51.40	193	81.09
H7	**15**	**3.73**	**44**	**4.94**	**4**	**1.68**
H8	0	0	0	0	0	0
H9	0	0	5	0.56	2	0.84
H10	28	6.97	9	1.01	17	7.14
H11	0	0	6	0.67	3	1.26

Answer to RQ1

We used the same full set of mutation operators of Judy for 3 projects. Because of each project has different source code including different operators. So, during the process generating HOMs, there are 3 different sets of operators used. Table 6 lists mutation operators used to generate HOMs for each project under test and 15 is the maximum mutation order in our experiment. MSI for each project under test after is presented in Table 7. JWBF is project with the highest MSI.

Table 6. Mutation operators used

Project	Operators
BeanBin	JIR_Ifle;JIR_Ifeq;JIR_Iflt;JIR_Ifne;JIR_Ifge;JIR_Iflt;JIR_Ifnull; AIR_Div;AIR_LeftOperand;AIR_RightOperand;AIR_Rem;AIR_Mul; EAM;EOC;PLD;CCE;LSF;REV;ISI;JTD;JTI;OAC;DUL;
Barbecue	JIR_Ifge;JIR_Ifeq;JIR_Ifgt;JIR_Ifne;JIR_Iflt;JIR_Ifle;JIR_Ifnull; AIR_Add;AIR_Div;AIR_Sub;AIR_LeftOperand;AIR_RightOperand; AIR_Mul;AIR__Rem; JTI;JTD;JID;JDC;EAM;OAC;EOC;FBD;IPC;EGE;EMM;CCE;LSF;REV
JWBF	JIR_Ifge;JIR_Ifeq;JIR_Ifgt;JIR_Ifne;JIR_Iflt;JIR_Ifle; EAM;EOA;JTD;JTI;JID;PRV;OAC;PLD;

Table 7. MSI for each project under test

Project	Total HOMs	Killed HOMs	MSI
BeanBin	402	146	36.32 %
Barbecue	891	629	70.59%
JWBF	238	225	94.54%

Answer to RQ2

RQ2 is designed to evaluate percentage of Non-Subsuming and De-coupled HOMs. These HOMs are "not good" because they are the HOMs which are easier to kill than their constituent FOMs and the set of test cases which kill HOMs cannot kills simultaneously their constituent FOMs. Our results indicate that the percentage of these HOMs is not large. The percentage of Non-Subsuming and De-coupled HOMs corresponding to the BeanBin, Barbecue, and JWBF projects are 1.74%, 10.21%, 2.52%, respectively.

Answer to RQ3

The ratio of subsuming HOMs to total generated HOMs is fairly high. This indicates that we can find the mutants that are harder to kill and more realistic (reflecting complex faults) than FOMs by applying multi objectives optimization algorithm. Ratio of strongly subsuming HOMs to subsuming HOMs is smaller, but they can be used to replace a large number of FOMs without loss of test effectiveness. In our study, the number of FOMs corresponding to number of Strongly Subsuming HOMs which can be used to replace FOMs for the JWBF, BeanBin and Barbecue projects are 11-4, 40-15 and 143-44, respectively.

Table 8. The percentage of subsuming HOMs and strongly subsuming HOMs

Project	**Ratio of Subsuming HOMs to total HOMs**	**Ratio of Strongly Subsuming to Subsuming HOMs**
BeanBin	26.9%	14%
Barbecue	60%	8.2%
JWBF	83.6%	2%

5 Conclusion and Future Work

We proposed a new, extended classification of HOMs based on the combination of set of test cases (TCs) that kill HOM and sets of TCs which kill FOMs. Furthermore, we presented the objectives and the fitness function to apply multi-objective optimization algorithm NSGA-II to search for strongly subsuming HOMs and ten other types of HOMs. The results indicate that our approach can be useful in searching for strongly subsuming HOMs. However, number of equivalent HOMs still is large. In this case, equivalent HOMs cannot be killed by any test case of the given set of test cases T (see Section 2). It means that they could be killed by some test cases that not belong to T. So, in the future, we will research to improve our approach to assess whether HOMs are equivalent or not. In addition, we will apply other search-based algorithms based on our proposed objectives and fitness function in order to compare the effectiveness of different algorithms for searching for strongly subsuming HOMs.

References

1. DeMillo, R.A., Lipton, R.J., Sayward, F.G.: Hints on test data selection: help for the practicing programmer. IEEE Computer 11(4), 34–41 (1978)
2. Hamlet, R.G.: Testing programs with the aid of a compiler. IEEE Transactions on Software Engineering SE-3(4), 279–290 (1977)
3. Jia, Y., Harman, M.: Higher order mutation testing. Information and Software Technology 51, 1379–1393 (2009)
4. Harman, M., Jia, Y., Langdon, W.B.: A Manifesto for Higher Order Mutation Testing. In: Third International Conf. on Software Testing, Verification, and Validation Workshops (2010)
5. Langdon, W.B., Harman, M., Jia, Y.: Efficient multi-objective higher order mutation testing with genetic programming. The Journal of Systems and Software 83 (2010)
6. Jia, Y., Harman, M.: Constructing Subtle Faults Using Higher Order Mutation Testing. In: Proc. Eighth Int'l Working Conf. Source Code Analysis and Manipulation (2008)
7. Langdon, W.B., Harman, M., Jia, Y.: Multi Objective Higher Order Mutation Testing with Genetic Programming. In: Proc. Fourth Testing: Academic and Industrial Conf. Practice and Research (2009)
8. Madeyski, L.: On the effects of pair programming on thoroughness and fault-finding effectiveness of unit tests. In: Münch, J., Abrahamsson, P. (eds.) PROFES 2007. LNCS, vol. 4589, pp. 207–221. Springer, Heidelberg (2007)
9. Madeyski, L.: The impact of pair programming on thoroughness and fault detection effectiveness of unit tests suites. Wiley, Software Process: Improvement and Practice 13(3), 281–295 (2008), doi:10.1002/spip.382
10. Madeyski, L.: The impact of test-first programming on branch coverage and mutation score indicator of unit tests: An experiment. Information and Software Technology 52(2), 169–184 (2010), doi:10.1016/j.infsof.2009.08.007
11. Madeyski, L., Orzeszyna, W., Torkar, R., Józala, M.: Overcoming the Equivalent Mutant Problem: A Systematic Literature Review and a Comparative Experiment of Second Order Mutation. IEEE Transactions on Software Engineering 40(1), 23–42 (2014), doi:10.1109/TSE.2013.44

12. Mresa, E.S., Bottaci, L.: Efficiency of mutation operators and selective mutation strategies: An empirical study. Software Testing, Verification and Reliability (1999)
13. Papadakis, M., Malevris, N.: An empirical evaluation of the first and second order mutation testing strategies. In: Proceedings of the 2010 Third International Conference on Software Testing, Verification, and Validation Workshops, ICSTW 2010, pp. 90–99. IEEE Computer Society (2010)
14. Vincenzi, A.M.R., Nakagawa, E.Y., Maldonado, J.C., Delamaro, M.E., Romero, R.A.F.: Bayesian-learning based guide-lines to determine equivalent mutants. International Journal of Soft. Eng. and Knowledge Engineering 12(6), 675–690 (2002)
15. Polo, M., Piattini, M., Garcia-Rodriguez, I.: Decreasing the Cost of Mutation Testing with Second-Order Mutants. Software Testing, Verification, and Reliability 19(2), 111–131 (2008)
16. Madeyski, L., Radyk, N.: Judy - a mutation testing tool for Java. IET Software 4(1), 32–42 (2010), doi:10.1049/iet-sen.2008.0038
17. Deb, K., Pratap, A., Agarwal, S., Meyarivan, T.: A Fast and Elitist Multiobjective Genetic Algorithm: NSGA-II. IEEE Transactions on Evolutionary Computation 6(2) (April 2002)
18. Nguyen, Q.V., Madeyski, L.: Problems of Mutation Testing and Higher Order Mutation Testing. In: van Do, T., Le Thi, H.A., Nguyen, N.T. (eds.) Advanced Computational Methods for Knowledge Engineering. AISC, vol. 282, pp. 157–172. Springer, Heidelberg (2014), doi:10.1007/978-3-319-06569-4_12
19. Papadakis, M., Yue, J., Harman, M., Traon, Y.L.: Trivial Compiler Equivalence: A Large Scale Empirical Study of a Simple, Fast and Effective Equivalent Mutant Detection Technique. In: 37th International Conference on Software Engineering, ICSE (2015)

Some Practical Aspects on Modelling Server Clusters

Nam H. Do and Thanh-Binh V. Lam

Analysis, Design and Development of ICT systems (AddICT) Laboratory,
Budapest University of Technology and Economics
H-1117 Budapest, Magyar tudósok körútja 2., Hungary

Abstract. In this paper we investigate the impact of some aspects that may be not considered in the performance evaluation of server clusters. The concerned practical aspects are S1) the distribution of switching times, S2) powering off the setup servers is forbidden and S3) the shutdown time.

The simulation results show that the impact of these aspects on the mean response time is negligible. However, relaxing S2 has a great impact on the accuracy of the mean setup of servers, and the average energy consumption of the cluster. On the contrary, a best trade-off between the complexity and the accuracy regarding to the mean response time and the average energy consumption of the cluster can be reached by relaxing only S3, and the accuracy of the model can be improved with a knowledge of the mean shutdown time.

Keywords: Data center, power-saving, shutdown time, analytical model.

1 Introduction

The efficient management and operation of data centers are one of the foremost interests of both providers and researchers. The key problem is the balancing between the energy consumption which formed in operation costs and the response time which presented in term of QoS.

The Dynamic Power Management (DPM) or Dynamic Voltage/Frequency Management (DVFM) algorithms [1, 2] are proposed by adapting the number of active servers to the arrival rate of jobs. However, this solution only solves a part of the problem as the servers still consume around 60% of the peak power in the idle state [3]. Moreover, both DPM and DVFM mainly focus on the power consumption of CPUs, meanwhile, the CPU no longer dominates platform power at peak usage in modern servers thanks to the advanced energy-efficiency techniques [3, 16]. Furthermore, DPM should be applied carefully in order to reduce the energy consumption [5].

In [9–11, 14], the authors state the running cost can be reduced with less harm to QoS by applying enhanced operation policies for turning on/off servers. Several analytical models [4, 6–8, 13, 14] were proposed for modelling data centers with controlling the number of idle servers. Some practical aspects are relaxed

H.A. Le Thi et al. (eds.), *Advanced Computational Methods for Knowledge Engineering*,
Advances in Intelligent Systems and Computing 358, DOI: 10.1007/978-3-319-17996-4_36

to make these models mathematically tractable. It is worth emphasizing that a mathematical analysis may require a low computational effort than a stochastic simulation [12].

In [4, 8, 13, 14], the authors assume that the shutdown phase can be ignored. However, it is obvious that, in switching off periods, the servers still consume energy without serving any jobs. Moreover, the number of switching off servers will reduce the number of available servers for serving jobs. Another assumption is that servers can be turned off in the setup phase, which is not practical. Note that this assumption ignores the shutdown phase. These observations motivate us to investigate the impacts of these aspects on the accuracy of the abstract models. Furthermore, we show that

- the assumption of the exponential distribution of the setup time and the relaxing of shutdown phase in the analytical models may not lead to the loss of accuracy. However, the accuracy of the performance evaluation can be improved with a knowledge of the mean shutdown time,
- turning off setup servers should be treated carefully as it greatly affects the accuracy regarding the computation of the system's average energy consumption.

The rest of the paper is organized as follows. In Section 2, the dynamics of the investigated system and the reference abstract model will be presented. The performance measures are provided in Section 3. The simulation experiments are carried out in Section 4. Finally, Section 5 concludes our work.

2 System Modelling

We consider a server cluster that consists of some servers, in which a server is able to serve only one job at a time. Power-saving modes are used to save the running cost, but require time to transition into and out of the power-saving mode, and the servers can not serve jobs during the state transitions. In order to reduce the energy consumption, operation policies can be used for turning on and off the servers. From a practical point of view, powering off a server should be avoided during its transition (into and out of the power-saving mode) periods.

Table 1 lists some analytical models proposed to model such server clusters [4, 8, 13, 14]. In order to have an analytically tractable model, two practical aspects are relaxed in these models: the shutdown phase is ignored and servers can be turned off during the powering up phase. Our aim in this paper is investigating the impact of these simplifications on the accuracy of the abstract models.

A reference abstract model is constructed which includes all concerned practical aspects. By relaxing one or more aspects, we are able to investigate their impacts on the abstract models' accuracy. It is important to note that we are only focus on the accuracy of the models, therefore, the operation policies are not taken into account.

In following, we present the description of the reference abstract model: the system consists of K homogeneous servers and a buffer with an infinite waiting

Table 1. Assumptions in some analytical models

	Setup time (S1)	Can't turn off setup servers (S2)	Has shutdown time (S3)	Policy
Artalejo et al.[4]	Exp.	No	No	Staggered On/Off
Gandhi et al.[8]	Exp.	No	No	On/Off, DelayedOff
Tian et al. [13, p. 231]	Exp.	No	No	On/Off
Mitrani[14]	Exp.	No	No	On/Off with block of servers
Reference model	Uni./Fix/Exp.	Yes	Yes, Uni./Fix/Exp.	On/Off

space. Jobs arrive according to a Poisson process with rate λ, and will be served according to First Come First Served (FCFS) principle. Jobs will enter the queue if there are not any free servers. A server serves one job at a time in the power on state, service times are distributed exponentially with mean $1/\mu$. Let ρ denote the system load with $\rho = \lambda/(K\mu)$. Time is required to transition into and out of the power-saving mode with the mean switching on and off time are $1/\alpha$ and $1/\omega$, respectively. The servers can not serve jobs during the state transitions. In addition, powering off a setup server is forbidden.

The simple On/Off policy is applied with the dynamics of the system are driven by the arrivals and the departures of jobs: The servers are switched into the power-saving mode after becoming idle as soon as the queue is empty. Otherwise they pick a job from the queue to serve according to the FCFS principle. The power-off servers will be turned on again upon on arrivals of new jobs, unless the number of the setup servers already exceeds the number of jobs in the queue.

Based on this reference abstract model, the following aspects will be considered:

1. Distribution of switching times can be Uniform, Fix or Exponential (S1).
2. Servers can not be powered off in the setup phase (S2).
3. The shutdown time is taken into account (S3).

Note that the assumption of turning off setup servers imply the simplification of ignoring the powering off phase.

3 Performance Measures

Following [5], let w_j be the waiting time in queue of job j before service and s_j be the service time needed to process job j. The response time r_j of job j is the time period between its arrival instant and departure instant. Therefore $r_j = w_j + s_j$. The mean waiting time $WT(n)$, the mean service time $ST(n)$ and the mean response time $RT(n)$ of n completed jobs are calculated as follows:

$$WT(n) = \frac{1}{n}\sum_{l=1}^{n} w_l, \quad ST(n) = \frac{1}{n}\sum_{l=1}^{n} s_l, \quad RT(n) = \frac{1}{n}\sum_{l=1}^{n} r_l.$$

The long term average service time, the long term average waiting time and the long term average response time are defined as

$$ST = \lim_{n\to\infty} ST(n), \quad WT = \lim_{n\to\infty} WT(n), \quad RT = \lim_{n\to\infty} RT(n).$$

Servers consume energy in busy, setup and switching off periods. When the server is busy, the consumed energy is used to process jobs. In the setup and switching off periods, consumed energy is needed due to the dynamic of the system, but can not be used for processing jobs, therefore, it should be minimized.

Let the departure time of job n be t_n. Let $\tau_i(t)$, $\zeta_i(t)$ and $\kappa_i(t)$ denote the total time server i spent in the busy, setup and switching off periods within a time interval t, respectively. Define $AE_{\mathrm{p}}(t_n)$ and $AE_{\mathrm{w}}(t_n)$ as average useful energy consumption per job and "*wasted*" energy consumption per job up to time t_n, respectively. We have

$$AE_{\mathrm{p}}(t_n) = \frac{P_{\max}\sum_{i=1}^{K}\tau_i(t_n)}{n},$$

$$AE_{\mathrm{w}}(t_n) = \frac{P_{\max}\sum_{i=1}^{K}(\zeta_i(t_n)+\kappa_i(t_n))}{n}.$$

with $P_{\max}$ is the power consumption of one server. The average energy consumption per job up to t_n is calculated as

$$AE(t_n) = AE_{\mathrm{p}}(t_n) + AE_{\mathrm{w}}(t_n).$$

The long term average energy consumptions per job are defined as

$$AE = \lim_{n\to\infty} AE(t_n), \quad AE_{\mathrm{p}} = \lim_{n\to\infty} AE_{\mathrm{p}}(t_n), \quad AE_{\mathrm{w}} = \lim_{n\to\infty} AE_{\mathrm{w}}(t_n).$$

Let $N_{\mathrm{se}}(t)$ denote the average number of setup servers within a time interval t and $n_{\mathrm{se}}(t)$ denote the number of setup servers at time t. We have

$$N_{\mathrm{se}}(t) = \frac{1}{t}\int_0^t n_{\mathrm{se}}(s)ds. \tag{1}$$

The long term average number of setup servers is defined as

$$N_{\mathrm{se}} = \lim_{T\to\infty}\frac{1}{T}\int_0^T n_{\mathrm{se}}(t)dt.$$

We consider the events of a server entering and leaving the setup phase. Let $0 < t_1 < t_2 < \dots$ be the times of the occurrences of these consecutive events and $t_0 = 0$.

Let define $L(t) = \max\{l : t_l \leq t\}$ and $t_{L(t)+1} = t$. It is trivial that $n_{\rm se}$ is invariable within the time interval $[t_l, t_{l+1})$, $0 \leq t_l \leq L(t)$. Hence, (1) can be rewritten as follows:

$$N_{\rm se}(t) = \frac{1}{t}\sum_{l=0}^{L(t)} n_{\rm se}(t_l)(t_{l+1} - t_l). \tag{2}$$

In the other hand, we can rewrite $n_{\rm se}$ as follows:

$$n_{\rm se}(s) = \sum_{i=1}^{K} \varepsilon(i, s),$$

with

$$\varepsilon(i,s) = \begin{cases} 1 & \text{If server } i \text{ is in the setup phase at time } s, \\ 0 & \text{otherwise.} \end{cases}$$

Therefore, (2) can be rewritten as follows:

$$\begin{aligned} N_{\rm se}(t) &= \frac{1}{t}\sum_{l=0}^{L(t)}\sum_{i=1}^{K} \varepsilon(i, t_l)(t_{l+1} - t_l) \\ &= \frac{1}{t}\sum_{i=1}^{K}\sum_{l=0}^{L(t)} \varepsilon(i, t_l)(t_{l+1} - t_l) \\ &= \frac{1}{t}\sum_{i=1}^{K} \zeta_i(t). \end{aligned}$$

Let $N_{\rm s}(t)$ and $N_{\rm sh}(t)$ denote the average number of busy and switching off servers within a time interval t, respectively. Similarly, we have

$$N_{\rm s}(t) = \frac{\sum_{i=1}^{K} \tau_i(t)}{t}, \quad N_{\rm sh}(t) = \frac{\sum_{i=1}^{K} \kappa_i(t)}{t}.$$

Their long term average values can be defined as

$$N_{\rm s} = \lim_{t\to\infty} N_{\rm s}(t), \quad N_{\rm sh} = \lim_{t\to\infty} N_{\rm sh}(t).$$

Define $AE_{\rm p}^*(t)$, $AE_{\rm w}^*(t)$ and $AE^*(t)$ as average useful, "*wasted*" and total energy consumption of the system up to time t, respectively, and can be calculated as follows:

$$\begin{aligned} AE_{\rm p}^*(t) &= \frac{\sum_{i=1}^{K} \tau_i(t) P_{\max}}{t} = N_{\rm s}(t) * P_{\max}, \\ AE_{\rm w}^*(t) &= \frac{\sum_{i=1}^{K} (\zeta_i(t) + \kappa_i(t)) P_{\max}}{t} = (N_{\rm se}(t) + N_{\rm sh}(t)) * P_{\max}, \\ AE^*(t) &= AE_{\rm p}^*(t) + AE_{\rm w}^*(t). \end{aligned}$$

The long term average energy consumption AE_{p}^{*}, AE_{w}^{*} and AE^{*} are defined as follows:

$$AE_{\mathrm{p}}^{*} = \lim_{t\to\infty} AE_{\mathrm{p}}^{*}(t) = N_{\mathrm{s}} * P_{\mathrm{max}},$$
$$AE_{\mathrm{w}}^{*} = \lim_{t\to\infty} AE_{\mathrm{w}}^{*}(t) = (N_{\mathrm{se}} + N_{\mathrm{sh}}) * P_{\mathrm{max}},$$
$$AE^{*} = \lim_{t\to\infty} AE^{*}(t) = (N_{\mathrm{se}} + N_{\mathrm{s}} + N_{\mathrm{sh}}) * P_{\mathrm{max}}.$$

4 Simulation Experiments

In analytical models [4, 8, 10, 13, 14], it is common assumption that the setup time is exponentially distributed. However, the switching on and off phases are time-bound in realistic systems. We took some experiments on a Linux server with Intel® Core™ i5-4670 3.40GHz processor, 16GB DDR3 1600 MHz RAM and 1TB 7200 RPM Hard disk running Ubuntu 14.04 64-bit OS. We manually turned on and turned off the server using Hibernate power-saving mode and recorded the required time of switching on and switching off phases using a stopwatch. The measured data is listed in Table 2.

Table 2. Experiment data

Phase	Min (s)	Max (s)	Mean (s)
Switch on	24.44	26.55	25.4
Switch off	8.93	10.92	9.7

When the system load is small, it is likely that we need to turn on an server for each new request and turn off it immediately after serving one request. Therefore, the system can be approximated by the $M/M/k$ queue with the mean service time of $(\frac{1}{\alpha} + \frac{1}{\mu} + \frac{1}{\omega})$. On the contrary, turning off servers is very rarely happens when the load is high enough. Hence, in this case the system can be approximated by the $M/M/k$ queue with the mean service time of $1/\mu$.

Similarly, it can be also applied when the mean service time is very large comparing to the mean setup time. In this case, both the setup time and the shutdown time can be ignored. We used $1/\mu = 50$s and $1/\mu = 300$s in the simulation in order to investigate the impact of the switch on and switch off time. The average power consumption of the HP Moonshot m300/m350/m710 server cartridges is chosen with $P_{\mathrm{max}} = 44.0$W [16], and $K = 45$ as the maximum number of cartridges of one Moonshot 1500 chassis [15].

The following 5 abstract models were compared:

1) a model with the switching times are uniform distributed (*Uni.,S2,S3*).
2) a model with the fix switching times (*Fix,S2,S3*).
3) a model with the switching times are exponential distributed (*Exp.,S2,S3*).
4) a model without aspect S3, the switching times are exponential distributed (*Exp.,S2*).

5) a model without aspect S2, the switching times are exponential distributed (*Exp.*). Note that this assumption also implies that S3 is neglected!

Simulation runs were performed with the confidence level of 99.9%. The confidence interval is $\pm 0.6\%$ of the collected data.

Figures 1 and 2 illustrate the mean response time against the system load for all scenarios. The response time consists of 2 components: the mean service time and the mean waiting time. The results show that the mean service times are not impacted by the simplification of some aspects, and are always equal to $1/\mu$.

When keeping the aspects S2 and S3 (*Uni./Fix/Exp.,S2,S3*), it can be observed that the uniform and fix switching times give the same mean response time, while the observed mean response time is always smaller in the case of the exponential distributed switching times. It is interesting that aspects S2 and S3

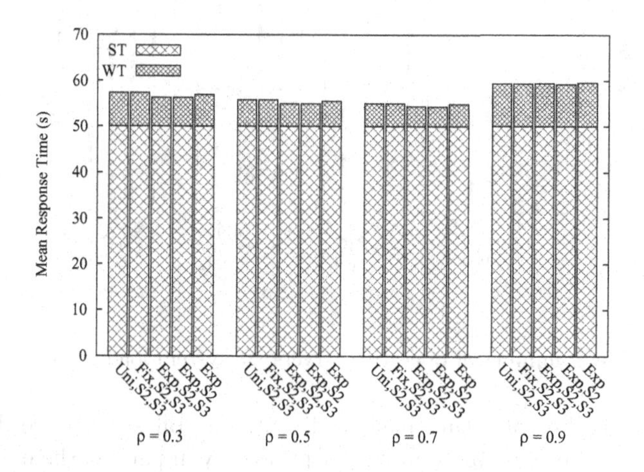

Fig. 1. Mean response time vs. system load for $1/\mu = 50$s

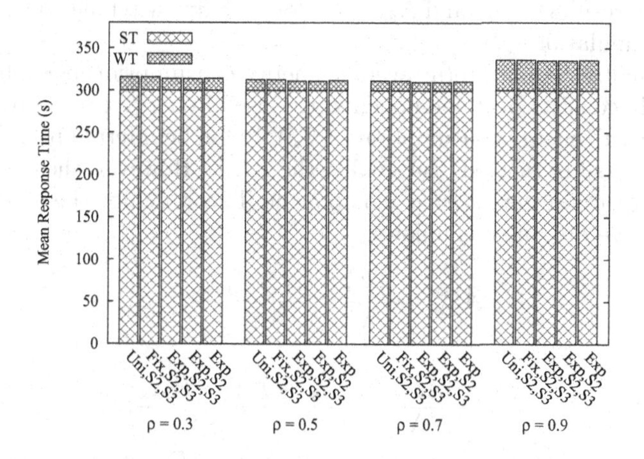

Fig. 2. Mean response time vs. system load for $1/\mu = 300$s

have a small impact on the accuracy regarding the computation of the mean response time. Usually relaxing only S3 gives the worst results, but they are approximately equal to the observed results of the scenario *Exp.,S2,S3*. However, the deviation is acceptable in the most all cases. For example, the deviations in the case of $1/\mu = 50$s are 1.04s (1.8%), 0.85s (1.5%), 0.37s (0.6%) for $\rho = 0.3, 0.5, 0.9$, respectively. The deviations are even smaller in case of higher $1/\mu$, for example the largest gap in case of $1/\mu = 300$s is 1.87s (0.6%) for $\rho = 0.3$.

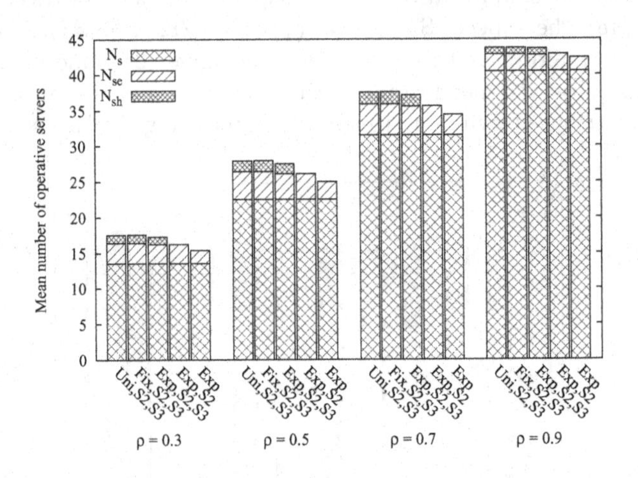

Fig. 3. Mean number of operative servers vs. system load for $1/\mu = 50s$

The mean number of setup, busy and switching off servers are depicted in Fig. 3. It is observed that the concerned aspects only impact on the mean number of setup and switching off servers, and the mean number of busy servers is always equal to $K\rho$. The other observation is that the uniform and fix switching times give the same results of N_{se} and N_{sh}, therefore the fix switching times should be used in the simulation.

Figure 4 plots the long term average energy consumption per job (AE) for $1/\mu = 50$s. It consists of 2 components: the useful energy consumption (AE_{p}) and the "*wasted*" energy consumption (AE_{w}). It can be inferred from Fig. 4 that the more aspects are relaxed, the smaller the AE is. However, the concerned aspects only impact on the "*wasted*" component. Theoretically, these performance measures can be obtained as follows:

$$AE_{\text{p}} = \frac{AE_{\text{p}}^*}{\lambda} = \frac{N_{\text{s}} * P_{\text{max}}}{\lambda},$$

and

$$AE_{\text{w}} = \frac{AE_{\text{w}}^*}{\lambda} = \frac{(N_{\text{se}} + N_{\text{sh}}) * P_{\text{max}}}{\lambda}.$$

It explains why the useful energy consumption (AE_{p}) remains constant in all scenarios.

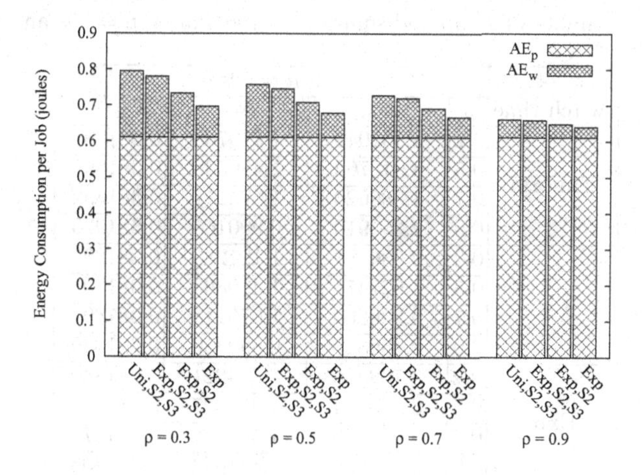

Fig. 4. Mean energy consumption per job vs. system load for $1/\mu = 50$s

The mean number of switching on and off servers are plotted again in Fig. 5 for $1/\mu = 50$s. Note that relaxing S3 means $N_{\text{sh}} = 0$. It can be observed that relaxing S2 (and also S3) has a great impact on the number of setup servers (N_{se}). For example, the deviations are $3.93 - 2.48 = 1.45$ and $2.36 - 1.88 = 0.88$ for $\rho = 0.5$ and $\rho = 0.9$, respectively.

The results from Fig. 5 also show that the mean number of setup servers in case of relaxing only S3 (*Exp.,S2*) approximately equals to the case of the reference abstract model with exponential distributed switching times (*Exp.,S2,S3*).

However, the rate of turning on servers must be equal to the rate of turning off servers in order to ensure the stability of the abstract models. Therefore, the

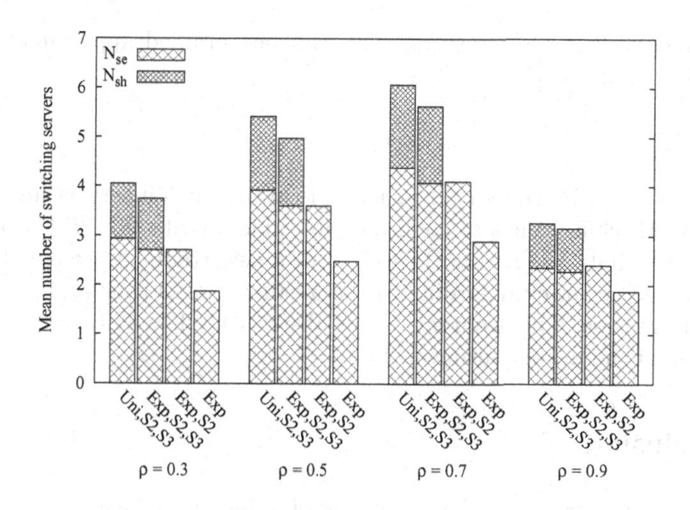

Fig. 5. Mean number of switching servers vs. system load for $1/\mu = 50$s

Table 3. Mean number of setup and shutdown servers for $1/\mu = 50s$ and $1/\mu = 300s$

Dist. of switch time	ρ	$1/\mu = 50s$ N_{se}	N_{sh}	N_{sh}/N_{se}
Uniform	0.2	2.932049	1.119708	0.381886
	0.5	3.918762	1.496538	0.381891
	0.7	4.386620	1.675217	0.381892
	0.9	2.356910	0.900166	0.381926
Fix	0.2	2.936141	1.121283	0.38189
	0.5	3.927475	1.499863	0.38189
	0.7	4.396860	1.679118	0.38189
	0.9	2.356310	0.899850	0.381889
Exp.	0.2	2.708456	1.034221	0.381849
	0.5	3.598524	1.374097	0.38185
	0.7	4.075039	1.556168	0.381878
	0.9	2.280834	0.871034	0.381893

Dist. of switch time	ρ	$1/\mu = 300s$ N_{se}	N_{sh}	N_{sh}/N_{se}
Uniform	0.2	0.911331	0.348031	0.381893
	0.5	1.306139	0.498796	0.381886
	0.7	1.585482	0.605479	0.38189
	0.9	1.065862	0.407044	0.381892
Fix	0.2	0.911614	0.348136	0.38189
	0.5	1.306596	0.498976	0.38189
	0.7	1.586221	0.605762	0.38189
	0.9	1.067366	0.407617	0.381891
Exp.	0.2	0.868859	0.331857	0.381946
	0.5	1.237465	0.472663	0.381961
	0.7	1.499902	0.572756	0.381862
	0.9	1.023362	0.390961	0.382036

mean number of shutdown servers (N_{sh}) can be obtained approximately with the following formula:

$$N_{sh} = N_{se} * \frac{\alpha}{\omega}.$$

This observation is true for all policies of turning on/off servers and the distribution of switching times as confirmed in Table 3 (with $9.7/25.4 = 0.38189$). Hence, relaxing only S3 (*Exp.,S2*) can give us a best trade-off between the complexity and the accuracy regarding to the mean response time and the average energy consumption. The accuracy of this abstract model can be increased with a knowledge of the mean shutdown time.

5 Conclusion

In this paper we investigated the impact of some simplifications assumed in several analytical models for modelling data centers. The concerned aspects are

S1) The distribution of switching times, S2) Turning off setup servers is forbidden and S3) The shutdown time.

The simulation results show that the impact of these aspects on the mean response time is small. However, relaxing S2 has a great impact on the accuracy of the mean setup of servers, and the average energy consumption. On the contrary, the knowledge of mean shutdown time can improve the accuracy of the model and a best trade-off between the complexity and the accuracy regarding to the mean response time and the average energy consumption can be reached by relaxing the condition of the shutdown time.

Acknowledgment. The authors would like to thank Prof. Tien Van Do for his continuing support during the preparation of this work.

References

1. Chen, Y., Das, A., Qin, W., Sivasubramaniam, A., Wang, Q., Gautam, N.: Managing server energy and operational costs in hosting centers. SIGMETRICS Perform. Eval. Rev. 33(1), 303–314 (2005), http://doi.acm.org/10.1145/1071690.1064253
2. Wang, Y., Wang, X., Chen, M., Zhu, X.: Power-efficient response time guarantees for virtualized enterprise servers. In: Real-Time Systems Symposium, pp. 303–312 (November 2008)
3. Barroso, L.A., Hölzle, U.: The case for energy-proportional computing. Computer 40(12), 33–37 (2007)
4. Artalejo, J.R., Economou, A., Lopez-Herrero, M.J.: Analysis of a multiserver queue with setup times. Queueing Syst. Theory Appl. 51(1-2), 53–76 (2005), http://dx.doi.org/10.1007/s11134-005-1740-6
5. Do, T.V., Vu, B.T., Tran, X.T., Nguyen, A.P.: A generalized model for investigating scheduling schemes in computational clusters. Simulation Modelling Practice and Theory 37, 30–42 (2013), http://www.sciencedirect.com/science/article/pii/S1569190X13000828
6. Do, T.V.: Comparison of allocation schemes for virtual machines in energy-aware server farms. Comput. J. 54(11), 1790–1797 (2011)
7. Do, T.V., Rotter, C.: Comparison of scheduling schemes for on-demand IaaS requests. Journal of Systems and Software 85(6), 1400–1408 (2012)
8. Gandhi, A., Harchol-Balter, M.: How data center size impacts the effectiveness of dynamic power management. In: 2011 49th Annual Allerton Conference on Communication, Control, and Computing (Allerton), pp. 1164–1169 (September 2011)
9. Gandhi, A., Gupta, V., Harchol-Balter, M., Kozuch, M.A.: Optimality analysis of energy-performance trade-off for server farm management. Performance Evaluation 67(11), 1155–1171 (2010), http://www.sciencedirect.com/science/article/pii/S0166531610001069, performance 2010
10. Gandhi, A., Harchol-Balter, M., Adan, I.: Server farms with setup costs. Perform. Eval. 67(11), 1123–1138 (2010), http://dx.doi.org/10.1016/j.peva.2010.07.004
11. Guenter, B., Jain, N., Williams, C.: Managing cost, performance, and reliability tradeoffs for energy-aware server provisioning. In: 2011 Proceedings IEEE INFOCOM, pp. 1332–1340 (April 2011)

12. Kleinrock, L.: Queueing Systems. Theory, vol. I. John Wiley & Sons, Inc. (1975)
13. Tian, N., Zhang, Z.G.: Vacation Queueing Models: Theory and Applications. International Series in Operations Research & Management Science. Springer (2006)
14. Mitrani, I.: Managing performance and power consumption in a server farm. Annals of Operations Research 202(1), 121–134 (2013), `http://dx.doi.org/10.1007/s10479-011-0932-1`
15. HP MoonShot 1500 chassis (December 2014), `http://h10032.www1.hp.com/ctg/Manual/c03728406.pdf`
16. HP MoonShot system - hewlett-packard (December 2014), `http://www8.hp.com/us/en/products/servers/moonshot/`

Erratum to: A Direct Method for Determining the Lower Convex Hull of a Finite Point Set in 3D

Thanh An Phan[1,2(✉)] and Thanh Giang Dinh[2,3]

[1] Institute of Mathematics, Vietnam Academy of Science and Technology, Hanoi, Vietnam
[2] CEMAT, Instituto Superior Técnico, University of Lisbon, Lisbon, Portugal
dtgiang@math.ist.utl.pt
[3] Department of Mathematics, Vinh University, Vinh City, Vietnam

Erratum to: Chapter "A Direct Method for Determining the Lower Convex Hull of a Finite Point Set in 3D" in: H.A. Le Thi et al. (eds.), *Advanced Computational Methods for Knowledge Engineering*, Advances in Intelligent Systems and Computing 358, DOI: 10.1007/978-3-319-17996-4_2

In the original version of the book, the revised Acknowledgement section has to be incorporated in Chapter "A Direct Method for Determining the Lower Convex Hull of a Finite Point Set in 3D". The erratum chapter and the book have been updated with the change.

The updated online version of this chapter can be found at
http://dx.doi.org/10.1007/978-3-319-17996-4_2

H.A. Le Thi et al. (eds.), *Advanced Computational Methods for Knowledge Engineering*,
Advanced in Intelligent System and Computing 358, DOI: 10.1007/978-3-319-17996-4_37

Author Index

Aïssani, Djamil 119
Akgüller, Ömer 209
Alkaya, Ali Fuat 83
Avril, François 369
Azizi, Nabiha 175

Balcı, Mehmet Ali 209
Bareche, Aicha 119
Bouabana-Tebibel, Thouraya 311
Bui, Alain 369
Bui, Thach V. 185

Cao, Son Thanh 243, 355
Carabas, Mihai 381
Chebba, Asmaa 311
Cherfaoui, Mouloud 119

Dang, Hai-Van 185
Dinh, Thanh Giang 15
Do, Nam H. 403
Do, Thanh-Nghi 231, 255
Doan, Thi-Huyen-Trang 141
Duong, Trong Hai 291

Ghazali, Najah 267
Gheorghe, Laura 381
Guiyassa, Yamina Tlili 175

Hamzah, Nor Hazadura 95
Hamzah, Norhizam 95
Hilaire, Vincent 27

Jahanshahloo, Almas 3

Koukam, Abderrafiaa 27

Lauri, Fabrice 27
Le, Bac 279
Le, Ba Cuong 151
Le, Duc Thuan 151
Le, Hoai Minh 37
Le, Hoang-Quynh 141
Le, Thi Hong Van 151
Le Thi, Hoai An 37, 57, 129

Madeyski, Lech 391
Manea, Valentina 381
Mogosanu, Lucian 381
Muthusamy, Hariharan 95

Nguyen, Canh Nam 49
Nguyen, Dang 163
Nguyen, Dinh-Thuc 185
Nguyen, Duc Anh 291
Nguyen, Duc-Than 185
Nguyen, Ha Hung Chuong 219
Nguyen, Hung Son 335
Nguyen, Huy 279
Nguyen, Linh Anh 321
Nguyen, Loan T.T. 197
Nguyen, Maja 335
Nguyen, Ngoc Thanh 197, 301
Nguyen, Nhu Tuan 151
Nguyen, Quang Thuan 69
Nguyen, Quang Vu 391
Nguyen, Thanh Binh 345
Nguyen, Thanh-Long 163
Nguyen, Thi Bich Thuy 37
Nguyen, Van Du 301
Nguyen, Van Duy 219

Pham, Thi Hoai 49
Pham, Van Huong 151
Pham Dinh, Tao 57
Phan, Nguyen Ba Thang 69
Phan, Thanh An 15
Poulet, François 255

Ramli, Dzati Athiar 267
Rubin, Stuart H. 311

Sohier, Devan 369
Świeboda, Wojciech 335

Ta, Minh Thuy 129
Tonyali, Samet 83
Tran, Mai-Vu 141
Tran, Thai-Son 185
Tran, Thi Thuy 57
Tran, Van-Hien 141
Tran, Van Huy 49

V. Lam, Thanh-Binh 403
Vo, Bay 163
Vo, Xuan Thanh 37
Vu, Ngoc-Trinh 141

Yaacob, Sazali 95

Zhu, Jiawei 27
Ziani, Amel 175
Zohrehbandian, Majid 3
Zufferey, Nicolas 107

Printed in the United States
By Bookmasters